Dairy Products for Human Health

Dairy Products for Human Health

Editor

Dennis Savaiano

MDPI • Basel • Beijing • Wuhan • Barcelona • Belgrade • Manchester • Tokyo • Cluj • Tianjin

Editor
Dennis Savaiano
Department of Nutrition
Science, College of Health
and Human Sciences, Purdue
University
USA

Editorial Office
MDPI
St. Alban-Anlage 66
4052 Basel, Switzerland

This is a reprint of articles from the Special Issue published online in the open access journal *Nutrients* (ISSN 2072-6643) (available at: https://www.mdpi.com/journal/nutrients/special_issues/Dairy_Products_Human_Health).

For citation purposes, cite each article independently as indicated on the article page online and as indicated below:

LastName, A.A.; LastName, B.B.; LastName, C.C. Article Title. *Journal Name* **Year**, *Volume Number*, Page Range.

ISBN 978-3-0365-4939-2 (Hbk)
ISBN 978-3-0365-4940-8 (PDF)

© 2022 by the authors. Articles in this book are Open Access and distributed under the Creative Commons Attribution (CC BY) license, which allows users to download, copy and build upon published articles, as long as the author and publisher are properly credited, which ensures maximum dissemination and a wider impact of our publications.

The book as a whole is distributed by MDPI under the terms and conditions of the Creative Commons license CC BY-NC-ND.

Contents

About the Editor vii

Preface to "Dairy Products for Human Health" ix

Tsz-Ning Mak, Imelda Angeles-Agdeppa, Marie Tassy, Mario V. Capanzana and
Elizabeth A. Offord
Contribution of Milk Beverages to Nutrient Adequacy of Young Children and Preschool
Children in the Philippines
Reprinted from: *Nutrients* **2020**, *12*, 392, doi:10.3390/nu12020392 1

Liya Anto, Sarah Wen Warykas, Moises Torres-Gonzalez and Christopher N. Blesso
Milk Polar Lipids: Underappreciated Lipids with Emerging Health Benefits
Reprinted from: *Nutrients* **2020**, *12*, 1001, doi:10.3390/nu12041001 19

William Chey, William Sandborn, Andrew J. Ritter, Howard Foyt, M. Andrea Azcarate-Peril
and Dennis A. Savaiano
Galacto-Oligosaccharide RP-G28 Improves Multiple Clinical Outcomes in Lactose-Intolerant
Patients
Reprinted from: *Nutrients* **2020**, *12*, 1058, doi:10.3390/nu12041058 53

Cristina Santurino, Bricia López-Plaza, Javier Fontecha, María V. Calvo, Laura M. Bermejo,
David Gómez-Andrés and Carmen Gómez-Candela
Consumption of Goat Cheese Naturally Rich in Omega-3 and Conjugated Linoleic Acid
Improves the Cardiovascular and Inflammatory Biomarkers of Overweight and Obese Subjects:
A Randomized Controlled Trial
Reprinted from: *Nutrients* **2020**, *12*, 1315, doi:10.3390/nu12051315 67

Marta Sajdakowska, Jerzy Gębski, Dominika Guzek, Krystyna Gutkowska and
Sylwia Żakowska-Biemans
Dairy Products Quality from a Consumer Point of View: Study among Polish Adults
Reprinted from: *Nutrients* **2020**, *12*, 1503, doi:10.3390/nu12051503 81

Kate P. Palmano, Alastair K. H. MacGibbon, Caroline A. Gunn and Linda M. Schollum
In Vitro and In Vivo Anti-inflammatory Activity of Bovine Milkfat Globule (MFGM)-derived
Complex Lipid Fractions
Reprinted from: *Nutrients* **2020**, *12*, 2089, doi:10.3390/nu12072089 97

Yifan Duan, Xuehong Pang, Zhenyu Yang, Jie Wang, Shan Jiang, Ye Bi, Shuxia Wang,
Huanmei Zhang and Jianqiang Lai
Association between Dairy Intake and Linear Growth in Chinese Pre-School Children
Reprinted from: *Nutrients* **2020**, *12*, 2576, doi:10.3390/nu12092576 119

Victoria M. Taormina, Allison L. Unger, Morgan R. Schiksnis, Moises Torres-Gonzalez and
Jana Kraft
Branched-Chain Fatty Acids—An Underexplored Class of Dairy-Derived Fatty Acids
Reprinted from: *Nutrients* **2020**, *12*, 2875, doi:10.3390/nu12092875 131

Claire M. Timon, Aileen O'Connor, Nupur Bhargava, Eileen R. Gibney and Emma L. Feeney
Dairy Consumption and Metabolic Health
Reprinted from: *Nutrients* **2020**, *12*, 3040, doi:10.3390/nu12103040 147

Hermine Pellay, Corinne Marmonier, Cécilia Samieri and Catherine Féart
Socio-Demographic Characteristics, Dietary, and Nutritional Intakes of French Elderly Community Dwellers According to Their Dairy Product Consumption: Data from the Three-City Cohort
Reprinted from: *Nutrients* **2020**, *12*, 3418, doi:10.3390/nu12113418 173

Min-Yu Tu, Kuei-Yang Han, Gary Ro-Lin Chang, Guan-Da Lai, Ku-Yi Chang, Chien-Fu Chen, Jen-Chieh Lai, Chung-Yu Lai, Hsiao-Ling Chen and Chuan-Mu Chen
Kefir Peptides Prevent Estrogen Deficiency-Induced Bone Loss and Modulate the Structure of the Gut Microbiota in Ovariectomized Mice
Reprinted from: *Nutrients* **2020**, *12*, 3432, doi:10.3390/nu12113432 191

Justyna Janiszewska, Joanna Ostrowska and Dorota Szostak-Węgierek
Milk and Dairy Products and Their Impact on Carbohydrate Metabolism and Fertility—A Potential Role in the Diet of Women with Polycystic Ovary Syndrome
Reprinted from: *Nutrients* **2020**, *12*, 3491, doi:10.3390/nu12113491 207

Moises Torres-Gonzalez, Christopher J. Cifelli, Sanjiv Agarwal and Victor L Fulgoni III
Association of Milk Consumption and Vitamin D Status in the US Population by Ethnicity: NHANES 2001–2010 Analysis
Reprinted from: *Nutrients* **2020**, *12*, 3720, doi:10.3390/nu12123720 225

Monica Ramakrishnan, Tracy K. Eaton, Omer M. Sermet and Dennis A. Savaiano
Milk Containing A2 β-Casein ONLY, as a Single Meal, Causes Fewer Symptoms of Lactose Intolerance than Milk Containing A1 and A2 β-Caseins in Subjects with Lactose Maldigestion and Intolerance: A Randomized, Double-Blind, Crossover Trial
Reprinted from: *Nutrients* **2020**, *12*, 3855, doi:10.3390/nu12123855 241

Stine B. Smedegaard, Maike Mose, Adam Hulman, Ulla R. Mikkelsen, Niels Møller, Gregers Wegener, Niels Jessen and Nikolaj Rittig
β-Lactoglobulin Elevates Insulin and Glucagon Concentrations Compared with Whey Protein—A Randomized Double-Blinded Crossover Trial in Patients with Type Two Diabetes Mellitus
Reprinted from: *Nutrients* **2021**, *13*, 308, doi:10.3390/nu13020308 257

Kristine Bach Korsholm Knudsen, Christine Heerup, Tine Røngaard Stange Jensen, Xiaolu Geng, Nikolaj Drachmann, Pernille Nordby, Palle Bekker Jeppesen, Inge Ifaoui, Anette Müllertz, Per Torp Sangild, Marie Stampe Ostenfeld and Thomas Thymann
Bovine Milk-Derived Emulsifiers Increase Triglyceride Absorption in Newborn Formula-Fed Pigs
Reprinted from: *Nutrients* **2021**, *13*, 410, doi:10.3390/nu13020410 271

Naomi M.M.P. de Hart, Ziad S. Mahmassani, Paul T. Reidy, Joshua J. Kelley, Alec I. McKenzie, Jonathan J. Petrocelli, Michael J. Bridge, Lisa M. Baird, Eric D. Bastian, Loren S. Ward, Michael T. Howard and Micah J. Drummond
Acute Effects of Cheddar Cheese Consumption on Circulating Amino Acids and Human Skeletal Muscle
Reprinted from: *Nutrients* **2021**, *13*, 614, doi:10.3390/nu13020614 287

About the Editor

Dennis Savaiano

Dennis Savaiano is the Virginia Meredith Professor in the Department of Nutrition Science and Dean Emeritus at Purdue University in West Lafayette, Indiana, USA. He earned MS and PhD degrees in Nutrition from the University of California at Davis, and a BA in Biology from Claremont McKenna College. For the past four decades, his research group has studied numerous factors which influence lactose digestion and tolerance, including lactose load, gastric and intestinal transit, the use of lactose digestive aids, the colon fermentation of lactose, and the consumption of fermented dairy foods and lactic acid bacteria. Major findings from these studies include: (1) The identification of a microbial lactase in yogurts that assists lactose digestion in the intestinal tract following the consumption of yogurt. (2) The characterization of the amount of lactose required to cause symptoms in lactose maldigesters as 12 g or more of lactose (one cup of milk). (3) The finding that lactose consumed with a meal is tolerated about three times better than lactose consumed in a fasting state. (4) Identifying the colonic flora as key to determining tolerance to lactose. The colonic flora readily adapt to lactose in the diet of maldigesters. Thus, maldigesters who routinely consume lactose have less symptoms due to the more efficient metabolism of lactose by their colon microflora. (5) The identification of a population of digesters and maldigesters who believe that they are extremely intolerant to lactose, but who tolerate lactose quite well in double-blinded clinical trials. (6) The characterization of the ability of lactic acid bacteria, including acidophilus and bifidus, to improve lactose digestion in vivo in the gastrointestinal system.

Preface to "Dairy Products for Human Health"

The consumption of dairy foods has changed dramatically over the past sixty years in the United States, with much less fluid milk consumed and an increased consumption of cheeses and yogurts. Coffee, sugar-containing beverages, and plant-based milks have replaced much of the fluid milk consumption in the American diet. At the same time, growth in both the international dairy industry and the global consumption of dairy foods has been substantial. Dairy foods are under considerable scrutiny, with concerns relating to the environment and the biological effects of dairy components, including their protein fractions, lipids, lactose, and other nutrients. At the same time, the high nutrient content of dairy foods, including protein, calcium, potassium, and riboflavin, make them a significant contributor to diet quality. Dairy product consumption can influence gut health, weight, cardiometabolic health, diabetes, bone mineral density, and many types of cancers.

This Special Issue of Nutrients aims to collect new scientific evidence addressing health concerns and opportunities related to dairy product consumption. Dairy products play an important role in diet quality and are associated with human health and disease. Our goal is to provide a stronger base of scientific information for the consumer as well as the professionals who advise them on their diet. Both professionals and consumers are undoubtedly confused about the values and risks of dairy foods in the diet due to the limited scientific evidence behind many of the claims made.

Dennis Savaiano
Editor

Article

Contribution of Milk Beverages to Nutrient Adequacy of Young Children and Preschool Children in the Philippines

Tsz-Ning Mak [1,*], Imelda Angeles-Agdeppa [2], Marie Tassy [1], Mario V. Capanzana [2] and Elizabeth A. Offord [1]

1. Nestlé Research, Route du Jorat 57, 1000 Lausanne, Switzerland; Marie.Tassy@rd.nestle.com (M.T.); elizabeth.offord-cavin@rdls.nestle.com (E.A.O.)
2. Food and Nutrition Research Institute-Department of Science and Technology, Taguig City 1631, Philippines; iangelesagdeppa@yahoo.com.ph (I.A.-A.); mar_v_c@yahoo.com (M.V.C.)
* Correspondence: TszNing.Mak@rd.nestle.com

Received: 19 December 2019; Accepted: 19 January 2020; Published: 1 February 2020

Abstract: Malnutrition is a major public health concern in the Philippines. Milk and dairy products are important sources of energy, protein, and micronutrients for normal growth and development in children. This study aims to assess the contribution of different types of milk to nutrient intakes and nutrient adequacy among young and preschool children in the Philippines. Filipino children aged one to four years ($n = 2992$) were analysed while using dietary intake data from the 8th National Nutrition Survey 2013. Children were stratified by age (one to two years and three to four years) and by milk beverage consumption type: young children milk (YCM) and preschool children milk (PCM), other milks (mostly powdered milk with different degrees of fortification of micronutrients), and non-dairy consumers (no milks or dairy products). The mean nutrient intakes and the odds of meeting nutrient adequacy by consumer groups were compared, percentage of children with inadequate intakes were calculated. Half (51%) of Filipino children (all ages) did not consume any dairy on a given day, 15% consumed YCM or PCM, and 34% consumed other milks. Among children one to two years, those who consumed YCM had higher mean intakes of iron, magnesium, potassium, zinc, B vitamins, folate, and vitamins C, D, and E (all $p < 0.001$) when compared to other milk consumers. Non-dairy consumers had mean intakes of energy, total fat, fibre, calcium, phosphorus, iron, potassium, zinc, folate, and vitamins D and E that were far below the recommendations. Children who consumed YCM or PCM had the highest odds in meeting adequacy of iron, zinc, thiamin, vitamin B6, folate, and vitamins C, D, and E as compared to other milks or non-dairy consumers, after adjusting for covariates. This study supports the hypothesis that dairy consumers had higher intakes of micronutrients and higher nutrient adequacy than children who consumed no milk or dairy products. Secondly, YCM or PCM have demonstrated to be good dairy options to achieve nutrient adequacy in Filipino children.

Keywords: milk; dairy; nutrient adequacy; Philippines; young children; preschool children

1. Introduction

A double burden of malnutrition exists in the Philippines. The prevalence of undernutrition in children is high, where one in three children under five years are stunted and 7% are wasted [1]. Micronutrient deficiency continues to be a key public health concern; vitamin A deficiency affects one in five children, while 14% and 18% of children < 5 years are anemic and zinc deficient, respectively [2].

A previous study found that children in the Philippines have poor diet diversity, and the prevalence of inadequate intakes of micronutrients is high [3]. Seventy-eight percent, 75% and 90% of Filipino children aged one, two, and three to four years, respectively, had inadequate intakes in iron; 62%, 66%, and 84% were inadequate in calcium; 52%, 46%, and 47% were inadequate in zinc, 60%, 41%, and 43% were inadequate in vitamin A, respectively. More than 40% of children aged one year were inadequate in B vitamins (thiamine, riboflavin, niacin, B6, folate, and B12), and in children three to four years, 72% were inadequate in folate and 60% in vitamin C [3]. The diet of children in the Philippines typically composed of refined rice and energy-dense foods such as cookies and sugar sweetened beverages, low in fruit, vegetables, and protein-rich foods [3]. The consumption of milk and dairy products is generally low among children in the Philippines with a decreasing trend with age [3]. At age one year, 37% of Filipino children were having breastmilk, 35% consuming cow's milk and 23% consuming toddler milk; at two years, only 11% were receiving breastmilk, 38% of children were consuming cow's milk, and 15% consuming toddler milk; and, at age three to four years, only 32% of children would consume milk on a given day [3].

Milk and dairy are part of an important food group for preschool children, as it greatly contributes to total energy, protein, and calcium in relation to children's daily requirements for normal growth and development [4]. In the Philippines, powdered milk is the most common type of milk available. It is a milk usually intended for the whole family, and is typically fortified with calcium and vitamin A. Young children milk (YCM) and preschool children milk (PCM), are milks tailored for children's nutritional needs at different age stages, fortified with vitamins and minerals, some with DHA and probiotics. A number of observational studies and randomized controlled trials (RCT) in Europe and Australia and New Zealand have demonstrated that, when compared to cow's milk, YCM improved nutrient intakes among children above 12 months, particularly in iron and vitamin D [5–9]. Several studies have shown that children who consumed YCM had better iron and vitamin D status than non-consumers [7,10,11], and it had a lower percentage of body fat than children who consumed cow's milk [12]. This is likely due to the lower protein of YCM than cow's milk, as evidence supports the early protein hypothesis that high protein intake in early years might promote weight gain and higher risk of obesity in later life [13,14].

Beyond developed countries, there is a lack of studies on the impact of YCM or PCM in children. A RCT on Indian children found that those who consumed fortified milk had improved weight and height gain and iron status than those who consumed unfortified milk [15]. In the Philippines, only one study can be found to measure the impact of fortified milk among young children. The clinical study showed that preschool children (three to five years) from a small town in Philippines, who consumed two servings of fortified milk a day for 12 weeks significantly increased the intakes of energy, protein, iron, vitamin A, calcium, thiamin, riboflavin, niacin, and vitamin C when compared to baseline [16]. Therefore, more evidence is needed to examine the role of milks in the diet of children across the Philippines, where nutrient inadequacy is highly prevalent.

The aim of the current study was therefore to assess the contribution of different milk types to nutrient intakes and nutrient adequacy among young children and preschool children in the Philippines. We hypothesized that children who consumed milk (YCM/ PCM/ other milks) would have higher intakes of micronutrients and higher nutrient adequacy than children who consumed no milk or dairy products. Furthermore, we also hypothesized that YCM/ PCM consumers may have higher nutrient adequacy than other milk consumers, given the high level of fortification of micronutrients in YCM/PCM.

2. Materials and Methods

2.1. Study Population

Dietary intake data of children aged one to two years ($n = 1461$) and three to four years ($n = 1531$) from the 2013 National Nutrition Survey (NNS) were analysed. The details of the NNS have been reported previously [17]. The 2013 NNS was a cross-sectional nationally representative health and nutrition survey of the Filipino population. Filipino households ($n = 35{,}825$) were sampled with a response rate of 91% [3]. The Ethics Committee of Food Nutrition Research Institute (FNRI) approved the survey protocol and data collection instruments. All of the surveyed households provided informed consent prior to participation with a registry number FIERC—2013—001 [17].

2.2. Data Collection

Trained dietitians conducted face-to-face 24-h dietary recalls with a parent or caregiver of each child during household visits, wherein the dietitian recorded all food and beverages that the child consumed the previous day. A first 24-h recall was performed on all children and a second 24-h recall was repeated in 50% of randomly selected households, typically two days after the first recall. The amount of each food item or beverage was estimated while using common household measures, such as cups, tablespoons, by size, or number of pieces. The information was then converted to grams while using a portion-to-weight list for common foods compiled by FNRI or through weighing of the food samples. For powdered milk, a conversion factor of 6.7 was multiplied to estimate the gram weight of fluid milk as consumed, based on the preparation instruction of approximately 36 g milk powder reconstituted to approximately 240 g of fluid milk (in line with recommendation from the Philippines Nutritional Guide Pyramid—four tablespoons of milk powder to one glass of water) [18]. All of the reported foods and drinks were reviewed, to ensure that all of the codes and quantities were correctly entered. The food and beverages consumed were converted to energy and nutrient intakes while using the Philippines Food Composition Tables (PFCT) [19]. The energy intake distribution of the full sample (n = 3016) was assessed and the outliers were defined as energy intakes three standard deviations above and below (±3SD) the mean per age group (one to two year and three to four years) and were removed from analyses. A final sample size of 2992 was retained for analysis.

The 888 unique food items reported were categorised into nine major food groups and further subgroups. The food grouping system was adapted from What We Eat in America Food Categories [20] and US Feeding Infants and Toddlers Study (FITS) adjusted for the Philippines local food culture [3].

Demographic characteristics, including dwelling location, wealth quintiles, mother's education, occupation status, and civil status were collected. Children's height-for-age and weight-for-age were obtained and the BMI z-scores were calculated to define their nutritional status (wasting, at risk of overweight, and overweight).

2.3. Data Processing

The three consumer groups were defined as follow. Children who reported to have consumed "Formula", which included subcategories "Toddler/Pre-school formula" and "Infant formula" were considered as "YCM or PCM consumers". We identified only 6% of children at age 1 year who were still consuming infant formula. Children who had reported to consume "Milk" which included subcategories "Milk, powdered", "Milk, fluid", "Other Milk", and "Dairy products" were categorized as "Other milk consumers". Dairy products (cheese and yogurt) consumption was only represented by 2.7% of the sample and was therefore grouped together with the milk subcategories. Children who had consumed both YCM/PCM and other milks were considered as "YCM/PCM consumers". All of the children who did not consume any of the above (YCM/PCM, other milk or dairy) were considered "non-dairy consumers". The nutritional compositions of YCM/PCM, milk (fluid), and milk (powdered) considered in the analyses can be found in Supplementary Materials Table S1.

Twenty-seven nutrients were available in the PFCT. After checking for completeness, 22 nutrients, including total energy (kcal), total carbohydrates (g), fibre (g), protein (g), total fat (g), sodium (mg), calcium (mg), phosphate (mg), iron (mg), magnesium (mg), potassium (mg), selenium (mg), zinc (mg), thiamine (mg), riboflavin (mg), niacin (mg), vitamin B6 (mg), folate (µg), vitamin B12 (mg), vitamin C (mg), vitamin D (µg), and vitamin E (mg) were analysed. Vitamin A, total sugars, saturated fat, monounsaturated fat, and polyunsaturated fat, while available in the PFCT, had high percentage of missing values for the food and beverages of interest, which were deemed too incomplete for the current analysis. The total daily intake of each nutrient based on food consumption was calculated per person per day. An average of nutrient intakes over the two days were considered in the subsequent analyses for children who had a second 24 h recall day.

2.4. Statistical Analysis

All analyses were stratified by age groups: one to two years, and three to four years. The relationships between sociodemographic variables (gender, age group, wealth quintiles, dwelling location, mother's education, civil status, and occupation status, and BMI z-scores) and consumer groups were tested for significance while using chi-squared tests and Kruskal–Wallis tests. The mean intake per nutrient per consumer group was calculated. Non-parametric Kruskal–Wallis tests were used to compare the mean ranks of each nutrient between three consumer groups due to the skewness of the nutrient intake data. Post hoc Dunn's tests were then performed to compare mean nutrient intakes between YCM/PCM consumers and other milk consumers, and YCM/PCM consumers and non-dairy consumers.

Inadequate intakes were defined while using the estimated average requirements (EAR) cut-off method. Children with intake below the EAR for a given nutrient were considered to be inadequate. For nutrients that did not have an established EAR cut-off, e.g., vitamin D and vitamin E, adequate intake (AI) values were used. The Philippines Nutrient Reference Intakes (PNRI) table was used to establish EAR/AI for protein, calcium, phosphorus, iron, selenium, zinc, thiamin, riboflavin, niacin, vitamins B6, B12, folate, and vitamins C, D, and E. For macronutrients (except for protein where EAR is available), Acceptable Macronutrient Distribution Ranges (AMDR) were used to evaluate carbohydrates and total fat, as a percentage of energy. The proportions of inadequate intakes were classified as %E less than the AMDR lower range. The proportion of children with inadequate intakes were calculated per consumer group.

Logistic regression analysis was used to estimate the odds ratios of children meeting the EAR/AI/AMDR of each nutrient with respect to the consumer group, to test the hypothesis that YCM/PCM consumers (reference group) had higher odds of meeting the EAR/AI/AMDR of nutrients when compared to other milk consumers and/or non-dairy consumers. Adjustment of potential confounding factors—wealth quintiles, dwelling location, mother's education, and mother's occupation was included in the final model. R version x64.3.6.1 and R-Studio version 1.2.1335 were used for all of the statistical analysis.

3. Results

3.1. Descriptive Statistics

Table 1 shows a description of the sample population by consumption group. There was an equal split in gender within the sample. Eleven percent of children one to two years suffered from wasting, while 13% were at risk of overweight or overweight. The majority came from the rural area and in poor to poorest wealth quintiles. Most mothers were not working, were married, and were largely high school graduates. There were no significant differences between consumption groups and gender, BMI z-scores, and mother's civil status. However, wealth quintiles, mother's education, and mother's occupation status were significantly related to if children consumed YCM/PCM, other milk or no dairy (all $p < 0.001$).

Of the 1461 children aged one to two years, 43% had no dairy (other milk, YCM or other dairy products), 35% were other milk consumers, and 22% YCM consumers. The mean consumption of YCM before reconstitution was 101 g/d (±SD 69 g) and in fluid weight was 641 g/d (±SD 432 g); mean consumption of other milk before reconstitution was 61 g/d (±SD 58 g) and in fluid weight was 354 g/d (±SD 330 g).

Among children aged three to four years ($n = 1531$), the majority (58%) did not consume any dairy, 34% consumed other milks, and 8% of children consumed PCM. Mean consumption of PCM before reconstitution was 85 g/d (±SD 73 g) and, in fluid weight, was 454 g/d (±SD 355 g); mean consumption of other milks before reconstitution was 46 g/d (±SD 53 g) and in fluid weight 224 g/d (±SD 233 g) for this age group.

Table 1. Sample characteristics by consumer group (toddlers and young children aged one to four years).

		YCM/PCM Consumers		Other milk Consumers		Non-Dairy Consumers		Total	p-Value
		n	%	n	%	n	%	n	
Gender	Boys	235	16	527	35	745	49	1507	
	Girls	210	14	502	34	773	52	1485	0.362
Age group	1 to 2 years	328	22	506	35	627	43	1461	
	3 to 4 years	117	8	523	34	891	58	1531	<0.001
Wealth quintiles	Poorest	29	3	187	22	648	75	864	
	Poor	37	6	222	36	357	58	616	
	Middle	75	13	233	41	266	46	574	
	Rich	116	24	230	47	142	29	488	
	Richest	175	48	124	34	68	19	367	<0.001
Dwelling location	Rural	169	10	519	31	991	59	1679	
	Urban	276	21	510	39	527	40	1313	<0.001
Mother's Education	No Grades Completed	1	3	2	6	31	91	34	
	Elementary Level	13	2	137	26	375	71	525	
	High School Level	66	7	324	34	554	59	944	
	Vocational Level	20	23	31	35	37	42	88	
	College Level	134	36	124	34	111	30	369	
	Others	0	0	0	0	5	100	5	<0.001
Mother's Civil Status	Single	9	14	22	34	34	52	65	
	Married	172	12	434	31	792	57	1398	
	Live-in	43	10	128	31	243	59	414	
	Separated/Annulled/Divorced	1	5	8	40	11	55	20	
	Widow/Widower	2	13	6	40	7	47	15	
	Unknown	0	0	0	0	1	100	1	0.917
Mother's occupation status	With Job/Business	99	20	165	33	238	47	502	
	Housekeeper/No Occupation/Pensioner/Student	128	9	432	31	850	60	1410	<0.001
BMI z-scores	<−2 (Wasting)	41	16	93	36	121	47	255	
	−2 to −1	54	9	187	33	334	58	575	
	−1 to 0	119	11	371	34	605	55	1095	
	0 to 1	125	17	257	35	342	47	724	
	1 to 2 (At risk of overweight)	66	29	79	34	85	37	230	
	>2 (Overweight)	40	35	42	37	31	27	113	0.495

3.2. Comparison of Mean Nutrient Intakes by Consumer Group

3.2.1. YCM/PCM Consumers vs. Other Milk Consumers

Table 2 shows the comparison of mean nutrient intakes by consumer group. YCM/PCM consumers and other milk consumers had similar macronutrient intakes, except for protein (one to two years only), where YCM consumers had lower intake. YCM/PCM consumers had higher total fat (three to four years only) and higher micronutrient intakes than other milk consumers, except for sodium, phosphorus, riboflavin, and selenium (one to two years only). No significant differences were found for calcium intakes in children aged one to two years. No significant differences were found for sodium, magnesium, selenium, and B12 in the older children.

YCM/PCM and other milk consumers both had mean intakes of energy, fibre, and potassium lower than the recommended intake levels. In general, YCM/PCM consumers had mean intakes of micronutrients closer to the recommendations compare to other milk consumers. In children aged one to two years, other milk consumers had mean intakes of iron, magnesium, folate, and vitamins D and E far below the recommendations.

Table 2. Mean and distribution of nutrient intakes by consumer group for children aged one to two years, three to four years.

		Reference Intake	1 to 2 Years			Mean and Distribution of Intakes					Reference Intake	3 to 4 Years			Mean and Distribution of Intakes				
		M/F	Mean	SD	a	10th	25th	50th	75th	90th	M/F	Mean	SD	a	10th	25th	50th	75th	90th
Energy (kcal)	All	1000/920	690.2	383.0		253.9	404.7	629.6	901.8	1236.5	1350/1260	915.1	395.0		455.6	616.1	860.5	1151.3	1459.5
	YCM		798.4	377.4		343.3	492.1	774.8	1047.8	1347.2		1058.6	408.5		502.6	817.5	1011.8	1351.4	1538.4
	Other milk		802.2	378.5	N.S.	380.3	532.8	741.1	1004.1	1338.4		1031.0	392.1	N.S.	574.0	727.8	974.8	1265.1	1607.6
	Non-dairy		543.3	338.1	*	175.0	290.4	474.2	727.9	989.8		828.1	371.6	*	410.7	552.6	772.3	1036.0	1318.2
Protein (g)	All	15/14	22.3	14.6		6.6	11.6	19.7	30.0	40.6	18/17 [1]	28.8	14.8		12.6	18.3	26.2	36.6	47.1
	YCM		24.6	13.3		9.6	14.9	22.9	33.8	42.4		34.8	18.8		16.9	22.3	33.8	42.8	50.2
	Other milk		28.7	15.9	‡	11.8	17.4	25.5	36.7	50.1		34.5	14.6	N.S.	18.8	24.2	32.0	42.8	54.6
	Non-dairy		16.0	11.2	*	4.5	7.7	13.4	22.2	31.3		24.7	12.7	*	11.0	15.3	22.6	31.2	41.8
Total fat (g)	All	25–35%E	19.0	15.9		2.8	7.0	14.9	27.1	40.8	15–30%E	21.1	16.7		4.4	9.0	16.6	29.1	44.0
	YCM		25.8	16.4		6.7	13.3	23.6	35.3	45.6		34.2	19.5		12.9	21.2	29.6	44.8	58.6
	Other milk		25.3	16.5	N.S.	8.1	13.5	20.9	33.1	48.3		28.8	16.8	#	11.1	15.9	24.7	39.3	50.9
	Non-dairy		10.3	9.8	*	1.3	3.4	7.5	14.0	22.1		14.9	12.9	*	2.9	5.9	11.6	19.4	31.5
Carbohydrates (g)	All	50–69%E	107.4	58.5		40.3	66.4	97.4	139.8	185.1	55–79%E	152.6	66.0		77.9	104.5	143.7	190.8	240.7
	YCM		116.5	54.2		53.6	76.1	110.0	152.8	188.3		153.2	62.7	N.S.	80.0	114.6	142.9	189.5	237.7
	Other milk		114.7	54.8	N.S.	56.0	75.6	102.1	141.4	189.4		158.6	64.3	N.S.	85.3	110.2	149.3	195.0	248.6
	Non-dairy		96.7	61.9	*	30.9	51.8	84.9	127.4	175.8		148.9	67.2		73.3	100.1	139.2	185.1	236.6
Fibre (g)	All	6–7	2.9	2.3		0.7	1.3	2.3	3.8	5.7	8–10	4.5	2.7		1.8	2.7	4.0	5.8	8.0
	YCM		2.6	2.0		0.5	1.1	2.1	3.7	5.4		4.3	2.8		1.6	2.6	3.5	5.5	7.6
	Other milk		2.8	2.0	N.S.	0.8	1.3	2.2	3.7	5.4		4.4	2.6	N.S.	1.6	2.6	4.0	5.7	7.8
	Non-dairy		3.1	2.5	#	0.8	1.5	2.4	3.9	6.1		4.7	2.8	N.S.	1.9	2.7	4.0	5.8	8.2
Sodium (mg)	All	225 [2]	512.6	499.2		70.2	195.6	405.4	685.1	1036.2	300 [2]	686.5	590.8		125.7	290.5	553.0	923.1	1369.2
	YCM		478.1	407.6	*	85.3	215.6	403.0	621.4	892.3		756.5	557.5	N.S.	243.3	410.5	601.5	994.4	1435.9
	Other milk		636.4	491.5	*	175.2	313.0	531.3	834.4	1160.5		833.8	622.9	N.S.	256.5	427.6	687.4	1043.2	1527.3
	Non-dairy		430.8	528.8		21.5	108.0	294.2	564.5	906.3		590.9	556.3	*	83.4	215.1	443.2	794.2	1266.9
Calcium (mg)	All	440	374.6	428.2		36.6	82.5	209.6	518.2	947.5	440	283.1	275.6		73.5	116.7	199.2	342.8	572.3
	YCM		622.3	534.3		57.3	204.7	510.2	890.5	1268.4		419.8	401.3		67.0	160.6	320.1	546.2	856.9
	Other milk		540.1	425.2	N.S.	134.8	239.3	407.8	697.5	1111.9		446.5	310.1	*	187.8	242.4	357.5	524.5	819.2
	Non-dairy		111.5	94.9	*	20.2	44.1	86.6	156.1	220.8		169.3	148.7	*	61.7	91.6	138.5	203.8	302.2
Phosphorus (mg)	All	380	395.7	304.2		99.6	183.4	318.9	516.7	788.0	405	460.4	241.2		199.9	289.4	416.2	582.7	768.4
	YCM		407.6	282.6	*	90.6	206.1	343.1	595.6	800.8		501.6	275.8	*	176.0	298.6	487.4	641.1	890.1
	Other milk		574.7	350.0	*	221.5	332.9	488.6	721.8	1074.3		593.3	262.0	*	321.9	411.4	536.9	711.1	956.7
	Non-dairy		245.1	163.1		69.3	123.0	214.8	328.7	456.8		377.0	179.6	*	174.3	242.9	346.6	478.2	621.5
Iron (mg)	All	6.4/7.0	4.4	5.1		0.9	1.6	3.0	5.4	9.7	7.5/7.4	5.2	3.8		1.8	2.7	4.2	6.6	9.1
	YCM		8.7	6.8	*	2.1	4.2	7.4	11.9	16.5		8.7	4.8	*	3.3	5.6	8.3	10.6	15.3
	Other milk		3.4	4.7	*	1.0	1.5	2.6	4.0	6.1		5.3	3.9	*	2.0	2.8	4.4	6.8	9.3
	Non-dairy		3.0	2.4		0.6	1.3	2.4	4.3	6.1		4.6	3.2		1.6	2.6	3.9	5.8	8.1

Table 2. Cont.

			1 to 2 Years									3 to 4 Years							
	Reference Intake		Mean and Distribution of Intakes								Reference Intake	Mean and Distribution of Intakes							
	M/F		Mean	SD	a	10th	25th	50th	75th	90th	M/F		Mean	SD	a	10th	25th	50th	75th	90th
Magnesium (mg)	60/60 [2]	All	59.3	48.0		15.7	28.0	47.7	76.9	114.0	70/70 [2]	All	88.0	60.1		37.6	52.5	76.5	107.9	142.1
		YCM	73.0	50.3	*	21.0	37.4	62.0	100.2	136.3		PCM	92.6	53.6	N.S.	34.9	56.5	87.7	121.3	153.3
		Other milk	56.7	53.4	*	16.3	26.0	41.9	68.6	109.2		Other milk	92.0	66.4	N.S.	34.7	52.4	75.8	111.0	154.7
		Non-dairy	54.3	40.2		13.6	26.3	45.4	72.2	102.9		Non-dairy	85.0	56.8		39.6	52.7	75.9	103.9	132.2
Potassium (mg)	1000 [2]	All	481.2	400.3		118.1	218.1	370.4	605.1	994.8	1400 [2]	All	627.3	396.4		252.8	357.1	536.5	795.8	1085.5
		YCM	707.7	541.9	*	156.1	314.5	560.7	1051.0	1351.9		PCM	812.4	525.4	#	280.2	464.9	718.2	1042.5	1441.4
		Other milk	471.0	355.3	*	164.7	247.1	380.1	567.8	840.6		Other milk	691.6	425.7	*	292.9	406.3	591.6	876.9	1145.8
		Non-dairy	370.8	281.6		88.5	164.0	300.9	506.5	769.1		Non-dairy	565.3	342.2		239.2	339.8	477.1	704.8	975.8
Selenium (µg)	13.6/13.0	All	30.7	23.4		7.9	14.2	25.8	40.9	58.9	16.1/15.6	All	47.1	27.0		17.6	27.1	41.5	61.2	83.5
		YCM	27.8	21.4		5.2	12.1	22.8	39.1	53.0		PCM	48.1	24.3	N.S.	19.5	30.5	44.9	65.2	79.3
		Other milk	33.7	24.3	*	10.9	16.9	27.6	43.0	62.4		Other milk	50.6	28.6	N.S.	20.0	30.5	44.6	65.1	87.8
		Non-dairy	29.8	23.4	N.S.	7.2	13.3	24.1	39.9	59.0		Non-dairy	44.9	26.1		16.5	25.9	39.6	58.7	79.6
Zinc (mg)	2.8/2.6	All	3.0	2.8		0.7	1.2	2.2	3.7	6.3	3.3/3.2	All	3.3	2.2		1.2	1.8	2.8	4.3	6.1
		YCM	5.4	3.8	*	1.3	2.6	4.8	7.4	10.1		PCM	5.6	2.6	*	2.6	3.8	5.4	6.9	9.0
		Other milk	2.7	1.8	*	1.0	1.5	2.3	3.5	4.8		Other milk	3.7	2.1	*	1.6	2.1	3.2	4.5	6.5
		Non-dairy	1.9	1.8		0.4	0.8	1.4	2.5	3.7		Non-dairy	2.8	1.9		1.0	1.6	2.4	3.6	4.8
Thiamin (mg)	0.4/0.4	All	0.4	0.4		0.1	0.2	0.3	0.6	0.9	0.5/0.4	All	0.5	0.4		0.2	0.3	0.4	0.7	1.0
		YCM	0.7	0.5	*	0.2	0.3	0.6	0.9	1.2		PCM	0.7	0.5	*	0.3	0.5	0.7	0.9	1.1
		Other milk	0.4	0.4	*	0.1	0.2	0.3	0.5	0.7		Other milk	0.6	0.4	*	0.2	0.3	0.5	0.7	1.0
		Non-dairy	0.3	0.3		0.1	0.1	0.2	0.4	0.7		Non-dairy	0.4	0.3		0.1	0.2	0.4	0.6	0.9
Riboflavin (mg)	0.4/0.4	All	0.7	0.8		0.1	0.2	0.4	1.0	1.6	0.5/0.4	All	0.6	1.8		0.1	0.2	0.4	0.7	1.1
		YCM	0.9	0.8	*	0.2	0.3	0.7	1.3	1.9		PCM	0.7	0.5	*	0.2	0.3	0.6	0.9	1.2
		Other milk	1.1	0.9	*	0.3	0.5	0.9	1.4	2.3		Other milk	1.1	3.1	*	0.4	0.5	0.8	1.1	1.7
		Non-dairy	0.2	0.2		0.1	0.1	0.2	0.3	0.5		Non-dairy	0.4	0.3		0.1	0.2	0.3	0.5	0.7
Niacin (mg)	5/5	All	5.8	4.1		1.5	2.7	4.9	7.9	11.2	5/5	All	8.4	5.1		3.3	5.0	7.4	10.7	14.4
		YCM	7.9	4.5	*	2.7	4.6	7.2	10.3	14.2		PCM	10.7	5.0	*	4.8	6.8	10.7	14.0	16.5
		Other milk	5.2	3.8	*	1.8	2.6	4.2	6.5	9.5		Other milk	8.6	6.1	*	3.4	5.2	7.5	10.4	14.4
		Non-dairy	5.1	3.8		1.2	2.3	4.2	7.1	10.2		Non-dairy	7.9	4.3		3.1	4.8	7.1	10.2	13.7
Vitamin B6 (mg)	0.4/0.5	All	0.5	0.5		0.1	0.2	0.3	0.6	0.9	0.5/0.5	All	0.6	0.5		0.2	0.4	0.5	0.8	1.1
		YCM	0.7	0.5	*	0.2	0.4	0.6	1.0	1.3		PCM	0.9	0.6	*	0.4	0.5	0.8	1.1	1.5
		Other milk	0.4	0.5	*	0.1	0.2	0.3	0.5	0.7		Other milk	0.7	0.6	*	0.2	0.4	0.5	0.8	1.2
		Non-dairy	0.4	0.3		0.1	0.2	0.3	0.5	0.7		Non-dairy	0.6	0.3		0.2	0.3	0.5	0.7	1.0
Vitamin B12 (mg)	0.8/0.9	All	1.2	1.8		0.0	0.2	0.7	1.6	2.7	0.9/1.0	All	1.9	3.8		0.2	0.5	1.1	2.1	3.8
		YCM	1.5	1.5	*	0.2	0.6	1.2	2.0	3.0		PCM	1.9	1.7	N.S.	0.4	0.8	1.5	2.4	3.5
		Other milk	1.2	2.2	*	0.0	0.2	1.0	1.5	2.9		Other milk	1.9	5.0	#	0.2	0.5	1.2	2.1	3.8
		Non-dairy	1.0	1.6		0.0	0.1	0.6	1.3	2.4		Non-dairy	1.8	3.2		0.1	0.4	1.0	2.1	4.0

Table 2. Cont.

		1 to 2 Years									3 to 4 Years								
	Reference Intake	Mean and Distribution of Intakes								Reference Intake	Mean and Distribution of Intakes								
	M/F	Mean	SD	a	10th	25th	50th	75th	90th	M/F	Mean	SD	a	10th	25th	50th	75th	90th	
Folate (µg)	120/120									160/160									
All		101.1	106.4		12.6	31.3	69.0	133.4	232.8		126.6	104.8		24.6	53.2	99.7	167.8	263.3
YCM		184.1	153.9	‡	36.7	66.6	140.0	266.1	388.5		212.7	131.1	*	75.3	121.2	180.2	292.4	404.0
Other milk		74.9	66.1	*	11.7	26.5	58.3	98.1	165.4		115.9	89.1		26.6	52.2	93.1	158.9	234.4
Non-dairy		78.9	76.5		9.6	24.5	55.6	109.7	175.9		121.5	104.5		22.1	48.4	95.0	158.5	255.2
Vitamin C (mg)	12/11									17/17								
All		21.2	33.3		0.0	2.0	9.2	24.9	58.4		19.2	30.4		0.0	2.6	8.6	21.9	47.4
YCM		50.8	43.9	‡	2.0	18.3	40.1	74.0	109.4		40.7	43.4	*	5.5	13.6	28.1	53.2	90.1
Other milk		14.2	18.2	*	2.0	4.4	9.4	16.8	28.7		18.8	25.5	*	2.6	5.0	10.2	20.4	42.4
Non-dairy		11.4	26.8		0.0	0.0	3.4	13.0	29.6		16.6	29.9		0.0	0.6	6.3	18.1	42.0
Vitamin D (µg)	5[3]									5[3]								
All		2.1	3.7		0.0	0.1	0.8	2.5	5.9		1.7	2.3		0.0	0.3	1.1	2.2	4.1
YCM		5.9	5.1	‡	0.6	2.3	4.8	8.1	12.4		4.3	3.6	*	1.3	2.1	3.3	5.2	8.1
Other milk		1.0	2.1	*	0.0	0.0	0.5	1.3	2.6		1.4	1.5	*	0.0	0.3	1.0	2.0	3.3
Non-dairy		1.1	2.2		0.0	0.0	0.5	1.4	2.6		1.5	2.3		0.0	0.3	0.9	1.9	3.6
Vitamin E (mg)	4[3]									4[3]								
All		1.8	2.3		0.1	0.4	1.0	2.2	4.6		2.0	2.1		0.4	0.7	1.5	2.6	4.3
YCM		4.2	3.3	‡	0.5	1.7	3.5	5.8	9.0		4.7	3.1	*	1.6	2.5	4.3	6.2	8.9
Other milk		1.2	1.5	*	0.1	0.3	0.8	1.4	2.5		2	1.9	*	0.4	0.8	1.5	2.7	4.1
Non-dairy		1.1	1.1		0.1	0.3	0.8	1.5	2.6		1.7	1.8		0.3	0.6	1.3	2.2	3.3

A represents test for significance between mean intakes of YCM/PCM consumers and Other milk consumers, and YCM/PCM consumers and Non-dairy consumers. N.S. non-significant; # $p < 0.05$; ‡ $p < 0.01$; * $p < 0.001$. M = Male; F = Female. [1] Recommended Energy Intake (REI); [2] Recommended Nutrient Intakes (RNI); [3] Adequate Intake (AI); Reference intakes w/o superscripts = EAR.

3.2.2. YCM/PCM Consumers vs. Non-Dairy Consumers

Non-dairy consumers had significantly lower intakes than YCM consumers in most macro- and micronutrients, except for fibre in children aged one to two years ($p < 0.05$). No significant differences were seen for carbohydrates, fibre, and magnesium and selenium between the PCM and non-dairy consumer groups in children aged three to four years. Non-dairy consumers had mean nutrient intakes far below the recommendation. The only nutrients where non-dairy consumers were in line or above the recommended levels were protein, sodium, selenium, and niacin (all ages); magnesium, B6 and B12 (three to four years only).

3.3. Percentage of Children with Inadequate Intakes

Table 3 shows the proportion of children with nutrient intakes below the EAR or AI across the three consumer groups. Among non-dairy consumers aged one to two years, except for carbohydrates and selenium, more than 50% of children had intakes below the EAR/AI. In particular, almost all children in this group were inadequate in calcium, vitamin D, and iron (99%, 97% and 91%, respectively). The highest level of inadequate intakes among other milk consumers (one to two years) were vitamin D (98%), vitamin E (97%), iron (92%), folate (82%), and vitamin B6 (73%). YCM consumers (1 to 2 years) had the lowest percentage of inadequate intakes when compared to the other groups. The highest levels of inadequacy in this group were vitamin E (68%), phosphorus (55%), vitamin D (52%), and folate (49%).

Among preschool children, the proportions with inadequate intakes were slightly lower than the younger children, except for calcium in PCM (67%) and other milk consumers (65%), and vitamin D (72%) in PCM consumers. Non-dairy consumers still had the highest proportion of children with inadequate intakes across most nutrients.

3.4. Odds of Nutrient Adequacy Per Consumer Group

Table 4 illustrates the odds ratios of other milk consumers and non-dairy consumers meeting nutrient adequacy when compared to YCM/PCM consumers (reference group), after adjusting for potential confounding factors (wealth quintiles, dwelling location, mother' education level, mother's occupation). Among children aged one to two years, other milk consumers were significantly less likely to meet the AMDR for carbohydrates, as well as the EARs for iron, zinc, thiamin, niacin, vitamin B6, folate, vitamins C, D, and E (all $p < 0.001$), and vitamin B12 ($p = 0.002$), than YCM consumers (reference group). Other milk consumers were more likely to meet the EAR of protein ($p = 0.003$), phosphorus, selenium, and riboflavin ($p \leq 0.001$) than YCM consumers. Non-dairy consumers had significantly lower odds of meeting the adequacy of all nutrients when compared to YCM consumers, except for carbohydrates and selenium (both non-significant).

In preschool children, other milk consumers were less likely to reach nutrient adequacy in iron, folate, vitamins C, D, E (all $p < 0.001$), zinc, thiamin, and vitamin B6 (all $p < 0.05$), but were more likely to meet adequacy in phosphorus and riboflavin when compared to PCM consumers. Non-dairy consumers had lower likelihood of meeting adequacy in total fat, calcium, iron, zinc, folate, vitamins C, D, E (all $p < 0.001$), and thiamin ($p = 0.005$) and vitamins B6 ($p = 0.024$), but more likely of being within AMDR of carbohydrates ($p = 0.016$) than PCM consumers.

Table 3. Percentage of children with inadequate intakes per consumer group by age groups one to two years, three to four years, and all ages.

	1 to 2 Years				3 to 4 Years					All Children			
	% Children Below EAR/AI/AMDR				% Children Below EAR/AI/AMDR					% Children Below EAR/AI/AMDR			
	YCM Consumers	Other milk Consumers	Non-Dairy Consumers	All	PCM Consumers	Other milk Consumers	Non-Dairy Consumers	All		YCM/PCM Consumers	Other milk Consumers	Non-Dairy Consumers	All
Protein	24%	17%	54%	22%	10%	8%	32%	22%		20%	12%	41%	28%
Total fat	36%	40%	83%	37%	4%	13%	56%	37%		28%	27%	67%	47%
Carbohydrates	15%	23%	5%	14%	32%	25%	6%	14%		19%	24%	6%	14%
Calcium	44%	53%	99%	84%	67%	65%	97%	84%		50%	59%	98%	77%
Phosphorus	55%	33%	82%	48%	41%	24%	63%	48%		51%	29%	71%	54%
Iron	44%	92%	92%	81%	43%	80%	87%	81%		44%	86%	89%	81%
Selenium	27%	15%	24%	8%	6%	6%	9%	8%		22%	10%	16%	15%
Zinc	26%	61%	78%	60%	19%	51%	70%	60%		24%	56%	73%	60%
Thiamin	34%	61%	73%	54%	24%	47%	62%	54%		31%	54%	67%	57%
Riboflavin	28%	17%	82%	52%	35%	16%	75%	52%		30%	16%	78%	50%
Niacin	28%	59%	57%	25%	12%	23%	27%	25%		24%	40%	40%	38%
Vitamin B6	34%	73%	68%	48%	21%	49%	51%	48%		31%	61%	58%	55%
Vitamin B12	37%	58%	61%	44%	34%	42%	47%	44%		36%	50%	53%	49%
Folate	49%	82%	83%	73%	45%	76%	75%	73%		48%	79%	78%	74%
Vitamin C	18%	60%	73%	68%	29%	68%	74%	68%		21%	64%	73%	62%
Vitamin D	52%	98%	97%	93%	72%	96%	94%	93%		58%	97%	96%	90%
Vitamin E	56%	96%	97%	88%	64%	95%	97%	94%		58%	96%	97%	91%

Table 4. Odds ratios of children meeting the estimated average requirements/adequate intake/Acceptable Macronutrient Distribution Ranges (EAR/AI/AMDR) of each nutrient by consumer group compared to YCM/PCM consumers (reference) in age groups one to two years and three to four years.

	Other milk Consumers				Factors Associated with Increased Odds of Meeting EAR/AI/AMDR of Nutrient ($p < 0.05$)	Factors Associated with Lower Odds of Meeting EAR/AI/AMDR of Nutrient ($p < 0.05$)
	OR	95% CI		p-Value		
Protein	2.27	1.33	3.88	0.003	Living in urban dwelling	Education levels: Elementary, high school, No grades completed
Total fat	1.41	0.89	2.24	0.146		Education level: Elementary
Carbohydrates	0.35	0.20	0.63	<0.001	Wealth quintiles: rich, richest	Wealth quintiles: richest
Calcium	1.05	0.67	1.66	0.828	Wealth quintile: richest	Wealth quintiles: poorest
Phosphorus	3.07	1.94	4.85	<0.001	Living in urban dwelling	-
Iron	0.09	0.05	0.16	<0.001	-	-
Selenium	2.60	1.49	4.53	0.001	-	Wealth quintiles: poor, poorest; Education level: No grades completed
Zinc	0.37	0.23	0.59	<0.001	Wealth quintile: richest; Living in urban dwelling	Education level: Elementary
Thiamin	0.43	0.27	0.68	<0.001	Wealth quintile: richest; Living in urban dwelling	Education level: Elementary
Riboflavin	3.07	1.83	5.17	<0.001	-	Education level: Elementary
Niacin	0.35	0.22	0.56	<0.001	Living in urban dwelling	Education levels: Elementary, high school
Vitamin B6	0.27	0.17	0.42	<0.001	Wealth quintile: richest; Living in urban dwelling	Education levels: Elementary, high school
Vitamin B12	0.51	0.33	0.78	0.002	-	Education levels: Elementary, high school
Folate	0.22	0.14	0.36	<0.001	-	Wealth quintiles: poor, poorest
Vitamin C	0.17	0.10	0.29	<0.001	-	-
Vitamin D	0.02	0.01	0.06	<0.001	-	-
Vitamin E	0.05	0.02	0.11	<0.001	-	Education level: Elementary

	Non-Dairy Consumers				Factors Associated with Increased Odds of Meeting EAR/AI/AMDR of Nutrient ($p < 0.05$)	Factors Associated with Lower Odds of Meeting EAR/AI/AMDR of Nutrient ($p < 0.05$)
	OR	95% CI		p-Value		
Protein	0.39	0.23	0.63	<0.001	Living in urban dwelling	Education levels: Elementary, high school, No grades completed
Total fat	0.22	0.13	0.35	<0.001	Wealth quintiles: rich, richest	Education level: Elementary
Carbohydrates	1.69	0.83	3.43	0.145		Wealth quintiles: richest
Calcium	0.01	0.01	0.04	<0.001	Wealth quintile: richest	Wealth quintiles: poorest
Phosphorus	0.34	0.21	0.54	<0.001	Living in urban dwelling	-
Iron	0.08	0.05	0.15	<0.001	-	-
Selenium	1.70	0.99	2.92	0.054	-	Wealth quintiles: poor, poorest; Education level: No grades completed
Zinc	0.17	0.11	0.28	<0.001	Wealth quintile: richest; Living in urban dwelling	Education level: Elementary
Thiamin	0.25	0.16	0.41	<0.001	Wealth quintile: richest; Living in urban dwelling	Education level: Elementary
Riboflavin	0.14	0.08	0.22	<0.001	-	Education level: Elementary
Niacin	0.38	0.24	0.60	<0.001	Living in urban dwelling	Education levels: Elementary, high school
Vitamin B6	0.36	0.23	0.56	<0.001	Wealth quintile: richest; Living in urban dwelling	Education levels: Elementary, high school
Vitamin B12	0.49	0.32	0.77	0.002	-	Education levels: Elementary, high school
Folate	0.21	0.13	0.35	<0.001	-	Wealth quintiles: poor, poorest
Vitamin C	0.10	0.06	0.16	<0.001	-	-
Vitamin D	0.03	0.02	0.08	<0.001	-	-
Vitamin E	0.04	0.02	0.09	<0.001	-	Education level: Elementary

	Other milk Consumers				Factors Associated with Increased Odds of Meeting EAR/AI/AMDR of Nutrient ($p < 0.05$)	Factors Associated with Lower Odds of Meeting EAR/AI/AMDR of Nutrient ($p < 0.05$)
	OR	95% CI		p-Value		
Protein	1.65	0.57	4.79	0.358	-	-
Total fat	0.53	0.15	1.86	0.320	-	Wealth quintiles: poor, poorest; Education level: Vocational
Carbohydrates	0.79	0.39	1.60	0.510	-	Wealth quintile: richest; Living in urban dwelling; Mother currently working
Calcium	1.48	0.76	2.88	0.244	Living in urban dwelling	-
Phosphorus	2.54	1.32	4.92	0.006	Wealth quintile: richest; Living in urban dwelling	Education level: Elementary
Iron	0.24	0.12	0.46	<0.001	Wealth quintile: richest; Living in urban dwelling	Education level: Vocational
Selenium	2.30	0.55	9.58	0.251	Living in urban dwelling	Education levels: Elementary, no grades completed
Zinc	0.43	0.20	0.89	0.024	Wealth quintiles: rich, richest; Living in urban dwelling	-
Thiamin	0.46	0.22	0.93	0.031	Living in urban dwelling	Education level: No grades completed
Riboflavin	6.24	3.03	12.84	<0.001	Wealth quintile: richest; Living in urban dwelling	Education level: Elementary
Niacin	0.95	0.40	2.25	0.912	Living in urban dwelling	Education level: Elementary
Vitamin B6	0.42	0.21	0.86	0.017	-	-
Vitamin B12	0.89	0.48	1.67	0.721	-	Education level: No grades completed
Folate	0.28	0.15	0.53	<0.001	-	-
Vitamin C	0.19	0.09	0.37	<0.001	Wealth quintile: richest	-
Vitamin D	0.11	0.04	0.27	<0.001	-	-
Vitamin E	0.12	0.05	0.27	<0.001	-	-

	Non-dairy Consumers				Factors Associated with Increased Odds of Meeting EAR/AI/AMDR of Nutrient ($p < 0.05$)	Factors Associated with Lower Odds of Meeting EAR/AI/AMDR of Nutrient ($p < 0.05$)
	OR	95% CI		p-Value		
Protein	0.51	0.18	1.44	0.204	-	-
Total fat	0.09	0.03	0.32	<0.001	-	
Carbohydrates	2.57	1.19	5.54	0.016	Wealth quintiles: poor, poorest	
Calcium	0.11	0.05	0.24	<0.001		
Phosphorus	0.65	0.34	1.23	0.186		
Iron	0.21	0.11	0.41	<0.001		
Selenium	2.24	0.54	9.25	0.266		
Zinc	0.26	0.12	0.55	<0.001		
Thiamin	0.36	0.18	0.73	0.005		
Riboflavin	0.55	0.28	1.09	0.087		
Niacin	1.28	0.54	3.06	0.576		
Vitamin B6	0.44	0.22	0.90	0.024		
Vitamin B12	0.87	0.46	1.63	0.656		
Folate	0.30	0.16	0.56	<0.001		
Vitamin C	0.14	0.07	0.28	<0.001		
Vitamin D	0.14	0.06	0.34	<0.001		
Vitamin E	0.09	0.04	0.22	<0.001		

YCM/PCM consumer group was considered as reference group (OR = 1.0) for the respective age groups (one to two years; three to four years).

The results also highlight the sociodemographic factors that were significantly related to increased or reduced odds of meeting nutrient adequacy. In children one to two years, living in urban areas increased the likelihoods of meeting protein, phosphorus, zinc, thiamin, niacin, and vitamin B6 adequacies; being in the richest wealth quintile were associated with adequacy in total fat, calcium, zinc, thiamin, and vitamin B6 (all $p < 0.05$). On the other hand, being in poor or poorest wealth quintiles, and/or mothers with education at or below high school levels reduced the odds of meeting adequacy in most nutrients, including protein, total fat, calcium, selenium, zinc, B vitamins, and vitamin E (all $p < 0.05$). Similar trends were also observed in children aged three to four years. The only nutrient that had an opposite trend than the rest of the nutrients was carbohydrate. In children aged one to two years, being in the richest wealth quintile lowered the odds of meeting the AMDR for carbohydrates, and similarly among three to four years, those who were the richest, living in urban dwelling, and with mother currently working reduced the likelihood of meeting AMDR for carbohydrates, while being in the poor or poorest wealth quintile increased the odds.

None of the sociodemographic factors were associated with iron, vitamins C and D adequacies in children aged one to two years, and vitamin B6, folate, vitamins D and E adequacies in children three to four years.

4. Discussion

The current study assessed the contribution of different milks to daily nutrient intakes among young and preschool children in the Philippines, by comparing the mean nutrient intakes and percentage of children with inadequate intakes between three consumer groups: YCM/PCM consumers, other milk consumers, and non-dairy consumers. Our analysis has shown that children in the Philippines who consumed dairy products, including YCM/PCM and other milks, had higher intake of most nutrients and lower nutrient inadequacy than non-dairy consumers. Non-dairy consumers, in particular, had intakes of energy, total fat, calcium, phosphorus, iron, potassium, folate, vitamin D, and vitamin E far below recommendations. On the other hand, YCM/PCM consumers had mean intakes of micronutrients that were closer to the recommendations when compared to other consumer groups. While other milk consumers had similar macronutrient intakes to YCM/PCM consumers, a higher percentage of other milk consumers had inadequate micronutrient intakes as compared to YCM/PCM consumers. These findings support the fact that dairy is an important food group that significantly contributes to the intakes of children in these two age groups. Moreover, the choice of dairy products (e.g., YCM/PCM vs. other milks) contributes differently to total daily nutrient intakes, and that YCM/PCM is a good option in terms of reducing micronutrient inadequacy.

4.1. Nutrient Inadequacy in Filipino Children

The alarming prevalence of inadequate intakes, particularly for energy and micronutrients, among young children in the Philippines deserves further attention. Overall, the energy intakes of the three consumer groups were below the age and gender specific Recommended Energy Intakes in the Philippines. Moreover, 77% of children (one to four years) were inadequate in calcium, 81% inadequate in iron, 74% in folate, 60% in zinc, 62% in vitamin C, and 90% in vitamin D. Previous studies have found that the diets of young Filipino children lacked diversity, which contributed to high levels of nutrient inadequacy [21,22]. While improving diet diversity (increasing number of food groups consumed) could incrementally increase micronutrient adequacy, a more recent study found that, even with high diet diversity, Filipino school children had difficulty in achieving adequacy in calcium, folate, iron, vitamin A, and vitamin C. This is likely due to the low quantities of food consumed, or that the current food supply in the Philippines might not contain enough of these nutrients to fulfil the needs of children [23]. Indeed, our logistic regression analysis provided interesting insights that those living in urban areas and wealthier socio-economic quintiles were more likely to achieve total fat and micronutrient adequacies, while those from poorer socio-economic backgrounds and having mothers with lower education levels had reduced odds of meeting micronutrient adequacies, except for carbohydrates. It is possible that those from higher socio-economic quintiles ate better diets and that living in urban areas meant better accessibility and availability of more nutritious foods (e.g., meat) and fortified beverages; and, those from lower socio-economic quintiles had a predominately carbohydrate-rich (e.g., rice) diet that lack variety. Our findings support the literature on key strategies to reduce stunting in Southeast Asia [24] and in Filipino children [23]. There is an urgent need for better access to a variety of nutrient-rich foods, particularly in the lower socio-economic groups, increase availability of fortified foods that are tailored for young children, in tandem with increased parental education on dietary intake, to close nutrient gaps among young children.

4.2. Large Proportion of Filipino Young Children Did Not Consume Dairy on a Given Day

The Daily Nutritional Guide Pyramid for Filipino Children aged one to six years recommends consuming one glass of milk & milk products per day, which includes one glass of whole milk, or 1/2 cup evaporated milk diluted with 1/2 glass water, or four tablespoons of powdered whole milk diluted with one glass of water [18]. However, the current study found half of Filipino children (43% aged one to two years and 58% aged three to four years) were not consuming any milk or milk products on a given day. Non-dairy consumers had mean energy intakes approximately 40% lower than the Recommended Energy Intakes (REI) (one to two years: mean intake of 543 kcal/d vs. REI of 1000 kcal/d for boys and 920 kcal/d for girls; three to four years: mean intake of 828 kcal/d vs. REI of 1350 kcal/d (boys) and 1260 kcal/d (girls)), suggesting these young children were at serious risk of undernutrition.

Indeed, the consumption of milk and animal-sourced foods is known to be limited among low-income countries. It is estimated that animal-source foods, such as milk, provide between 5% to 15% of total daily energy in Asian countries, when compared to over 20% daily energy in western countries, such as the U.S. and Europe [4]. A study on South East Asian countries, including Thailand, Malaysia, Vietnam, and Indonesia also observed sub-optimal milk and dairy intake in children aged one to 12 years. Only around half of Indonesian (52%) and Vietnamese (47%) children consumed dairy products on a daily basis. The study found that children who consumed less than one portion of dairy per day had significantly lower nutrient intakes and higher prevalence of underweight and stunting than children who consumed ≥1 portion of dairy per day [25].

Our analysis also suggests that a high percentage of non-dairy consumers came from poorer wealth quintiles, from rural areas, and with mothers who had lower education levels; while, the opposite sociodemographic characteristics were observed for those who consumed YCM/PCM. Financial constraints, limited product availability, and lack of parental awareness of the importance of nutrition and dairy food could be the reasons for the high proportion of children not consuming YCM/PCM or other milks.

4.3. The Role of YCM/PCM in the Diet of Filipino Children

A previous study has highlighted that both cow's milk and YCM/PCM were two of the top five contributors of energy, carbohydrate, protein, fat, thiamine, riboflavin, vitamins A, C, calcium, iron and zinc for Filipino children aged 12 to 35.9 months [3]. Particularly, YCM was a top contributor of iron and zinc in children aged 12–23.9 months. Other studies in Europe and Oceania found that YCM/PCM consumers had higher intakes of nutrients, such as iron, vitamins C, and D than cow's milk consumers [5–8]. Our study adds to the evidence that YCM/PCM and other milk consumers had nutrient intakes that were more in line with recommendations than non-dairy consumers, and that the dairy food group can contribute hugely to the daily intake of nutrients in young children. Furthermore, the fact that lower percentages of YCM/PCM consumers had inadequate intakes, and that they had higher odds of achieving adequacy in iron, zinc, thiamine, niacin, folate, vitamins B6, B12, C, D when compared to other milk consumers, suggests that YCM/PCM might be a good choice of dairy product for Filipino children.

In contrast to the expert opinions and recommendations on toddler milk and YCM from Europe and the U.S. [13,26–28], the routine use of YCM is not deemed a necessity, as the missing nutrients in the diet of young children can also be provided by other dietary sources in these countries, the findings of the current study did support the role of YCM in the diet of young children in the Philippines. Furthermore, our study also supported that the use of YCM/PCM can increase the intake of iron and vitamin D and decrease the intake of protein when compared to other milk, in accordance to the European Food Safety Authority (EFSA) scientific opinion [26], conclusion from the European Society for Paediatric Gastroenterology Hepatology and Nutrition (ESPGHAN) committee [13], and findings from previous studies [5–11].

However, one caution to highlight is that, while our data supports that, overall, YCM/PCM consumers have lower nutrient inadequacies than other consumer groups, not all YCM/PCMs on the market have favorable nutritional compositions for children. Table S1 highlights the minimum, maximum, and mean nutrient compositions across 18 brands of YCM/PCMs. Parents and health care professionals should consider the nutritional profiles of YCM/PCMs when choosing YCM/PCM for children, as, evidently, YCM/PCMs have highly variable macro- and micronutrient content, and not all are fortified with essential micronutrients to the same extent.

4.4. Strengths and Limitations

The large sample of children ($n = 2992$) from a nationally representative survey of the Philippines covering the young toddlers one to four years is a strength of this study. The nutrient intakes of this age group have not been extensively studied in this country or in the Southeast Asia region, and this study has added to the existing literature that dairy is a key food group to the diet of young Filipino children beyond infancy. This study, to our knowledge, is the first to identify the high percentage of young children who are not consuming any dairy in the Philippines. This highlights the opportunity for the public health authority in the Philippines to improve dairy consumption in children.

One limitation of the current study is the method for estimating total quantity of powdered milk and YCM/PCM consumed. Powdered milk and YCM/PCM consumption were recorded as gram weight in powder form in the survey and a conversion factor of 6.7 was applied to estimate fluid weight (g), as consumed. While we had checked several milk brands' on-pack preparation instructions, and the conversion the factor used was deemed appropriate, it is possible that the total quantity of other milk and YCM/PCM as consumed was overestimated. Indeed, the estimated total YCM consumed by children aged one to two years (641 g/d) was higher than anticipated. A report of an expert panel recommended two to three servings, with a total of 400–600 mL of YCM/PCM per day are appropriate for children age one to six years [29]. The nutrient composition of the milk or YCM/PCM did not change and would not affect the calculation of mean nutrient intakes or percentage of children with inadequate intakes, despite the conversion to weight as consumed.

Another limitation is that the high proportion of missing data for some of the nutrient variables (vitamin A, saturated fat, monounsaturated, and polyunsaturated fats) meant that they were excluded for analyses. Given that vitamin A inadequacy is known to be high in the Philippines from the literature and that the fatty acids also are of public health importance, it was unfortunate that they were not investigated in the current analysis.

5. Conclusions

This study highlights the high prevalence of inadequate intakes, particularly for energy and micronutrients, among young children in the Philippines, regardless of milk consumption type. Secondly, the current study also provides novel insights on the importance of dairy food group including YCM/PCM and other milk to energy and macronutrient intakes, and meeting micronutrient adequacy in the diet of Filipino children. YCM/PCM have demonstrated to be a more superior option than other milks to achieve adequacy in key nutrients, such as iron, zinc, vitamins C, D, E, and some B vitamins in this population. Furthermore, expert opinions and recommendations on the role of YCM/PCM from Europe or the U.S may not apply to countries, such as the Philippines, where nutrient inadequacies and deficiencies are much more common. Finally, strategies targeting specific socio-demographic segments (e.g., by wealth status, dwelling location, and mother's education) should be adopted to increase the intake of YCM/PCM or other fortified nutrient-dense foods and beverages among young children in the Philippines to improve nutrient intakes and reduce nutrient inadequacy.

Supplementary Materials: The following are available online at http://www.mdpi.com/2072-6643/12/2/392/s1, Table S1: Nutritional compositions of YCMBs, Milk (fluid), Milk (powdered) considered in the current study.

Author Contributions: T.-N.M. conceptualized and designed the study, interpreted the data, and drafted the manuscript. M.T. conducted the data analysis, interpreted the data, and drafted the manuscript. I.A.-A. and M.V.C. guided the data analysis and critically reviewed the draft of the manuscript. E.A.O. conceptualized the study, oversaw the project and critically reviewed the manuscript. All authors proofread and approved the final manuscript.

Funding: This research and the APC were funded by Nestlé Research.

Acknowledgments: We would like to thank Emma F. Jacquier, Lolita Bazarova, Delphine Egli, Hervé Vancheri and Ryan Carvalho for their guidance of the manuscript. We would like to thank FNRI for granting access to the data.

Conflicts of Interest: The authors declare no conflicts of interest. T.N.M., M.T., E.O. are employees of Nestlé Research, Vers-chez-les-Blanc, Lausanne, Switzerland. This research study is made possible to Nestlé Research via virtual access of the requested data of the 2013 National Nutrition Survey, funded by Nestlé Research, Lausanne, Switzerland.

References

1. Food and Nutrition Research Institute-Department of Science and Technology (FNRI-DOST). *2015 Updating of the Nutritional Status of Filipino Children and Other Population Group: Food Security Survey*; FNRI-DOST: Metro Manila, Philippines, 2016.
2. Food and Nutrition Research Institute-Department of Science and Technology (FNRI-DOST). *Philippine Nutrition Facts and Figures 2013: Biochemical Survey.*; FNRI-DOST: Metro Manila, Philippines, 2015.
3. Denney, L.; Angeles-Agdeppa, I.; Capanzana, M.V.; Toledo, M.B.; Donohue, J.; Carriquiry, A. Nutrient Intakes and Food Sources of Filipino Infants, Toddlers and Young Children are Inadequate: Findings from the National Nutrition Survey 2013. *Nutrients* **2018**, *10*, 1730. [CrossRef] [PubMed]
4. Dror, D.K.; Allen, L.H. The importance of milk and other animal-source foods for children in low-income countries. *Food Nutr. Bull.* **2011**, *32*, 227–243. [CrossRef] [PubMed]
5. Walton, J.; Flynn, A. Nutritional adequacy of diets containing growing up milks or unfortified cow's milk in Irish children (aged 12–24 months). *Food Nutr. Res.* **2013**, *57*. [CrossRef] [PubMed]
6. Ghisolfi, J.; Fantino, M.; Turck, D.; de Courcy, G.P.; Vidailhet, M. Nutrient intakes of children aged 1-2 years as a function of milk consumption, cows' milk or growing-up milk. *Public Health Nutr.* **2013**, *16*, 524–534. [CrossRef] [PubMed]
7. Sidnell, A.; Pigat, S.; Gibson, S.; O'Connor, R.; Connolly, A.; Sterecka, S.; Stephen, A.M. Nutrient Intakes and Iron and Vitamin D status Differ Depending on Main Milk Consumed by UK Children Aged 12–18 Months—Secondary Analysis from the Diet and Nutrition Survey of Infants and Young Children. *J. Nutr. Sci.* **2016**, *5*, e32. [CrossRef]
8. Lovell, A.L.; Davies, P.S.; Hill, R.J.; Milne, T.; Matsuyama, M.; Jiang, Y.; Chen, R.X.; Grant, C.C.; Wall, C.R. A comparison of the effect of a Growing Up Milk Lite versus Cow's Milk on longitudinal dietary patterns and nutrient intakes in children aged 12 to 23 months: The Growing Up Milk Lite (GUMLi) randomised controlled trial. *Br. J. Nutr.* **2018**, 1–23. [CrossRef]
9. Chouraqui, J.-P.; Turck, D.; Tavoularis, G.; Ferry, C.; Dupont, C. The Role of Young Child Formula in Ensuring a Balanced Diet in Young Children (1–3 Years Old). *Nutrients* **2019**, *11*, 2213. [CrossRef]
10. Akkermans, M.D.; Eussen, S.R.; van der Horst-Graat, J.M.; van Elburg, R.M.; van Goudoever, J.B.; Brus, F. A micronutrient-fortified young-child formula improves the iron and vitamin D status of healthy young European children: A randomized, double-blind controlled trial. *Am. J. Clin. Nutr.* **2017**, *105*, 391–399. [CrossRef]

11. Lovell, A.L.; Davies, P.S.W.; Hill, R.J.; Milne, T.; Matsuyama, M.; Jiang, Y.; Chen, R.X.; Wouldes, T.A.; Heath, A.M.; Grant, C.C.; et al. Compared with Cow Milk, a Growing-Up Milk Increases Vitamin D and Iron Status in Healthy Children at 2 Years of Age: The Growing-Up Milk-Lite (GUMLi) Randomized Controlled Trial. *J. Nutr.* **2018**, *148*, 1570–1579. [CrossRef]
12. Wall, C.R.; Hill, R.J.; Lovell, A.L.; Matsuyama, M.; Milne, T.; Grant, C.C.; Jiang, Y.; Chen, R.X.; Wouldes, T.A.; Davies, P.S.W. A multicenter, double-blind, randomized, placebo-controlled trial to evaluate the effect of consuming Growing Up Milk "Lite" on body composition in children aged 12-23 mo. *Am. J. Clin. Nutr.* **2019**, *109*, 576–585. [CrossRef]
13. Hojsak, I.; Bronsky, J.; Campoy, C.; Domellof, M.; Embleton, N.; Fidler Mis, N.; Hulst, J.; Indrio, F.; Lapillonne, A.; Molgaard, C.; et al. Young Child Formula: A Position Paper by the ESPGHAN Committee on Nutrition. *J. Pediatric Gastroenterol. Nutr.* **2018**, *66*, 177–185. [CrossRef]
14. Luque, V.; Closa-Monasterolo, R.; Escribano, J.; Ferre, N. Early Programming by Protein Intake: The Effect of Protein on Adiposity Development and the Growth and Functionality of Vital Organs. *Nutr. Metab. Insights* **2015**, *8*, 49–56. [CrossRef]
15. Sazawal, S.; Dhingra, U.; Dhingra, P.; Hiremath, G.; Sarkar, A.; Dutta, A.; Menon, V.P.; Black, R.E. Micronutrient fortified milk improves iron status, anemia and growth among children 1-4 years: A double masked, randomized, controlled trial. *PLoS ONE* **2010**, *5*, e12167. [CrossRef] [PubMed]
16. Cervo, M.M.C.; Mendoza, D.S.; Barrios, E.B.; Panlasigui, L.N. Effects of Nutrient-Fortified Milk-Based Formula on the Nutritional Status and Psychomotor Skills of Preschool Children. *J Nutr Metab* **2017**, *2017*, 6456738. [CrossRef] [PubMed]
17. Food and Nutrition Research Institute. *Philippine Nutrition Facts and Figures 2013. 8th National Nutrition Survey Overview*; Department of Science and Technology-Food and Nutrition Research Institute (DOST-FNRI): Metro Manila, Philippines, 2015.
18. Food and Nutrition Research Institute-Department of Science and Technology (FNRI-DOST). Daily Nutritional Guide Pyramid. Available online: https://www.fnri.dost.gov.ph/index.php/tools-and-standard/nutritional-guide-pyramid#toddler (accessed on 9 December 2019).
19. Food and Nutrition Research Institute-Department of Science and Technology (FNRI-DOST). *The Philippine Food Composition Tables*; Food and Nutrition Research Institute-Department of Science and Technology (FNRI-DOST): Metro Manila, Philippines, 1997.
20. USDA. What We Eat in America Food Categories. Available online: https://www.ars.usda.gov/ARSUserFiles/80400530/pdf/1314/food_category_list.pdf (accessed on 31 January 2020).
21. Daniels, M.C.; Adair, L.S.; Popkin, B.M.; Truong, Y.K. Dietary diversity scores can be improved through the use of portion requirements: An analysis in young Filipino children. *Eur. J. Clin. Nutr.* **2009**, *63*, 199–208. [CrossRef] [PubMed]
22. Kennedy, G.L.; Pedro, M.R.; Seghieri, C.; Nantel, G.; Brouwer, I. Dietary diversity score is a useful indicator of micronutrient intake in non-breast-feeding Filipino children. *J. Nutr.* **2007**, *137*, 472–477. [CrossRef] [PubMed]
23. Mak, T.N.; Angeles-Agdeppa, I.; Lenighan, Y.M.; Capanzana, M.V.; Montoliu, I. Diet Diversity and Micronutrient Adequacy among Filipino School-Age Children. *Nutrients* **2019**, *11*, 2197. [CrossRef] [PubMed]
24. Bloem, M.W.; de Pee, S.; Le Hop, T.; Khan, N.C.; Laillou, A.; Minarto; Moench-Pfanner, R.; Soekarjo, D.; Soekirman; Solon, J.A.; et al. Key Strategies to Further Reduce Stunting in Southeast Asia: Lessons from the ASEAN Countries Workshop. *Food Nutr. Bull.* **2013**, *34*, S8–S16. [CrossRef]
25. Nguyen Bao, K.L.; Sandjaja, S.; Poh, B.K.; Rojroongwasinkul, N.; Huu, C.N.; Sumedi, E.; Aini, J.N.; Senaprom, S.; Deurenberg, P.; Bragt, M.; et al. The Consumption of Dairy and Its Association with Nutritional Status in the South East Asian Nutrition Surveys (SEANUTS). *Nutrients* **2018**, *10*, 759. [CrossRef]
26. EFSA NDA Panel (EFSA Panel on Dietetic Products, NDA). Scientific Opinion on nutrient requirements and dietary intakes of infants and young children in the European Union. *EFSA J.* **2013**, *11*, 3408.
27. Lott, M.; Callahan, E.; Welker Duffy, E.; Story, M.; Daniels, S. Healthy Beverage Consumption in Early Childhood: Recommendations from Key National Health and Nutrition Organizations. In *Technical Scientific Report*; Healthy Eating Research: Durham, NC, USA, 2019.

28. Suthutvoravut, U.; Abiodun, P.O.; Chomtho, S.; Chongviriyaphan, N.; Cruchet, S.; Davies, P.S.W.; Fuchs, G.J.; Gopalan, S.; van Goudoever, J.B.; Nel, E.R.; et al. Composition of Follow-Up Formula for Young Children Aged 12–36 Months: Recommendations of an International Expert Group Coordinated by the Nutrition Association of Thailand and the Early Nutrition Academy. *Ann. Nutr. Metab.* **2015**, *67*, 119–132. [CrossRef] [PubMed]
29. Lippman, H.E.; Desjeux, J.-F.; Ding, Z.-Y.; Tontisirin, K.; Uauy, R.; Pedro, R.A.; Van Dael, P. Nutrient Recommendations for Growing-up Milk: A Report of an Expert Panel. *Crit. Rev. Food Sci. Nutr.* **2016**, *56*, 141–145. [CrossRef] [PubMed]

© 2020 by the authors. Licensee MDPI, Basel, Switzerland. This article is an open access article distributed under the terms and conditions of the Creative Commons Attribution (CC BY) license (http://creativecommons.org/licenses/by/4.0/).

Review

Milk Polar Lipids: Underappreciated Lipids with Emerging Health Benefits

Liya Anto [1], Sarah Wen Warykas [1], Moises Torres-Gonzalez [2] and Christopher N. Blesso [1,*]

1. Department of Nutritional Sciences, University of Connecticut, Storrs, CT 06269, USA; liya.anto@uconn.edu (L.A.); sarah.warykas@uconn.edu (S.W.W.)
2. National Dairy Council, Rosemont, IL 60018-5616, USA; moises.torres-gonzalez@dairy.org
* Correspondence: christopher.blesso@uconn.edu; Tel.: +1-860-486-9049

Received: 6 March 2020; Accepted: 2 April 2020; Published: 4 April 2020

Abstract: Milk fat is encased in a polar lipid-containing tri-layer milk fat globule membrane (MFGM), composed of phospholipids (PLs) and sphingolipids (SLs). Milk PLs and SLs comprise about 1% of total milk lipids. The surfactant properties of PLs are important for dairy products; however, dairy products vary considerably in their polar lipid to total lipid content due to the existence of dairy foods with different fat content. Recent basic science and clinical research examining food sources and health effects of milk polar lipids suggest they may beneficially influence dysfunctional lipid metabolism, gut dysbiosis, inflammation, cardiovascular disease, gut health, and neurodevelopment. However, more research is warranted in clinical studies to confirm these effects in humans. Overall, there are a number of potential effects of consuming milk polar lipids, and they should be considered as food matrix factors that may directly confer health benefits and/or impact effects of other dietary lipids, with implications for full-fat vs. reduced-fat dairy.

Keywords: polar lipids; dairy; sphingomyelin; heart disease; gut health; cancer; inflammation

1. Introduction

Polar lipids are essential components of all biological membranes and found in the human diet as phospholipids (PLs) and sphingolipids (SLs). Consumption of dietary polar lipids is relatively common in the Western dietary pattern and estimated to be in the range of 2–8 g/day for PLs (~1–10% of daily fat intake) [1] and 50–400 mg/day for SLs [2,3]. In milk, polar lipids are primarily located within the milk fat globule membrane (MFGM), which is a trilayered biological membrane that surrounds the fat globule. MFGM are typically derived from cell membranes of lactating cells and the endoplasmic reticulum membranes [4]. Animal cell membranes, in general, have PLs as the major structural lipids and SLs, such as sphingomyelin (SM), are typically components of lipid rafts in association with cholesterol [5]. Of the total milk lipids, polar lipids account for approximately 1% of total lipids in milk. SM content encompasses approximately 25% of total milk polar lipids, and SM is found at ~3:1 ratio to cholesterol by mass [6]. While health effects of PLs from eggs [7] and SLs [8] have been reviewed previously, the health effects of consuming milk polar lipids have not been reviewed extensively. Since the SLs found within MFGM are known to impact various aspects of lipid metabolism [9,10], gut microbiota [10], and inflammation [9,11], milk polar lipids may be considered as food matrix factors that may confer health benefits and/or impact effects of other dietary lipids, with implications for full-fat vs. lower-fat dairy varieties. This review summarizes the recent basic science and clinical research examining food sources and health effects of milk polar lipids, as well as to identify gaps in the scientific literature related to milk polar lipids research.

2. Polar Lipids

2.1. Classes of Polar Lipids

Phospholipids and sphingolipids are the two major classes of polar lipids in milk. PLs in milk includes SM (considered both a PL and a SL) and glycerophospholipids. The glycerophospholipids consist of phosphatidylethanolamine (PE), phosphatidylserine (PS), phosphatidylcholine (PC), and phosphatidylinositol (PI). Sphingomyelin is a phosphosphingolipid and is the major SL found in milk. Other SLs in milk include glucosylceramide (GluCer) and lactosylceramide (LacCer). Classification of milk polar lipids is given in Figure 1.

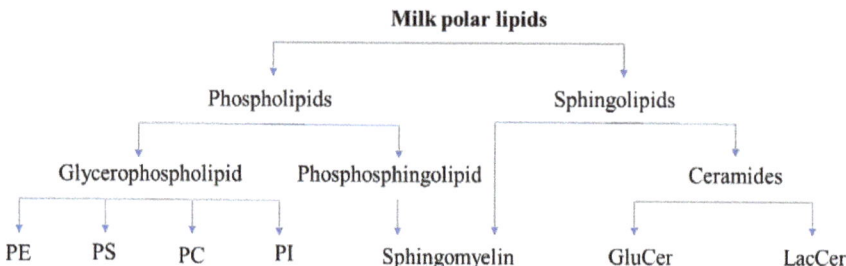

Figure 1. Classification of milk polar lipids. Abbreviations: GluCer, glucosylceramide; LacCer, lactosylceramide; PC, phosphatidylcholine; PE, phospatidylethanolamine; PI, phosphatidylinositol; and PS, phosphatidylserine.

Glycerophospholipids are amphipathic molecules comprised of a glycerol backbone, two ester-linked fatty acids (FAs) which form the hydrophobic tail (at sn-1 and sn-2 positions of glycerol), and a polar head group linked by a phosphate residue (at sn-3 position). The polar head group can be ethanolamine, serine, choline, or inositol, which makes it PE, PS, PC, or PI, respectively [12]. Fatty acids seen in glycerophospholipids are usually unsaturated and long chain. However, PC is reported to have more saturated FAs when compared with other PLs seen in milk [6]. Esterification of very-long-chain fatty acids in glycerophospholipids is also reported, although to a lesser extent than long-chain FAs [13].

Sphingolipids are polar molecules that differ from glycerophospholipids in that instead of glycerol, they contain a long-chain amino alcohol, known as a sphingoid base, as their backbone [14]. The predominant mammalian sphingoid base is the 18-carbon sphingosine. Ceramide is formed when the amino group of sphingosine is linked to a FA via an amide bond. Sphingomyelin (or ceramide phosphocholine) is an amphipathic sphingolipid in which a phosphorylcholine head group is linked to ceramide [15]. In glucosylceramide (GluCer), the 1-position of ceramide is linked to a glucose residue, whereas in lactosylceramide (LacCer), it is linked to lactose [8,16].

Previous literature has cited the presence of other minor PLs in milk, including lysophosphatidylethanolamine (LPE) and lysophosphatidylcholine (LPC) [17–19]. However, the origin of these lipids in milk is unclear, as they are the hydrolytic products of PE and PC and can be formed by hydrolysis during milk handling [12]. Another minor class of PLs seen in milk are the plasmalogens, which are structurally characterized as a glycerophospholipid having a vinyl-ether linkage with a fatty alcohol at the sn-1 position of the glycerol backbone [20,21].

2.2. Biological Functions

Phospholipids are present in all biological membranes, where they display amphipathic properties. The amphipathic nature of PLs is characterized by two hydrophobic tails and a hydrophilic, polar head group. This feature helps them to aggregate spontaneously in the aqueous phase by exposing their hydrophilic head and hiding their hydrophobic tails away from the aqueous phase, thereby attaining an energetically favorable conformation [22]. Phospholipids aggregate in two different ways; they can either bury their hydrophobic tails in the interior forming a spherical structure, as seen in micelles, or they can sandwich their tails between the hydrophilic head groups forming sheets or bilayers, as seen in cell membranes [23].

Almost 50% of the total mass of the animal cell membrane is made up of lipids, with PLs being the most abundant. The major PLs in cell membranes include PC, PE, PS, and SM. They serve a wide variety of structural and biological functions in the cell membrane [23]. Phospholipid bilayers in cell membranes help in maintaining a semi-permeable barrier between the cells and the organelles [24,25]. Due to their amphipathic nature and cylindrical structure, the PLs in cell membrane can also demonstrate a self-healing property. A tear in the membrane can produce an energetically unfavorable form, which compels the PLs to rearrange spontaneously to heal/close the tear [23]. This specific property of PLs is fundamental to the existence of a cell. Within the lipid bilayers, individual lipids can move freely, which provides fluidity to the membrane [26]. The presence of PS in the cytosolic surface of the bilayer is important for the enzyme activity of cytosolic protein kinase C. Phospholipids are important intracellular mediators. Extracellular signals can activate various membrane phospholipases that can cleave PLs and the fragments produced can act as intracellular mediators [24,27,28]. Thus, PLs are integral parts of cellular membranes and are also critically involved in vital functions of the cell.

During lipid digestion in the small intestine, mixed micelles are formed, which also contain PLs on their surface. As previously mentioned, the spherical structure of the micelle is due to the amphipathic property of PLs. Phospholipids are arranged in micelle in such a way that the hydrophilic head group face to the exterior, while the hydrophobic tails are arranged on the inside. This particular arrangement of PLs, along with bile salts, helps in the packing of highly hydrophobic triglyceride (TG) and cholesteryl esters in the core of mixed micelle, so that they can be transported through the aqueous environment in the intestine where they can be digested and absorbed. Likewise, the emulsifying function of PL monolayers in lipoproteins derived from the intestine and liver is similar to what is observed in micelles.

2.3. Milk Polar Lipids: Classes and Quantity

Not all dairy products are created equal when it comes to PL and SL contents. There is a considerable amount of variability when comparing polar lipids to total lipid content in dairy products. For example, some fat-rich products like butter and cream have relatively low levels compared to other products. For the content of SLs, which is mostly SM, dairy products vary from trace levels of SM in anhydrous milk fat to higher levels in buttermilk (~12–21.5 g/100 g of PL) and butter serum (also called beta serum) (23.8–28.92 g/100 g of PL) [29]. The major PL and SM content of raw milk and other dairy products is summarized in Table 1. Whole milk contains about twice the PLs of skim milk, while some dairy co-products can become relatively enriched in PLs, including buttermilk and butter serum [29]. Phospholipid content in whole milk can vary between 0.7 to 2.3 g/100 g of fat [30–34]. This variation in the level of reported PL content in milk can be attributed to the different analytical techniques used, as well as the diet [35], season of the year [36], age, and stage of lactation of the animal [37,38]. The most abundant PL in milk is reported to be PE, followed by PC and SM [20,29,31,33,34,37,39–41]. Fat-rich dairy products like butter and cream are found to be less enriched in PL content as a proportion of total fat, when compared with their aqueous co-products. Cream is rich in fat (~32%) and protein (~2%) [31] and has 0.3–5.65 g of PL/100 g of fat. Most authors reported PE being the most abundant PL found in cream, followed by PC and SM [30,32,42,43]. Butter (~78% fat) [32] has a PL content ranging from 0.195–5.31 g/100 g fat. The order of abundance of PL in butter is the same as cream, with PE being the richest followed by PC and SM [30–32,44]. Buttermilk (4.485–35.32 g/100 g of fat) and butter serum (46.69–48.39 g/100 g of fat), which are the co-products of butter and anhydrous milk fat production, respectively, are rich sources of polar lipids, yet are low in fat [32]. Cheese whey is also a good source of milk PL (5.3–23.66 g/100 g fat).

Table 1. Milk PL and SM content of raw milk and dairy products.

Product	PL (g/100 g DM)	PL (g/100 g Fat)	PE (% of Total PL)	PI (% of Total PL)	PS (% of Total PL)	PC (% of Total PL)	SM (% of Total PL)	Reference
Whole milk	0.2–0.3	0.7–2.3	23.2–72.2	1.4–7.5	3.4–24.5	8.0–46.4	4.0–29.5	[20,30–34,39–42,45]
Skim milk	0.1	10.7–11.1	26.7–38.2	5.5–8.4	8.4–9.9	19.6–35.2	16.7–21.2	[30,32]
Cream	0.2–0.4	0.3–5.6	17.7–45.6	6.8–15.4	6.7–14.8	14.6–33.7	11.9–28.6	[19,30,32,42–44,46]
Butter	0.3	0.2–5.31	17.7–43.3	4.3–15.8	7.0–15.3	19.9–35.6	16.6–21.8	[30–32,43,44,46]
Buttermilk	1.1–2.0	4.5–35.3	17.0–44.8	2.4–17.3	8.0–18.0	17.3–46.0	12.1–21.5	[29,30,37,42,44,47–49]
Butter serum	11.5	46.7–48.4	26.7–31.4	9.0–11.2	6.9–10.1	24.9–27.2	23.8–28.9	[30,32,47]
Cheese whey	0.3–1.8	5.3–23.7	27.4–41.1	2.8–3.7	3.9–9.3	19.0–32.2	9.9–16.4	[29,30,48,50]
Yogurt (skimmed)	0.2	5.5	31.1	6.3	7.9	19.9	24.9	[30]
Ricotta cheese	1.16	2.7	45.4	4.4	5.8	15.8	14.2	
Mozzarella cheese	0.28	0.5	42.5	5.7	5.6	19.4	14.6	[30]
Cheddar cheese	0.25	0.5	38.0	7.7	8.5	20.3	16.3	

Abbreviations: DM, dry matter; PC, phosphatidylcholine; PE, phosphatidylethanolamine; PI, phosphatidylinositol; PL, phospholipid; PS, phosphatidylserine; and SM, sphingomyelin.

It has been reported that various dairy production methods (homogenization, heating, etc.) can disrupt the MFGM [30]. They can also affect the MFGM protein content in the final product and will enhance the milk PL content in the serum phase [51–53]. In addition, several reports have shown that milk products which are rich in PL are also rich in MFGM proteins [6,30,54–56]. As a consequence, it is suggested that, due to the close association of milk PL with MFGM proteins, these fractions may be migrating together during dairy processing [6]. Churning, centrifuging, homogenization, and spray-drying are some of the dairy processing techniques which are known to affect the composition of milk PLs and SLs [30]. This could be one probable reason behind this unexpected inverse relationship between fat and milk PL content in dairy ingredients and co-products. Due to the high PL content on a dry matter and fat base, buttermilk, butter serum, and cheese whey are the most suitable sources of milk PL purification. However, these ratios are much higher for butter serum when compared with buttermilk and cheese whey, which makes it the most attractive source for milk PL purification [49]. However, researchers have also successfully used buttermilk [57,58] and cheese whey [56] for the extraction of milk PLs.

3. Health Effects of Milk Polar Lipids

3.1. Effects on Intestinal Lipid Absorption

Dietary PLs can inhibit intestinal lipid absorption when added to the diet in significant amounts by interfering with lipid mobilization from mixed micelles (as reviewed by Cohn J et al. [1]). Dietary SM, in particular, is known to dose-dependently reduce the intestinal absorption of cholesterol, TG, and fatty acids in rodents [10]. Products of SM digestion, such as ceramides and sphingosine, also inhibit cholesterol and fatty acid absorption [59–61]. Compared to SM derived from egg yolk, SM from milk has been shown to have stronger effects on inhibiting the intestinal absorption of fat and cholesterol in rats, potentially due to stronger hydrophobic interactions between milk SM and other lipids [62]. Evidence of dietary SM and its hydrolytic products inhibiting intestinal lipid absorption through cell-dependent and cell-independent mechanisms is provided in previous preclinical studies [63]. Enrichment of the dietary PL pool by milk SM has shown to decrease intestinal cholesterol absorption by decreasing the active concentration of cholesterol monomers in mixed micelles [64]. In vitro studies suggest that SM in mixed micelles can reduce TG hydrolysis by inhibiting human pancreatic lipase-colipase activity [65,66]. Long-chain bases of SLs and long-chain fatty acids also have been reported to compete with each other for cellular uptake, since they both utilize acyl-CoA synthetases [61].

Other milk polar lipids such as PC, PE, and gangliosides are also known to reduce the intestinal absorption of dietary lipids. Phosphatidylcholine, which when present in the bile can facilitate intestinal absorption of dietary lipids, can inhibit lipid absorption when present in large amounts in the diet. The presence of PC in taurocholate containing mixed micelles reduced the uptake of cholesterol by Caco-2 cells [67]. Incubation of Caco-2 cells with micelles containing 200 µM of PC reduced cholesterol absorption, accompanied by reduced cellular esterification and secretion of cholesteryl esters. In contrast, the presence of lysophosphatidylcholine showed only a minor effect. Additionally, previous research also revealed the capacity of intact PC in mixed micelles to inhibit the absorption of cholesterol and FAs using in vivo (isolated jejunal segment technique) and in vitro (everted sac experimental model from rat jejuna) studies [68]. Similar effects of PC are also reported by Rampone and Long (1977) using the same in vitro model [69]. In human trials, the administration of PC via intraduodenal infusion attenuated cholesterol absorption when compared to the placebo group, which received the same amount of safflower oil with similar FA composition [70]. Later studies showed that phospholipase A2 hydrolysis of surface PL in lipid emulsions prior to pancreatic lipase/co-lipase-mediated TG hydrolysis is necessary for the cellular uptake of cholesterol and FAs. The cellular uptake of cholesterol from lipid emulsions with high PL/TG molar ratio (>0.3) was significantly lower [71]. However, the exact mechanism by which high PC concentration reduces intestinal cholesterol absorption remains unclear. Although, a plausible mechanism can be attributed to an increased solubility of cholesterol in the micelle and, thus, shifting the partition coefficient away from the cell membrane [72]. Additionally, a higher concentration of PC in lipid emulsions may also lead to a decreased availability of cholesterol for absorption, due to the increased packing density of the micellar surface [71].

Phosphatidylethanolamine is shown to have a hypocholesterolemic effect in animal studies [73,74]. Supplementation with 2% PE for two weeks in rats fed with a 1% cholesterol-containing diet reduced serum cholesteryl ester compared to control animals, with the cholesteryl ester concentration being inversely related to the level of hepatic PE [74]. A similar effect of PE in rats was also reported by Imaizumi and co-workers [73]. It is also known that mono- and di-unsaturated PE can exhibit a similar affinity to cholesterol as that of SM and can influence cholesterol absorption [75,76]. Therefore, PE found in milk may also have the capability to reduce intestinal lipid absorption, due to its affinity for cholesterol; however, further studies are needed to confirm this.

3.2. Anti-Inflammatory Effects

Dietary SM and its hydrolytic products (i.e., sphingosine and ceramide) have shown promising anti-inflammatory action in preclinical studies. Mazzei et al. [77] showed that sphingosine can activate a peroxisome proliferator-activated receptor-gamma (PPARγ) reporter in macrophages. PPARγ is a nuclear receptor that represses the transcriptional activation of inflammatory response genes in mouse macrophages [78]. Similarly, consumption of milk SM effectively decreased disease activity and colonic inflammatory lesions in mice with chemically induced colitis, partly through PPARγ [77]. Norris et al. [11] found that milk SM was not cytotoxic to RAW264.7 macrophages at physiological dosages tested but strongly decreased LPS-stimulated pro-inflammatory gene expression. The major bioactive component in these experiments was sphingosine, as only sphingosine and sphingosine-containing SLs recapitulated the anti-inflammatory effects of milk SM. Consistent with this finding, sphingosine reduced TNF-α production from macrophages stimulated with LPS [79]. The choline moiety of milk PLs may also contribute some protection against macrophage inflammation. Choline dose-dependently reduced TNF-α release from macrophages stimulated with endotoxin by inhibiting NF-κB activation [80]. In mice, intraperitoneal injection of choline (50 mg/kg) prior to the endotoxin treatment significantly improved the survival rate and decreased plasma TNF-α level [80]. This anti-inflammatory effect of choline was found to be nicotinic acetylcholine receptor subunit α7 (α7nAChR)-dependent, which is an essential component of the cholinergic anti-inflammatory pathway [80]. Collectively, these data demonstrate that milk SM has potential to be anti-inflammatory in macrophages.

Milk polar lipids and their metabolic products also appear effective against endotoxemia. Endotoxemia is a persistent sub-clinical, low-grade inflammation due to circulating endotoxins, primarily LPS, which may be absorbed into circulation due to defects in the gut barrier. Milard et al. [81] observed gut barrier effects when Caco-2/TC7 cells were treated with milk SM incorporated into mixed micelles. They observed significant inductions in the gene expression of tight junction proteins, which was related to a specific induction of interleukin-8 (IL-8) by milk SM. While IL-8 has been reported to have pro-inflammatory effects in some conditions, this study showed recombinant IL-8 specifically increased tight junction protein expression. More research is warranted to investigate the role of milk SM on IL-8 and gut barrier function. While these effects were observed with milk SM, some studies have been conducted with MFGM as a source of polar lipids. Snow et al. [82] studied the effect of MFGM on gut-barrier and systemic inflammation in LPS-challenged mice by feeding them with 10% MFGM containing diet (6 g of milk polar lipids/kg of the diet). The MFGM supplemented diet significantly attenuated LPS inducted systemic inflammation, partly by improving gut barrier integrity. Moreover, a four-week parallel intervention study in healthy adults showed that consumption of 10 g of a commercially-available MFGM-rich milk protein concentrate (Lacprodan PL-20, containing 16% PL by weight) twice daily can provide in vivo resistance to food-borne infections [83]. Norris et al. [10] was the first to report that circulating LPS activity was reduced in high-fat diet (HFD)-fed C57BL/6J mice supplemented with milk SM (0.25% w/w) for 4 weeks. Additionally, supplementation of milk SM (0.1% w/w) to an obesogenic diet dramatically reduced serum inflammatory cytokines/chemokines and mRNA expression of inflammation markers in adipose tissue in C57BL/6J mice [9,11]. Accordingly, Li et al. [84] reported that supplementation of MFGM (200 mg/kg body weight) for eight weeks attenuated HFD-induced intestinal inflammation, improved gut barrier tight junction protein expression, and reduced LPS activity and inflammation biomarkers in the circulation of C57BL/6 mice.

3.3. Modulation of Gut Microbiota

An emerging area of importance is the potential modulation of gut microbiome by dietary polar lipids [63]. Humans harbor trillions of bacteria in their gastrointestinal (GI) tract as part of the natural gut microbiota [85]. The residency of these microbes in the GI tract is closely linked with human physiology and plays a vital role in the function of the gut. The three most abundant phyla in human gut microbiome are Bacteroidetes, Firmicutes, and Actinobacteria [86]. Both Firmicutes and Bacteroidetes make up about 90% of phyla in the human gut microbiota [87]. The significance of the Firmicutes/Bacteroidetes ratio to human health has yet to be fully elucidated, yet it appears that lean individuals have a greater proportion of gut microbiota as Bacteroidetes compared to obese individuals, whereas the opposite is seen for Firmicutes [88]. Thus, the Firmicutes/Bacteroidetes ratio is often used as a marker of gut dysbiosis related to obesity and HFD consumption. Bifidobacterium belongs to the phylum Actinobacteria, and it is the most abundant bacteria seen in the gut of breastfed infants [89,90]. Lower levels of gut bifidobacteria are commonly associated with many disease conditions, including hepatitis B [91], cystic fibrosis [92], Type 1 and Type 2 diabetes [91,93], and obesity [94,95].

The SL fraction of milk polar lipids are reported to possess antibacterial effects. Sprong et al. [96] were the first to report the antibacterial effect of milk SLs. In their study, they showed the antibacterial effects of galactosylsphingosine and lysoSLs against Gram-positive and Gram-negative bacteria. In addition, Fischer et al. [97] showed that sphingoid bases (e.g., sphingosine, phytosphingosine, and dihydrosphingosine) also have broad-spectrum antimicrobial activity, which was supported by the induction of ultra-structural damage subsequent to pathogen uptake [98]. Nejrup et al. [99] conducted a 24 h in vitro fermentation study in the fecal sample of healthy infants to determine the effect of digestive products of milk lipids on modulating gut bacteria. They found that long-chain non-esterified fatty acids (LC-NEFA) with 10% sphingosine can increase bifidobacteria relative abundance in fecal content, whereas the LC-NEFA alone did not have an influence on bifidobacteria populations [99].

These in vitro findings suggest that milk SLs and their metabolic products may potentially exert changes in the gut microbiota when consumed regularly. Milk SM is digested and absorbed in the middle and distal part of the small intestine in rats and, presumably, in humans [100]. However, a large fraction of dietary SM and its digestive products reach the colon [101], where they may exert their bactericidal and gut-modulating effects. The effects of milk PLs on gut microbiota composition in both pre-clinical and clinical studies are summarized in Table 2.

Norris et al. [10] first reported the gut microbiota modulating effects of purified milk SM. Supplementation of a HFD (45% kcal as fat) with 0.25% (w/w) milk SM for four weeks increased the relative abundances of fecal Bifidobacterium, Firmicutes, and Actinobacteria, as well as reduced Bacteroidetes in C57BL/6J mice [10]. Supplementation of 0.25% (w/w) milk SM to a semi-purified low-fat diet also increased fecal Bifidobacterium in C57BL/6J mice [63]. However, a longer supplementation (10 weeks) of a lower dose of SM (0.1% milk SM) in mice fed an obesogenic diet (60% kcal as fat) had weaker effects on gut microbiota, with little change except for an increase in Acetatifactor relative abundance [11].

With SM being the most bioactive component in milk PLs, effects of the total milk polar lipids fraction on gut microbiota have shown similar results to the animal studies using purified SM. We have recently observed that feeding 2% milk PLs (containing 0.4% milk SM) to HFD-fed LDL-receptor knockout mice resulted in a similar increase in fecal Bifidobacterium relative abundance, as previously observed by Norris et al. [102]. Interestingly, there was also an increase in the Bacteroidetes phylum, which significantly reduced the Firmicutes to Bacteroidetes ratio in the 2% milk PL-supplemented group [102]. Milard et al. [103] found that supplementing an HFD with 1.6% milk PLs (0.38% SM) induced a reduction in fecal Lactobacillus in C57BL/6 mice, while 1.1% milk PLs (0.25% SM) induced an increase in Bifidobacterium compared with the HFD-fed controls. Akkermansia muciniphila, which is classified under the Verrucomicrobia phylum, was also significantly higher in milk PL-fed mice. Akkermansia muciniphila is noted for its positive metabolic effects, which includes improving insulin sensitivity and protecting against metabolic endotoxemia-induced inflammation [104–106]. In addition, there was a significant positive correlation between Bifidobacterium animalis and Akkermansia muciniphila for the milk PL-supplemented mice [103]. However, a study conducted by Reis et al. [107] did not find any effect on the gut microbiota composition by supplementing total polar lipids, PLs, or SLs in HFD-fed C57BL/6J mice. This particular experiment by Reis et al. [107] differed from other experimental designs, as mice were first fed with HFD for five weeks and then later supplemented milk polar lipids along with the HFD for an additional five weeks. As the HFD consumption likely resulted in altered gut microbiota after five weeks, the delay in polar lipid treatment may have contributed to the insignificant effects in this case. Modulation of the gut microbiota by MFGM, i.e., the milk PL-rich fraction of milk, has also been tested in rodents. The pup-in-a-cup model was used in five-day-old rats to examine the feeding of formula with fat from vegetable sources only, formula with supplemented MFGM, or mother's milk on intestinal development and the gut microbiota [108]. After 10 days, the pups fed a formula supplemented with MFGM had a more similar intestinal development and gut microbiota to those fed mother's milk when compared to those fed formula only. In another study, Li et al. [84] reported that supplementation of MFGM (200 mg/kg body weight) to C57BL/6 mice for eight weeks attenuated gut dysbiosis that occurs with HFD, including increasing the Bacteroidetes/Firmicutes ratio. Recently, modulation of the gut microbiome by ethanolamine, which is the base constituent of PE, was tested by supplementing 0, 250, 500, and 1000 µM ethanolamine to the drinking water of rats [109]. Ethanolamine supplementation at 500 and 1000 µM significantly increased Bacteroidetes and decreased Proteobacteria, Elusimicrobia, and Tenericutes. In addition, a reduction in Spirochetes was also noticed in the mice provided with 500 µM ethanolamine [109]. Overall, most pre-clinical studies in mice have noted some impact of milk polar lipids on gut microbiota composition; however, there are differences in microbiota profiles across studies. The observed differences between studies may be related to varying dosages, forms of milk polar lipids, base diet composition, or the use of prevention or treatment models.

Table 2. Animal and human studies examining the effects of milk polar lipids on gut microbiota.

Authors	Model	Control	Treatment	Duration	Results	Reference
Reis et al. (2013)	C57BL/6J mice	HFD (n = 13)	HFD followed by supplementation of total polar lipids (TPL), phospholipids (PL), or sphingolipids (SPL) through HFD (n = 13)	5 weeks on HFD followed by 5 weeks on TPL/PL/SM	Little effect of the polar lipid dietary supplementation on the composition of cecal microbiota was observed ($p > 0.05$).	[107]
Nejrup et al. (2015)	Fecal samples from nine healthy infants (aged 2–5 months)		Medium chained and long chain NEFA with and without 10 mol% sphingosine	24 h in vitro	LC-NEFA with sphingosine: increased bifidobacteria	[99]
Zhou et al. (2018)	21-d-old Sprague–Dawley rats	0 µM Ethanolamine in drinking water	250, 500 and 1000 µM Ethanolamine from milk in drinking water for 2 weeks	2 weeks	Increased: Bacteroidetes (500 and 1000 µM) Decreased: Proteobacteria, Elusimicrobia and Tenericutes (500 and 1000 µM) Spirochetes (500 µM)	[109]
Norris et al. (2016)	Male C57BL/6 mice	HFD (21% added milk fat by weight) (n = 3)	0.25% (w/w) milk SM in HFD (n = 10)	4 weeks	Increased: Firmicutes, bifidobacteria, Actinobacteria and Gram-positives Decreased: Bacteroidetes, Tenericutes and Gram-negatives	[9]
Norris et al. (2017)	Male C57BL/6 mice	HFD (31% lard; 0.15% cholesterol by weight) (n = 14)	0.1% (w/w) milk SM in HFD. (n = 14)	10 weeks	Increased Acetatifactor	[11]
Bhinder et al. (2017)	5 to 15 days old Rats (Used pup in a cup model)	Fed with mothers' milk (MM)	Formula with MFGM comprising part of the fat component or Formula with fat derived entirely from vegetable source	15 days	MFGM formula: microbial richness and evenness similar to MM. Similar abundances of Firmicutes and Proteobacteria compared to MM MFGM formula: Increased Lactobacilli, Enterococcus, Clostridiales, Streptococcus, and Morganella vs. vegetable fat formula.	[108]

Table 2. Cont.

Authors	Model	Control	Treatment	Duration	Results	Reference
Li et al. (2018)	5 weeks old C57BL/6J mice	Chow diet ($n = 10$)	HFD ($n = 10$) or HFD + MFGM (Lacprodan® MFGM-10) at 200 mg/kg BW ($n = 10$)		MFGM diet increased the relative abundance of Porphyromonadaceae, S24-7, norank_f_Bacteroidates, S24-7_group, unclassified_f_Lachnospiraceae, and Odoribacter compared with the HFD group Increased ACE index compared with HFD. MFGM supplementation recovered 13 key genera found enriched in control group Simpson's index showed no difference among three group	[84]
Milard et al. (2019)	Male C57BL/6J mice	HFD (21% w/w palm oil in chow)	8 weeks on HFD with 1.1% (w/w) milk PL or 1.6% (w/w) of milk PL	8 weeks	Increased: *Bifidobacterium*, in particular *Bifidobacterium animalis* in 1.1% of milk PL group Decreased: Lactobacillus in 1.6% of milk PL group Positive correlation between *Bifidobacterium animalis* and *Akkermansia muciniphila*	[103]
Vors et al. (2019)	Double-blind, parallel clinical trial in 58 Overweight postmenopausal women	No milk PL via butter serum ($n = 19$)	3 mg ($n = 19$) or 5 mg ($n = 20$) of milk PL via butter serum	4 weeks	No change in major phylogenetic groups and bacterial species of gut microbiota Increased fecal coprostanol/cholesterol ratio	[110]
Millar et al. (2020)	LDLr−/− mice	HFD (45%) for ($n = 15$)	HFD (45%) with 1% or 2% milk PL (MPL) ($n = 15$)	14 weeks	2% MPL: Increased Actinobacteria, Bacteroidetes, *Bifidobacterium*, Bacteriodales_unclassified. Reduced Firmicutes/Bacteroidetes ratio 1% MPL: Increased Shannon diversity	[102]

Abbreviations: ACE, abundance based coverage estimator; BW, body weight; HFD, high-fat diet; LC-NEFA, long-chain non-esterified fatty acids; LDLr, low-density lipoprotein receptor; MM, mother's milk; MFGM, milk fat globular membrane; NEFA, non-esterified fatty acids; PL, phospholipid; SM, sphingomyelin, SPL, sphingolipids; and TPL, total polar lipids.

While the pre-clinical findings reporting modulation of the gut microbiome by milk polar lipids are promising, only one human clinical trial has been conducted investigating this area. In a recent study by Vors et al. [110], post-menopausal women who supplemented their diets for four weeks with either 3 g or 5 g of milk polar lipids daily through butter serum showed no significant changes in major phylogenetic groups or bacterial species of gut microbiota when compared with the control group fed only butter oil. This may be attributed to the much lower dosage used in this study when compared to animal studies. A simple allometric approach considering the body surface area can be used to convert mice dose in mg/kg to human equivalent dose (mg/kg) by multiplying by 0.081 [111]. For example, the animal dose of milk PLs in Millar et al. [102] shown to modulate gut microbiota was ~1.25–2.5 g PLs/kg of body weight, which is equivalent to ~0.1–0.2 mg/kg of bodyweight in humans. This would equate to a dose of 7–14 g of milk PLs in a 70 kg human. However, Vors et al. [110] reported there was a significantly greater amount of fecal coprostanol, as well as a higher coprostanol/cholesterol ratio in those supplemented with milk polar lipids compared to control. While not examined further, these effects suggest there were changes in gut microbiome metabolism specific to increased coprostanol conversion in the gut with milk polar lipid supplementation. Some gut microbes are known to have the ability to convert cholesterol to coprostanol [112,113]. Research has shown an inverse relationship between blood cholesterol concentrations and the coprostanol/cholesterol ratio in human feces, suggesting the ability of coprostanol to modulate cholesteremia [114]. Thus, metagenomic effects of milk polar lipids and their influence on coprostanoligenic bacteria warrant more investigation.

Although comparisons of studies investigating gut microbiota composition are often challenging, five out of eight studies described above, including in vitro studies, showed an increase in fecal bifidobacteria by supplementing milk polar lipids at different concentrations. Strains of Bifidobacterium are commonly used as probiotic agents [115–118] and have been shown to have preventive and therapeutic effects in infant gut diseases [119] and in respiratory and gastrointestinal disorders in adults [120]. If supported in human studies, milk PLs may have potential as prophylactic or therapeutic agents against these diseases. It is quite interesting to note the differences in the changes in Firmicutes/Bacteroidetes ratio by the supplementation of milk PLs and milk SM. Norris et al. [10] reported a decrease in Bacteroidetes and increase in Firmicutes (hence an increased Firmicutes/Bacteroidetes ratio) by supplementing 0.25% milk SM for four weeks to HFD-fed mice. It is noteworthy that SLs can be produced in the gut by a small fraction of bacteria belonging to the Bacteroidetes phylum. A plausible justification may be that chronic exogenous SM supplementation may be triggering a feedback signaling pathway that is lethal to Bacteroidetes [63]. Another possibility may be related to greater amount of lipids getting to the colon of animals supplemented with high amounts of purified milk SM due to its noted inhibitory effects on lipid absorption. However, supplementing the same or higher amount of milk SM in the presence of other milk PLs did not change the Firmicutes/Bacteroidetes ratio [103] or decreased it [102]. The varying responses of the gut microbiota across studies may be due to differences in animal models or diets used, as well as the duration of milk polar lipid supplementation. While effects on gut microbiota composition observed at the phylum level are intriguing, more research is warranted in this area to investigate genus- and species-level compositional changes, as well as the metagenomic effects.

Recent research also suggests that choline-containing PLs can be metabolized by gut bacteria to generate trimethylamine (TMA) and, subsequently, oxidized to trimethylamine-N-oxide (TMAO) in the liver after absorption. Many observational and metabolomic studies have reported TMAO as a predictive risk factor for cardiovascular disease (CVD) [121,122] and colorectal cancer [123]. Choline, as one of the precursors for TMA, is found in many foods as free choline or as part of phosphatidylcholine, phosphocholine, or SM. Since PC is one of the most abundant polar lipids found in milk and milk products, researchers have analyzed the association between milk consumption and TMAO production. A cross-sectional study conducted in a German adult population reported a positive association between elevated plasma TMAO levels and milk consumption [124]. In contrast, another study conducted on healthy adults (KarMeN study) found no association between milk consumption and plasma TMAO [125], while intervention studies by Zheng et al. reported lower urinary TMAO levels in overweight women on high dairy diets [126] and adult men on both high milk and high cheese diets [127].

Current evidence appears insufficient to associate milk polar lipids and milk to high plasma or urinary TMAO levels. The direct effects of elevated plasma TMAO levels in promoting CVD risk is also controversial. Fish and seafoods are the rich sources of TMAO and TMA [128]. Many reports associate high fish intake to elevated plasma TMAO levels [128,129]. If Plasma TMAO can increase the risk of CVD, it should be speculated that high fish intake can increase the risk of CVD. However, epidemiological and observational studies report a protective effect of fish consumption on CVD risk [130–133]. On the other hand, fish oil has reported to ameliorate the adverse effects caused by TMAO in HFD fed mice [134]. A recent Mendelian randomization trial suggests that type 2 diabetes and kidney disease can increase circulating TMAO and evidence for the association between TMAO and CVD in observational studies may be due to reverse causality or confounding [135]. Overall, current evidence suggest choline-containing milk PLs may be used as substrates for TMA generation by gut microbiota, but the effect on disease risk is unclear. Fortunately, a number of studies have investigated the effects of milk PLs on other CVD risk factors.

3.4. Cardiovascular Disease

Cardiovascular disease (CVD) remains the most prominent contributor to mortality in the United States [136]. Atherosclerosis, which is characterized by the deposition of fatty plaque in the inner walls of the arteries, is a key player in the development of CVD. Dietary modification is recommended as a primary prevention strategy for managing blood lipid levels to reduce the risk for CVD [137]. The health effects of milk PLs on serum lipid levels are summarized in Tables 3 and 4. Genetically-obese KK-Ay mice displayed significant reductions in plasma LDL-cholesterol when fed a diet supplemented with SL-concentrated butter serum (0.35% SLs in diet by weight) or milk-derived ceramides (0.35% w/w) [138]. The feeding of milk SM (~0.25% w/w of diet) has been shown to significantly reduce serum cholesterol by ~15–25% when fed to C57BL/6 mice consuming both milk fat-enriched [10,139] and low-fat diets [63]. In humans, milk SM also shows potential to improve serum lipids. In a single-blind, randomized, controlled isocaloric parallel study, Rosqvist et al. [140] observed that an eight-week consumption of 40 g milk fat/day as whipping cream (rich in MFGM) in overweight adults resulted in lower plasma LDL-C, non-HDL-C, and apoB:apoA-I ratio compared to the same amount of milk fat as butter oil (free of MFGM). Conway et al. [141] reported that ingestion of 45 g/day of buttermilk for four weeks resulted in reductions in plasma cholesterol and TG compared to placebo, in a double-blind randomized study of healthy adults. The lower plasma LDL-C concentrations observed with buttermilk were associated with changes in plasma β-sitosterol, a marker of cholesterol absorption. In a clinical trial in postmenopausal women by Vors et al. [110], a dose of 5 g/day of milk PLs (1.3 g/day milk SM) via a butter serum concentrate lowered total and LDL-C, as well as decreased total/HDL-cholesterol and decreased apoB:apoA-I ratio, compared to a control cream cheese devoid of milk PL. Overall, it appears that the hypolipidemic effects of milk PLs observed in animal studies were also observed in several human studies, even when using lower dosages than in animal studies.

We have previously reported that supplementing purified milk SM at both 0.25% and 0.1% (w/w of diet) attenuated dyslipidemia and inflammation in HFD-fed C57BL/6J mice. Our research group [142] and others [143] have previously reported that dietary egg SM could attenuate atherosclerosis development in apoE$^{-/-}$ mice, even in the absence of changes in serum lipids. In a recent study, we report that supplementation of 2% (w/w) milk PLs to LDLr$^{-/-}$ mice fed a milk fat-rich diet strongly reduced atherogenic lipoprotein cholesterol, modulated gut microbiota, modestly lowered inflammatory markers, and markedly attenuated atherosclerosis development [102]. Milk polar lipids were provided by supplementing diets with butter serum, a dairy co-product rich in both PLs and SLs. Thus, due to potential beneficial effects on both serum lipids and inflammation, milk polar lipids may be important to consider when choosing foods for the prevention of CVD.

3.5. Non-Alcoholic Fatty Liver Disease (NAFLD)

NAFLD is the most common cause of chronic liver disease worldwide [144]. In Western societies, it affects 20%–30% of the general population and over 75% of obese individuals [145]. The "two-hit hypothesis" model of NAFLD states that the disease progresses in a stepwise manner, with a "first hit" from obesity and insulin resistance resulting in hepatic lipid accumulation, i.e., hepatic steatosis. Subsequently, a "second hit" in the form of oxidative stress and inflammation promotes liver injury and fibrosis [146]. In laboratory animals, dietary SM appears useful in preventing hepatic lipid accumulation (as extensively reviewed previously by Norris and Blesso [15]). The health effects of milk PLs on hepatic lipid metabolism are summarized in Table 3. Previous research in our laboratory has reported that HFD-fed mice supplemented with milk SM (0.25% w/w) had reduced hepatic TG after fouur weeks, compared to HFD-fed control animals [10]. In addition, our experiments investigating 10-week supplementation of a lard-based HFD (31% lard, 0.15% cholesterol by weight) with either 0.1% (w/w) milk SM or egg SM significantly attenuated the development of hepatic steatosis and adipose tissue inflammation in C57BL/6J mice [9]. Moreover, gene expression analysis revealed lower hepatic mRNA expression of stearoyl-CoA desaturase-1, indicating a reduced capacity for hepatic TG synthesis. Likewise, Cohn and colleagues have also noted that chronic supplementation of Western-type diets (21% butter fat, 0.15% cholesterol by weight) with various milk PL extracts (0.25–0.35% SM w/w of diet) significantly attenuated hepatic cholesterol and TG accumulation in mice. Recently, we have reported that supplementing milk fat-based diets (21% milk fat by weight) with milk PLs (at 1% and 2% w/w) resulted in significantly lower hepatic cholesterol concentrations in LDLr$^{-/-}$ mice, although no effects were seen in hepatic TG content [102]. For human studies, there is limited data available evaluating the effects of milk PLs on NAFLD-related markers, although one study by Weiland et al. [147] found beneficial effects of 2–3 g/day of milk PLs for 7–8 weeks on serum γ-glutamyl transferase (GGT), a marker of fatty liver disease, with no changes in alanine transaminase and aspartate transaminase (markers of liver injury) in two separate clinical trials of overweight or obese men.

Table 3. Animal studies examining the effect of milk polar lipids on serum and hepatic lipids.

Authors	Animal Model	Control	Treatment	Duration	Results	Reference
Nyberg et al. (2000)	Male Sprague-Dawley rats ($n = 5$–8)	Cholesterol mixed in soybean oil (without PL)	2.6:1, 1:1 or 0.5:1 molar ratio of cholesterol:SM	3 days	Decreased intestinal cholesterol absorption (lowest in cholesterol:SM ratio 1:1)	[148]
Eckhardt et al. (2002)	Male C57BL/6 mice ($n = 6$)	Chow	Chow diet enriched in PL (containing 0.1%, 0.5% or 5% of milk SM by weight)	4 days	Decreased intestinal cholesterol absorption	[64]
Wat et al. (2009)	Male C57BL/6 mice ($n = 10$)	LFD or HFD without milk PL	LFD or Western-type diet with 1.2% (w/w) PL from phospholipid-rich dairy milk extract (PLRDME)	8 weeks	Serum lipids: PLRDME with western-type diet group: Decreased TG (−20%), phospholipids (−21%) and HDL-C (−19%) PLRDME with LFD group: No change in TC, TG, phospholipids and HDL-C Hepatic Lipids: PLRDME with western-type diet group: Decreased total lipid (−33%), TG (−44%), TC (−48%) and phospholipids (−16%) PLRDME with LFD group: No change in total lipid, TG, TC and phospholipids	[139]
Kamili et al. (2010)	Male C57BL/6 mice ($n = 10$)	Western-type diet without milk PL	Western-type diet (21% AMF; 0.15% cholesterol by weight) with 1.2% (w/w) PL from PLRDME or milk phospholipid concentrate (PC-700)	3, 5 or 8 weeks	Plasma lipids: PLRDME after 8 weeks: Decreased plasma TC (−23%) Hepatic lipids: PLRDME after 5 weeks: Decreased total lipid (−41%), TG (−47%) and TC (−39%) PLRDME after 8 weeks: Decreased total lipid (−18%) and TG (−28%) PC–700 after 5 weeks: Decreased total lipid (−45%), TG (−63%) and TC (−57%)	[149]
Watanabe et al. (2011)	Female KK-A^y mice ($n = 7$)	AIN-93G diet	AIN-93G diet with 1.7% (w/w) of lipid-concentrated butter serum (LC-BS), or 0.5% (w/w) of ceramide-rich fraction (Cer-fr) or 0.5% (w/w) of SM-rich fraction (SM-fr)	4 weeks	Plasma lipids: SM-fr: no change LC-BS: Decreased TC (−18%) and LDL-C (−45%) Cer-fr: Change only in TC (−25%) Hepatic lipids: SM-fr: No change LC-BS: Decreased TG (−27%) Cer-fr: Decreased TG (−38%) and TC (−47%)	[138]
Zhou et al. (2012)	Fischer-344 rats ($n = 3$–4)	AIN-76A diet with corn oil or AMF (0.5% w/w)	2.5% (w/w) MFGM, 2.5% (w/w) AMF in AIN-76A diets	12 weeks	Decreased esterified cholesterol and increased TG in liver	[150]

Table 3. Cont.

Authors	Animal Model	Control	Treatment	Duration	Results	Reference
Reis et al. (2013)	Male C57BL/6 ($n = 13$)	HFD	HFD (~20% lard by weight) with 1.7% (w/w) total polar lipids extracts or 1.4% (w/w) phospholipids-rich extract or 0.4% (w/w) SM-rich extract	5 weeks	Decreased FA synthesis in liver by total PL extract and PL-rich extract Decreased 16:1n-7/16:0 in liver by SM-rich extract	[107]
Lecomte et al. (2015)	Female Swiss mice ($n = 7$)	Emulsion with soybean PL (gavaged)	Emulsion with 5.7 mg milk PL (gavaged)	1, 2 or 4 h	After 1 h: Increased plasma NEFA and a trend to increase TG After 4 h: Decreased plasma TG and NEFA associated with a decreased duodenal gene expression of APOB 48 and Sar1b	[151]
Norris et al. (2016a)	Male C57BL/6 mice ($n = 10$)	HFD (21% AMF by weight)	0.25% (w/w) milk SM in HFD (21% AMF by weight)	4 weeks	Decreased serum TC and hepatic TG No change in serum TG and hepatic TC	[10]
Norris et al. (2017)	Male C57BL/6 mice ($n = 14$)	HFD (31% lard; 0.15% cholesterol by weight)	0.1% (w/w) milk SM in HFD	10 weeks	No change in serum lipids Decreased hepatic TC (~23%) and TG (~30%)	[9]
Lecomte et al. (2016)	Male C57BL/6J mice ($n = 10-12$)	HFD (17% w/w palm oil) + soybean PL	1.2 % (w/w) milk PL or SPL in HFD (17% w/w palm oil)	8 weeks	No change in plasma and hepatic lipids Increased fecal VLCFA such as C22:0, C24:0 and C22:4(n-6)	[152]
Yamauchi et al. (2016)	Obese/diabetic KK-A^y ($n = 7$) and male C57BL/6 mice ($n = 6$)	HFD (lard, soybean, linseed or fish)	1% (w/w) milk SM in HFD (lard, soybean, linseed or fish)	4 weeks	No effect on wild type mice In KK-A^y mice: Soybean + SM: decreased serum LDL-C and non-HDL-C. Increased hepatic total lipids, cholesterol, bile acid. Linseed + SM: Decreased serum LDL-C. Decreased hepatic total FA and increased fecal total lipid and cholesterol. Lard + SM: Increased fecal total lipids, cholesterol, and decreased hepatic total FA.	[153]
Zhou et al. (2019)	Male ob/ob mice ($n = 11-18$)	Moderately high-fat AIN-93G diet (34% kcal as fat) without milk PL or gangliosides	(0.2% (w/w) milk gangliosides (GG) or 1% (w/w) milk PL (PL) in moderately high-fat AIN-93G diet (34% kcal as fat)	2 weeks	No change in plasma and hepatic lipids by milk GG. PL increased plasma NEFA, PL, SM and DAG and decreased hepatic CE.	[154]

Table 3. Cont.

Authors	Animal Model	Control	Treatment	Duration	Results	Reference
Millar et al. (2020)	LDLr−/− mice	HFD (45%) for ($n = 15$)	HFD (45%) with 1% or 2% milk PL (MPL) ($n = 15$)	14 weeks	2% MPL: Decreased serum cholesterol (−51%), with dose-dependent reduction in VLDL-C and LDL-C. Decreased hepatic TC (−55%) 1% MPL: Decreased hepatic TC (−53%)	[102]

Abbreviations: AMF, anhydrous milk fat; ApoB, apolipoprotein B; Cer-fr, ceramide-rich fraction; CE, cholesteryl ester; DAG, diacylglycerol; FA, fatty acid; GG, gangliosides; HDL-C, high-density lipoprotein cholesterol; HFD, high-fat diet; LC-BS, lipid-concentrated butter serum; LDL-C, low-density lipoprotein cholesterol; LFD, low-fat diet; MFGM, milk fat globular membrane; NEFA, non-esterified fatty acids; PL, phospholipids; PLRDME, phospholipid-rich dairy milk extract; Sar1B, secretion-associated; SM, sphingomyelin; SM-fr, sphingomyelin-rich fraction; SPL, soybean polar lipids; TC, total cholesterol; TG, triglyceride; and VLCFA, very long-chain fatty acids.

Table 4. Human clinical trials examining the effects of milk polar lipids on serum lipids.

Authors	Population and Study Design	Control	Treatment	Duration	Results	Reference
Ohlsson et al. (2009)	Parallel group study with 33 healthy men and 15 healthy women	119 mg of total SL (isocaloric)	2 drinks/day totaling 975 mg SL, containing 700 mg SM, 180 mg GC and 95 mg GS	4 weeks	No change in plasma lipids. Trend for decreasing LDL-C (only in women)	[155]
Ohlsson et al. (2010)	Human ileostomy contents from 6 men and 6 women		1. Milk SM (250 mg) mixed in skimmed milk 2. Milk SM (50,100 or 200 mg) mixed in milk-like oat drink	Collected after 8 h	Increased the out-put of VLCFA specific of milk SM (22:0, 23:0, 24:0)	[156]
Ohlsson et al. (2010)	Crossover study in 18 healthy adult males	High-fat (40 g) standard breakfast together with a milk-like formulation lacking polar milk lipids	High-fat (40 g) standard breakfast together with a milk-like formulation containing 975 mg of milk SL	1 to 7 h	No change in plasma lipids after 1 h Trend for decreasing cholesterol in large TG-rich lipoproteins.	[157]
Keller et al. (2013)	Parallel study in 14 healthy women	Baseline	2 supplementation cycle–3 g milk PL/day followed by 6 g milk PL/day	10 days each	3 g milk PL: Decreased plasma TC, HDL-C After 6 g milk PL supplementation: Increased plasma TC and LDL-C	[158]
Conway et al. (2013)	Double-blinded crossover study in 34 healthy adults	45 g/day of a macro/micronutrient matched placebo	45 g buttermilk powder/day	4 weeks	Decreased serum cholesterol (−3.1%), TG (−10.7%) and trend for decreasing LDL-C ($p = 0.057$) Decreased LDL-C (−5.6%) in participants with highest (top 50%) baseline LDL-C	[141]
Baumgartner et al. (2013)	Single-blind parallel study in 97 healthy adults	One or two eggs a week ($n = 20$)	1. One egg/day ($n = 57$) 2. 100 mL/day of buttermilk drink containing one egg yolk ($n = 20$)	12 weeks	No difference in serum lipids, liver inflammatory markers, Apo-A1, Apo-B100, campesterol, or lathosterol between the two treatment groups	[159]
Rosqvist et al. (2015)	Single-blind, parallel study in 57 overweight adults	Butter oil (1.3 mg total PL), matched for calories, macronutrients, and calcium	40 g milkfat/day as whipping cream (198 mg total PL)	8 weeks	Decreased plasma cholesterol, LDL-C, non-HDL-C, and apoB:apoA1 ratio	[140]
Severins et al. (2015)	Single-blind, parallel study in 92 mildly hypercholesterolemic adults	80 mL of skim-milk powder ($n = 25$)	1. 80 mL skim-milk with lutein enriched egg yolk (28 g from 1.5 eggs providing 323 mg cholesterol) 2. Buttermilk (72 mg PL) 3. Buttermilk with lutein enriched egg yolk (28 g from 1.5 eggs providing 323 mg cholesterol)	12 weeks	Buttermilk addition could not change the increased serum lipids levels due to of egg yolk Buttermilk group showed a trend for decreasing TC ($p = 0.077$), but not for LDL-C	[160]

Table 4. *Cont.*

Authors	Population and Study Design	Control	Treatment	Duration	Results	Reference
Weiland et al. (2016)	Double-blind parallel-group intervention trials in overweight or obese males.	Milk enriched with 2 g milk fat (n = 31)	Milk enriched with 2 g milk PL (n = 31)	8 weeks	Decreased GGT and waist circumference. No change in plasma lipids (total, HDL- and LDL-cholesterol, total cholesterol: HDL-cholesterol ratio, TAG, PL), ALT, AST, apoB, apoA1, glucose, insulin, insulin sensitivity index, C-reactive protein, IL-6, soluble intracellular adhesion molecule and total homocysteine (tHcy).	[147]
		Milk enriched with 2.8 g soy PL (n = 57)	Milk enriched with 3 g milk PL (n = 57)	7 weeks	Decreased only GGT. No change in plasma lipids (TC, HDL-C, LDL-C, TG, phospholipids, TC:HDL-C ratio), apoA1, apoB, glucose, insulin, HOMA-IR, hs-CRP, IL-6, sICAM, ALT, AST	
Grip et al. (2018)	Double blinded study in formula fed infants.	Breast fed infants (n = 80)	Formula without MFGM (n = 160)	4, 6 and 12 months	Decreased plasma PC and SM	[161]
Vors et al. (2019)	Double blinded parallel study in 58 postmenopausal women	No milk PL via butter serum (n = 19)	3 g (n = 19) or 5 g (n = 20) of milk PL via butter serum	4 weeks	Decreased fasting total cholesterol, LDL-C, TC/HDL-C ratio, ApoB/ApoA1 ratio, post-prandial total cholesterol, chylomicron lipids.	[110]
	Double blind cross-over study in 4 ileostomized subjects	No milk PL via butter serum (n = 19) with ^{2}H-cholesterol tracer	3 g (n = 19) or 5 g (n = 20) of milk PL via butter serum with ^{2}H-cholesterol tracer	Acute post-prandial	Decreased ^{2}H-cholesterol tracer in plasma and chylomicrons. Increased ileal output of total cholesterol and of milk SM	

Abbreviations: ALT, alanine transaminase; ApoA-I, apolipoprotein A-I; ApoB, apolipoprotein B; AST, aspartate transaminase; GC, glucosylceramide; GGT, gamma-glutamyl transferase; GS, gangliosides; HDL-C, high-density lipoprotein cholesterol; HOMA-IR, homeostatic model assessment of insulin resistance; hs-CRP, high-sensitivity C-reactive protein; IL-6, interleukin -6; LDL-C, low-density lipoprotein cholesterol; MFGM, milk fat globule membrane; PC, phosphatidylcholine; PL, phospholipids; sICAM, soluble intercellular adhesion molecule; SL, sphingolipids; SM, sphingomyelin; TAG, triacylglycerol; TC, total cholesterol; TG, triglyceride; and VLCFA, very long chain fatty acid.

3.6. Insulin Resistance and Type 2 Diabetes

Insulin resistance occurs when the body's cells cannot effectively import glucose in response to the release of endogenous or exogenous insulin within the bloodstream [162]. Insulin resistance can cause hyperglycemia, which makes it a risk factor for the development of type 2 diabetes mellitus (T2DM). Diabetes is a major cause of death in the U.S. and, also, contributes to significant comorbidities related to micro- and macrovascular complications, including CVD, NAFLD, kidney disease, and blindness [163]. While studies specifically examining the effects of milk polar lipids in T2DM models are lacking, the health effects of milk polar lipids on insulin resistance and glycemia are summarized in Table 5.

Nagasawa et al. [164] observed the effects of dihydrosphingosine (DHS) on activating GPR120, a receptor expressed by enteroendocrine cells that promote the secretion of the incretin GLP-1. DHS, along with phytosphingosine, was shown to strongly activate GPR120 in vitro, although sphingosine did not. Interestingly, milk SM is known to have more saturated sphingoid backbones than other SM sources [165] and, thus, would provide DHS as a hydrolytic digestive product. GLP-1 is known to have inhibitory effects on insulin resistance and T2DM; thus, more research should investigate the effects of milk polar lipids on regulating incretin production.

In contrast to in vitro effects, results from in vivo studies have been less promising. Yamauchi et al. [153] studied the effects of SM supplementation in male obese/diabetic KK-Ay mice. Mice that were supplemented with 1% (w/w) milk SM in several low-fat diets (7% by weight as lard, soybean oil, or linseed oil) for four weeks showed no significant effects in body weight, adiposity, or blood glucose compared to control groups. Correspondingly, there were no significant differences in blood glucose concentrations when measured at various time points (days 0, 7, 14, 21, and 28). Similarly, Norris et al. [9] reported that blood glucose and HOMA-IR were not significantly affected by 0.1% (w/w) milk SM supplementation in C57BL/6J mice fed an obesogenic HFD for 10 weeks, although egg SM was shown to significantly reduce fasting glucose in the same study. Weiland et al. [147] observed the effects of three different milk interventions administered to overweight and obese men. Within this report, there were two double-blind parallel-group trials that occurred involving PL-enriched milk supplementation in overweight/obese men. Trial 1 consisted of administering milk enriched with 2 g milk PL or 2 g milk fat (control) to 62 male participants over an eight-week time period. Trial 2 consisted of administering milk enriched with 3 g milk PL or 2.8 g soy PL to 57 male participants over a seven-week time period. The overall results showed a reduction in waist circumference in participants that received 2 g of milk-PL (intervention) in Trial 1 when compared to those that received the control. However, there were no differences in fasting glucose, insulin, or the insulin sensitivity index in both trials.

Table 5. Studies examining effects of milk polar lipids on insulin resistance and type 2 diabetes.

Author	Model	Control	Treatment	Duration	Results	Reference
Nagasawa et al. (2018)	293 T cells		Dihydrosphingosine or phytosphingosine or sphingosine	24 h	Significant upregulation of GPR120 (a receptor for long chain fatty acids) by dihydrosphingosine and phytosphingosine. Sphingosine—no effect.	[164]
Yamauchi et al. (2016)	obese/diabetic KK-A^y mice ($n = 7$)	Lard or soybean oil or linseed oil	Lard + 1% SM or soybean oil +1% SM or linseed oil + 1% SM	4 weeks	No difference in blood glucose level	[153]
	wild-type C57BL/6J mice ($n = 7$)	Linseed oil or fish oil or lard + soybean oil	Linseed oil + 1% SM or fish oil + 1% SM or lard + soybean oil + 1% SM	4 weeks	No difference in blood glucose level	
Weiland et al. (2016)	Double-blind parallel-group intervention trials in overweight or obese males.	Milk enriched with 2 g milk fat ($n = 31$)	Milk enriched with 2 g milk PL ($n = 31$)	8 weeks	No difference in blood glucose, insulin and HOMA-IR between groups	[147]
		Milk enriched with 2.8 g soy PL ($n = 57$)	Milk enriched with 3 g milk PL ($n = 57$)	7 weeks	No difference in blood glucose, insulin and HOMA-IR between groups	
Norris et al. (2017)	Male C57BL/6 mice	HFD (31% lard; 0.15% cholesterol by weight) ($n = 14$)	0.1% (w/w) milk SM ($n = 14$) in HFD	10 weeks	No difference in fasting serum insulin, glucose concentrations and HOMA-IR between groups	[9]

Abbreviations: HFD, high-fat diet; HOMA-IR, homeostatic model assessment of insulin resistance; PL, phospholipids; and SM, sphingomyelin.

3.7. Cognitive Function and Neurodevelopment

There is a growing interest in the potential health benefits of milk PLs on neurodevelopment. The development of an infant's brain starts at two weeks of conception and stops when the individual has reached about 20 years of age or early adulthood [166]. As galactosylceramide (cerebroside) content of myelin is important for central nervous system function [167], it is hypothesized that milk polar lipids may influence cognition when introduced into the diet at a young age. The health effects of cow's milk polar lipids on cognitive function and neurodevelopment are summarized in Table 6.

Oshida et al. [167] studied the effects of cow's milk SM supplementation on l-cycloserine (LCS) (an inhibitor of myelination) treatment in male Wister rat pups. Rat pups were divided into two treatment groups that were either administered LCS treatment only (control) or LCS + cow's milk SM (0.81% *w/w* of diet) for 28 days. This study found an increase in brain weight and CNS myelin dry weight in the LCS + SM group when compared to the LCS only group. Gurnida et al. [168] examined the effects of infant formula supplemented with complex milk lipids and gangliosides on Griffith Scale values and serum ganglioside levels in infants. The treatment group consisted of infants that were given a supplemented infant formula with complex milk lipids and gangliosides (11–12 µg/mL) (derived from cow's milk), while the control group consisted of infants that were given unsupplemented, standard infant formula. The reference group consisted of infants given breastmilk only. The infants received the treatment or control products starting at two to eight weeks of age and consumed them until six months of age. Overall, findings showed an increase in Griffith Scale values within the treatment group when compared to the control group. Correspondingly, the Griffith scale values were comparable to the values of a reference group of breastfed infants. Additionally, an increase in serum GM3, GD3, and total ganglioside levels were observed in the treatment group when compared to the control group.

3.8. Colorectal Cancer and Colitis

While milk polar lipids, particularly SM and other SLs, have shown promise in controlling inflammation and modulation of gut microbiota, there has also been interest in their effects on chronic diseases of the GI tract, such as colorectal cancer and inflammatory bowel disease (IBD). Colorectal cancer is the second leading cause of cancer-related deaths in the United States [169]. Meanwhile, IBD has rapidly expanded in both Western civilizations and newly industrialized nations in the 21st century [170]. In North America, over 1.5 million people suffer from IBD, which includes ulcerative colitis (UC) and Crohn's disease (CD) [170]. These two diseases of the gut are linked, as patients with IBD have a greater risk of developing colorectal cancer [171]. The health effects of milk PLs on colorectal cancer and colitis are summarized in Table 7.

Table 6. Studies examining effects of milk polar lipids on cognitive/brain development.

Authors	Model/Population and Study Design	Control	Treatment	Duration	Results	Reference
Oshida et al. (2003)	Male Wistar rat pups ($n = 30$)	No l-Cycloserine (LCS) treatment or dietary SM (non-LCS)	Daily s/c injection of 100 mg/kg of LCS from 8 days old + diet without (LCS group) or with 810 mg/100 g of SM (SM-LCS group) from 17 days old	Until 28 days old	Significantly high myelin dry weight, myelin total lipid content, and cerebroside content in the SM-LCS group than in the LCS group. Axon diameter, nerve fiber diameter, myelin thickness, and g value of optic nerve were similar in SM-LCS and non-LCS groups.	[167]
Tanaka et al. (2013)	Randomized, double-blind controlled trial in 28 premature infants with birth weight less than 1500 g	Milk (13 g SM/100 g PL) ($n = 14$)	Sphingomyelin fortified milk (20 g SM/100 g PL) ($n = 14$)	18 months	Significantly better Behavior Rating Scale of the BSID-II, Fagan test scores, latency of VEP, and sustained attention test scores	[172]
Gurnida et al. (2012)	Double-blind, parallel study in infants 2 to 8 weeks of age	Standard infant formula (0.22% milk PL and 0.006% gangliosides) ($n = 30$)	Complex lipid-supplemented formula (0.235% milk PL and 0.009% gangliosides) ($n = 29$)	From 2–8 weeks of age to until 24 weeks of age	Increased Hand and Eye coordination IQ score ($p < 0.006$), Performance IQ score ($p < 0.001$) and General IQ score ($p = 0.041$).	[168]

Abbreviations: BSID, Bayley Scales of Infant Development; LCS, 1-Cycloserine; PL, phospholipids; and SM, sphingomyelin.

Table 7. Studies examining effects of milk polar lipids on colon cancer and colitis.

Authors	Model	Control	Treatment	Duration	Results	Reference
Kutchta-Noctor et al. (2016)	SW480 colon cancer cells and FHC cells (normal human colon cells)	Controls contained only media. Sodium butyrate (5 mM), a potent apoptotic fatty acid, served as a positive control.	Buttermilk between 0 and 0.94 mg/mL of media	3 days at 37 °C in CO_2 incubator	Inhibited growth of SW480 colon cancer cells in dose-dependent manner with selective antiproliferative activity toward cancer cells. Downregulated growth signaling pathways mediated by Akt, ERK1/2, and c-myc.	[173]
Schmelz et al. (2000)	5 weeks old female CF1 mice	i/p injection of 1,2-DMH (DMH)@ 30 mg/kg body weight for 6 weeks + sphingolipid free AIN 76A diet	i/p injection of 1,2-dimethylhydrazine (DMH) at 30 mg/kg body weight for 6 weeks + AIN 76A diet with 0.025 or 0.1 g/100 g of milk GluCer, LacCer or ganglioside GD3 after 1 week	4 weeks	Glycosphingolipid groups: >40% reduction ($p < 0.001$) in appearance of aberrant crypt foci, reduced proliferation (up to 80%; $p < 0.001$) in colonic crypts.	[174]
Dillehay et al. (1994)	CF1 mice	Injection of 1,2-DMH + diet without SM	Injection of DMH + diets with 0.025 to 0.1 g/100 g of SM for 28 weeks followed by diet without SM	52 weeks	SM fed groups: 20% incidence of colon tumors (vs 47% in controls)	[175]
Schmelz et al. (1996)	5 weeks old female CF1 mice	i/p injection of 0.5 mL/kg of DMH once weekly for 6 weeks followed by diet without SM	i/p injection of 0.5 mL/kg of DMH once weekly for 6 weeks followed by diet supplemented with 0 to 0.1% (w/w) buttermilk or powdered milk SM	34 weeks	0.1% SM: Reduced appearance of aberrant colonic crypt foci ($p < 0.001$) and significantly fewer aberrant crypts per colonic focus	[176]
Snow et al. (2010)	Fischer-344 rats	i/p injection of 1,2-DMH (25 mg/kg BW) once weekly for 2 weeks followed by AIN-76A diet corn oil	i/p injection of 1,2-dimethylhydrazine (25 mg/kg BW) once weekly for 2 weeks followed by AIN-76A diet with AMF or with 50% MFGM, 50% AMF	9 weeks	MFGM group had significantly fewer aberrant crypt foci	[177]
Mazzei et al. (2011)	PPARγ+/+ and PPARγ−/− mice	Semi-purified sphingolipid-free AIN76A diet for 7 weeks followed by single injection of azoxymethane (10 mg/kg BW).	0.1% SM (w/w) supplemented diet for 7 weeks followed by single injection of azoxymethane (10 mg/kg BW).	9 weeks	SM group of both genotypes: Decreased disease activity and colonic inflammatory lesions (more efficiently in PPARγ+/+ mice).	[77]

Abbreviations: AMF, anhydrous milk fat; BW, bodyweight; DMH, dimethylhydrazine; GluCer, glucosylceramide; LacCer, lactosylceramide; and SM, sphingomyelin.

Kutchta-Noctor et al. [173] observed the effects of buttermilk, containing SM, lactosylceramide (LacCer), and ceramide, on growth inhibition of SW480 human colon cancer cells and noncancerous fetal human colon (FHC) cells. They reported that buttermilk containing SM and LacCer led to growth inhibition of SW480 cells and was selective towards cancer cells, with no effect on FHC cell growth. Early experiments reported decreased aberrant crypt foci formation in CF1 mice treated with 1,2-dimethylhydrazine (DMH) when supplemented with milk SLs in the diet [174–176]. In terms of the control groups, CF1 mice were fed a semi-purified diet (A1N76A) consisted of 0–0.005% SL content throughout the various studies [174–176]. A 66–70% decrease in aberrant colonic crypt appearance was observed by Schmelz et al. [176] when mice were supplemented with 0.1% SM by weight. Similar findings were observed by Dillehay et al. [175], who reported decreased incidence of DMH-induced colon tumors in mice fed 0.05% milk SM (w/w) when compared to the control group. Schmelz et al. [174] showed that DMH-injected CF1 mice supplemented with 0.025%–0.1% (w/w) of glucosylceramide (GluCer), LacCer, or GD3 had a measurable decrease in aberrant crypt foci formation of >40%. When compared to milk SM supplementation, GluCer, LacCer, and GD3 supplementation yielded similar decreases in aberrant crypt foci formation in DMH-injected CF1 animals. Snow et al. [177] reported a significant reduction in aberrant crypt foci when AMF-containing diets were supplemented with MFGM in Fischer-344 rats. Mazzei et al. [77] studied the effects of 0.1% (w/w) milk SM supplementation on azoxymethane (AOM) and dextran sulfate sodium (DSS)-induced colitis and colon tumor formation in PPARγ$^{+/+}$ and PPARγ$^{-/-}$ C57BL/6 mice. Compared to AOM + DSS control animals, those fed milk SM had reductions in disease activity index and colonic inflammatory lesions, with greater effects in PPARγ-expressing animals. Additional findings show decreased AOM-induced colon tumors only in PPARγ$^{-/-}$ mice fed SM. Thus, pre-clinical studies show promising effects of milk SLs on attenuating colitis and colorectal cancer. Clinical studies are necessary to find out if these favorable effects also translate to humans.

4. Gaps in Scientific Literature and Future Directions

Significant research has been conducted to understand the health benefits of phytochemicals, whereas less is known about zoochemicals, such as SM found in milk polar lipids. Our research laboratory as well as others have shown in animal studies that milk polar lipids may impart health benefits through lowering blood cholesterol, inflammation, and altering gut bacteria. However, very little research has been conducted to investigate how milk polar lipids affect lipoprotein profiles, inflammation, and gut health in men and women at risk for CVD. Vors et al. [110] recently reported that postmenopausal women who consumed 5 g/day of milk polar lipids had beneficial effects on plasma lipids and increased the coprostanol/cholesterol ratio in feces. While crucial data were revealed with this important study, there are many questions that remain. For example, (1) what are the impacts of chronic intake of milk polar lipids on systemic and intestinal inflammation? Additionally, (2) how are lipoprotein particle characteristics that are known predictors of CVD risk (e.g., LDL size and HDL particle number) affected by milk polar lipids? While Vors et al. [110] did not observe changes in major bacterial phyla composition after four weeks (i.e., who is there), the increased coprostanol conversion in feces with milk polar lipid consumption suggests significant changes in gut microbiome metabolic capacity; thus, (3) are there effects of milk polar lipids on the metagenome of gut bacteria (i.e., what they are doing)? Finally, while the study by Vors et al. [110] was conducted in moderately hyperlipidemic, overweight post-menopausal women, (4) do milk polar lipid health effects differ between men and women? Further research should address these important unresolved questions.

It is quite challenging to arrive at a comparative assessment and extrapolate in vivo experiments across different mammalian models. Apart from the category of polar lipid used in different studies, some of the confounding factors could be partly due to the differences in diet, dosage, and duration of the studies across different species. A significant number of studies reviewed in this paper used SM, SLs, milk PLs, total polar lipids, or MFGM as the source of milk polar lipids. When interpreting the results of studies that used total polar lipids or milk PL, the likelihood of the synergistic effects of different polar lipids should be considered. Unlike SM studies, the effect of SLs in animal models could also be attributed to the various fractions involved. Likewise, in studies that used MFGM the effect of membrane proteins, glycoproteins, and gangliosides should be considered as potential bioactive compounds influencing the results.

5. Conclusions

In this review, we evaluated the potential health effects of milk PLs in humans by examining in vitro and in vivo studies. Milk PLs were shown to favorably influence health in relation to inflammation, CVD, NAFLD, gastrointestinal diseases, and neurodevelopment, with most effects observed in pre-clinical studies (Figure 2). As described above, this can be attributed to the much lower dosages used in human studies when compared to animal studies and more clinical studies with higher doses are needed to confirm these effects in humans. Evidence from such studies may further support the development of "designer" dairy products rich in milk PLs and SLs to enhance value and promote health. Additionally, inexpensive dairy co-products rich in PLs, such as butter serum or buttermilk, could be promoted and utilized as value-added sources of milk PLs. Overall, Milk PLs are emerging as commonly consumed dairy matrix components that may be important to consider when planning diets for the prevention of chronic disease.

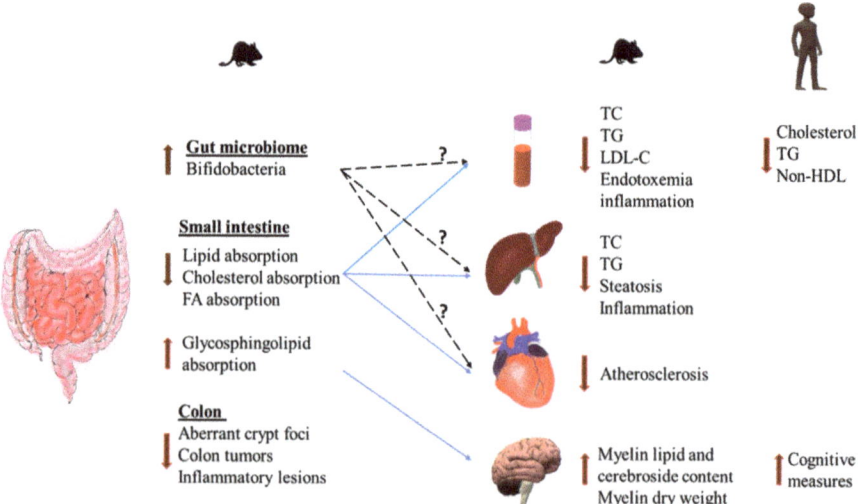

Figure 2. Functional properties of dietary milk polar lipids on various organs. Dietary milk polar lipids appear to have local effects in the GI tract on gut microbiome, colon health, and lipid absorption. Although the reported effects of milk polar lipids on gut microbiome are quite variable, a consistent finding in most studies is an increase in bifidobacterial population. Broken black lines indicate the hypothetical contribution of gut modulating effect of milk polar lipids on changes seen in other organs. Solid blue lines indicate the known underlying mechanisms by which milk polar lipids exert their systemic effects. Abbreviations: FA, fatty acids; TC, total cholesterol; TG, triglycerides; LDL-C, low-density lipoprotein-cholesterol; non-HDL, non-high-density lipoproteins.

Author Contributions: Conceptualization, M.T.-G. and C.N.B.; methodology and literature search, L.A. and S.W.W.; writing—original draft preparation, L.A. and S.W.W.; writing—review and editing, M.T.-G. and C.N.B.; supervision, M.T.-G. and C.N.B.; and project administration, M.T.-G. All authors have read and agreed to the published version of the manuscript.

Funding: This work was supported by a research grant (#3085) to C.N.B. by National Dairy Council.

Conflicts of Interest: M.T.-G. is a National Dairy Council employee. C.N.B. has received research funding from National Dairy Council. L.A. and S.W.W. declare no conflict of interest.

References

1. Cohn, J.S.; Kamili, A.; Wat, E.; Chung, R.W.S.; Tandy, S. Dietary phospholipids and intestinal cholesterol absorption. *Nutrients* **2010**, *2*, 116–127. [CrossRef] [PubMed]
2. Vesper, H.; Schmelz, E.-M.; Nikolova-Karakashian, M.N.; Dillehay, D.L.; Lynch, D.V.; Merrill, A.H. Sphingolipids in Food and the Emerging Importance of Sphingolipids to Nutrition. *J. Nutr.* **1999**, *129*, 1239–1250. [CrossRef] [PubMed]
3. Yunoki, K.; Ogawa, T.; Ono, J.; Miyashita, R.; Aida, K.; Oda, Y.; Ohnishi, M. Analysis of sphingolipid classes and their contents in meals. *Biosci. Biotechnol. Biochem.* **2008**, *72*, 222–225. [CrossRef] [PubMed]
4. Bourlieu, C.; Michalski, M.C. Structure-function relationship of the milk fat globule. *Curr. Opin. Clin. Nutr. Metab. Care* **2015**, *18*, 118–127. [CrossRef] [PubMed]
5. Lopez, C.; Briard-Bion, V.; Ménard, O.; Beaucher, E.; Rousseau, F.; Fauquant, J.; Leconte, N.; Robert, B. Fat globules selected from whole milk according to their size: Different compositions and structure of the biomembrane, revealing sphingomyelin-rich domains. *Food Chem.* **2011**, *125*, 355–368. [CrossRef]
6. Dewettinck, K.; Rombaut, R.; Thienpont, N.; Le, T.T.; Messens, K.; Van Camp, J. Nutritional and technological aspects of milk fat globule membrane material. *Int. Dairy J.* **2008**, *18*, 436–457. [CrossRef]
7. Blesso, C.N. Egg phospholipids and cardiovascular health. *Nutrients* **2015**, *7*, 2731–2747. [CrossRef]
8. Norris, G.H.; Blesso, C.N. Dietary and Endogenous Sphingolipid Metabolism in Chronic Inflammation. *Nutrients* **2017**, *9*, 1180. [CrossRef]
9. Norris, G.H.; Porter, C.M.; Jiang, C.; Millar, C.L.; Blesso, C.N. Dietary sphingomyelin attenuates hepatic steatosis and adipose tissue inflammation in high-fat-diet-induced obese mice. *J. Nutr. Biochem.* **2017**, *40*, 36–43. [CrossRef]
10. Norris, G.H.; Jiang, C.; Ryan, J.; Porter, C.M.; Blesso, C.N. Milk sphingomyelin improves lipid metabolism and alters gut microbiota in high fat diet-fed mice. *J. Nutr. Biochem.* **2016**, *30*, 93–101. [CrossRef]
11. Norris, G.; Porter, C.; Jiang, C.; Blesso, C.; Norris, G.H.; Porter, C.M.; Jiang, C.; Blesso, C.N. Dietary Milk Sphingomyelin Reduces Systemic Inflammation in Diet-Induced Obese Mice and Inhibits LPS Activity in Macrophages. *Beverages* **2017**, *3*, 37. [CrossRef]
12. Contarini, G.; Povolo, M.; Contarini, G.; Povolo, M. Phospholipids in Milk Fat: Composition, Biological and Technological Significance, and Analytical Strategies. *Int. J. Mol. Sci.* **2013**, *14*, 2808–2831. [CrossRef] [PubMed]
13. Fong, B.Y.; Norris, C.S.; MacGibbon, A.K.H. Protein and lipid composition of bovine milk-fat-globule membrane. *Int. Dairy J.* **2007**, *17*, 275–288. [CrossRef]
14. Beare-Rogers, J.; Dieffenbacher, A.; Holm, J.V. Lexicon of Lipid Nutrition (IUPAC Technical Report). *Pure Appl. Chem.* **2001**, *73*, 685–744. [CrossRef]
15. Norris, G.H.; Blesso, C.N. Dietary sphingolipids: Potential for management of dyslipidemia and nonalcoholic fatty liver disease. *Nutr. Rev.* **2017**, *75*, 274–285. [CrossRef] [PubMed]
16. Sandhoff, K.; Kolter, T. Biosynthesis and degradation of mammalian glycosphingolipids. *Philos. Trans. R. Soc. Lond. Ser. B Biol. Sci.* **2003**, *358*, 847–861. [CrossRef] [PubMed]
17. Hay, J.D.; Morrison, W.R. Polar lipds in bovin milk. III. Isomeric cis and trans monoenoic and dienoic fatty acids, and alkyl and alkenyl ethers in phosphatidyl choline and phosphatidyl ethanolamin. *Biochim. Biophys. Acta (BBA)/Lipids Lipid Metab.* **1971**, *248*, 71–79. [CrossRef]
18. Murgia, S.; Mele, S.; Monduzzi, M. Quantitative characterization of phospholipids in milk fat via 31P NMR using a monophasic solvent mixture. *Lipids* **2003**, *38*, 585–591. [CrossRef]
19. Gallier, S.; Gragson, D.; Cabral, C.; Jiménez-Flores, R.; Everett, D.W. Composition and fatty acid distribution of bovine milk phospholipids from processed milk products. *J. Agric. Food Chem.* **2010**, *58*, 10503–10511. [CrossRef]
20. Garcia, C.; Lutz, N.W.; Confort-Gouny, S.; Cozzone, P.J.; Armand, M.; Bernard, M. Phospholipid fingerprints of milk from different mammalians determined by 31P NMR: Towards specific interest in human health. *Food Chem.* **2012**, *135*, 1777–1783. [CrossRef]

21. Nagan, N.; Zoeller, R.A. Plasmalogens: Biosynthesis and functions. *Prog. Lipid Res.* **2001**, *40*, 199–229. [CrossRef]
22. Mayor, S.; Rao, M. Rafts: Scale-dependent, active lipid organization at the cell surface. *Traffic* **2004**, *5*, 231–240. [CrossRef] [PubMed]
23. Alberts, B.; Johnson, A.; Lewis, J.; Raff, M.; Roberts, K.; Walter, P. Molecular Biology of the Cell - The Lipid Bilayer. In *New York: Garland Science*; Garland Science: New York, NY, USA, 2002; Available online: https://www.ncbi.nlm.nih.gov/books (accessed on 13 August 2019).
24. Grecco, H.E.; Schmick, M.; Bastiaens, P.I.H. Signaling from the living plasma membrane. *Cell* **2011**, *144*, 897–909. [CrossRef] [PubMed]
25. Dowhan, W.; Bogdanov, M. Chapter 1 Functional roles of lipids in membranes. In *Biochemistry of lipids, lipoproteins Membranes*; Elsevier: Amsterdam, The Netherlands, 2008; pp. 1–37.
26. Zimmerberg, J.; Gawrisch, K. The physical chemistry of biological membranes. *Nat. Chem. Biol.* **2006**, *2*, 564–567. [CrossRef] [PubMed]
27. Simons, K.; Toomre, D. Lipid rafts and signal transduction. *Nat. Rev. Mol. Cell Biol.* **2000**, *1*, 31–39. [CrossRef] [PubMed]
28. Van Meer, G.; Voelker, D.R.; Feigenson, G.W. Membrane lipids: Where they are and how they behave. *Nat. Rev. Mol. Cell Biol.* **2008**, *9*, 112–124. [CrossRef] [PubMed]
29. Rombaut, R.; Camp, J.V.; Dewettinck, K. Phospho- and sphingolipid distribution during processing of milk, butter and whey. *Int. J. Food Sci. Technol.* **2006**, *41*, 435–443. [CrossRef]
30. Rombaut, R.; Dewettinck, K.; Van Camp, J. Phospho- and sphingolipid content of selected dairy products as determined by HPLC coupled to an evaporative light scattering detector (HPLC–ELSD). *J. Food Compos. Anal.* **2007**, *20*, 308–312. [CrossRef]
31. Le, T.T.; Miocinovic, J.; Nguyen, T.M.; Rombaut, R.; van Camp, J.; Dewettinck, K. Improved Solvent Extraction Procedure and High-Performance Liquid Chromatography–Evaporative Light-Scattering Detector Method for Analysis of Polar Lipids from Dairy Materials. *J. Agric. Food Chem.* **2011**, *59*, 10407–10413. [CrossRef]
32. Barry, K.M.; Dinan, T.G.; Murray, B.A.; Kelly, P.M. Comparison of dairy phospholipid preparative extraction protocols in combination with analysis by high performance liquid chromatography coupled to a charged aerosol detector. *Int. Dairy J.* **2016**, *56*, 179–185. [CrossRef]
33. Ferreiro, T.; Gayoso, L.; Rodríguez-Otero, J.L. Milk phospholipids: Organic milk and milk rich in conjugated linoleic acid compared with conventional milk. *J. Dairy Sci.* **2015**, *98*, 9–14. [CrossRef] [PubMed]
34. Rodríguez-Alcalá, L.M.; Castro-Gómez, P.; Felipe, X.; Noriega, L.; Fontecha, J. Effect of processing of cow milk by high pressures under conditions up to 900 MPa on the composition of neutral, polar lipids and fatty acids. *LWT - Food Sci. Technol.* **2015**, *62*, 265–270. [CrossRef]
35. Lopez, C.; Briard-Bion, V.; Menard, O.; Rousseau, F.; Pradel, P.; Besle, J.-M. Phospholipid, Sphingolipid, and Fatty Acid Compositions of the Milk Fat Globule Membrane are Modified by Diet. *J. Agric. Food Chem.* **2008**, *56*, 5226–5236. [CrossRef] [PubMed]
36. Puente, R.; García-Pardo, L.A.; Rueda, R.; Gil, A.; Hueso, P. Seasonal variations in the concentration of gangliosides and sialic acids in milk from different mammalian species. *Int. Dairy J.* **1996**, *6*, 315–322. [CrossRef]
37. Castro-Gómez, M.P.; Rodriguez-Alcalá, L.M.; Calvo, M.V.; Romero, J.; Mendiola, J.A.; Ibañez, E.; Fontecha, J. Total milk fat extraction and quantification of polar and neutral lipids of cow, goat, and ewe milk by using a pressurized liquid system and chromatographic techniques. *J. Dairy Sci.* **2014**, *97*, 6719–6728. [CrossRef]
38. Bitman, J.; Wood, D.L. Changes in Milk Fat Phospholipids During Lactation. *J. Dairy Sci.* **1990**, *73*, 1208–1216. [CrossRef]
39. Fagan, P.; Wijesundera, C. Liquid chromatographic analysis of milk phospholipids with on-line pre-concentration. *J. Chromatogr. A* **2004**, *1054*, 241–249. [CrossRef]
40. Donato, P.; Cacciola, F.; Cichello, F.; Russo, M.; Dugo, P.; Mondello, L. Determination of phospholipids in milk samples by means of hydrophilic interaction liquid chromatography coupled to evaporative light scattering and mass spectrometry detection. *J. Chromatogr. A* **2011**, *1218*, 6476–6482. [CrossRef]
41. Rodríguez-Alcalá, L.M.; Fontecha, J. Major lipid classes separation of buttermilk, and cows, goats and ewes milk by high performance liquid chromatography with an evaporative light scattering detector focused on the phospholipid fraction. *J. Chromatogr. A* **2010**, *1217*, 3063–3066. [CrossRef]

42. MacKenzie, A.; Vyssotski, M.; Nekrasov, E. Quantitative Analysis of Dairy Phospholipids by 31P NMR. *J. Am. Oil Chem. Soc.* **2009**, *86*, 757–763. [CrossRef]
43. Verardo, V.; Gómez-Caravaca, A.M.; Gori, A.; Losi, G.; Caboni, M.F. Bioactive lipids in the butter production chain from Parmigiano Reggiano cheese area. *J. Sci. Food Agric.* **2013**, *93*, 3625–3633. [CrossRef] [PubMed]
44. Avalli, A.; Contarini, G. Determination of phospholipids in dairy products by SPE/HPLC/ELSD. *J. Chromatogr. A* **2005**, *1071*, 185–190. [CrossRef] [PubMed]
45. Kiełbowicz, G.; Micek, P.; Wawrzeńczyk, C. A new liquid chromatography method with charge aerosol detector (CAD) for the determination of phospholipid classes. Application to milk phospholipids. *Talanta* **2013**, *105*, 28–33. [CrossRef] [PubMed]
46. Costa, M.R.; Elias-Argote, X.E.; Jiménez-Flores, R.; Gigante, M.L. Use of ultrafiltration and supercritical fluid extraction to obtain a whey buttermilk powder enriched in milk fat globule membrane phospholipids. *Int. Dairy J.* **2010**, *20*, 598–602. [CrossRef]
47. Britten, M.; Lamothe, S.; Robitaille, G. Effect of cream treatment on phospholipids and protein recovery in butter-making process. *Int. J. Food Sci. Technol.* **2008**, *43*, 651–657. [CrossRef]
48. Rombaut, R.; Camp, J.V.; Dewettinck, K. Analysis of Phospho- and Sphingolipids in Dairy Products by a New HPLC Method. *J. Dairy Sci.* **2005**, *88*, 482–488. [CrossRef]
49. Rombaut, R.; Dewettinck, K. Properties, analysis and purification of milk polar lipids. *Int. Dairy J.* **2006**, *16*, 1362–1373. [CrossRef]
50. Zhu, D.; Damodaran, S. Dairy Lecithin from Cheese Whey Fat Globule Membrane: Its Extraction, Composition, Oxidative Stability, and Emulsifying Properties. *J. Am. Oil Chem. Soc.* **2013**, *90*, 217–224. [CrossRef]
51. Kim, H.-H.Y.; Jimenez-Flores, R. Heat-Induced Interactions between the Proteins of Milk Fat Globule Membrane and Skim Milk. *J. Dairy Sci.* **1995**, *78*, 24–35. [CrossRef]
52. Ye, A.; Singh, H.; Taylor, M.W.; Anema, S. Characterization of protein components of natural and heat-treated milk fat globule membranes. *Int. Dairy J.* **2002**, *12*, 393–402. [CrossRef]
53. Cano-Ruiz, M.E.; Richter, R.L. Effect of Homogenization Pressure on the Milk Fat Globule Membrane Proteins. *J. Dairy Sci.* **1997**, *80*, 2732–2739. [CrossRef]
54. Corredig, M.; Roesch, R.R.; Dalgleish, D.G. Production of a Novel Ingredient from Buttermilk. *J. Dairy Sci.* **2003**, *86*, 2744–2750. [CrossRef]
55. Roesch, R.R.; Rincon, A.; Corredig, M. Emulsifying Properties of Fractions Prepared from Commercial Buttermilk by Microfiltration. *J. Dairy Sci.* **2004**, *87*, 4080–4087. [CrossRef]
56. Rombaut, R.; Dejonckheere, V.; Dewettinck, K. Filtration of Milk Fat Globule Membrane Fragments from Acid Buttermilk Cheese Whey. *J. Dairy Sci.* **2007**, *90*, 1662–1673. [CrossRef] [PubMed]
57. Sachdeva, S.; Buchheim, W. Recovery of phospholipids from buttermilk using membrane processing. *Kiel. Milchwirtsch. Forsch.* **1997**, *1*, 47–68.
58. Smith, L.M.; Jack, E.L. Isolation of Milk Phospholipids and Determination of Their Polyunsaturated Fatty Acids. *J. Dairy Sci.* **1959**, *42*, 767–779. [CrossRef]
59. Garmy, N.; Taïeb, N.; Yahi, N.; Fantini, J. Interaction of cholesterol with sphingosine: Physicochemical characterization and impact on intestinal absorption. *J. Lipid Res.* **2005**, *46*, 36–45. [CrossRef]
60. Feng, D.; Ohlsson, L.; Ling, W.; Nilsson, Å.; Duan, R.-D. Generating Ceramide from Sphingomyelin by Alkaline Sphingomyelinase in the Gut Enhances Sphingomyelin-Induced Inhibition of Cholesterol Uptake in Caco-2 Cells. *Dig. Dis. Sci.* **2010**, *55*, 3377–3383. [CrossRef] [PubMed]
61. Narita, T.; Naganuma, T.; Sase, Y.; Kihara, A. Long-chain bases of sphingolipids are transported into cells via the acyl-CoA synthetases. *Sci. Rep.* **2016**, *6*, 25469. [CrossRef]
62. Noh, S.K.; Koo, S.I. Milk Sphingomyelin Is More Effective than Egg Sphingomyelin in Inhibiting Intestinal Absorption of Cholesterol and Fat in Rats. *J. Nutr.* **2004**, *134*, 2611–2616. [CrossRef]
63. Norris, G.H.; Milard, M.; Michalski, M.-C.; Blesso, C.N. Protective Properties of Milk Sphingomyelin against Dysfunctional Lipid Metabolism, Gut Dysbiosis, and Inflammation. *J. Nutr. Biochem.* **2019**, *73*, 108224. [CrossRef] [PubMed]
64. Eckhardt, E.R.M.; Wang, D.Q.-H.; Donovan, J.M.; Carey, M.C. Dietary sphingomyelin suppresses intestinal cholesterol absorption by decreasing thermodynamic activity of cholesterol monomers. *Gastroenterology* **2002**, *122*, 948–956. [CrossRef] [PubMed]

65. Mathiassen, J.H.; Nejrup, R.G.; Frøkiær, H.; Nilsson, Å.; Ohlsson, L.; Hellgren, L.I. Emulsifying triglycerides with dairy phospholipids instead of soy lecithin modulates gut lipase activity. *Eur. J. Lipid Sci. Technol.* **2015**, *117*, 1522–1539. [CrossRef]
66. Patton, J.S.; Carey, M.C. Inhibition of human pancreatic lipase-colipase activity by mixed bile salt-phospholipid micelles. *Am. J. Physiol. - Gastrointest. Liver Physiol.* **1981**, *4*, 328–336. [CrossRef]
67. Homan, R.; Hamelehle, K.L. Phospholipase A2 relieves phosphatidylcholine inhibition of micellar cholesterol absorption and transport by human intestinal cell line Caco-2. *J. Lipid Res.* **1998**, *39*, 1197–1209.
68. Rodgers, J.B.; O'Connor, P.J. Effect of phosphatidylcholine on fatty acid and cholesterol absorption from mixed micellar solutions. *Biochim. Biophys. Acta (BBA)/Lipids Lipid Metab.* **1975**, *409*, 192–200. [CrossRef]
69. Rampone, A.J.; Long, L.R. The effect of phosphatidylcholine and lysophosphatidylcholine on the absorption and mucosal metabolism of oleic acid and cholesterol in vitro. *Biochim. Biophys. Acta Lipids Lipid Metab.* **1977**, *486*, 500–510. [CrossRef]
70. Beil, F.U.; Grundy, S.M. Studies on plasma lipoproteins during absorption of exogenous lecithin in man. *J. Lipid Res.* **1980**, *21*, 525–536.
71. Young, S.C.; Hui, D.Y. Pancreatic Lipase/Colipase-Mediated Triacylglycerol Hydrolysis Is Required for Cholesterol Transport from Lipid Emulsions to Intestinal Cells. *Biochem. J.* **1999**, *339*, 615–620. [CrossRef]
72. Hollander, D.; Morgan, D. Effect of plant sterols, fatty acids and lecithin on cholesterol absorption in vivo in the rat. *Lipids* **1980**, *15*, 395–400. [CrossRef]
73. Imaizumi, K.; Mawatari, K.; Murata, M.; Ikeda, I.; Sugano, M. The Contrasting Effect of Dietary Phosphatidylethanolamine and Phosphatidylcholine on Serum Lipoproteins and Liver Lipids in Rats. *J. Nutr.* **1983**, *113*, 2403–2411. [CrossRef] [PubMed]
74. Imaizumi, K.; Sekihara, K.; Sugano, M. Hypocholesterolemic action of dietary phosphatidylethanolamine in rats sensitive to exogenous cholesterol. *J. Nutr. Biochem.* **1991**, *2*, 251–254. [CrossRef]
75. Grzybek, M.; Kubiak, J.; Łach, A.; Przybyło, M.; Sikorski, A.F. A raft-associated species of phosphatidylethanolamine interacts with cholesterol comparably to sphingomyelin. A Langmuir-Blodgett monolayer study. *PLoS ONE* **2009**, *4*, e5053. [CrossRef] [PubMed]
76. Shaikh, S.R.; Brzustowicz, M.R.; Gustafson, N.; Stillwell, W.; Wassall, S.R. Monounsaturated PE does not phase-separate from the lipid raft molecules sphingomyelin and cholesterol: Role for polyunsaturation? *Biochemistry* **2002**, *41*, 10593–10602. [CrossRef] [PubMed]
77. Mazzei, J.C.; Zhou, H.; Brayfield, B.P.; Hontecillas, R.; Bassaganya-Riera, J.; Schmelz, E.M. Suppression of intestinal inflammation and inflammation-driven colon cancer in mice by dietary sphingomyelin: Importance of peroxisome proliferator-activated receptor γ expression. *J. Nutr. Biochem.* **2011**, *22*, 1160–1171. [CrossRef] [PubMed]
78. Pascual, G.; Fong, A.L.; Ogawa, S.; Gamliel, A.; Li, A.C.; Perissi, V.; Rose, D.W.; Willson, T.M.; Rosenfeld, M.G.; Glass, C.K. A SUMOylation-dependent pathway mediates transrepression of inflammatory response genes by PPAR-γ. *Nature* **2005**, *437*, 759–763. [CrossRef]
79. Józefowski, S.; Czerkies, M.; Łukasik, A.; Bielawska, A.; Bielawski, J.; Kwiatkowska, K.; Sobota, A. Ceramide and Ceramide 1-Phosphate Are Negative Regulators of TNF-α Production Induced by Lipopolysaccharide. *J. Immunol.* **2010**, *185*, 6960–6973. [CrossRef]
80. Parrish, W.R.; Rosas-Ballina, M.; Gallowitsch-Puerta, M.; Ochani, M.; Ochani, K.; Yang, L.H.; Hudson, L.Q.; Lin, X.; Patel, N.; Johnson, S.M.; et al. Modulation of TNF release by choline requires α7 subunit nicotinic acetylcholine receptor-mediated signaling. *Mol. Med.* **2008**, *14*, 567–574. [CrossRef]
81. Milard, M.; Penhoat, A.; Durand, A.; Buisson, C.; Loizon, E.; Meugnier, E.; Bertrand, K.; Joffre, F.; Cheillan, D.; Garnier, L.; et al. Acute effects of milk polar lipids on intestinal tight junction expression: Towards an impact of sphingomyelin through the regulation of IL-8 secretion? *J. Nutr. Biochem.* **2019**, *65*, 128–138. [CrossRef]
82. Snow, D.R.; Ward, R.E.; Olsen, A.; Jimenez-Flores, R.; Hintze, K.J. Membrane-rich milk fat diet provides protection against gastrointestinal leakiness in mice treated with lipopolysaccharide. *J. Dairy Sci.* **2011**, *94*, 2201–2212. [CrossRef]
83. Ten Bruggencate, S.J.; Frederiksen, P.D.; Pedersen, S.M.; Floris-Vollenbroek, E.G.; Lucas-van de Bos, E.; van Hoffen, E.; Wejse, P.L. Dietary Milk-Fat-Globule Membrane Affects Resistance to Diarrheagenic Escherichia coli in Healthy Adults in a Randomized, Placebo-Controlled, Double-Blind Study. *J. Nutr.* **2016**, *146*, 249–255. [CrossRef] [PubMed]

84. Li, T.; Gao, J.; Du, M.; Mao, X. Milk fat globule membrane supplementation modulates the gut microbiota and attenuates metabolic endotoxemia in high-fat diet-fed mice. *J. Funct. Foods* **2018**, *47*, 56–65. [CrossRef]
85. Cho, I.; Blaser, M.J. The human microbiome: At the interface of health and disease. *Nat. Rev. Genet.* **2012**, *13*, 260–270. [CrossRef] [PubMed]
86. Cresci, G.A.; Bawden, E. Gut Microbiome. *Nutr. Clin. Pract.* **2015**, *30*, 734–746. [CrossRef] [PubMed]
87. Jonsson, A.L.; Bäckhed, F. Role of gut microbiota in atherosclerosis. *Nat. Rev. Cardiol.* **2017**, *14*, 79–87. [CrossRef] [PubMed]
88. Ley, R.E.; Turnbaugh, P.J.; Klein, S.; Gordon, J.I. Human gut microbes associated with obesity. *Nature* **2006**, *444*, 1022–1023. [CrossRef] [PubMed]
89. Roger, L.C.; Costabile, A.; Holland, D.T.; Hoyles, L.; McCartney, A.L. Examination of faecal Bifidobacterium populations in breast- and formula-fed infants during the first 18 months of life. *Microbiology* **2010**, *156*, 3329–3341. [CrossRef] [PubMed]
90. Mariat, D.; Firmesse, O.; Levenez, F.; Guimarăes, V.; Sokol, H.; Doré, J.; Corthier, G.; Furet, J.-P. The Firmicutes/Bacteroidetes ratio of the human microbiota changes with age. *BMC Microbiol.* **2009**, *9*, 123. [CrossRef]
91. Wu, X.; Ma, C.; Han, L.; Nawaz, M.; Gao, F.; Zhang, X.; Yu, P.; Zhao, C.; Li, L.; Zhou, A.; et al. Molecular Characterisation of the Faecal Microbiota in Patients with Type II Diabetes. *Curr. Microbiol.* **2010**, *61*, 69–78. [CrossRef] [PubMed]
92. Duytschaever, G.; Huys, G.; Bekaert, M.; Boulanger, L.; De Boeck, K.; Vandamme, P. Dysbiosis of bifidobacteria and Clostridium cluster XIVa in the cystic fibrosis fecal microbiota. *J. Cyst. Fibros.* **2013**, *12*, 206–215. [CrossRef]
93. Murri, M.; Leiva, I.; Gomez-Zumaquero, J.M.; Tinahones, F.J.; Cardona, F.; Soriguer, F.; Queipo-Ortuño, M.I. Gut microbiota in children with type 1 diabetes differs from that in healthy children: A case-control study. *BMC Med.* **2013**, *11*, 46. [CrossRef]
94. Kalliomäki, M.; Carmen Collado, M.; Salminen, S.; Isolauri, E. Early differences in fecal microbiota composition in children may predict overweight. *Am. J. Clin. Nutr.* **2008**, *87*, 534–538. [CrossRef] [PubMed]
95. Gao, X.; Jia, R.; Xie, L.; Kuang, L.; Feng, L.; Wan, C. Obesity in school-aged children and its correlation with Gut E.coli and Bifidobacteria: A case–control study. *BMC Pediatr.* **2015**, *15*, 64. [CrossRef] [PubMed]
96. Sprong, R.C.; Hulstein, M.F.E.; Van Der Meer, R. Bactericidal Activities of Milk Lipids. *Antimicrob. Agents Chemother.* **2001**, *45*, 1298–1301. [CrossRef]
97. Fischer, C.L.; Drake, D.R.; Dawson, D.V.; Blanchette, D.R.; Brogden, K.A.; Wertz, P.W. Antibacterial activity of sphingoid bases and fatty acids against Gram-positive and Gram-negative bacteria. *Antimicrob. Agents Chemother.* **2012**, *56*, 1157–1161. [CrossRef]
98. Fischer, C.L.; Walters, K.S.; Drake, D.R.; Blanchette, D.R.; Dawson, D.V.; Brogden, K.A.; Wertz, P.W. Sphingoid bases are taken up by Escherichia coli and Staphylococcus aureus and induce ultrastructural damage. *Skin Pharmacol. Physiol.* **2013**, *26*, 36–44. [CrossRef]
99. Nejrup, R.G.; Bahl, M.I.; Vigsnæs, L.K.; Heerup, C.; Licht, T.R.; Hellgren, L.I. Lipid hydrolysis products affect the composition of infant gut microbial communities in vitro. *Br. J. Nutr.* **2015**, *114*, 63–74. [CrossRef]
100. Nilsson, Å.; Duan, R.-D. Pancreatic and mucosal enzymes in choline phospholipid digestion. *Am. J. Physiol. Gastrointest. Liver Physiol.* **2019**, *316*, G425–G445. [CrossRef]
101. Zhang, Y.; Cheng, Y.; Hansen, G.H.; Niels-Christiansen, L.L.; Koentgen, F.; Ohlsson, L.; Nilsson, Å.; Duan, R.D. Crucial role of alkaline sphingomyelinase in sphingomyelin digestion: A study on enzyme knockout mice. *J. Lipid Res.* **2011**, *52*, 771–781. [CrossRef]
102. Millar, C.L.; Jiang, C.; Norris, G.H.; Garcia, C.; Seibel, S.; Anto, L.; Lee, J.-Y.; Blesso, C.N. Cow's milk polar lipids reduce atherogenic lipoprotein cholesterol, modulate gut microbiota and attenuate atherosclerosis development in LDL-receptor knockout mice fed a Western-type diet. *J. Nutr. Biochem.* **2020**, *79*, 108351. [CrossRef]
103. Milard, M.; Laugerette, F.; Durand, A.; Buisson, C.; Meugnier, E.; Loizon, E.; Louche-Pelissier, C.; Sauvinet, V.; Garnier, L.; Viel, S.; et al. Milk Polar Lipids in a High-Fat Diet Can Prevent Body Weight Gain: Modulated Abundance of Gut Bacteria in Relation with Fecal Loss of Specific Fatty Acids. *Mol. Nutr. Food Res.* **2019**, *63*, 1801078. [CrossRef]
104. Li, J.; Lin, S.; Vanhoutte, P.M.; Woo, C.W.; Xu, A. Akkermansia Muciniphila Protects Against Atherosclerosis by Preventing Metabolic Endotoxemia-Induced Inflammation in Apoe-/- Mice. *Circulation* **2016**, *133*, 2434–2446. [CrossRef] [PubMed]

105. Dao, M.C.; Everard, A.; Aron-Wisnewsky, J.; Sokolovska, N.; Prifti, E.; Verger, E.O.; Kayser, B.D.; Levenez, F.; Chilloux, J.; Hoyles, L.; et al. Akkermansia muciniphila and improved metabolic health during a dietary intervention in obesity: Relationship with gut microbiome richness and ecology. *Gut* **2016**, *65*, 426–436. [CrossRef]
106. Schneeberger, M.; Everard, A.; Gómez-Valadés, A.G.; Matamoros, S.; Ramírez, S.; Delzenne, N.M.; Gomis, R.; Claret, M.; Cani, P.D. Akkermansia muciniphila inversely correlates with the onset of inflammation, altered adipose tissue metabolism and metabolic disorders during obesity in mice. *Sci. Rep.* **2015**, *5*, 16643. [CrossRef]
107. Reis, M.G.; Roy, N.C.; Bermingham, E.N.; Ryan, L.; Bibiloni, R.; Young, W.; Krause, L.; Berger, B.; North, M.; Stelwagen, K.; et al. Impact of Dietary Dairy Polar Lipids on Lipid Metabolism of Mice Fed a High-Fat Diet. *J. Agric. Food Chem.* **2013**, *61*, 2729–2738. [CrossRef] [PubMed]
108. Bhinder, G.; Allaire, J.M.; Garcia, C.; Lau, J.T.; Chan, J.M.; Ryz, N.R.; Bosman, E.S.; Graef, F.A.; Crowley, S.M.; Celiberto, L.S.; et al. Milk Fat Globule Membrane Supplementation in Formula Modulates the Neonatal Gut Microbiome and Normalizes Intestinal Development. *Sci. Rep.* **2017**, *7*, 45274. [CrossRef] [PubMed]
109. Zhou, J.; Xiong, X.; Wang, K.-X.; Zou, L.-J.; Ji, P.; Yin, Y.-L. Ethanolamine enhances intestinal functions by altering gut microbiome and mucosal anti-stress capacity in weaned rats. *Br. J. Nutr.* **2018**, *120*, 241–249. [CrossRef] [PubMed]
110. Vors, C.; Joumard-Cubizolles, L.; Lecomte, M.; Combe, E.; Ouchchane, L.; Drai, J.; Raynal, K.; Joffre, F.; Meiller, L.; Le Barz, M.; et al. Milk polar lipids reduce lipid cardiovascular risk factors in overweight postmenopausal women: Towards a gut sphingomyelin-cholesterol interplay. *Gut* **2019**, *69*, 487–501. [CrossRef]
111. Nair, A.; Jacob, S. A simple practice guide for dose conversion between animals and human. *J. Basic Clin. Pharm.* **2016**, *7*, 27. [CrossRef]
112. Gérard, P.; Lepercq, P.; Leclerc, M.; Gavini, F.; Raibaud, P.; Juste, C. Bacteroides sp. Strain D8, the First Cholesterol-Reducing Bacterium Isolated from Human Feces. *Appl. Environ. Microbiol.* **2007**, *73*, 5742–5749. [CrossRef]
113. Gérard, P. Metabolism of Cholesterol and Bile Acids by the Gut Microbiota. *Pathogens* **2013**, *3*, 14–24. [CrossRef]
114. Sekimoto, H.; Shimada, O.; Makanishi, M.; Nakano, T.; Katayama, O. Interrelationship between serum and fecal sterols. *Jpn. J. Med.* **1983**, *22*, 14–20. [CrossRef]
115. Picard, C.; Fioramonti, J.; Francois, A.; Robinson, T.; Neant, F.; Matuchansky, C. Review article: Bifidobacteria as probiotic agents - physiological effects and clinical benefits. *Aliment. Pharmacol. Ther.* **2005**, *22*, 495–512. [CrossRef]
116. Kailasapathy, K.; Chin, J. Survival and therapeutic potential of probiotic organisms with reference to *Lactobacillus acidophilus* and *Bifidobacterium* spp. *Immunol. Cell Biol.* **2000**, *78*, 80–88. [CrossRef]
117. Grill, J.P.; Schneider, F.; Ballongue, J. Bifidobacteria and probiotic effects: Action of Bifidobacterium species on conjugated bile salts. *Curr. Microbiol.* **1995**, *31*, 23–27. [CrossRef]
118. Pellissery, A.J.; Radhakrishnan Nair, U. Pellissery and Uma (2013). Lactic Acid Bacteria as Mucosal Delivery Vaccine Review Article ARTICLE HISTORY ABSTRACT. *Adv. Anim. Vet. Sci.* **2013**, *1*, 183–187.
119. Di Gioia, D.; Aloisio, I.; Mazzola, G.; Biavati, B. Bifidobacteria: Their impact on gut microbiota composition and their applications as probiotics in infants. *Appl. Microbiol. Biotechnol.* **2014**, *98*, 563–577. [CrossRef] [PubMed]
120. Tojo, R.; Suárez, A.; Clemente, M.G.; de los Reyes-Gavilán, C.G.; Margolles, A.; Gueimonde, M.; Ruas-Madiedo, P. Intestinal microbiota in health and disease: Role of bifidobacteria in gut homeostasis. *World J. Gastroenterol.* **2014**, *20*, 15163–15176. [CrossRef] [PubMed]
121. Wang, Z.; Klipfell, E.; Bennett, B.J.; Koeth, R.; Levison, B.S.; Dugar, B.; Feldstein, A.E.; Britt, E.B.; Fu, X.; Chung, Y.M.; et al. Gut flora metabolism of phosphatidylcholine promotes cardiovascular disease. *Nature* **2011**, *472*, 57–65. [CrossRef]
122. Randrianarisoa, E.; Lehn-Stefan, A.; Wang, X.; Hoene, M.; Peter, A.; Heinzmann, S.S.; Zhao, X.; Königsrainer, I.; Königsrainer, A.; Balletshofer, B.; et al. Relationship of serum trimethylamine N-oxide (TMAO) levels with early atherosclerosis in humans. *Sci. Rep.* **2016**, *6*, 1–9. [CrossRef]
123. Bae, S.; Ulrich, C.M.; Neuhouser, M.L.; Malysheva, O.; Bailey, L.B.; Xiao, L.; Brown, E.C.; Cushing-Haugen, K.L.; Zheng, Y.; Cheng, T.Y.D.; et al. Plasma choline metabolites and colorectal cancer risk in the women's health initiative observational study. *Cancer Res.* **2014**, *74*, 7442–7452. [CrossRef] [PubMed]

124. Rohrmann, S.; Linseisen, J.; Allenspach, M.; von Eckardstein, A.; Müller, D. Plasma Concentrations of Trimethylamine-N-oxide Are Directly Associated with Dairy Food Consumption and Low-Grade Inflammation in a German Adult Population. *J. Nutr.* **2016**, *146*, 283–289. [CrossRef] [PubMed]
125. Krüger, R.; Merz, B.; Rist, M.J.; Ferrario, P.G.; Bub, A.; Kulling, S.E.; Watzl, B. Associations of current diet with plasma and urine TMAO in the KarMeN study: Direct and indirect contributions. *Mol. Nutr. Food Res.* **2017**, *61*, 1700363. [CrossRef] [PubMed]
126. Zheng, H.; Lorenzen, J.; Astrup, A.; Larsen, L.; Yde, C.; Clausen, M.; Bertram, H. Metabolic Effects of a 24-Week Energy-Restricted Intervention Combined with Low or High Dairy Intake in Overweight Women: An NMR-Based Metabolomics Investigation. *Nutrients* **2016**, *8*, 108. [CrossRef] [PubMed]
127. Zheng, H.; Yde, C.C.; Clausen, M.R.; Kristensen, M.; Lorenzen, J.; Astrup, A.; Bertram, H.C. Metabolomics investigation to shed light on cheese as a possible piece in the French paradox puzzle. *J. Agric. Food Chem.* **2015**, *63*, 2830–2839. [CrossRef] [PubMed]
128. Zhang, A.Q.; Mitchell, S.C.; Smith, R.L. Dietary precursors of trimethylamine in man: A pilot study. *Food Chem. Toxicol.* **1999**, *37*, 515–520. [CrossRef]
129. Cho, C.E.; Taesuwan, S.; Malysheva, O.V.; Bender, E.; Tulchinsky, N.F.; Yan, J.; Sutter, J.L.; Caudill, M.A. Trimethylamine-N-oxide (TMAO) response to animal source foods varies among healthy young men and is influenced by their gut microbiota composition: A randomized controlled trial. *Mol. Nutr. Food Res.* **2017**, *61*, 1600324. [CrossRef]
130. Daviglus, M.L.; Stamler, J.; Orencia, A.J.; Dyer, A.R.; Liu, K.; Greenland, P.; Walsh, M.K.; Morris, D.; Shekelle, R.B. Fish Consumption and the 30-Year Risk of Fatal Myocardial Infarction. *N. Engl. J. Med.* **1997**, *336*, 1046–1053. [CrossRef]
131. Dewailly, É.; Blanchet, C.; Gingras, S.; Lemieux, S.; Holub, B.J. Fish consumption and blood lipids in three ethnic groups of Québec (Canada). *Lipids* **2003**, *38*, 359–365. [CrossRef]
132. Kromhout, D.; Bosschieter, E.B.; Cor de Lezenne, C. The Inverse Relation between Fish Consumption and 20-Year Mortality from Coronary Heart Disease. *N. Engl. J. Med.* **1985**, *312*, 1205–1209. [CrossRef]
133. Shekelle, R.B.; Missell, L.; Paul, O.; Shryock, A.M.; Stamler, J.; Vollset, S.E.; Heuch, I.; Bjelke, E.; Curb, J.D.; Reed, D.M.; et al. Fish Consumption and Mortality from Coronary Heart Disease. *N. Engl. J. Med.* **1985**, *313*, 820–824.
134. Gao, X.; Xu, J.; Jiang, C.; Zhang, Y.; Xue, Y.; Li, Z.; Wang, J.; Xue, C.; Wang, Y. Fish oil ameliorates trimethylamine N-oxide-exacerbated glucose intolerance in high-fat diet-fed mice. *Food Funct.* **2015**, *6*, 1117–1125. [CrossRef] [PubMed]
135. Jia, J.; Dou, P.; Gao, M.; Kong, X.; Li, C.; Liu, Z.; Huang, T. Assessment of causal direction between gut microbiota-dependent metabolites and cardiometabolic health: A bidirectional mendelian randomization analysis. *Diabetes* **2019**, *68*, 1747–1755. [CrossRef] [PubMed]
136. Santulli, S. Santulli Gaetano Epidemiology of Cardiovascular Disease in the 21st Century: Updated Numbers and Updated Facts. *J. Cardiovasc. Dis.* **2013**, *1*, 1–2.
137. Expert Panel on Detection, Evaluation, and Treatment of High Blood Cholesterol in Adults. Executive summary of the third report of the National Cholesterol Education Program (NCEP) expert panel on detection, evaluation, and treatment of high blood cholesterol in adults (Adult Treatment Panel III). *JAMA* **2001**, *285*, 2486–2497.
138. Watanabe, S.; Takahashi, T.; Tanaka, L.; Haruta, Y.; Shiota, M.; Hosokawa, M.; Miyashita, K. The effect of milk polar lipids separated from butter serum on the lipid levels in the liver and the plasma of obese-model mouse (KK-Ay). *J. Funct. Foods* **2011**, *3*, 313–320. [CrossRef]
139. Wat, E.; Tandy, S.; Kapera, E.; Kamili, A.; Chung, R.W.S.; Brown, A.; Rowney, M.; Cohn, J.S. Dietary phospholipid-rich dairy milk extract reduces hepatomegaly, hepatic steatosis and hyperlipidemia in mice fed a high-fat diet. *Atherosclerosis* **2009**, *205*, 144–150. [CrossRef]
140. Rosqvist, F.; Smedman, A.; Lindmark-Månsson, H.; Paulsson, M.; Petrus, P.; Straniero, S.; Rudling, M.; Dahlman, I.; Risérus, U. Potential role of milk fat globule membrane in modulating plasma lipoproteins, gene expression, and cholesterol metabolism in humans: A randomized study1. *Am. J. Clin. Nutr.* **2015**, *102*, 20–30. [CrossRef]
141. Conway, V.; Couture, P.; Richard, C.; Gauthier, S.F.; Pouliot, Y.; Lamarche, B. Impact of buttermilk consumption on plasma lipids and surrogate markers of cholesterol homeostasis in men and women. *Nutr. Metab. Cardiovasc. Dis.* **2013**, *23*, 1255–1262. [CrossRef]

142. Millar, C.L.; Norris, G.H.; Vitols, A.; Garcia, C.; Seibel, S.; Anto, L.; Blesso, C.N. Dietary egg sphingomyelin prevents aortic root plaque accumulation in apolipoprotein-E knockout mice. *Nutrients* **2019**, *11*, 1124. [CrossRef]
143. Chung, R.W.S.; Wang, Z.; Bursill, C.A.; Wu, B.J.; Barter, P.J.; Rye, K.-A. Effect of long-term dietary sphingomyelin supplementation on atherosclerosis in mice. *PLoS ONE* **2017**, *12*, e0189523. [CrossRef] [PubMed]
144. Loomba, R.; Sanyal, A.J. The global NAFLD epidemic. *Nat. Rev. Gastroenterol. Hepatol.* **2013**, *10*, 686–690. [CrossRef] [PubMed]
145. Bellentani, S.; Scaglioni, F.; Marino, M.; Bedogni, G. Epidemiology of Non-Alcoholic Fatty Liver Disease. *Dig. Dis.* **2010**, *28*, 155–161. [CrossRef] [PubMed]
146. Anstee, Q.M.; Goldin, R.D. Mouse models in non-alcoholic fatty liver disease and steatohepatitis research. *Int. J. Exp. Pathol.* **2006**, *87*, 1–16. [CrossRef]
147. Weiland, A.; Bub, A.; Barth, S.W.; Schrezenmeir, J.; Pfeuffer, M. Effects of dietary milk- and soya-phospholipids on lipid-parameters and other risk indicators for cardiovascular diseases in overweight or obese men—Two double-blind, randomised, controlled, clinical trials. *J. Nutr. Sci.* **2016**, *5*, e21. [CrossRef]
148. Nyberg, L.; Duan, R.D.; Nilsson, Å. A mutual inhibitory effect on absorption of sphingomyelin and cholesterol. *J. Nutr. Biochem.* **2000**, *11*, 244–249. [CrossRef]
149. Kamili, A.; Wat, E.; Chung, R.W.; Tandy, S.; Weir, J.M.; Meikle, P.J.; Cohn, J.S. Hepatic accumulation of intestinal cholesterol is decreased and fecal cholesterol excretion is increased in mice fed a high-fat diet supplemented with milk phospholipids. *Nutr. Metab. (Lond.)* **2010**, *7*, 90. [CrossRef]
150. Zhou, A.L.; Hintze, K.J.; Jimenez-Flores, R.; Ward, R.E. Dietary fat composition influences tissue lipid profile and gene expression in fischer-344 rats. *Lipids* **2012**, *47*, 1119–1130. [CrossRef]
151. Lecomte, M.; Bourlieu, C.; Meugnier, E.; Penhoat, A.; Cheillan, D.; Pineau, G.; Loizon, E.; Trauchessec, M.; Claude, M.; Ménard, O.; et al. Milk Polar Lipids Affect In Vitro Digestive Lipolysis and Postprandial Lipid Metabolism in Mice. *J. Nutr.* **2015**, *145*, 1770–1777. [CrossRef]
152. Lecomte, M.; Couëdelo, L.; Meugnier, E.; Plaisancié, P.; Létisse, M.; Benoit, B.; Gabert, L.; Penhoat, A.; Durand, A.; Pineau, G.; et al. Dietary emulsifiers from milk and soybean differently impact adiposity and inflammation in association with modulation of colonic goblet cells in high-fat fed mice. *Mol. Nutr. Food Res.* **2016**, *60*, 609–620. [CrossRef]
153. Yamauchi, I.; Uemura, M.; Hosokawa, M.; Iwashima-Suzuki, A.; Shiota, M.; Miyashita, K. The dietary effect of milk sphingomyelin on the lipid metabolism of obese/diabetic KK-Ay mice and wild-type C57BL/6J mice. *Food Funct.* **2016**, *7*, 3854–3867. [CrossRef] [PubMed]
154. Zhou, A.L.; Ward, R.E. Milk polar lipids modulate lipid metabolism, gut permeability, and systemic inflammation in high-fat-fed C57BL/6J ob/ob mice, a model of severe obesity. *J. Dairy Sci.* **2019**, *102*, 4816–4831. [CrossRef] [PubMed]
155. Ohlsson, L.; Burling, H.; Nilsson, A. Long term effects on human plasma lipoproteins of a formulation enriched in butter milk polar lipid. *Lipids Health Dis.* **2009**, *8*, 44. [CrossRef] [PubMed]
156. Ohlsson, L.; Hertervig, E.; Jönsson, B.A.; Duan, R.-D.; Nyberg, L.; Svernlöv, R.; Nilsson, Å. Sphingolipids in human ileostomy content after meals containing milk sphingomyelin. *Am. J. Clin. Nutr.* **2010**, *91*, 672–678. [CrossRef]
157. Ohlsson, L.; Burling, H.; Duan, R.-D.; Nilsson, Å. Effects of a sphingolipid-enriched dairy formulation on postprandial lipid concentrations. *Eur. J. Clin. Nutr.* **2010**, *64*, 1344–1349. [CrossRef] [PubMed]
158. Keller, S.; Malarski, A.; Reuther, C.; Kertscher, R.; Kiehntopf, M.; Jahreis, G. Milk phospholipid and plant sterol-dependent modulation of plasma lipids in healthy volunteers. *Eur. J. Nutr.* **2013**, *52*, 1169–1179. [CrossRef] [PubMed]
159. Baumgartner, S.; Kelly, E.R.; van der Made, S.; Berendschot, T.T.J.M.; Husche, C.; Lütjohann, D.; Plat, J. The influence of consuming an egg or an egg-yolk buttermilk drink for 12 wk on serum lipids, inflammation, and liver function markers in human volunteers. *Nutrition* **2013**, *29*, 1237–1244. [CrossRef]
160. Severins, N.; Mensink, R.P.; Plat, J. Effects of lutein-enriched egg yolk in buttermilk or skimmed milk on serum lipids & lipoproteins of mildly hypercholesterolemic subjects. *Nutr. Metab. Cardiovasc. Dis.* **2015**, *25*, 210–217.

161. Grip, T.; Dyrlund, T.S.; Ahonen, L.; Domellöf, M.; Hernell, O.; Hyötyläinen, T.; Knip, M.; Lönnerdal, B.; Orešič, M.; Timby, N. Serum, plasma and erythrocyte membrane lipidomes in infants fed formula supplemented with bovine milk fat globule membranes. *Pediatr. Res.* **2018**, *84*, 726–732. [CrossRef]
162. Lebovitz, H. Insulin resistance: Definition and consequences. *Exp. Clin. Endocrinol. Diabetes* **2001**, *109*, S135–S148. [CrossRef]
163. American Diabetes Association. Standards of medical care in diabetes. *Diabetes Care* **2012**, *35*, S11. [CrossRef] [PubMed]
164. Nagasawa, T.; Nakamichi, H.; Hama, Y.; Higashiyama, S.; Igarashi, Y.; Mitsutake, S. Phytosphingosine is a novel activator of GPR120. *J. Biochem.* **2018**, *164*, 27–32. [CrossRef] [PubMed]
165. Ramstedt, B.; Leppimäki, P.; Axberg, M.; Slotte, J.P. Analysis of natural and synthetic sphingomyelins using high-performance thin-layer chromatography. *Eur. J. Biochem.* **1999**, *266*, 997–1002. [CrossRef] [PubMed]
166. Tierney, A.L.; Nelson, C.A. Brain Development and the Role of Experience in the Early Years. *Zero Three* **2009**, *30*, 9–13. [PubMed]
167. Oshida, K.; Shimizu, T.; Takase, M.; Tamura, Y.; Shimizu, T.; Yamashiro, Y. Effects of dietary sphingomyelin on central nervous system myelination in developing rats. *Pediatr. Res.* **2003**, *53*, 589–593. [CrossRef] [PubMed]
168. Gurnida, D.A.; Rowan, A.M.; Idjradinata, P.; Muchtadi, D.; Sekarwana, N. Association of complex lipids containing gangliosides with cognitive development of 6-month-old infants. *Early Hum. Dev.* **2012**, *88*, 595–601. [CrossRef] [PubMed]
169. Siegel, R.L.; Miller, K.D.; Jemal, A. Cancer statistics, 2018. *CA. Cancer J. Clin.* **2018**, *68*, 7–30. [CrossRef]
170. Ng, S.C.; Shi, H.Y.; Hamidi, N.; Underwood, F.E.; Tang, W.; Benchimol, E.I.; Panaccione, R.; Ghosh, S.; Wu, J.C.Y.; Chan, F.K.L.; et al. Worldwide incidence and prevalence of inflammatory bowel disease in the 21st century: A systematic review of population-based studies. *Lancet* **2017**, *390*, 2769–2778. [CrossRef]
171. Kim, E.R.; Chang, D.K. Colorectal cancer in inflammatory bowel disease: The risk, pathogenesis, prevention and diagnosis. *World J. Gastroenterol.* **2014**, *20*, 9872–9881. [CrossRef]
172. Tanaka, K.; Hosozawa, M.; Kudo, N.; Yoshikawa, N.; Hisata, K.; Shoji, H.; Shinohara, K.; Shimizu, T. The pilot study: Sphingomyelin-fortified milk has a positive association with the neurobehavioural development of very low birth weight infants during infancy, randomized control trial. *Brain Dev.* **2013**, *35*, 45–52. [CrossRef]
173. Kuchta-Noctor, A.M.; Murray, B.A.; Stanton, C.; Devery, R.; Kelly, P.M. Anticancer Activity of Buttermilk Against SW480 Colon Cancer Cells is Associated with Caspase-Independent Cell Death and Attenuation of Wnt, Akt, and ERK Signaling. *Nutr. Cancer* **2016**, *68*, 1234–1246. [CrossRef] [PubMed]
174. Schmelz, E.M.; Sullards, M.C.; Dillehay, D.L.; Merrill, A.H. Colonic Cell Proliferation and Aberrant Crypt Foci Formation Are Inhibited by Dairy Glycosphingolipids in 1,2-Dimethylhydrazine-Treated CF1 Mice. *J. Nutr.* **2000**, *130*, 522–527. [CrossRef] [PubMed]
175. Dillehay, D.L.; Webb, S.K.; Schmelz, E.-M.; Merrill, A.H. Dietary Sphingomyelin Inhibits 1,2-Dimethylhydrazine-Induced Colon Cancer in CF1 Mice. *J. Nutr.* **1994**, *124*, 615–620. [CrossRef] [PubMed]
176. Schmelz, E.M.; Dillehay, D.L.; Webb, S.K.; Reiter, A.; Adams, J.; Merrill, A.H. Sphingomyelin consumption suppresses aberrant colonic crypt foci and increases the proportion of adenomas versus adenocarcinomas in CF1 mice treated with 1,2-dimethylhydrazine: Implications for dietary sphingolipids and colon carcinogenesis. *Cancer Res.* **1996**, *56*, 4936–4941. [PubMed]
177. Snow, D.R.; Jimenez-Flores, R.; Ward, R.E.; Cambell, J.; Young, M.J.; Nemere, I.; Hintze, K.J. Dietary Milk Fat Globule Membrane Reduces the Incidence of Aberrant Crypt Foci in Fischer-344 Rats. *J. Agric. Food Chem* **2010**, *58*, 2157. [CrossRef] [PubMed]

© 2020 by the authors. Licensee MDPI, Basel, Switzerland. This article is an open access article distributed under the terms and conditions of the Creative Commons Attribution (CC BY) license (http://creativecommons.org/licenses/by/4.0/).

Protocol

Galacto-Oligosaccharide RP-G28 Improves Multiple Clinical Outcomes in Lactose-Intolerant Patients>

William Chey [1,*], William Sandborn [2], Andrew J. Ritter [3], Howard Foyt [3], M. Andrea Azcarate-Peril [4] and Dennis A. Savaiano [5]

1. Gastroenterology and Nutrition Sciences 3912 Taubman Center, SPC 5362, Ann Arbor, MI 48109, USA
2. Division of Gastroenterology, University of California San Diego, San Diego, CA 92093, USA; wsandborn@ucsd.edu
3. Ritter Pharmaceuticals, Inc., Los Angeles, CA 90067, USA; andrew@ritterpharma.com (A.J.R.); HFoyt@viacyte.com (H.F.)
4. Department of Medicine, Division of Gastroenterology and Hepatology and UNC Microbiome Core, School of Medicine, University of North Carolina at Chapel Hill, NC 27514, USA; azcarate@med.unc.edu
5. Meredith Professor of Nutrition Science, Institution: Purdue University, West Lafayette, IN 47907, USA; savaiano@purdue.edu
* Correspondence: wchey@umich.edu; Tel.: +1-734-936-4780

Received: 25 February 2020; Accepted: 3 April 2020; Published: 10 April 2020

Abstract: Background and Aims: Lactose intolerance (LI) is a global problem affecting more than half of the world's population. An ultra-purified, high-concentration galacto-oligosaccharide, RP-G28, is being developed as a treatment for patients with LI. The efficacy and safety of RP-G28 in reducing symptoms of lactose intolerance were assessed in a blinded, randomized, placebo-controlled trial. **Methods:** In this multiclinical site, double-blinded, placebo-controlled trial, 377 patients with LI were randomized to one of two doses of orally administered RP-G28 or placebo for 30 days. A LI test and symptom assessment were performed at baseline and on day 31. The primary endpoint was a ≥4-point reduction or a score of zero on LI composite score on day 31. Voluntary milk and dairy intake and global outcome measures assessed patients' overall treatment satisfaction and quality of life before therapy and 30 days after therapy. This study received Institutional Review Board (IRB) approval. **Results:** For the primary endpoint, 40% in the RP-G28 groups reported a ≥4-point reduction or no symptoms on LI symptom composite score compared to 26% with placebo ($p = 0.016$). Treatment with RP-G28 also led to significantly higher levels of milk and dairy intake and significant improvements in global assessments compared to placebo. RP-G28 but not placebo led to significant increases in five *Bifidobacterium* taxa. **Conclusions:** RP-G28 for 30 days significantly reduced symptoms and altered the fecal microbiome in patients with LI. Treatment with RP-G28 also improved milk/dairy consumption and quality of life and was safe and well tolerated.

Keywords: lactose intolerance; lactase non-persistence; galacto-oligosaccharide; gut microbiome; abdominal pain; bloating; gas; diarrhea

---

1. Introduction

Non-persistence of the lactase enzyme in the small intestinal mucosa (LNP) affects 65–70% percent of the population globally, impacting health and potentially causing distressing gastrointestinal (GI) symptoms, which are commonly referred to as lactose intolerance (LI) [1]. LNP results in lactose malabsorption, which allows undigested lactose to move into the colon. Fermentation of lactose in the colon can produce carbon dioxide, hydrogen gas, methane and short-chain fatty acids, leading to a range of abdominal and bowel-related symptoms that can include abdominal pain, cramping, discomfort, bloating, distension, flatulence, increased stool frequency and/or loose or watery stools [2].

Patients with LI often avoid dairy products. Prolonged restriction of dairy can result in insufficient dietary calcium and vitamin D, which can, in turn, lead to osteoporosis, osteomalacia, and hypertension [3,4]. The daily calcium intake in patients with LI, ranging from 320 to 388 mg/day, is significantly less than the recommended amount of 1000–1200 mg/day [5,6].

There is accumulating evidence that treatments which alter the gut microbiome can improve a wide range of diseases, either by reversing disease-associated alterations in the microbiome or "dysbiosis" or by modifying the "normal" gut microbiome. One example of this latter strategy includes galacto-oligosaccharides (GOSs), which pass intact through to the colon, where they stimulate the growth of lactose-metabolizing bacteria. Elevated populations of *Lactobacillus* and *Bifidobacterium* enhance β-galactosidase activity and GOS utilization [7,8], resulting in enhanced fermentation of lactose to glucose, galactose, and short-chain fatty acids as well as reduced lactose-derived gas production, which could be beneficial to the symptoms of LI [9].

RP-G28, an ultra-purified, high-concentration (>95%) galacto-oligosaccharide, has been evaluated in a Phase 2a study in 85 patients [2]. RP-G28 reduced abdominal pain in 50 percent of patients after treatment and 30 days post-treatment ($p = 0.019$). Additionally, RP-G28-treated patients were 6 times less likely than patients given placebo to report LI 30 days post-treatment after reintroduction of dairy foods ($p = 0.039$).

The aim of the current study was to further evaluate the efficacy and safety of two doses of RP-G28 in patients with LI in a larger, multicenter, randomized, double-blind, placebo-controlled, parallel group trial.

2. Materials and Methods

2.1. Study Overview

A multicenter, randomized, double-blind, placebo-controlled, parallel group clinical trial was conducted to determine the efficacy, safety, and tolerability of two doses of RP-G28 in subjects with moderate to severe LI. The study took place between March and October 2016 at 15 investigative centers throughout the U.S. and included a 7 day screening, a 30 day treatment, and a 30 day post-treatment "real-world" observation, during which dairy was re-introduced to patients' diets (Figure 1).

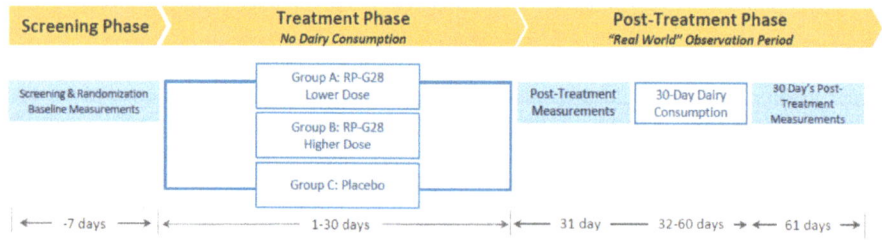

Figure 1. Protocol Design.

The clinical trial consisted of 3 phases: a screening phase, a treatment phase and a post-treatment phase. Patients were screened for LI for 7 days prior to study. Patients were then stratified into higher and lower-dose RP-G28 treatment or placebo for a 30 day treatment phase, during which patients did not consume lactose. On day 31, post-treatment, LI symptom assessments were made. Following this, "real-world" dairy intake began, and LI symptoms were assessed over a 30 day period.

Study patients, investigators, study site staff, the sponsor, the medical monitor and the study monitors were blinded to the treatment during the trial. All sites obtained Institutional Review Board (IRB) approval. This investigation was carried out in accordance with the Declaration of Helsinki of the World Medical Assembly and its revisions, as well as the rules of Good Clinical Practice (GCP) of the United States FDA (Protection of Human Subjects, 21 CFR 50; IRB, 21 CFR 56; and IND, 21 CFR 312).

IRB approval for protocol G28-003 was approved on 26 January 2016, with approval of Amendment #1 on 5 May 2016, and Amendment #2 on 17 June 2016. This clinical trial can be found on the clinical trial registry website (www.clinicaltrials.gov), trial number NCT02673749.

The primary and secondary endpoints provided a comprehensive evaluation of the effect of treatment on the symptoms of LI, milk and dairy consumption, quality of life experiences, and fecal microbiome to assess meaningful treatment benefits.

2.2. Screening Period

To establish eligibility, a pre-screening questionnaire was administered to confirm patient perception of LI. Eligible patients were required to have a moderate to severe symptom severity score, and a positive Hydrogen Breath Test (HBT) following a standardized in-clinic lactose challenge. The criteria for diagnosis of LI, and the main criteria for inclusion into the study, are provided in Table 1.

Table 1. Diagnosis and Main Criteria for Inclusion.

Male or female subjects, and female subjects were to be non-pregnant, and non-lactating.
Aged 18 to 75 years, inclusive.
A Hydrogen Breath Test (HBT) result that was positive for lactase deficiency.
A total abdominal pain score of at least 5 and at least 1 time point rated 3 or higher on the 11-point Numerical Rating Scale (NRS) over the 5 h screening.
At least 2 individual symptom scores present, measured over a 5 h screening: abdominal cramping, bloating, movement of gas (stomach rumbling), release of gas (flatulence), and bowel urgency.

A 5 h HBT at baseline confirmed LNP in patients. Hydrogen, methane, and carbon dioxide concentrations were measured in exhaled breath following a single-blinded lactose challenge three times during the study. Anhydrous food-grade lactose was administered at 0.35/kilogram body weight on day 0 (baseline, in order to confirm that the patient had symptoms of LI for eligibility), on day 31 (end-of-treatment, primary endpoint), and on day 61 (end of post-treatment period). After fasting overnight, patients were assessed for symptoms of LI using a validated symptom questionnaire (LI Symptom Questionnaire) to create a composite symptom score (comprised of abdominal pain, cramping, bloating and gas) at 6 time points starting at 30 minutes and hourly thereafter for 5 h.

Subjects who met the eligibility criteria completed a global assessment questionnaire and a 7 day lactose consumption assessment recall, provided stool samples, and were randomized to 1 of 2 dose regimens (higher or lower) of RP-G28 or placebo in a 1:1:1 ratio via a centralized, randomized Interactive Response (IXR) system using the subject ID number assigned at the beginning of screening. The IXR system was then used to assign drug kit numbers.

Subjects were asked to refrain from ingesting lactose-containing beverages/foods during the treatment period and record LI symptoms and adverse events (AEs) on a daily basis. The first dose of the study drug was administered on day 1 and the final dose on day 30. Subjects were followed for an additional 30 days after the final dose, during which time subjects were encouraged to ingest lactose-containing foods and record symptoms of LI daily. The demographics and characteristics of the study population are provided in Table 2.

2.3. Treatment Dose and Period

The lower dose of RP-G28 was 5 grams twice daily on days 1–10 followed by 7.5 grams twice daily on days 11–30. The higher dose of RP-G28 was 7.5 g twice daily for days 1–10, followed by 10 g twice daily on days 11–30. The placebo (powdered corn syrup that matched the consistency, color, sweetness, and taste of the drug) was administered in a blinded matching packet.

Table 2. Demographics and Baseline Characteristics (mITT Population).

Characteristic	Placebo (N = 121)	Lower Dose (N = 126)	Higher Dose (N = 121)	All Randomized (N = 368)
Age (year)				
Mean (SD)	39.9 (13.03)	42.8 (12.75)	40.8 (12.96)	41.2 (12.93)
Median	37	45	40	41
Min, Max	19, 74	18, 73	18, 70	18, 74
Gender (N [%])				
Male	43 (35.5)	48 (38.1)	57 (47.1)	148 (40.2)
Female	78 (64.5)	78 (61.9)	64 (52.9)	220 (59.8)
Ethnicity/Race (N [%])				
Hispanic or Latino				
African American	0	4 (3.2)	5 (4.1)	9 (2.4)
American Indian or Alaska Native	0	1 (0.8)	0	1 (0.3)
Asian	1 (0.8)	0	0	1 (0.3)
White	42 (34.7)	48 (38.1)	33 (27.3)	123 (33.4)
Other	0	3 (2.4)	1 (0.8)	4 (1.1)
Not Hispanic or Latino				
African American	55 (45.5)	51 (40.5)	57 (47.1)	163 (44.3)
American Indian or Alaska Native	1 (0.8)	1 (0.8)	2 (1.7)	4 (1.1)
Asian	7 (5.8)	5 (4.0)	2 (1.7)	14 (3.8)
White	15 (12.4)	13 (10.3)	17 (13.5)	45 (12.2)
Other	0	0	4 (3.3)	4 (1.1)
Height (cm)				
Mean (SD)	167.0 (8.97)	168.5 (8.82)	170.0 (10.60)	168.5 (9.54)
Median	167	168	168.9	167.8
Min, Max	147, 188	147, 194	146, 199	146, 199
Weight (kg)				
Mean (SD)	82.3 (21.97)	87.7 (24.28)	86.3 (19.08)	85.4 (21.91)
Median	77.2	85.3	85.4	83.1
Min, Max	41, 166	51, 163	45, 138	41, 166
BMI (kg/m)				
Mean (SD)	29.5 (7.82)	30.8 (8.09)	29.9 (6.37)	30.1 (7.48)
Median	27.2	29.6	29.3	29
Min, Max	18, 58	17, 57	18, 50	17, 58

BMI = body mass index; max = maximum; min = minimum; mITT = modified intent to treat; N = number of subjects; SD = standard deviation; kg = kilogram.

2.4. At 30 Days Post-Treatment "Real-World" Observation Period

The amount of lactose consumed each day for 7 days pre-treatment, and for 30 days post-treatment, was assessed via a food diary. Further, during the 30 day post-treatment period, global patient assessment questionnaires measured feelings, experiences and dietary changes resulting from treatment (Table 3). Global assessments are widely accepted as qualitative tools to evaluate the efficacy of treatment for functional disorders of the gastrointestinal tract [10].

2.5. Gut Microbiome Analysis

The 16S rRNA amplicon sequencing was performed utilizing patient stool samples collected on days 0, 31 and 61. Amplification using universal primers targeting the V4 region of the bacterial 16S rRNA gene was performed on 12.5 nanograms of total DNA from collected samples. [10] Each 16S rRNA gene amplicon was purified using the AMPure XP reagent (Beckman Coulter, Indianapolis, IN). Next, each sample was amplified using a limited cycle Polymerase Chain Reaction (PCR) program, adding Illumina sequencing adapters and dual-index barcodes (index 1(i7) and index 2(i5)) (Illumina) to the amplicon target. The final libraries were again purified using the AMPure XP reagent, quantified and normalized prior to pooling. The DNA library pool was then denatured with NaOH, diluted with hybridization buffer and heat denatured before loading on the MiSeq reagent cartridge and on the MiSeq instrument (Illumina). Automated cluster generation and paired-end sequencing with dual reads were performed according to the manufacturer's instructions.

Table 3. Global Patient Assessment Tool.

Assessment Questionnaires (Quality of Life Instrument)	Response Type	Scale	Description
Patient Global Impression of Severity	NRS[1]	5-point	1 = no symptoms, 1 = mild, 2 = moderate, 3 = severe, 4 = very severe
Patient Assessment of Satisfaction	NRS	5-point	1 = not at all satisfied, 2 = a little satisfied, 3 = somewhat satisfied, 4 = very satisfied, 5 = extremely satisfied
Patient Assessment of Adequate Relief	Binary	2-point	Yes/No
Patient Global Impression of Change	Likert-type scale	7-point	1 = very much improved, 2 = much improved, 3 = minimally improved, 4 = no change, 5 = minimally worse, 6 = much worse, 7 = very much worse

1. NRS = Numerical Rating Scale.

The 1050 DNA samples corresponding to 345 subjects receiving placebo, lower-dose or higher-dose treatments at three time points (days 0, 31, and 61) were analyzed by high-throughput quantitative PCR (qPCR) targeting *Bifidobacteria* and *Lactobacilli* using specific 16S rRNA gene and GroEL probes [11,12]. Microfluidic qPCR was performed using a BioMark HD reader (Fluidigm Corporation, San Francisco, CA) with a Dynamic Array 24.192 chip processed following manufacturer's instructions.

2.6. Statistical Analysis

2.6.1. Primary Efficacy Endpoint

A LI Symptom Questionnaire was developed, validated and applied during this clinical trial. This questionnaire rated individual symptoms of LI on an 11-point Numerical Response Scale (NRS) as well as a Verbal Rating Scale (VRS).

The primary efficacy endpoint was the proportion of responders on day 31. A responder was defined as a patient with a reduction from baseline in the composite score of 4 points or greater or a composite score of 0 (i.e., symptom resolution) on day 31. The composite score was calculated from the maximum symptom scores for each symptom (abdominal pain, cramping, bloating, and gas movement) after a lactose challenge test. Each symptom was rated on a 10-point Likert-type scale over 5 h after a lactose challenge. Each maximum symptom was then averaged into a composite score ranging from 0 to 10, where 0 indicated no symptoms and 10 indicated symptoms at their worst. A 4-point change was considered a meaningful improvement in symptoms of LI, based on psychometric analysis (blinded review of the clinical data prior to unblinding) and two rounds of cognitive interviews (N = 30 and 23). In the cognitive interviews, a 4-point change was meaningful for 84% of the subject responses and 53% of the subject responses in the first and second round of interviews, respectively. Symptom resolution was also considered meaningful. The 4-point threshold was further supported by empirical cumulative distribution functions using patient global severity anchor and was associated with sensitivity of 71%, specificity of 68%, positive predictive value of 63%, and negative predictive of 75%.

Statistical analysis used two-tailed tests at the α = 0.05 level of significance. The proportions of responders on day 31 were analyzed by a stratified (by quartile of baseline LI symptom composite score) Cochran–Mantel–Haenszel (CMH) test comparing both doses combined versus placebo, higher dose versus placebo, and lower dose versus placebo. Prior to unblinding, the protocol was modified to specify that the primary endpoint was a comparison of the 2 active arms combined versus placebo, using a two-sided test at the α = 0.05 level of significance. Pooling the data provided greater statistical power. The modified intent to treat (mITT) analysis population was all randomized subjects who received at least 1 dose of drug.

The sample size of 372 subjects was designed with the standard deviation of percentage abdominal pain reduction being 50%. It was hypothesized that the mean percentage reduction reported would be 50% for placebo-treated subjects and 70% for actively treated subjects. A simulation using Dunnett comparisons indicated that with 113 evaluable subjects in each arm, there would be 90% power to detect at least 1 of the active arms to be superior to placebo. The primary endpoint was based on a dichotomization of the distribution of the composite symptom score, which was thought to have at least as much power as percentage abdominal pain reduction.

The SAS software version 9.4 was used.

2.6.2. Secondary Efficacy Endpoints

A number of secondary analyses were conducted. The proportion of participants in each group reporting no symptoms by the lactose intolerance composite score as well as the individual symptoms of abdominal pain, cramping, bloating, and gas movement after lactose challenge on day 31 was determined. The amount of milk consumed by participants on day 61 after RP-G28 or placebo was assessed. Global endpoints as defined in Table 3 were also compared on day 61 after RP-G28 or placebo. Analysis was performed by a stratified Cochran–Mantel–Haenszel (CMH) test for 3 comparisons: both doses combined versus placebo, higher dose versus placebo, and lower dose versus placebo. Analyses were conducted using two-sided tests at the α = 0.05 level of significance.

2.7. Gut Microbiome Analysis

Paired-end fastq files were joined into a single multiplexed, single-end fastq using the software tool fastq-join. Demultiplexing and quality filtering were performed on the joined results. Quality analysis reports were produced using the FastQC software [13]. Bioinformatics analysis of bacterial 16S amplicon sequencing data was conducted using the Quantitative Insights into Microbial Ecology (QIIME) software at a 25,000 reads/sample depth [14]. Operational Taxonomic Unit (OTU) picking was performed on the quality filtered results using pick_de_novo_otus.py. Chimeric sequences were detected and removed using ChimeraSlayer [15]. Summary reports of taxonomic assignment by sample and all categories were produced using QIIME summarize_taxa_through_plots.py and summarize_otu_by_cat.py. The script group_significance.py was used to compare taxa frequencies in sample groups and to determine whether there were statistically significant differences between abundances in the different groups. The non-parametric Analysis of Variance (ANOVA) test (Kruskal–Wallis) with False Discovery Rate (FDR) correction was used to compare treatments and placebo groups. For high-throughput qPCR data, the relative proportion of *Bifidobacterium* and *Lactobacillus* species was computed based on the Livak method. Quantitation cycle (Cq) values for each sample were normalized against the Cq value for the universal primers. Fold differences were calculated by $2^{-\Delta\Delta Ct}$. Paired t-test and ANOVA with Tukey tests were conducted to assess statistically significant differences between groups.

2.8. Safety

All treated subjects were included in the safety analyses using summary statistics by treatment group of AEs, concomitant medications, vital signs, physical examinations, and clinical laboratory measurements. AEs were reported from initiation of treatment through 30 days of the post-treatment period. An AE was classified as a Severe Adverse Event if it interfered significantly with the patients' usual functions.

2.9. Irregular Site

A for-cause audit of one study site (out of 15 sites) was conducted due to significant data irregularities. The audit found significant deviations from the protocol and failure to comply with regulatory requirements for Good Clinical Practice (GCP), and identified protocol deviations including medical history and concomitant medications, inclusion/exclusion entry criteria, and administration of the Hydrogen Breath Test (HBT), lactose challenge, and patient diaries. Compared to the other sites, there were significant differences in multiple baseline symptom scores at this site. Patients reported two times more dairy intake prior to entry into the trial compared to subjects at other centers ($p = 0.04$). In addition, patients reported higher symptom severity scores during the in-clinic lactose challenge compared to the other site's subjects ($p = 0.035$). High milk consumption in addition to high symptom severity scores based on a blinded in-clinic lactose challenge is inconsistent with patients with LI. Further, the screen failure rate was also significantly lower at this center compared to other centers.

Thus, efficacy analyses were conducted for both the mITT population which included all randomized subjects that received at least one dose of the drug and a mITT Efficacy Subset population, which included all randomized subjects that received at least one dose of the drug, excluding those who were enrolled at the site with numerous GCP violations.

3. Results

A total of 1398 subjects were screened for LI, 377 subjects from 15 study sites were enrolled and randomized (127 lower dose, 123 higher dose and 127 placebo), and 344 (87% placebo, 92% lower dose, 94% higher dose) completed the study. A high screen failure rate (>70%) was expected based on the rigorous inclusion and exclusion criteria required, including testing positive for LNP from a HBT and meeting LI symptom thresholds (63% of screen failures were due to HBT or LI symptoms scores not being met).

3.1. Primary Endpoint Analysis

Symptom Reduction in RP-G28-Treated Patients

In the Efficacy Subset mITT group, significantly more patients in the pooled RP-G28 group responded as compared to patients in the placebo group—40% versus 26% ($p = 0.016$). In total, 41% of subjects treated with lower-dose RP-G28 ($p = 0.043$) and 38% of subjects treated with higher-dose RP-G28 ($p = 0.029$) responded. (Table 4) In the mITT population, the pooled RP-G28 group trended toward significance $p = 0.062$ (40% with treatment versus 31% with placebo).

3.2. Secondary Endpoint Analysis

Individual Assessed Symptom Response to RP-G28 Post-Treatment

In the Efficacy Subset mITT, RP-G28-treated patients were significantly more likely than patients treated with placebo to report complete elimination of LI symptoms. RP-G28 was more likely to lead to complete elimination of the LI symptom composite score ($p = 0.004$) and individual symptoms of abdominal pain ($p = 0.014$), abdominal cramping ($p = 0.002$), abdominal bloating ($p = 0.015$), and gas movement ($p = 0.001$) than placebo (Figure 2A).

Table 4. Primary Endpoint [1]

	Efficacy Subset mITT [2] (N = 296)		
	Higher dose	Lower dose	Pooled (Higher + Lower)
Number of subjects	97	102	199
RP-G28 Treatment	37 (38%)	42 (41%)	79 (40%)
Placebo	25 (26%)	25 (26%)	25 (26%)
CMH p-value versus placebo [4]	0.029	0.043	0.016
	mITT [3] (N = 368)		
	Higher dose	Lower dose	Pooled (Higher + Lower)
Number of subjects	121	126	247
RP-G28 treatment	46 (38%)	53 (42%)	99 (40%)
Placebo	38 (31%)	38 (31%)	38 (31%)
CMH p-value versus placebo [4]	0.096	0.117	0.062

1. Proportion of LI symptom composite score responders post-treatment (day 31). 2. Efficacy Subset (mITT)—mITT data in which observed inconsistent data from one study center was removed from analysis. 3. mITT—modified intent to treat (all patients who received at least one dose of drug). 4. CMH = Cochran–Mantel–Haenszel. p-value versus placebo. N = number of subjects.

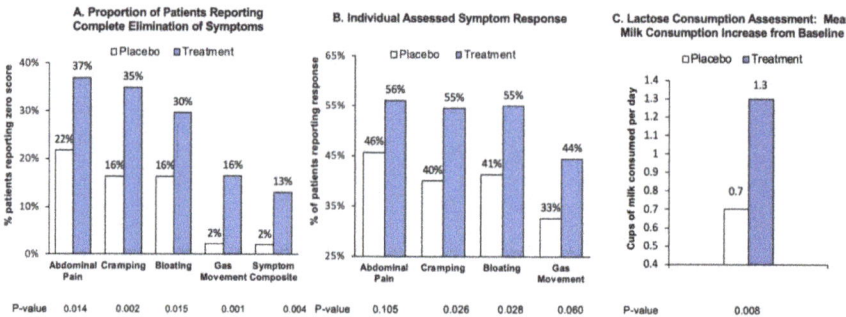

Figure 2. Secondary Endpoints Analysis. (A) The proportion of patients reporting complete elimination of lactose intolerance (LI) symptoms (abdominal pain, cramping, bloating, and gas movement) with RP-G28 treatment or placebo. (B) The proportion of patients reporting a response (≥4-point improvement from baseline or a score of 0 on day 31) in each key LI symptom (abdominal pain, cramping, bloating, and gas movement) with RP-G28 treatment or placebo. (C) RP-G28 led to a significantly greater increase in daily average milk consumption from baseline 30 days after treatment (day 61) compared to placebo.

Patients treated with RP-G28 exhibited a consistent and significantly greater decrease in LI symptoms including cramping ($p = 0.026$) and bloating ($p = 0.028$). Non-significant trends for improvement were seen for abdominal pain ($p = 0.105$) and gas movement ($p = 0.060$) (Figure 2B).

"Real-world" Observation Period—Milk and Dairy Intake 30 days Post-Treatment: After 30 days post-treatment (day 61), patients treated with RP-G28 reported drinking significantly more milk, drinking an average of 1.5 cups/day versus 0.2 cups/day prior to treatment. The mean increase in milk intake of 1.3 cups/day (SD 1.479) in treated patients was significantly higher than that of the placebo group, which was 0.7 cups/day (SD 1.591) ($p = 0.008$, Efficacy Subset mITT) (Figure 2C).

In addition, 59% of treatment patients consumed ≥1 cups/d of milk after being treated with RP-G28 in comparison to 42% with placebo ($p = 0.01$, Efficacy Subset mITT)). Patients treated with RP-G28 also consumed more dairy in general, ingesting 5.4 cups/day on day 61 versus only 1 cup prior to treatment, at baseline. In comparison, the placebo group ingested 4.5 cups/day on day 61 versus 1.7 cups/day at baseline. The mean increase in dairy intake of 4.3 cups/day trended toward significance compared to the placebo group ($p = 0.057$, Efficacy Subset mITT).

"Real-world" Observation Period—Global Patient Assessments 30 days Post-Treatment: Significantly more treated patients (82%) reported "no symptoms" or "mild symptoms", respectively) as compared to 64% in the placebo group after treatment ($p = 0.001$, Efficacy Subset mITT). In addition, significantly more treated patients reported satisfaction with the ability of RP-G28 to prevent or treat their LI symptoms, with 66% of treatment patients reporting "very satisfied" or "extremely satisfied" as compared to 52% in the placebo group ($p = 0.030$, Efficacy Subset mITT). Patients' perception of adequate relief from LI symptoms and patients' global impression of change were also improved with treatment ($p = 0.042$ and $P = 0.034$ respectively, Efficacy Subset mITT) (Figure 3).

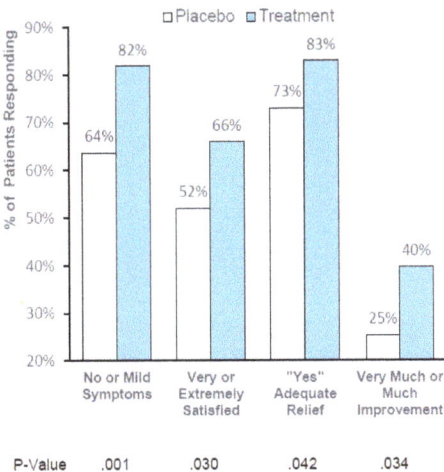

Figure 3. Global Patient Assessments. Patients' global assessments were evaluated 30 days after treatment (day 61) and were based on a patient questionnaire.

3.3. Gut Microbiome Analysis

Of 543 bacterial species identified overall, 28 were differentially represented (Kruskal–Wallis, FDR corrected $p < 0.05$). Of those, relative abundances of five *Bifidobacterium* taxa (*Bifidobacterium_Other*, *Bifidobacterium sp.*, *Bifidobacterium adolescentis*, *Bifidobacterium longum* and *Bifidobacterium pseudolongum*) were increased by both RP-G28 treatments. High-throughput qPCR quantitative data confirmed a significant ($p < 0.05$) increase in the abundance of the phylum Actinobacteria, the family *Bifidobacteriaceae*, and the genus *Bifidobacterium* on day 31 in the higher- and lower-dose treatment groups, but not the placebo group (Figure 4). Specifically, treatment with RP-G28 resulted in an elevated abundance of *Bifidobacterium longum*, *Bifidobacterium bifidum*, *Bifidobacterium breve*, *Bifidobacterium catenulatum*, *Bifidobacterium angulatum*, *Bifidobacterium gallicum*. In total, 78% (77/99) of patients treated with RP-G28 had elevated *Bifidobacteria* levels, as compared to 52% (49/94) of patients in the placebo group ($p < 0.001$). Data using Firmicutes, *Lactobacillaceae* and *Lactobacillus* species-specific 16S rRNA gene probes showed a significant increase in the relative abundance of the family *Lactobacillaceae* on day 31 in the lower-dose treatment and a non-significant increase in the higher-dose group, but not in the placebo group.

On day 31 after treatment with RP-G28, qPCR quantitative data revealed an increase in the relative abundance of the genus *Bifidobacterium* in the lower-dose galacto-oligosaccharide RP-G28 and higher-dose RP-G28 treatment groups, an effect not seen in the placebo group.

3.4. Safety

RP-G28 was well tolerated. In total, 6.9% of patients had treatment-emergent adverse events (TEAEs) with no differences between placebo and RP-G28. None of the serious adverse events (SAEs) were treatment related. No TEAEs resulted in hospitalization or death (Table 5).

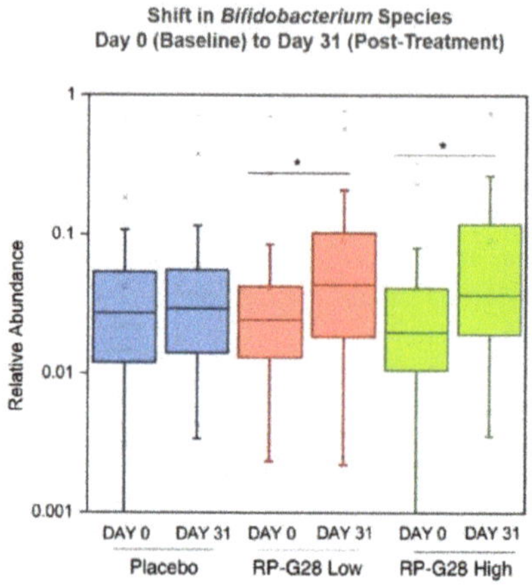

Figure 4. Gut Microbiome Analysis.

Table 5. Overall Summary of Treatment-Emergent Adverse Events (Safety Population).

Type of Adverse Event	Placebo (N = 121)	Lower Dose (N = 126)	Higher Dose (N = 121)	Pooled Dose (Lower + Higher) (N = 247)
Subject with at least 1 TEAE	38 (31.4%)	40 (31.7%)	36 (29.8%)	76 (30.8%)
Subjects with at least 1 treatment-related TEAE [1]	8 (6.6%)	11 (8.7%)	6 (5.0%)	17 (6.9%)
Subjects with at least 1 SAE [2]	3 (2.5%)	1 (0.8%)	0	1 (0.4%)
Subjects with at least 1 treatment-related SAE	0	0	0	0
Subjects with a TEAE leading to study drug withdrawal, interruption, or reduction	2 (1.7%)	1 (0.8%)	0	1 (0.4%)
Any AE resulting in death	0	0	0	0

AE = adverse event; N = Number of subjects; SAE = Serious adverse event; TEAE = treatment-emergent adverse event. 1. Treatment-related TEAT was defined as any TEAT that was possibly, probably, or definitely related to study drug. 2. One subject (a randomized placebo) experienced the SAE of spontaneous abortion.

4. Discussion

Though lactose intolerance is remarkably common, treatment options have not changed for decades [16,17]. The cornerstone of treatment, abstinence from consuming dairy products, is potentially nutritionally detrimental, inconvenient and limiting to patients. This study assessed a novel treatment strategy for lactose intolerance involving supplementation of RP-G28, which promoted the proliferation of lactose-fermenting bacteria, including *Lactobacillus* and *Bifidobacterium* [12]. Increased abundance and a higher diversity of these species is associated with enhanced levels of beta-galactosidase [17], reduced lactose-derived gas production in the colon [9,17], and mitigation of clinical symptoms in patients with LI [12,18].

To assess the efficacy and meaningfulness of treatment, three broad groups of LI patient experiences were evaluated—a LI composite score to measure specific symptoms, milk and dairy intake to evaluate patients' ability to increase dietary lactose with treatment, and global assessment measures to assess patients' overall satisfaction with therapy. This three-pronged assessment provided a comprehensive evaluation of how treatment influenced symptoms of LI and the day to day lives of patients.

A composite symptom score and a stringent threshold of meaningful change was established prior to unblinding the study in order to identify patients who reported a meaningful decrease in

LI symptoms. Psychometric analyses indicated that the four symptom severity items were distinct yet related enough to support the creation of a composite score, and this composite score was shown to be reliable, valid, and responsive to change in LI symptoms over time. This tool provides a valuable resource for accurately assessing symptoms from the patient perspective in future LI clinical trials, and for identifying subjects who have experienced a meaningful treatment benefit in terms of patient-reported LI symptom severity. This study is one of the first to define and measure a meaningful treatment benefit for LI patients. The tool was developed adhering to good measurement principles following the FDA PRO requirements [19], and with consultation with the FDA. The instrument content was developed through a literature review, patient surveys, and concept elicitation and was tested through several rounds of cognitive interviews.

In the Efficacy Subset mITT population, the primary endpoint of response on day 31 was achieved in the pooled and individual RP-G28 groups. Individual symptoms were consistently reduced. The most rigorous endpoint for symptom relief is complete elimination. In the Efficacy Subset mITT population, RP-G28 administration resulted in complete symptom relief in a significantly greater proportion of patients than placebo, for the measures of abdominal pain, cramping, bloating, gas movement and overall symptoms ($p < 0.05$).

RP-G28-treated patients drank significantly more milk after treatment compared to the control group, drinking an average of 1.3 cups/day, as compared to an average of 0.7 cups/day more by the placebo group. Milk is the primary source of lactose in the diet, and thus milk consumption is an important indicator of the clinical benefits of supplementation with RP-G28. The ability to drink milk without LI symptoms supports optimal nutrient intake, contributing to the USDA's recommended 2–3 cups of dairy per day [20]. The sustained improvement in dairy intake and symptom relief on day 61 are further validation of the durability of treatment.

Consistent, statistically significant improvement in global assessments support the clinical meaningfulness of RP-G28. When consuming dairy foods for 30 days after treatment, 82% of patients reported "no or mild symptoms", 66% reported being "very or extremely satisfied," 83% reported adequate relief and 40% reported "very much or much improvement".

Overall, RP-G28 was safe and well tolerated. There were no differences in adverse events in those receiving RP-G28 or placebo. There were no serious adverse events in either group. Of note, there were no increases in GI side effects with RP-G28. This is important as RP-G28 is a GOS and, thus, part of the family of fermentable, oligo, di, monosaccharides, and polyols referred to as FODMAPs. FODMAPs have been shown to trigger GI symptoms in patients with irritable bowel syndrome. It is reassuring that at the doses administered, RP-G28 did not induce any significant GI side effects.

Gut microbiome changes and a reduction in net hydrogen gas production support the hypothesized mode of action [18]. In total, 78% of individuals treated with RP-G28 had elevated *Bifidobacteria*, and *Lactobacillaceae*. Patient microbiomes adapted further with the reintroduction of lactose into patient diets during a 30 day period post-treatment, with an increase in lactose-fermenting *Roseburia* species [2,12]. Elevated abundance of *Bifidobacterium* and *Lactobacillus* resulting from RP-G28 treatment, known to enhance lactose fermentation, is likely instrumental in better clinical outcomes. Further, stool from patients adapted to lactose produce less hydrogen, due to an absolute reduction in hydrogen production, rather than an alteration in hydrogen uptake by the microbiome [21]. Additional studies to understand how the changes in gut microbiome induced by RP-G28 lead to clinical improvements are warranted.

5. Limitations

One of the 15 study sites was excluded from the Efficacy Subset mITT due to significant GCP violations. Nevertheless, this is the largest and most rigorously designed double-blinded randomized trial for the treatment of LI ever conducted. Another limitation is that the construct that identifies LI is subjective, depending on a single lactose dose and a subsequent time course. Real-world LI is likely intermittent, and depends on diet, dose, transit and other environmental and biological factors that are impossible to fully control within a clinical trial. However, in this study, the primary endpoint is part of a comprehensive assessment, including primary and secondary efficacy endpoints, global patient assessments, lactose consumption assessments, and correlation of treatment to changes in the microbiome—all of which are indicative of a beneficial effect of RP-G28 in improving symptoms of LI.

Another limitation of the study is the duration, which does not allow determination of long-term durability. While significant improvements in diet quality and lactose tolerance were evident on days 31 and 61, we cannot determine whether retreatment will be necessary or effective.

6. Summary

RP-G28 was safe and effective for reducing or eliminating symptoms of LI. Treatment with RP-G28 led to increased milk and dairy intake and improved quality of life. RP-G28 was safe and well tolerated at the doses administered. RP-G28 led to changes in the microbiome, which may be involved in the clinical benefits observed in patients with LI. These findings are relevant not only to lactose intolerance, but also usher in an era of using prebiotics to manipulate the microbiome to facilitate gut health.

Author Contributions: Conceptualization, W.C., W.S., A.J.R., H.F. and D.S.; methodology, W.C., W.S., A.J.R., H.F. and D.A.S.; formal analysis, W.C., W.S., A.R., H.F., M.A.A.-P. and D.A.S.; investigation, W.C., W.S., A.J.R., H.F., M.A.A.-P. and D.A.S.; data curation, W.C., W.S., A.R., H.F., M.A.A.-P. and D.A.S.; writing—original draft preparation, W.C. and D.A.S.; writing—review and editing, W.C., W.S., A.R., M.A.A.-P. and D.A.S.; funding acquisition, A.J.R. All authors have read and agreed to the published version of the manuscript.

Funding: Support for this research was provided by Ritter Pharmaceuticals.

Acknowledgments: We thank Dana Barberio, of Edge Bioscience Communications and Tara Nassari for their assistance in drafting and revising the manuscript.

Conflicts of Interest: W. Chey, W. Sandborn and D. Savaiano are on the Medical Advisory Board of Ritter Pharmaceuticals, H. Foyt; A. Azcarate-Peril are consultants of Ritter Pharmaceuticals; A. Ritter is an employee of Ritter Pharmaceuticals.

Abbreviations

Adverse Events (AE), Analysis of Variance (ANOVA), Code of Federal Regulations (CFR), Cochran–Mantel–Haenszel (CMH), False Discovery Rate (FDR), Food and Drug Administration (FDA), Gastrointestinal (GI), Galacto-Oligosaccharides (GOSs), Good Clinical Practice (GCP), Hydrogen Breath Test (HBT), Institutional Review Board (IRB), Investigational New Drug (IND), Lactose Non-Persistence (LNP), Lactose Intolerance (LI), Modified Intent to Treat (mITT), Numerical Response Scale (NRS), Polymerase Chain Reaction (PCR), Serious Adverse Event (SAE), Treatment-Emergent Adverse Events (TEAEs), and Verbal Rating Scale (VRS).

References

1. Bayless, T.M.; Brown, E.; Paige, D.M. Lactase Non-persistence and Lactose Intolerance. *Curr. Gastroenterol. Rep.* **2017**, *19*, 23. [CrossRef] [PubMed]
2. Savaiano, D.A.; Ritter, A.J.; Klaenhammer, T.R.; James, G.M.; Longcore, A.T.; Chandler, J.R.; Walker, W.A.; Foyt, H.L. Improving lactose digestion and symptoms of lactose intolerance with a novel galacto-oligosaccharide (RP-G28): A randomized, double-blind clinical trial. *Nutr. J.* **2013**, *12*, 160. [CrossRef] [PubMed]
3. McCarron, D.A.; Heaney, R.P. Estimated healthcare savings associated with adequate dairy food intake. *Am. J. Hypertens.* **2004**, *17*, 88–97. [CrossRef] [PubMed]
4. Standing Committee on the Scientific Evaluation of Dietary Reference Intakes, Food and Nutrition Board, Institute of Medicine. *Dietary Reference Intakes for Calcium, Phosphorus, Magnesium, Vitamin D, and Fluoride*; National Academies Press: Washington, DC, USA, 1997.
5. Carroccio, A.; Montalto, G.; Cavera, G.; Notarbatolo, A.; Lactase Deficiency Study Group. Lactose intolerance and self-reported milk intolerance: Relationship with lactose maldigestion and nutrient intake. Lactase Deficiency Study Group. *J. Am. Coll. Nutr.* **1998**, *17*, 631–636. [CrossRef] [PubMed]
6. Buchowski, M.S.; Semenya, J.; Johnson, A.O. Dietary calcium intake in lactose maldigesting intolerant and tolerant African-American women. *J. Am. Coll. Nutr.* **2002**, *21*, 47–54. [CrossRef] [PubMed]

7. Silk, D.B.; Davis, A.; Vulevic, J.; Tzortzis, G.; Gibson, G.R. Clinical trial: The effects of a trans-galactooligosaccharide prebiotic on faecal microbiota and symptoms in irritable bowel syndrome. *Aliment. Pharmacol. Ther.* **2009**, *29*, 508–518. [CrossRef] [PubMed]
8. Depeint, F.; Tzortzis, G.; Vulevic, J.; I'Anson, K.; Gibson, G.R. Prebiotic evaluation of a novel galactooligosaccharide mixture produced by the enzymatic activity of *Bifidobacterium bifidum* NCIMB 41171, in healthy humans: A randomized, double-blind, crossover, placebo-controlled intervention study. *Am. J. Clin. Nutr.* **2008**, *87*, 785–791. [CrossRef] [PubMed]
9. Jiang, T.; Mustapha, A.; Savaiano, D.A. Improvement of lactose digestion in humans by ingestion of unfermented milk containing *Bifidobacterium longum*. *J. Dairy Sci.* **1996**, *79*, 750–757. [CrossRef]
10. Drossman, D.A. The functional gastrointestinal disorders and the Rome III process. *Gastroenterology* **2006**, *130*, 1377–1390. [CrossRef] [PubMed]
11. Monteagudo-Mera, A.; Arthur, J.C.; Jobin, C.; Keku, T.; Bruno-Barcena, J.M.; Azcarate-Peril, M.A. High purity galacto-oligosaccharides enhance specific *Bifidobacterium* species and their metabolic activity in the mouse gut microbiome. *Benef. Microbes* **2016**, *7*, 247–264. [CrossRef] [PubMed]
12. Azcarate-Peril, M.A.; Ritter, A.J.; Savaiano, D.; Monteagudo-Mera, A.; Anderson, C.; Magness, S.T.; Klaenhammer, T.R. Impact of short-chain galactooligosaccharides on the gut microbiome of lactose-intolerant individuals. *Proc. Natl. Acad. Sci. USA* **2017**, *114*, E367–E375. [CrossRef] [PubMed]
13. *Babraham Bioinformatics: FastQC*; Babraham Institute: Babraham, UK, 2018.
14. Caporaso, J.G.; Kuczynski, J.; Stombaugh, J.; Bittinger, K.; Bushman, F.D.; Costello, E.K.; Fierer, N.; Pena, A.G.; Goodrich, J.K.; Gordon, J.I.; et al. QIIME allows analysis of high-throughput community sequencing data. *Nat. Methods* **2010**, *7*, 335–336. [CrossRef] [PubMed]
15. *Microbiome Utilities Portal of the Broad Institute*; Broad Institute: Cambridge, MA, USA, 2011.
16. Briet, F.; Flourie, B.; Achour, L.; Maurel, M.; Rambaud, J.C. Bacterial adaptation in patients with short bowel and colon in continuity. *Gastroenterology* **1995**, *109*, 1446–1453. [CrossRef]
17. Hertzler, S.R.; Savaiano, D.A. Colonic adaptation to daily lactose feeding in lactose maldigesters reduces lactose intolerance. *Am. J. Clin. Nutr.* **1996**, *64*, 232–236. [CrossRef] [PubMed]
18. Forsgard, R.A. Lactose digestion in humans: Intestinal lactase appears to be constitutive whereas the colonic microbiome is adaptable. *Am. J. Clin. Nutr.* **2019**, *110*, 273–279. [CrossRef] [PubMed]
19. U.S. Department of Health and Human Services Food and Drug Administration CDER, CBER and CDRH. *Guidance for Industry Patient-Reported Outcome Measures: Use in Medical Product Development to Support Labeling Claims*; United States Department of Agriculture: Washington, DC, USA, 2009; December 2009 Clinical/Medical.
20. *USDA Choosemyplate.gov*; United States Department of Agriculture: Washington, DC, USA, 2017.
21. Hertzler, S.R.; Savaiano, D.A.; Levitt, M.D. Fecal Hydrogen Production and Consumption Measurements: Response to Daily Lactose Ingestion by Lactose Maldigesters. *Dig. Dis. Sci.* **1997**, *42*, 348–353. [CrossRef]

© 2020 by the authors. Licensee MDPI, Basel, Switzerland. This article is an open access article distributed under the terms and conditions of the Creative Commons Attribution (CC BY) license (http://creativecommons.org/licenses/by/4.0/).

Article

Consumption of Goat Cheese Naturally Rich in Omega-3 and Conjugated Linoleic Acid Improves the Cardiovascular and Inflammatory Biomarkers of Overweight and Obese Subjects: A Randomized Controlled Trial

Cristina Santurino [1], Bricia López-Plaza [1,*], Javier Fontecha [2], María V. Calvo [2], Laura M. Bermejo [1], David Gómez-Andrés [3,4] and Carmen Gómez-Candela [1,5]

1. Nutrition Research Group, Hospital La Paz Institute for Health Research (IdiPAZ), 28046 Madrid, Spain; cristina.santurino@idipaz.es (C.S.); laura.bermejol@salud.madrid.org (L.M.B.); cgcandela@salud.madrid.org (C.G.-C.)
2. Food Lipid Biomarkers and Health Group, Institute of Food Science Research (CIAL, CSIC), Campus of Autonomous University of Madrid, 28049 Madrid, Spain; j.fontecha@csic.es (J.F.); mv.calvo@csic.es (M.V.C.)
3. Department of Anatomy, Histology and Neuroscience, School of Medicine, Autonomous University of Madrid, 28049 Madrid, Spain; dgandres10@hotmail.com
4. Pediatric Neurology Unit, Hospital Universitari Vall d'Hebron, VHIR, 08035 Barcelona, Spain
5. Dietetic and Clinical Nutrition Department, La Paz, University Hospital, 28046 Madrid, Spain
* Correspondence: bricia.plaza@idipaz.es

Received: 17 March 2020; Accepted: 28 April 2020; Published: 5 May 2020

Abstract: This study examines the value of a goat cheese naturally enriched in polyunsaturated fatty acids (PUFA) (n-3 PUFA and conjugated linolenic acid (CLA)) as means of improving cardiovascular and inflammatory health. Sixty-eight overweight and obese subjects (BMI ≥ 27 and <40 kg/m^2), with at least two risk factors for cardiovascular disease (CVD) in a lipid panel blood tests, participated in a randomized, placebo-controlled, double-blind, parallel designed study. The subjects consumed for 12 weeks: (1) 60 g/d control goat cheese and (2) 60 g/d goat cheese naturally enriched in n-3 PUFA and CLA. Diet and physical activity were assessed. Anthropometric and dual-energy X-ray absorptiometry (DXA) tests were performed. Blood samples were collected at the beginning and at the end of the study period. Changes in health status, lifestyle and dietary habits, and daily compliance were recorded. The consumption of a PUFA-enriched goat cheese significantly increased plasma high-density lipoprotein (HDL)-cholesterol, as well as in apolipoprotein B, and it significantly decreased high-sensitivity C-reactive protein concentrations compared to the control goat cheese ($p < 0.05$). The significant improvement of the plasma lipid profile and inflammatory status of people with risk for CVD due to the consumption of PUFA-enriched cheese suggests a potential role of this dairy product as an alternative to develop high nutritional value food in a balanced diet comprising regular exercise.

Keywords: n-3 PUFA; CLA; cheese; blood lipids; dairy fat

1. Introduction

Cardiovascular disease (CVD) is the leading cause of death worldwide. The major risk factors are well-established and are mediated mainly by hypertension, dyslipidemia, and smoking, as well as others such as obesity, elevated cholesterol, poor diet, and physical activity. Due to the significant influence exerted by diet and lifestyle, the current nutritional recommendations like controlling the amount and quality and quality of the fats consumed in the diet and salt intake, as well as regular physical exercise, are key to the prevention and treatment of CVD [1].

Within this framework, the effect of polyunsaturated fatty acids (PUFA), such as α-linolenic acid (ALA) n-3, has been demonstrated through mechanisms involving anti-inflammatory, anti-arrhythmic, and anti-thrombotic properties, which reduce low-density lipoprotein cholesterol levels (LDL-C) and, to a lesser extent, high-density lipoprotein cholesterol levels (HDL-C) when they replace saturated fatty acids (SFA) [2]. Even though these beneficial effects are well known, in the last 100 years, the dietary ratio of n-6/n-3 PUFA in modern Western diets has dramatically increased to 15-17:1, which has been associated with an increase of many other illnesses including inflammatory diseases and cholesterolaemia [3]. In addition, some clinical trials in humans have indicated that conjugated linolenic acid (CLA) may have several beneficial effects for health, such as improving the blood lipid profile related to CVD and diabetes [4]. However, it is important to know the exact dose of CLA and the duration of the treatment to know the biological effects [4].

In recent decades, a wide variety of functional foods have been designed to reduce some of the factors that induce cardiovascular risks and to improve health [5]. Thus, one possible way to increase PUFA consumption is to enrich foods that are regularly consumed by the majority of the population such as dairy products. Even though full fat dairy products consumption has long been considered a risk factor for cardiovascular health, such products contribute to the mean daily intakes of energy (11%), protein (14%), fat (17%), calcium (48%), phosphorous (24%), and vitamin A (27%). Though further studies are needed, a recent meta-analysis has demonstrated that dairy product consumption is not associated with CVD [6]. Some studies have even proposed a distinction between dairy and other food sources of SFA based on their different effects on blood lipids [7] and the possible cardioprotective effect of eating fermented dairy products [8].

Modulating milk FA composition through the ruminant feeding, particularly with oilseeds rich in PUFA, has shown to be a valuable tool to improve milk nutritional value [9]. In particular, goat's milk possesses some inherent properties and a great nutritional quality determined by its lipid composition, which makes it an attractive alternative to developing dairy products with a high added value, like cheese. In this respect, our group developed and characterized a goat cheese naturally enriched in CLA and omega-3 [9] to be further employed in a clinical trial on cardiovascular risk prevention in humans.

Thus, a randomized controlled trial was performed in order to assess the effect of the consumption of that PUFA-enriched cheese in modulating blood lipids (total cholesterol (TC), HDL-C, LDL-C, triglycerides (TAG), apolipoprotein A1 (ApoA1), apolipoprotein B (ApoB), and free fatty acids (FFA), as well as other cardiovascular risk factors, such as inflammatory markers, in overweight and obese subjects.

2. Materials and Methods

The present study was registered at http://clinicaltrials.gov under the number NCT02630602.

2.1. Subjects

For the present study, the Clinical Nutrition Department of La Paz University Hospital (HULP) in Madrid (Spain) recruited 68 overweight and obese subjects (52 women and 16 men) between January and March 2014. The inclusion criteria were: aged 18–65 years living in the region of Madrid, Spain; body mass index (BMI) ≥ 27 < 40 Kg/m^2; to have a CVD risk score < 10% [10]; at least two atherogenic risk factors: TAG ≥ 150 mg/dL and <200 mg/dL, TC ≥ 200 mg/dL, HDL-C <40 mg/dL men or <50 mg/dL women, and/or LDL-C ≥130 mg/dL and <160 mg/dL, reflecting a risk for CVD [10]; having a suitable understanding of the clinical trial level; agreeing to voluntarily participate in the study; and signing the informed consent. Exclusion criteria were a diagnosis of diabetes mellitus, chronic degenerative diseases (e.g., liver or kidney), dyslipidemia, mental illness or diminished cognitive function, or the taking of antihypertension or lipid-lowering medication (e.g., statins, omega-3 supplements). Persons with lactose intolerance and dairy protein allergies were not enrolled. Pregnant or breastfeeding women were also excluded. The participants were individually allocated to one of the two study groups by randomization (Figure 1).

In addition, all groups followed the same balanced hypocaloric diet. All subjects gave their informed consent to take part in the study, which was approved by The Scientific Research and Ethics Committee of the Hospital Universitario La Paz (HULP 4092) and conformed to the ethical standards of the Declaration of Helsinki [11]; authorization for the disclosure of protected health information was obtained from all subjects before protocol-specific procedures. The participants were individually allocated to one of the two study groups, generated by a randomization procedure provided by the Biostatistics Unit of La Paz University Hospital. The allocation ratio of the study groups was 1:1.

2.2. Study Design

The controlled, randomized, double blind, parallel dietary intervention trial consisted of a 12-week investigation period (84 days). The control group (CG) received 60 g/day of a commercial goat cheese, and the experimental group (EG) received 60 g/day of the goat cheese that was naturally enriched with n-3 PUFA and CLA. Both control and enriched cheeses were produced as described by Santurino et al. (2017) [12]. Immediately after manufacture, the control and enriched cheeses were vacuum packed, refrigerated, and marked to maintain the conditions of blinding. Thus, neither the participants nor the researchers knew to which group the members belonged until the end of the study.

2.3. Dietetic, Physical Activity and Comorbidities' Data

Balanced hypocaloric and personalized diets were individually prescribed for all participants. An energy restriction of approximately 400 kcal/day was prescribed depending on gender, age, BMI, nutritional habits, physical activity, comorbidities and previous dietary treatments. Dietary intake was recorded using a food frequency questionnaire and a "3-day food and drink record" validated for the Spanish population [13] for computing energy, fat and protein intake. Two weekdays and one weekend day were included in the dietary record to take any differences in nutrient intake during weekdays and weekends into account. This was achieved by guidance from our dietitian. Subjects attended the department to collect the test food and for follow-up every three weeks throughout the intervention period. A questionnaire was fulfilled to collect the current use of medications and supplements, and the presence of relevant previous diseases and a physical activity metabolic equivalent of task (MET) score was determined based on self-reported energy-consuming activities during work, at home, while travelling, and at leisure time based on "Global Recommendations on Physical Activity for Health" by the WHO.

2.4. Anthropometric Variables

Blood pressure and heart rate were measured three times at 5-min intervals on the right arm using a Welch automatic monitor (Allyn Spot Vital Signs 420 series, Amsterdam, The Netherlands) (accuracy ±5 mmHg). The measurements were taken with subjects sitting, and the means were calculated. Dual-energy X-ray absorptiometry (DXA) was used to measure the total fat mass (TFM (%)), bone mineral density (BMD (g/cm^2)), android fat (AF (%)), gynoid (GF), and the lean mass (LM (%)), employing a GE Lunar Prodigy apparatus (GE Healthcare, Madison, WI, USA). Finally, anthropometric measurements as subject composition (TANITA BC-420MA, Biológica Tecnología Médica S.L. Barcelona, Spain), BMI, and waist and hip circumference were measured and recorded while adhering to international norms set out by the WHO.

2.5. Blood Collection

Blood samples were taken at baseline and at the end of the study period after a 12 h overnight fast at the Extraction Unit of the Hospital Universitario La Paz (Madrid, Spain). Samples were collected early in a 5 ml vacutainer tube with EDTA, and they were centrifuged at 4 °C over 7 min at 3500 rpm. Finally, samples were kept at −40 °C until analysis. A biochemical serum lipid profile (TC, HDL and

LDL cholesterol, triglycerides, apolipoprotein A1, apolipoprotein B, and free fatty acids), and glucose determinations were performed by an enzymatic-spectrophotometric assay using an Olympus AU 5400 apparatus (Izasa, CA, USA). C-reactive protein (CRP) concentrations were determined using a BNII nephelometer (Siemens Healthcare Diagnostics GmbH, Eschborn, Germany). Tumor necrosis factor-α (TNF-α) and interleukin 6 (IL-6) were determined using a Luminex ®-100 (Luminex Corporation. Texas City, TX, USA) multianalyte profiling system with commercially available immunoassay panels. Total lipid peroxides in plasma were determined as an indicator of oxidative stress by using the thiobarbituric acid reactive substances (TBARS) method21. The results were expressed as µmol MDAeq/mL. Data were analyzed using the xPONENT v.3.1 software (Merck Millipore, Burlington, VT, USA) and were determined using specific protocols of La Paz University Hospital.

2.6. Compliance and Adverse Events

Compliance was measured at the end of each experimental period using a specific questionnaire, and a subject was considered compliant when he/she consumed the contents of ≥70% of the product. Adverse events were recorded during the experimental periods. An adverse event was defined as any unfavorable, unintended effect reported by a subject or observed by the investigator. All were recorded along with the symptoms involved (nausea, vomiting, diarrhea, halitosis, and/or constipation). No participants showed any signs of intolerance to the supplement of the study diets. Subjects were informed of their right to withdraw from the study at any time.

2.7. Statistical Analysis

The sample size of 30 subjects in each group was calculated to provide 90% power at a 5% level of significance by the power analysis (nQuery Advisor Release 2.0, Statistical Solutions, Boston, MA, USA) based on LDL-C as a target effect size. The primary outcomes of the study were the changes from baseline to week 12 in the TC, low density lipoprotein cholesterol, and high-density lipoprotein cholesterol. Changes in triglycerides, free fatty acids, lipoproteins apoA-1 and apoB, fasting glucose, fasting insulin, body mass index, waist circumference, and the percent of fat tissue and its distribution assessed by android-to-gynoid fat percent ratio, as well as the total visceral adipose tissue, inflammatory markers (CRP, IL-6, TNF-α, oxidized low-density lipoprotein (OxLDL), and fibrinogen), calcium, phosphate, vitamin D, ghrelin, and leptin were considered as secondary outcomes. Baseline features in the intervention and control group were compared by a t-test (continuous variables) or by a chi-squared test (categorical variables). Changes in the primary and secondary outcomes from baseline to week 12 were defined by the absolute difference of the value of a parameter in week 12 minus the value at baseline. The statistical analysis of not normally distributed parameters were assessed by a Mann–Whitney U non-parametric test. The 95% confidence intervals of the absolute difference of the mean changes between the intervention and control groups were calculated by adjusted bootstrap percentile method after a 1000-replication bootstrap. Statistical calculations were performed in R (R Core Team (2013), Vienna, Austria).

3. Results

3.1. Recruitment and Study Population

Eighty possible patients were screened for enrolment in this study, but only sixty-eight met the inclusion and exclusion criteria and were randomized. The participant flow diagram is shown in Figure 1. Nine participants did not finish the study due to personal reasons, refusal to participate further, or relocations. Thus, fifty-nine subjects finished the 12-wk intervention period (control group: 31 subjects; experimental group: 28 subjects); only their data were included in analysis.

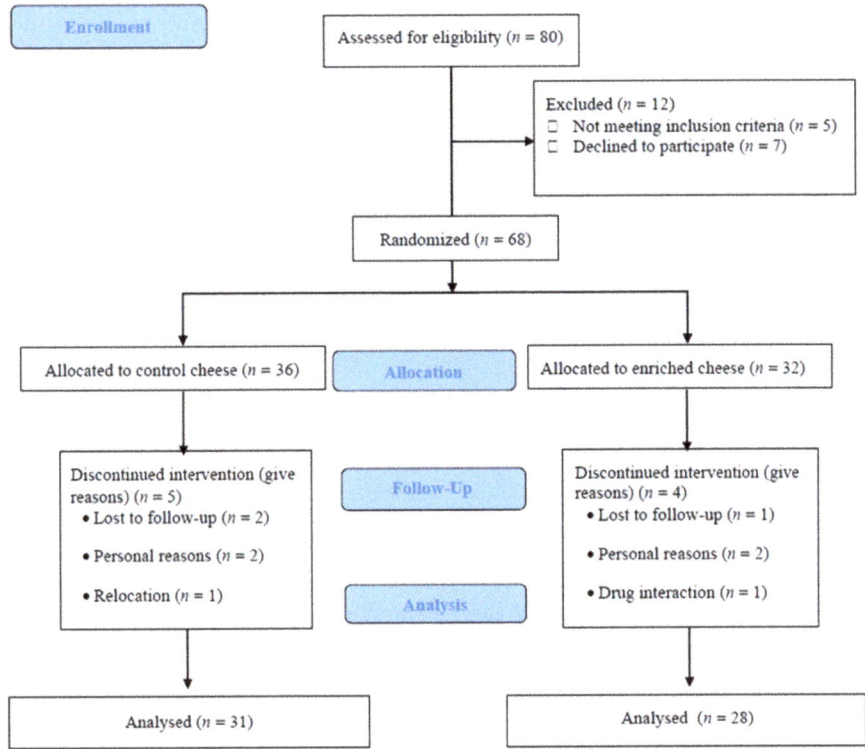

Figure 1. Flow chart describing the present trial.

3.2. Baseline Characteristics

The baseline characteristics of the fifty-nine subjects who completed the study were found to be comparable between the two groups are described in Table 1. The treatment compliance was high, and no differences were observed between groups (>85% of the scheduled doses consumed in the CG; >87% in the EG; $p < 0.374$) (Table 1).

Table 1. Baseline characteristics and anthropometric parameters of the study participants before and after the intervention.

Characteristic	Week 0		Week 12		Week 12–Week 0		p-Value
	CG	EG	CG	EG	CG	EG	
Age (years)	47.60 ± 9.40	48.50 ± 7.80	-	-	-	-	-
Men (n)	8	6	-	-	-	-	-
Women (n)	23	22	-	-	-	-	-
Weight (kg)	85.6 ± 11.30	86.80 ± 15.80	82.18 ± 11.77	83.13 ± 15.75	−3.41 ± 3.13	−3.66 ± 2.46	0.865
BMI (kg/m^2)	31.05 ± 3.30	30.74 ± 4.20	30.47 ± 3.69	30.54 ± 4.09	−1.12 ± 0.20	−0.93 ± 0.17	0.756
Waist circ. (cm)	105.0 ± 10.50	99.55 ± 10.60	96.29 ± 12.79	97.03 ± 11.09	−5.74 ± 6.63	−5.87 ± 3.18	0.889
BMD (g/cm^2)	1.17 ± 0.12	1.16 ± 0.13	1.17 ± 0.12	1.16 ± 0.14	−0.004 ± 0.01	−0.002 ± 0.14	0.767
Lean mass (%)	45.23 ± 7.77	46.15 ± 9.72	44.75 ± 8.00	45.72 ± 9.84	−0.49 ± 1.24	−0.43 ± 1.21	0.723
Android fat (%)	51.62 ± 7.19	50.72 ± 7.14	48.83 ± 9.29	48.92 ± 7.82	−2.80 ± 2.99	−1.80 ± 2.44	0.436
Gynoid fat (%)	47.24 ± 7.53	47.22 ± 8.24	45.64 ± 8.22	45.20 ± 8.14	−1.60 ± 1.99	−2.02 ± 1.90	0.356
Total fat mass (%)	46.94 ± 5.87	46.27 ± 6.29	44.55 ± 6.87	43.11 ± 6.67	−1.79 ± 1.80	−1.72 ± 1.68	0.645
Systolic BP (mm Hg)	110.3 ± 14.00	110.9 ± 13.10	110.7 ± 10.09	105.64 ± 22.05	0.45 ± 9.41	−5.21 ± 21.35	0.123
Diastolic BP (mm Hg)	77.50 ± 10.70	76.50 ± 9.50	76.90 ± 8.53	75.07 ± 8.38	−0.61 ± 7.99	−1.46 ± 7.79	0.385
HR (rate per minute)	75.30 ± 10.50	77.70 ± 10.50	68.10 ± 11.97	75.71 ± 13.70	−7.23 ± 9.39	−1.96 ± 10.97	0.259

Data are expressed as the means ± SDs. Abbreviations: BMI: body mass index; Waist circ.: waist circumference; BMD: bone mineral density; BP: blood pressure; and HR: heart rate.

3.3. Dietetic and Anthropometric Variables

In general, the diets followed by the volunteers showed a similar intake of macro- and micro-nutrients. No significant baseline differences in the basal diet were noted among the groups, except for the weekly rations of legumes ($p = 0.025$) and water ($p = 0.012$), which were higher in the EG compared to the CG. Instead, weekly rations of meat ($p = 0.021$) were higher in the CG. Regarding the low number of adverse events reported, no conclusion towards a relationship with a specific intervention could be drawn. There was no significant change in body weight in either treatment group, nor in the BMIs after the 12 weeks of study ($p > 0.05$) (Table 1). Additionally, there were no significant differences between treatments in all the parameters of the DXA analysis, which allowed us to obtain accurate values of the variation of body composition ($p > 0.05$) (Table 1). Regarding the waist circumference, no significant differences were found in the baseline GC and EG values ($p > 0.05$).

3.4. Blood Pressure and Biochemical Variables

On the other hand, at the end of the intervention period, both systolic and diastolic blood pressure remained within normal values for the general population (120/80 mmHg). Though systolic blood pressure decreased by -5.21 ± 21.35 mm Hg in the EG there were no significant differences among groups, possibly due to intragroup differences, nor were there any significant differences at baseline or after the intervention for 12 weeks ($p > 0.05$) (Table 1).

The subjects' blood lipids and apolipoproteins concentrations before and after intervention are shown in Table 2. At the end of the study, the level of TC increased significantly in the EG in comparison to the CG ($p > 0.05$), despite no significant difference from baseline observed in either group. Even though randomization, there was an imbalance between both groups in baseline HDL-C concentration ($p = 0.04$), with lower baseline HDL-C levels in the EG. However, this result was corrected, and a significant increase of HDL-C in favor of the EG occurred at the end of the intervention. The plasma levels of ApoA1 and ApoB remained within the reference values for the study population throughout the intervention period (Table 2). At the end of the intervention, no changes in ApoA1 levels (related to HDL-C, the most abundant apolipoprotein in plasma, which contributed to a good cardiovascular health [14]) were detected in any group. On the contrary, the plasma concentration of ApoB increased in the EG by the end of the intervention period (Table 2).

Additionally, the increase of TC in the EG could have been related to the significant increase of the HDL content in this group. Conversely, the consumption of cheese in both intervention groups did not significantly affect LDL-C values, leading to a significant improvement in the LDL/HDL ratio, a good lipid indicator of atherogenic risk along with the TC/HDL-C ratio (Table 3).

Table 2. Blood lipids and apolipoproteins concentrations before and after intervention.

(mg/dL)	CG			EG			Week 12−Week 0		
	Week 0	Week 12		Week 0	Week 12		CG	EG	CI 95%
TC	201.80 ± 40.41	197.16 ± 36.95		195.39 ± 37.38	201.46 ± 38.81		−2 (−90; 43)	6.5 (−24; 50)	10.72 (0.37; 22.74) #
HDL-C	54.97 ± 16.47	53.97 ± 11.08	*	47.89 ± 8.00	52.00 ± 9.24	*	2 (−48; 12)	4.5 (−5; 16)	5.11 (1.96; 11.46) #
LDL-C	128.97 ± 31.34	125.61 ± 29.29		127.32 ± 32.45	129.71 ± 31.59		3 (−11; 16)	2.5 (−9; 31)	1.39 (−2.15; 5.92)
TAG	89.00 ± 42.24	87.81 ± 47.65		100.71 ± 34.33	99.50 ± 31.59		6 (−65; 75)	1 (−56; 53)	−0.02 (−12.8; 14.62)
ApoA1	161.32 ± 26.62	153.94 ± 20.85		150.68 ± 16.30	152.14 ± 18.40		−1 (−28; 9.6)	0.3 (−9.8; 24)	2.74 (−0.3; 6.98)
ApoB	102.42 ± 21.45	101.65 ± 22.06		102.86 ± 23.33	108.00 ± 26.20		−2* (−19; 27)	3.5* (−20; 34)	5.92 (0.81; 11.77)
Glucose	93.06 ± 8.08	95.39 ± 8.49		90.82 ± 9.40	94.54 ± 7.54		2.32 (70.04; 94.06)	3.71 (67.05; 95.04)	0.05 (−0.1; 0.19)

Data are expressed as the means ± SDs. Statistical analysis was assessed by Mann–Whitney U test. * Significant difference between groups at the end of the intervention period (week 0–week 12) $p < 0.05$. # Significant difference between groups at week 0 $p < 0.05$. Abbreviations: TC: total cholesterol; HDL-C high-density lipoprotein cholesterol; LDL-C low-density lipoprotein cholesterol; and TAG triglycerides.

Table 3. Values of total cholesterol/high-density lipoprotein (TC/HDL), low-density lipoprotein (LDL)/HDL and apolipoprotein B/apolipoprotein A1 (ApoB/ApoA1) ratios before and after the intervention.

Ratio	CG (n = 31)		EG (n = 28)	
	Week 0	Week 12	Week 0	Week 12
TC/HDL-C	3.67 ± 2.5	3.67 ± 3.3	4.06 ± 4.6	3.87 ± 4.6
LDL-C/HDL-C	1.71 ± 1.9	1.72 ± 2.6	1.91 ± 4.0	1.79 ± 3.4 *
ApoB/ApoA1	0.63 ± 0.8	0.66 ± 1.1	0.68 ± 1.43	0.71 ± 1.42

Data are presented as mean ± s.d. * Significant difference between groups before and after the intervention $p < 0.05$. Abbreviations: TC: total cholesterol; HDL-C high-density lipoprotein cholesterol; LDL-C low-density lipoprotein cholesterol; and TAG: triglycerides.

3.5. Inflammation Variables

Systemic inflammation (TNF-α, IL-6, CRP, and others) is described in Table 4. At baseline, there was no difference between groups in any of these characteristics. After intervention, there was a significant decrease in CRP in the EG by 36%, taking into account the intragroup variation.

Plasma calcium and phosphorous remained within the ranges of normality described throughout the intervention period (Table 5). At baseline, there was no difference between groups in the plasma levels of vitamin D, but there were lower than the reference range for the study population (20–40 ng/mL) in both the CG and the EG, maybe linked to overweight and obesity status [15]. In contrast, at the end of the intervention period, vitamin D plasma levels were found within values considered in the reference range for the study population [16]. There was also no significant change in ghrelin and leptin after the intervention in either group.

Table 4. Inflammatory biomarker concentration before and after the intervention.

	CG		EG		Week 12–Week 0		CI 95%
	Week 0	Week 12	Week 0	Week 12	CG	EG	
CRP (mg/L)	2.67 ± 4.36	2.98 ± 7.62	2.95 ± 6.06	1.02 ± 6.33	0.03 (−18.12; 32.1)	−0.76 (−12.34; 3.38)	−10.72 (0.37; 22.74) #
TNF-α (pg/mL)	3.88 ± 1.45	4.47 ± 1.43	4.51 ± 1.57	4.85 ± 1.71	0.5 (−0.8; 18.3)	0.75 (−1.2; 2.5)	−0.46 (−246; 0.28)
IL-6 (pg/mL)	2.60 ± 2.37	3.70 ± 2.25	2.70 ± 1.53	3.65 ± 1.60	1.2 (−1.7; 3.2)	0.95 (−2; 5.2)	−0.12 (−0.82; 0.6)
Fibrinogen (mg/dL)	397 ± 102.82	413 ± 97.33	419 ± 124.14	346 ± 129.64	−13 (−333;425)	−17 (−202; 106)	−26.88 (−81.27; 22.96)
OxLDL (ng/mL)	72.71 ± 238.10	80.53 ± 245.14	135.09 ± 229.32	113.47 ± 218.01	0.01 (−164.2; 654.9)	−0.31 (−250.9; 90.4)	−13.3 (−87.2; 16.9)
FFA (mM)	0.28 ± 0.17	0.25 ± 0.08	0.28 ± 0.11	0.23 ± 0.10	−0.01 (−0.47; 0.21)	−0.06 (−0.41; 0.2)	−0.03 (−0.09; 0.04)

Data are expressed as the means ± SDs. Statistical analysis was assessed by Mann-Whitney U test. Significant difference between groups at week 0 $p < 0.05$. # Significant difference between groups at the end of the intervention period (week 0–week 12) $p < 0.05$. Abbreviations: CRP: high-sensitivity C reactive protein; TNF-α: tumor necrosis alpha factor; IL-6: interleukin 6; oxLDL: oxidized LDL; and FFA: free fatty acids.

Table 5. Mineral and hormone concentrations before and after the intervention.

	CG		EG		Week12–Week 0		CI 95%
	Week 0	Week 12	Week 0	Week 12	CG	EG	
Vitamin D (ng/mL)	14.26 ± 6.15	21.89 ± 6.94	13.95 ± 6.06	22.5 ± 6.33	7 (1; 23)	8 (0; 17)	0.39 (−2.2; 2.26)
Calcium (mg/dL)	9.16 ± 0.38	9.24 ± 0.30	9.16 ± 0.32	9.30 ± 0.35	0.1 (−0.6; 0.7)	0.15 (−0.3; 0.9)	0.05 (−0.1; 0.19)
Phosphorous (mg/dL)	3.35 ± 0.39	3.41 ± 0.52	3.46 ± 0.59	3.32 ± 0.47	0.1 (−0.8; 0.8)	−0.1 (−1.4; 0.8)	−0.19 (−0.41; 0.02)
Ghrelin (pg/mL)	9.9 ± 13.75	10 ± 14.42	8.45 ± 19.62	8.10 ± 20.69	0.6 (−35.7; 40.3)	0.05 (−41.8; 39.2)	−0.37 (−7.72; 6.78)
Leptin (pg/mL)	14.14 ± 10.74	11.01 ± 10.00	14.80 ± 10.80	11.15 ± 10.50	−1.8 (−23.03; 13.1)	−2.7 (−9.5; 2.5)	−0.12 (−0.82; 0.6)

Data are presented as mean ± SDs.

4. Discussion

This study was designed to evaluate the combined effect of a enriched cheese with a balanced hypocaloric diet and physical activity in overweight and obese subjects on cardiovascular risk factors. This diet enriched with n-3 PUFA and CLA did not significantly modify the body composition of either group. The lack of significant differences is in agreement with recent clinical trials in which the consumption of cheese naturally enriched with PUFA did not significantly modify the body composition of both healthy volunteers and subjects with altered lipid profiles [4,17]. Though the difference was not significant between both groups ($p > 0.05$), the reduction of the waist circumference in both intervention groups could be related to the good efficacy of the nutritional intervention and the guidelines for physical activity carried out in both intervention groups, thus decreasing the metabolic risk in relation to waist circumference specified by the WHO (>88 cm in women and >102 cm in men) (Table 1) [18]. These results are in accordance with those obtained in a recent clinical trial in which cheese consumption did not significantly modify anthropometric parameters related to metabolic risk among the different study groups [19]. Lastly, the slight non-significant decrease in heart rate observed at the end of the study in both groups may have been a consequence of weight loss, thus improving an important cardiovascular risk (CVR) factor [20] (Table 1). These results are in accordance with those obtained in a recent clinical trial in which a similar intervention period of 12 weeks has been previously shown to induce significant weight loss [21].

The DXA analysis revealed a baseline value of TFM that exceeded the typical values in overweight people (BMI 25–30 kg/m^2) in both intervention groups. Though no significant differences were obtained between treatments, probably due to the low-calorie diet received by all volunteers ($p > 0.05$), at the end of the intervention, both groups had a slightly decreased TFM, AF and GF (Table 1). Recent studies have shown that weight loss and/or muscle mass could lead to a loss of BMD, thus highlighting the importance of a good dietary strategy in the management of overweight and obesity and avoiding the loss of muscle mass by performing regular physical exercise [22]. Consequently, the results showed the good follow-up of the recommendations for the daily performance of physical activity by all volunteers.

Regarding the increase in TC (the sum of HDL-C and LDL-C) in the EG, TC provides limited information about cardiovascular risk, and it is not useful for diagnosing metabolic syndrome [23] because it cannot be associated with a circulating increase in atherogenic lipoprotein concentration [24]. However, the significant increase in HDL-C in the EG after the intervention period was in line with another clinical trial in which hypercholesterolemic volunteers consumed PUFA-enriched yogurt for 10 weeks [25]. Furthermore, de Goede et al. (2015) in a recent review, concluded that the cheese intake as compared to butter might have beneficial effects on certain plasma lipids that are directly related to the antiatherogenic properties of CLA. Though the differences in the LDL-C/HDL-C ratio between the CG and the EG were not statistically significant ($p > 0.05$), the results showed a slight decrease in the atherogenic risk only in the EG at the end of the intervention. These results were in line with a recent study that evaluated the effect of the consumption of a LC n-3-PUFA-enriched cheese on the lipid profile in hypercholesterolemic adults [26]. In the study, no significant differences were found between both groups on plasma TAG (Table 2). This approach is in line with a recent review and meta-analysis where de Goede et al. (2015) concluded that cheese consumption has no effect on TAG levels in humans, and this effect could also be dependent on the intervention time [27]. Furthermore, a recent large review of nine RCT suggests that CLA did not significantly affect TC, TAG, or LDL-C contents [28]. On the contrary, Carrero et al. (2007) [29] supplemented hyperlipidemic volunteers with a milk product containing EPA plus DHA, and they observed a significant reduction in TAG and TC after eight weeks. In terms of increasing the plasma ApoB concentration in the EG, previous studies have reported significant changes in plasma ApoB concentration or even increases when prescribing therapies with LC n-3-PUFA, underscoring the importance of treatment duration to attain consistent results [30]. However, ApoB plasma levels in both groups were in the reference range of Apo B levels in adults, and there were no significant changes in plasma LDL-C levels, so this increase in ApoB levels cannot be considered an increase of cardiovascular risk [31]. The ApoB/ApoA1 ratio reflects the balance between two processes: the transport of cholesterol to peripheral tissues and the reverse transport to the liver. Due to the results obtained for the ApoA1 and ApoB values in both intervention groups, significant differences between the CG and the EG in the ApoB/ApoA1 ratio between before and after the intervention were not observed, but this ratio slightly increased at the end of the clinical trial in both groups.

Weight loss is associated with reduced levels of pro-inflammatory cytokines responsible for inflammation, such as TNF-α and IL-6 [32]. Though recent studies have shown an increase in TNF-α and IL-6 levels due to the state of inflammation related to overweight and obesity [15], the consumption of 60 g/day of cheese, following a balanced and hypocaloric diet, kept these levels stable at the end of the intervention period (Table 4). These results were in line with those reported by Dawczynski et al. (2013) [7], where the consumption of PUFA-enriched yogurt did not significantly modify the values of inflammation markers studied in overweight and obese volunteers. Similarly, in our study at the end of 12 weeks of supplementation, there was no significant effect in plasma OxLDL, fibrinogen, and FFA compared with the control group, probably due to the wide variability among the volunteers in each group. These results were in line with those obtained by Joseph et al. (2011) [5] in an eight-week crossover clinical trial in which dietary supplementation with CLA-enriched oil did not modify plasma OxLDL values in overweight and hyperlipidemic subjects. Additionally, the plasma CRP concentration increases its levels in response to generalized inflammation, as in the case of overweight and obese individuals [33]. At the end of the intervention, the plasma CRP concentration increased by 37% in the CG, whereas this value, taking into account the intragroup variation, significantly decreased in the EG by 36%. This significant decrease in the plasma CRP concentration only in the EG did not coincide with previous studies in which the consumption of enriched dairy products in FA n-3 did not significantly modify the plasma levels of pro-inflammatory cytokines [33]. The lack of significant changes in levels of CRP has been also attributed to the duration of the intervention. Additionally, the reduction in CRP levels could be directly related to weight loss, decreasing 0.13 mg/dL CRP per kg of weight lost [33]. The consumption of FA n-3 increases their concentration in blood, cells, and tissues, and it alters the physical properties of cell membranes and the function of membrane proteins. FA n-3 is incorporated into cell membranes in competition with n-6 FA and AA. Considering that the replacement of n-6 FA with n-3 FA in membranes of the immune active cells may induces leucocytes to produce pro-inflammatory processes and lead to the reduced formation of pro-inflammatory compounds, the significant changes of plasma CRP levels in overweight and obese subjects only in the EG could have been due to the synergistic effect among the anti-inflammatory effect of the consumption of dietary FA n-3 and CLA, weight loss, and the consumption of a balanced diet, together with regular physical activity.

Calcium and phosphorus interact in numerous processes of the organism. Blood calcium values considered normal for a studied population are usually between 8 and 10.5 mg/dL, as well as between 2.4 and 4.5 mg/dL for phosphorus. For the population under study, the normal blood calcium and phosphorus values ranged between 8 and 10.5 mg/dL and between 2.4 and 4.5 mg/dL, respectively. In regards to vitamin D (which regulates mineral homeostasis, protects the integrity of the skeleton, and modulates cell growth and differentiation in a wide variety of tissues [34]), although there were no significant differences between both groups, the baseline levels of vitamin D in both the GC and the EG were lower than those considered normal for the study population (20–40 ng/mL), and this may be linked to overweight and obesity status [35]. On the other hand, hormonal regulators of satiety, such as ghrelin and leptin, are also related to body weight. Though no significant changes were seen in terms of intervention time and treatment group at the end of the clinical trial, an intragroup analysis revealed a slight decrease in leptin levels in both the GC and the EG. These results were expected after the hypocaloric diet and the consequent weight loss in both groups, as well as changes in the blood lipid profile [36].

5. Conclusions

Overall, the consumption of 60 g/day of cheese (both control and enriched), within the context of a balanced hypocaloric diet and recommendations for physical activity, was effective for the reduction of body weight, BMI and waist circumference in both the CG and the EG. Additionally, the healthy habits carried out by all subjects resulted in a slight decrease in heart rate, as well as maintenance of the BMD, resulting in a decrease in CVR.

On the other hand, the significant increase of HDL and the significant decrease in blood levels of CRP in the EG improved the plasma lipid profile and the inflammatory status, thus producing a decrease in the atherogenic risk. Therefore, the consumption of this PUFA n-3 and CLA naturally enriched goat cheese could have a potential role as a high nutritional value food to improve the state of health.

Author Contributions: Conceptualization, B.L.-P. and C.G.-C.; Methodology, L.M.B.; Formal analysis D.G.-A. and B.L.P.; Investigation, B.L.-P., L.M.B., C.G.-C., and C.S.; Resources, M.V.C.; Data curation, C.S.; Writing—Original draft preparation, C.S. and J.F.; Writing—Review and editing, J.F., B.L.-P., C.G.-C., and C.S.; Visualization, J.F.; Supervision, C.G.-C. and M.V.C.; Project administration, B.L.-P., J.F., and C.G.-C. All authors have read and agreed to the published version of the manuscript.

Funding: This study was supported by the LODYN S.L. group through the Centre for the Development of Industrial Technology CDTI of Economy and Competitiveness Ministry of Spain (MINECO).

Acknowledgments: We thank LODYN S.L. for the manufacture and supply of the experimental and control cheeses.

Conflicts of Interest: The authors declare no conflicts of interest.

References

1. Di Nicolantonio, J.; Lucan, S.C. The evidence of saturated fat and for sugar related to coronary heart disease. *Prog. Cardiovasc. Dis.* **2016**, *58*, 464–472. [CrossRef] [PubMed]
2. De Goede, J.; Geleijnse, J.; Ding, E.; Soedamah-Muthu, S. Effect of cheese consumption on blood lipids: A systematic review and meta-analysis of randomized controlled trials. *Nutr. Rev.* **2015**, *73*, 259–275. [CrossRef] [PubMed]
3. Dias, C.; Wood, L.; Garg, M. Effects of dietary saturated and n-6 polyunsaturated fatty acids on the incorporation of long-chain n-3 polyunsaturated fatty acids into blood lipids. *Eur. J. Clin. Nutr.* **2016**, *70*, 812–818. [CrossRef] [PubMed]
4. Klok, M.; Jakobsdottir, S.; Drent, M. The role of leptin and ghrelin in the regulation of food intake and body weight in humans: A review. *Obes. Rev.* **2007**, *8*, 21–34. [CrossRef]
5. Joseph, S.; Jacques, H.; Plourde, M.; Mitchell, P.; McLeod, R.; Jones, P. Conjugated linoleic acid supplementation for 8 weeks does not affect body composition, lipid profile, or safety biomarkers in overweight, hyperlipidemic men. *J. Nutr.* **2011**, *141*, 1286–1291. [CrossRef]
6. Fontecha, J.; Calvo, M.V.; Juarez, M.; Gil, A.; Martínez-Vizcaino, V. Milk and dairy product consumption and cardiovascular diseases: An overview of systematic reviews and meta-analyses. *Adv. Nutr.* **2019**, *10*, S164–S189. [CrossRef]
7. Dawczynski, C.; Massey, K.; Ness, C.; Kiehntopf, M.; Stepanow, S.; Platzer, M.; Grun, M.; Nicolau, A.; Jahreis, G. Randomized placebo-controlled intervention with n-3 LC-PUFA-supplemented yoghurt: Effects on circulating eicosanoids and cardiovascular risk factors. *Clin. Nutr.* **2013**, *32*, 686–696. [CrossRef]
8. De Oliveira Otto, M.; Nettleton, J.; Lemaitre, R.M.; Steffen, L.; Kromhout, D.; Rich, S.; Tsay, M.Y.; Jacobs, D.R.; Mozaffarian, D. Biomarkers of dairy fatty acids and risk of cardiovascular disease in the multi-ethnic study of atherosclerosis. *J. Am. Heart Assoc.* **2013**, *2*, e000092. [CrossRef]
9. Dittrich, M.; Jahreis, G.; Bothor, K.; Drechsel, C.; Kiehntopf, M.; Blüher, M. Benefits of foods supplemented with vegetable oils rich in α-linolenic, stearidonic or docosahexaenoic acid in hypertriglyceridemic subjects: A double-blind, randomized, controlled trail. *Eur. J. Nutr.* **2014**, *54*, 881–893. [CrossRef]
10. Houston, D.; Driver, K.; Bush, A.; Kritchevsky, S. The association between cheese consumption and cardiovascular risk factors among adults. *J. Hum. Nutr. Diet.* **2008**, *21*, 129–140. [CrossRef]
11. World Medical Association (WMA). Declaration of Helsinki. Ethical principles for medical research involving human subjects. *JAMA* **2013**, *310*, 2191–2194. [CrossRef] [PubMed]

12. Santurino, C.; Calvo, M.; Gómez-Candela, C.; Fontecha, J. Characterization of naturally goat cheese enriched in conjugated linoleic acid and omega-3 fatty acids for human clinical trial in overweight and obese subjects. *PharmaNutrition* **2017**, *5*, 8–17. [CrossRef]
13. Khandelwal, S.; Demonty, I.; Jeemon, P.; Lakshmy, R.; Mukherjee, R.; Gupta, R.; Snehi, U.; Niveditha, D.; Singh, Y.; van der Knaap, H.C.; et al. Independent and interactive effects of plant sterols and fish oil n-3 long-chain polyunsaturated fatty acids on the plasma lipid profile of mildly hyperlipidaemic Indian adults. *Br. J. Nutr.* **2009**, *102*, 722. [CrossRef] [PubMed]
14. Intorre, F.; Foddai, M.; Azzini, E.; Martin, B.; Montel, M.; Catasta, G. Differential effect of cheese fatty acid composition on blood lipid profile and redox status in normolipidemic volunteers: A pilot study. *Int. J. Food Sci. Nutr.* **2011**, *62*, 660–669. [CrossRef]
15. Stone, N.; Robinson, J.; Lichtenstein, A.; Bairey Merz, C.; Blum, C.; Eckel, R. 2013 ACC/AHA guideline on the treatment of blood cholesterol to reduce atherosclerotic cardiovascular risk in adults. *Circulation* **2013**, *129*, S1–S45. [CrossRef]
16. Kim, B.; Lim, H.; Lee, H.; Lee, H.; Kang, W.; Kim, E. The effects of conjugated linoleic acid (CLA) on metabolic syndrome patients: A systematic review and meta-analysis. *J. Funct. Foods* **2016**, *25*, 588–598. [CrossRef]
17. Klop, B.; Proctor, S.; Mamo, J.; Botham, K.; Castro Cabezas, M. Understanding postprandial inflammation and its relationship to lifestyle behaviour and metabolic diseases. *Int. J. Vasc. Med.* **2012**, *2012*, 947417. [CrossRef]
18. López Gómez, J.; Pérez Castrillón, J.; Romero Bobillo, E.; De Luis Román, D. Efecto del tratamiento dietoterápico de la obesidad sobre el metabolismo óseo. *Nutr. Hosp.* **2016**, *33*, 1452–1460.
19. Marrugat, J.; Solanas, P.; D'Agostino, R.; Sullivan, L.; Ordovas, J.; Cordón, F.; Ramos, R.; Sala, J. Estimación del riesgo coronario en España mediante la ecuación de Framingham calibrada. *Revista Esp. Cardiol.* **2003**, *56*, 253–261. [CrossRef]
20. Navarro-Alarcón, M.; Cabrera-Vique, C.; Ruiz-López, M.; Olalla, M.; Artacho, R.; Giménez, R.; Quintana, V.; Bergillos, T. Levels of Se, Zn, Mg and Ca in commercial goat and cow milk fermented products: Relationship with their chemical composition and probiotic starter culture. *Food Chem.* **2011**, *129*, 1126–1131. [CrossRef] [PubMed]
21. Intorre, F.; Venneria, E.; Finotti, E.; Foddai, M.; Toti, E.; Catasta, G. Fatty acid content of serum lipid fractions and blood lipids in normolipidaemic volunteers fed two types of cheese having different fat compositions: A pilot study. *Int. J. Food Sci. Nutr.* **2012**, *64*, 185–193. [CrossRef] [PubMed]
22. Nestel, P.; Mellett, N.; Pally, S.; Wong, G.; Barlow, C.; Croft, K.; Mori, T.A.; Meikle, P.K. Effects of low-fat or full-fat fermented and non-fermented dairy foods on selected cardiovascular biomarkers in overweight adults. *Br. J. Nutr.* **2013**, *110*, 2242–2249. [CrossRef] [PubMed]
23. Pannu, P.; Calton, E.; Soares, M. Calcium and vitamin d in obesity and related chronic disease. In *Advances in Food and Nutrition Research*; Academic Press: Cambridge, MA, USA, 2016; pp. 57–100.
24. Pariza, M. Perspective on the safety and effectiveness of conjugated linoleic acid. *Am. J. Clin. Nutr.* **2004**, *79*, 1132S–1136S. [CrossRef] [PubMed]
25. Nilsen, R.; Høstmark, A.; Haug, A.; Skeie, S. Effect of a high intake of cheese on cholesterol and metabolic syndrome: Results of a randomized trial. *Food Nutr. Res.* **2015**, *59*, 27–51. [CrossRef]
26. Obregón, O.; Gestne, A.; Lares, M.; Castro, J.; Stulin, I.; Rivas, K. Estatinas y factor de necrosis tumoral alfa. *Revista Latinoam. Hipertens.* **2010**, *5*, 6–10.
27. Piepoli, M.; Hoes, A.; Agewall, S.; Albus, C.; Brotons, C.; Catapano, A.; Cooney, M.T.; Corrà, U.; Bernard Cosyns, C.; Deaton, C.; et al. European guidelines on cardiovascular disease prevention in clinical practice (version 2016). *Eur. Heart J.* **2016**, *37*, 2315–2381. [CrossRef]
28. Pranger, I.; Muskiet, F.; Kema, I.; Singh-Povel, C.; Bakker, S.K. Potential biomarkers for fat from dairy and fish and their association with cardiovascular risk factors: Cross-sectional data from the lifelines biobank and cohort study. *Nutrients* **2019**, *11*, 1099. [CrossRef]
29. Salas-Salvadó, J.; Rubio, M.; Barbany, M.; Moreno, B. Consenso SEEDO 2007 para la evaluación del sobrepeso y la obesidad y el establecimiento de criterios de intervención terapéutica. *Med. Clín.* **2007**, *128*, 184–196.
30. Sánchez, F.; Albo Castaño, M.; Casallo Blanco, S.; Vizuete Calero, A.; Matías Salces, L. Importancia de las apoproteínas A1 y B como marcadores de riesgo cardiovascular. *An. Med. Interna* **2008**, *25*, 199–200. [CrossRef]

31. Shaikh, N.; Yantha, J.; Shaikh, S.; Rowe, W.; Laidlaw, M.; Cockerline, C.; Ali, A.; Holub, B.; Jackowski, G. Efficacy of a unique omega-3 formulation on the correction of nutritional deficiency and its effects on cardiovascular disease risk factors in a randomized controlled VASCAZEN®REVEAL Trial. *Mol. Cell. Biochem.* **2014**, *396*, 9–22. [CrossRef]
32. Stone, N.; Robinson, J.; Lichtenstein, A.; Goff, D.; Lloyd-Jones, D.; Smith, S.; Blum, C.; Schwartz, J.S. Treatment of blood cholesterol to reduce atherosclerotic cardiovascular disease risk in adults: Synopsis of the 2013 americcollege of cardiology/American heart association cholesterol guideline. *Ann. Int. Med.* **2014**, *160*, 339–343. [CrossRef] [PubMed]
33. Aguillón, G.J.; Cruzat, C.A.; Cuenca, M.J.; Cuchacovich, T.M. El polimorfismo genético del factor de necrosis tumoral alfa como factor de riesgo en patología. *Revista Méd Chile* **2012**, *130*, 1043–1050.
34. Warensjo, E.; Jansson, J.; Cederholm, T.; Boman, K.; Eliasson, M.; Hallmans, G.; Johansson, I.; Sjogren, P. Biomarkers of milk fat and the risk of myocardial infarction in men and women: A prospective, matched case-control study. *Am. J. Clin. Nut.* **2010**, *92*, 194–202. [CrossRef] [PubMed]
35. Aymé, S.; Rath, A.; Bellet, B. WHO International Classification of Diseases (ICD) Revision Process: Incorporating rare diseases into the classification scheme: State of art. *Orphanet J. Rare Dis.* **2018**, *5*, P1.
36. Castro, I.; Monteiro, V.; Barroso, L.; Bertolami, M. Effect of eicosapentaenoic/docosahexaenoic fatty acids and soluble fibers on blood lipids of individuals classified into different levels of lipidemia. *Nutrition* **2007**, *23*, 127–137. [CrossRef]

© 2020 by the authors. Licensee MDPI, Basel, Switzerland. This article is an open access article distributed under the terms and conditions of the Creative Commons Attribution (CC BY) license (http://creativecommons.org/licenses/by/4.0/).

Article

Dairy Products Quality from a Consumer Point of View: Study among Polish Adults

Marta Sajdakowska *, Jerzy Gębski, Dominika Guzek, Krystyna Gutkowska and Sylwia Żakowska-Biemans

Department of Food Market and Consumer Research, Institute of Human Nutrition Sciences, Warsaw University of Life Sciences (SGGW-WULS), 159C Nowoursynowska Street, 02-787 Warsaw, Poland; jerzy_gebski@sggw.edu.pl (J.G.); dominika_guzek@sggw.edu.pl (D.G.); krystyna_gutkowska@sggw.edu.pl (K.G.); sylwia_zakowska_biemans@sggw.edu.pl (S.Ż.-B.)
* Correspondence: marta_sajdakowska@sggw.edu.pl; Tel.: +48-225-937-145

Received: 10 April 2020; Accepted: 19 May 2020; Published: 21 May 2020

Abstract: The aims of the current study were (a) to deepen the understanding of food quality from animal origin with particular emphasis on dairy products, including yoghurt; (b) to determine the level of acceptance of methods and ingredients used to enhance the quality of food from animal origin; (c) to identify how the perception of animal products quality affects the acceptance of changes in production methods and (d) to identify the projective image of consumers purchasing high-quality yoghurt. The data were collected using a CAPI (Computer Assisted Personal Interview) survey on a sample of 983 consumers. The k-means clustering method (k-means clustering algorithm is an unsupervised algorithm that is used to segment the interest area from the background) was used to identify five clusters of consumers. Moreover, the logistic regression models were used in order to examine the impact of opinions related to the quality of product on acceptance of food production methods. The results showed that food quality is generally perceived by consumers using the following attributes: its freshness, naturalness, production method, as well as appearance, taste and smell, but when it comes to the quality of food from animal origin, convenience, connected with the availability, nutritional value and health benefits is of primary importance. The most accepted production method of high-quality food is animal production that takes into consideration the welfare of farm animals. Results also show that the increase in the level of education among the surveyed people contributed to the acceptance of ensuring welfare of farm animals as a method of increasing food quality while consumers' openness to new products favored the acceptance of adding health-promoting ingredients to livestock feed. As regards the assessment of the level of acceptance of enhancing food with beneficial ingredients, people for whom health aspects were important declared their willingness to accept such a method of increasing food quality. The research findings can be used to develop educational campaigns as well as marketing communication of enterprises operating on the food market. Furthermore, the results could be used to strengthen the competitive position of food enterprises searching for innovative solutions.

Keywords: consumer; quality; animal-derived food; yoghurt

1. Introduction

Consumers take various factors into consideration when choosing food; they include taste and freshness as well as naturalness [1,2]. In addition to taste, smell, freshness and naturalness of the product, the following factors also affect consumer choices: the method of food production and processing, ensuring welfare of farm animals and maintaining the health values of food, especially for consumers for whom it is most crucial [3,4]. Moreover, results of studies also indicated that relative advantage, naturalness, novelty and discomfort are the most important factors of the perception of some innovative food products [5]. The literature showed that from a consumer perspective, among other factors, food selection factors are seen as food quality [6–8]. This also applies to dairy products [6,7]. The results of studies confirm that consumers take into account a number of attributes associated with quality, so they expect the product to be safe, natural, healthy and generally of high quality [9]. Furthermore, some consumers underlined the role of quality signs, particularly in the field of positioning origin and organic products in the segment of premium prices, emphasizing the authenticity of these products [10].

Moreover, when it comes to functional food, consumer acceptance was also analyzed from the perspective of consumer quality perception of food products. Functional foods provide, from the consumers' perspective, synergies between healthiness and convenience but may, in the consumers' opinion, lead to trade-offs between healthiness on the one hand and taste and naturalness on the other hand [11].

The results of studies showed that acceptance of functional dairy products increases among consumers with higher diet/health-related knowledge, as well as with ageing. General interest in health, food-neophobia and perceived self-efficacy seem also to contribute to shaping the acceptance of functional dairy products [12]. Furthermore, products with "natural" matches between carriers and ingredients have the highest level of acceptance among consumers [12,13]. A review by Kaur and Singh indicated that a high level of education and high income greatly influence consumer uptake of functional food, as well as an increased personal health consciousness [14]. Furthermore, results of the research indicate that health benefits and ingredient naturalness are positively valued, but such preferences and valuations depend on an individual's education, income and food purchase behaviors; thus, naturally occurring nutrients are preferred over fortification [15].

As earlier mentioned, taking into account food safety and food naturalness, the method of food production is also a crucial point. When it comes to the method of food production, including animal welfare-friendly methods, the results of studies among European consumers indicate that public perceptions of farm-animal welfare represent a potentially important driver of consumption behaviors by European consumers [16]. However, some Europeans currently do not think there is sufficient choice of welfare-friendly animal food products in shops and supermarkets [17]. In addition, there is an increasing need to develop policies pertaining to animal production diseases, sustainable intensification and animal welfare, which incorporate consumer priorities as well as technical assessments of farm animal welfare. Consumers may have concerns about intensive production systems and whether animal production disease pose a barrier to consumer acceptance of their increased use [18].

With reference to yoghurt, its nutritional content varies depending on the processing method and the ingredients used. Similar to milk, it is a good source of protein and calcium and may be a source of iodine, potassium and B vitamins [19]. Moreover, some dairy products are fortified with vitamin D [20]. Furthermore, during the past years, interest in yoghurt manufacture has increased for scientific and commercial reasons [21]. Additionally, the functional food market has experienced a tremendous level of growth particularly in yoghurt in the last couple of decades, due to the ease of incorporating pre- and probiotics [22].

Yoghurt still plays an important role in the human diet today due to its pleasant taste and health benefits [22,23]. Moreover, yoghurt is the most-frequently consumed healthy and nutritious food around the world. Therefore, it offers an appropriate potential to provide nutritious ingredients to human diet [24]. Furthermore, the results of research indicate that with respect to the safety and health effects of food products, the probiotic yoghurt is recommended for consumption [25]. Considering the fast evolution of functional yoghurts either at research stage or marketplace, further development would require an accurate measure of quality, safety and efficacy to meet consumers' expectations on quality and claimable health benefits [26].

Therefore, the aims of the current study were (a) to deepen the understanding of the quality of food from animal origin with particular emphasis on dairy products, including yoghurt; (b) to determine the level of acceptance of methods and ingredients used to enhance the quality of food from animal origin; (c) to identify how the perception of animal products quality affects the acceptance of changes in production methods and (d) to identify the projective image of consumers purchasing high-quality yoghurt.

2. Material and Methods

2.1. Data Collection Process

The sample in our study ($N = 983$) was drawn from the Social Security addresses database and was representative of the national population in terms of age, gender and the region that consumers lived in. The survey was conducted in each of the 16 voivodships in Poland. After drawing the starting addresses, the random route method was used in the selection of the sample [27,28]. A good number of sampling points were drawn with a probability proportional to population size, for total coverage of the country and for population density. In order to achieve this, the sampling points were drawn systematically from each of the "administrative regional units", after stratification by individual unit and type of area. They thus represent the whole of Poland as well as the distribution of the resident population. In each of the selected sampling points, a starting address was drawn at random. Further addresses were selected by standard "random route" procedures from the initial address. In each household, a respondent was drawn at random (following the "closest birthday rule").

The interviews were conducted face-to-face at respondents' homes by a professional market research agency in accordance with the ESOMAR (European Society for Opinion and Marketing Research) code of conduct using the CAPI (Computer Assisted Personal Interview) technique. All respondents were aged 21+. Only those respondents who met the recruitment criteria, i.e., made their own or cooperative food purchases and declared dairy product consumption, participated in the study.

2.2. Description of Questionnaire

The questionnaire used in the study was structured in a few main blocks and covered aspects such as consumer opinion towards: (1) the quality of food, including quality of animal origin food and (2) production methods of animal origin food, formulated into various types of questions:

(A) An open question: *What, in your opinion, shows the quality of food? Please indicate one of the most important attributes;*

(B) Two questions related to the quality of animal origin food (I), including the dairy products (II):

(I) *Below are statements describing food of animal origin. For each statement, how much you agree are indicated on a 1–7 scale, where 1 is the lowest level of compliance and 7 is the highest level of compliance; High-quality animal food is food (1) with the right taste and traditional recipe; (2) preservative free and with a short shelf life, (3) having nutritional value and health benefits, (4) produced in an environmentally friendly area, including taking into account production ensuring welfare of farm animals, (5) of low processing level/derived from an organic production method, (6) which is easy to prepare and easily available in a wide range;*

(II) *Please indicate how much you agree with the following statements. Please provide answers on a scale of 1–7, where 1 means "strongly disagree" and 7 means "strongly agree"; (1) I buy dairy products because they have a positive effect on my figure, (2) I buy dairy products because they have a good effect on my children's health, (3) Quality is important to me when choosing dairy products, (4) I buy dairy products for those members of my family who have health issues;*

(C) Questions referring to methods of increasing the quality of food of animal origin are formulated as follows: *To what extent do you accept the following methods of increasing the quality of food of animal origin? Please provide answers on a scale of 1–7, where 1 means "definitely do not accept" and 7 "definitely accept"; (1) Adding health-promoting ingredients to livestock feed, (2) Production ensuring welfare of farm animals, (3) Enhancing food products with health-promoting ingredients at the processing stage;*

(D) Questions related to increasing the level of ingredients in dairy products are as follows: *Please specify if you think the content of the ingredients listed below should be increased in dairy products? Where 1 definitely should not be increased, 7 definitely should be increased; (1) Minerals; (2) Fibre, (3) Cholesterol-lowering ingredients, (4) Omega-3 acid, (5) Live bacterial cultures, (6) Protein, (7) Coenzyme Q_{10};*

(E) Questions that allow the determination of projective image of buyers purchasing high-quality yoghurt are formulated as follows: *Who do you think is the most willing to buy high-quality yoghurt? Please give your answer on a scale from 1–7, where: 1 means "Definitely no" and 7 "Definitely yes", (1) professionally active individuals, (2) sport doers, (3) those looking for nutritional news, (4) the young, (5) the overworked, (6) people with abnormal intestinal motility, (7) cooking lovers, (8) those oriented on the convenience of preparing a meal, (9) bargain hunters, (10) people who are particularly health-conscious.*

2.3. Data Analysis

Referring to analysis of the results collected using the open question, the χ^2 test was applied in order to determine statistically significant differences between the variables (part A of the questionnaire). Moreover, the k-means clustering method was used to identify segments of consumers. In the k-means method (k-means clustering algorithm is an unsupervised algorithm that is used to segment the interest area from the background), in order to increase its efficiency, the average values for individual clusters obtained using the hierarchical method were used as seeds. The statements about the characteristic of food of animal origin were used as segmentation variables (part B I of the questionnaire; Table 1).

Five well-separated clusters were obtained, which was confirmed by both statistics assessing the selection of clusters such as CCC (Cubic Clustering Criteria), pseudo T2 or ANOVA (Analysis of Variance) statistics comparing the average values of variables for individual clusters. Socio-demographic variables such as gender, age, education, subjective assessment of the financial situation and size of the place of residence were used to profile the clusters. The independence χ^2 test was used to assess the diversity of profile features between clusters.

In all statements analyzed, statistically significant ($p < 0.05$) differences between mean scores particularly clusters have been observed. Additionally, post-hoc test (Waller–Duncan K-ratio *t* Test) was used to compare mean values of opinions between pairs of clusters.

As mentioned, the segmentation analysis made it possible to identify five consumer segments. Clusters have been named according to consumers' opinions towards statements referring to high quality food of animal origin (Table 1):

(1) "Convenience-oriented" consumers with a high level of compliance with the statement referring to convenience associated with the easy preparation and availability of high-quality food (9.12);
(2) "Uninvolved" consumers with the lowest levels of compliance with most of the statements compared to other segments;
(3) "Health-oriented" consumers with a significantly high level of compliance with the statement describing the acceptance of nutritional value and health values (10.53) compared to other segments;
(4) "Particularly demanding in terms of quality", consumers with a significantly high level of compliance for most statements referring to high-quality food;
(5) "Neutral but valuing food quality", consumers declaring relatively high rating levels for most statements but lower rating level for people classified in segment No. 4.

Table 1. Statements used as segmentation variables regarding the characteristics of high-quality products of animal origin.

Attributes	Mean	Convenience-Oriented 1	Uninvolved 2	Health-Oriented 3	Particularly Demanding in Terms of Quality 4	Neutral but Valuing Food Quality 5	p-Value
Easy preparation and availability	6.79	9.12 a	2.20 d	6.79 c	6.50 c	8.30 b	<0.0001
Nutritional value and health benefits	5.95	2.82 d	2.37 e	10.53 a	7.84 b	5.76 c	<0.0001
Processing, organic production	4.51	2.73 d	2.99 d	3.66 c	7.18 a	6.23 b	<0.0001
Tradition and taste	4.17	2.64 c	2.71 c	2.71 c	8.84 a	4.81 b	<0.0001
Lack of preservatives and shelf life	4.12	2.72 c	3.07 c	3.01 c	8.10 a	4.47 b	<0.0001
Environment and animal rights	3.92	2.60 d	2.25 d	2.98 c	7.10 a	5.06 b	<0.0001

One-Way ANOVA (Analysis of Variance), $p < 0.05$; a, b, c, d, e—Means with the same letter are not significantly different in Waller-Duncan test.

In the second step of data analysis, logistic regression was performed to determine how the perception of animal products quality impacts on:

- The acceptance of adding health-promoting ingredients to livestock feed;
- The acceptance of production ensuring welfare of farm animals;
- The acceptance of enhancing food products with health-promoting ingredients at the processing stage.

Due to the dichotomous nature of dependent variables (accept/not accept), logistic regression models were used [29,30], where dependent variables (regressants) were declarations regarding the acceptance of the above-mentioned 3 methods, and explanatory variables (regressors) were opinions about yoghurts and dairy products expressed in questions, i.e.,: *How much do you agree with the statements describing the quality of dairy products, Who do you think is the most willing to buy high-quality yoghurt? Do you think the content of the ingredients listed below should be increased in dairy products?* The models were built with a stepwise selection of explanatory variables. Only statistically significant variables at the significance level $\alpha = 0.05$ were included in the models. The statistical analysis was carried out using IBM SPSS Statistics, version 25.0 (IBM Corp., Armonk, NY, USA) and SAS 9.4 statistical package (SAS Institute, Cary, NC, USA).

3. Results

3.1. Profile of the Total Sample and Perception of Food Quality

The detailed socio-demographic characteristic of the sample and segments identified is included in Table 2.

Table 2. Socio-demographic characteristics of the consumers surveyed ($N = 983$, Poland) (%).

Variables	Total Sample (%)	Convenience-Oriented N = 208; 21% 1	Uninvolved N = 172; 18% 2	Health-Oriented N = 218; 22% 3	Particularly Demanding in Terms of Quality N = 159; 16% 4	Neutral But Valuing Food Quality N = 226; 23% 5	p-Value
Gender							0.7462 *
Female	51.41	54.85	52.07	51.15	51.01	48.15	
Male	48.59	45.15	47.93	48.85	48.99	51.85	
Age							0.5216 *
21–27	16.30	15.05	15.98	13.36	19.46	18.52	
28–34	15.99	16.02	14.20	12.90	20.13	17.59	
35–44	18.18	17.48	15.98	17.97	19.46	19.91	
45–54	20.06	24.76	21.30	20.74	14.77	17.59	
55–64	18.81	18.45	21.30	21.66	17.45	15.28	
65–75	10.66	8.25	11.24	13.36	8.72	11.11	

Table 2. Cont.

Variables	Total Sample (%)	Convenience-Oriented N = 208; 21% 1	Uninvolved N = 172; 18% 2	Health-Oriented N = 218; 22% 3	Particularly Demanding in Terms of Quality N = 159; 16% 4	Neutral But Valuing Food Quality N = 226; 23% 5	p-Value
Education							
Primary, lower secondary, vocational	47.75	41.26	55.03	43.78	54.36	47.69	0.0102
Secondary	37.10	38.35	36.09	39.17	36.91	34.72	
Higher	15.15	20.39	8.88	17.05	8.73	17.59	

* Differences between groups not significant (χ^2 test, p-value > 0.05).

Results of the study show that the quality of food of animal origin is connected with the ease of preparation, availability, as well as nutritional value and health benefits (Table 1), but when it comes to the food quality in general, it is perceived by consumers mainly through the following attributes: its freshness, naturalness, production method, as well as appearance, taste and smell (Table 3).

Table 3. Attributes describing food quality in consumer reviews (%).

Attributes	Number of Indications	%	Convenience-Oriented N = 208; 21% 1	Uninvolved N = 172; 18% 2	Health-Oriented N = 218; 22% 3	Particularly Demanding in Terms of Quality N = 159; 16% 4	Neutral But Valuing Food Quality N = 226; 23% 5	p-Value
Freshness	197	20.04	19.71	19.19	27.98	7.55	22.12	
Naturalness, production method	162	16.48	20.19	15.70	18.35	7.55	18.14	
Appearance, taste, smell	121	12.31	17.31	10.47	8.26	13.84	11.95	<0.0001
Composition, nutritional values	113	11.50	10.10	12.79	7.34	15.09	13.27	
Preservative-free	100	10.17	7.69	10.47	12.84	8.81	10.62	
Price	62	6.31	3.85	4.65	1.38	13.84	9.29	
Producer	47	4.78	4.81	6.98	5.5	4.40	2.65	
Shelf life	42	4.27	2.88	4.65	5.96	5.66	2.65	
Quality mark	38	3.87	3.85	4.65	5.05	6.29	0.44	
Origin	37	3.76	6.25	2.91	3.67	3.77	2.21	
No answer/ I do not know	64	6.51	3.37	7.56	3.67	13.21	6.64	

Test of independence χ^2 $p < 0.05$.

Results also show (Tables 2 and 3) that in segment No. 1 ("Convenience-oriented"; N = 208; 21%) the largest share of opinions indicate that the quality of food of animal origin is evidenced by its naturalness, production method, but also its freshness. This segment was characterized by the largest share of middle-aged people (45–54 years).

In segment No. 2 ("Uninvolved"; N = 172; 18%), the largest share of the food quality is reflected in its freshness, naturalness and the production method. This segment had a relatively high share of people with low levels of education.

In segment No. 3 ("Health-oriented"; N = 218; 22%), the highest share of answers indicating that the quality of food is reflected in its freshness was recorded. There was also a relatively large share of opinions indicating that the quality of food of animal origin is reflected in its naturalness and production method. The additional attribute of food quality that was mentioned by people was lack of preservatives in the product.

In segment No. 4 ("Particularly demanding in terms of quality"; N = 159; 16%), there was a relatively large share of opinions indicating that the quality of food of animal origin is reflected in its composition and nutritional values as well as appearance, taste and smell. The segment had the largest share of indications that the determinant of food quality is its price. This segment had the relatively high share of people with low levels of education.

In segment No. 5 ("Neutral but valuing food quality"; N = 226; 23%), among the indications characterizing the quality of food of animal origin, mainly its freshness, naturalness, method of production as well as the aspects referring to composition and nutritional value of food were mentioned.

3.2. Methods of Improving Quality of Animal Origin among The Clusters of Consumers

The results show that among the methods of increasing the quality of food of animal origin, consumers scored the highest for the production method ensuring welfare of farm animals compared to the other two methods. Comparison of the segments show that respondents from segment No. 3 ("Health-oriented") displayed significantly lower acceptance, compared to other segments regarding the method of adding health-promoting ingredients to livestock feed (Table 4).

Table 4. The level of acceptance of methods to increase food quality of animal origin in the opinion of respondents.

Selected Methods of Increasing Food Quality	Mean	Convenience-Oriented N = 208; 21% 1	Uninvolved N = 172; 18% 2	Health-Oriented N = 218; 22% 3	Particularly Demanding in Terms of Quality N = 159; 16% 4	Neutral But Valuing Food Quality N = 226; 23% 5	p-Value
Animal production ensuring welfare of farm animals	5.90	6.21 a	6.32 a	6.18 a	5.13 c	5.60 b	<0.0001
Adding health-promoting ingredients to livestock feed	4.14	4.32 a	4.35 a	3.39 b	4.43 a	4.35 a	<0.0001
Enhancing food products with health-promoting ingredients at the processing stage	3.85	3.89 b	4.21 b	2.84 c	4.61 a	4.00 b	<0.0001

One-Way ANOVA, $p < 0.05$; a, b, c—Means with the same letter are not significantly different in Waller–Duncan test.

In addition, scores on production ensuring welfare of farm animals by respondents in segment No. 5 ("Neutral but valuing food quality") was significantly lower compared to consumers in segments 1, 2 and 3, while respondents from segment No. 4 significantly displayed lower acceptance of this type of production compared to other segments. Scores on enhancing food products with health-promoting ingredients at the processing stage by respondents from segment No. 4 ("Particularly demanding in terms of quality") was significantly higher compared to other segments, and respondents from segment No. 3 ("Health-oriented") had significantly lower acceptance of this type of enrichment compared to other segments (Table 4).

Regarding the increase of some ingredients in dairy products, live bacterial cultures and cholesterol lowering ingredients were the most important types of ingredients that should be increased in the consumers' opinion in dairy products (Table 5). Consumers in segment 2 ("Uninvolved"), which was also significantly higher compared to segments 3, 4 and 5, agreed that the content of cholesterol-lowering ingredients should be increased in dairy products. "Uninvolved" also declared a significantly higher level of acceptance compared to other segments, in terms of increasing the content of live bacterial cultures and coenzyme Q_{10} in dairy products. In the case of increasing the content of minerals and increasing the fiber content in dairy products, the "Uninvolved" significantly agreed with this opinion in comparison to "Health-oriented" and "Particularly Demanding in Terms of Quality". On the other hand, "Health-oriented" segment displayed the lowest degree of acceptance with regards to increasing the level of Omega-3 acid, proteins and Coenzyme Q_{10} in dairy products compared to other segments (Table 5).

Table 5. Consumers' opinion on increasing in dairy products the level of ingredients that have a positive impact on health.

Food Ingredients Whose Level Should Be Increased	Mean	Convenience-Oriented N = 208; 21% 1	Uninvolved N = 172; 18% 2	Health-Oriented N = 218; 22% 3	Particularly Demanding in Terms of Quality N = 159; 16% 4	Neutral But Valuing Food Quality N = 226; 23% 5	p-Value
Live bacterial cultures	4.88	5.03 b	5.51 a	4.51 c	4.78 bc	4.76 bc	<0.0001
Cholesterol lowering ingredients	4.84	4.97 ab	5.27 a	4.55 b	4.67 b	4.81 b	0.0101
Minerals	4.76	4.85 ab	5.20 a	4.37 c	4.70 bc	4.78 ab	0.0024
Fiber	4.69	4.85 ab	5.07 a	4.41 c	4.49 bc	4.69 abc	0.0142
Omega-3 acid	4.50	5.67 a	5.00 ab	4.01 c	4.48 b	4.51 b	0.0006
Protein	4.42	4.35 b	4.93 a	3.89 c	4.75 ab	4.40 b	<0.0001
Coenzyme Q_{10}	4.32	4.41 b	5.03 a	3.73 c	4.31 b	4.35 b	<0.0001

One-Way ANOVA, $p < 0.05$; a, b, c—Means with the same letter are not significantly different in Waller–Duncan test.

The next part of the study was aimed at determining the image of consumers of high-quality yoghurt (Table 6). Respondents perceived consumers of high quality yoghurts referring to two main aspects: (1) health and (2) physical activity.

Table 6. Projective image of high-quality yoghurt consumers.

High-Quality Yoghurts Are Purchased by	Mean	Convenience-Oriented N = 208; 21% 1	Uninvolved N = 172; 18% 2	Health-Oriented N = 218; 22% 3	Particularly Demanding in Terms of Quality N = 159; 16% 4	Neutral But Valuing Food Quality N = 226; 23% 5	p-Value
those who are particularly health-conscious	5.95	6.25 a	6.30 a	5.94 b	5.21 c	5.89 b	<0.0001
those with abnormal intestinal motility	5.82	6.03 a	6.07 a	6.04 a	5.15 c	5.71 b	<0.0001
sport doers	5.76	6.03 a	5.89 a	5.94 a	5.18 c	5.62 b	<0.0001
the young	5.55	5.57 b	5.86 a	5.69 ab	5.08 c	5.45 b	<0.0001
professionally active	5.53	5.68 a	5.71 a	5.10 b	5.67 a	5.44 a	0.0005
those looking for nutritional novelties	5.36	5.48 ab	5.73 a	5.17 bc	5.09 c	5.34 bc	0.0019
those oriented on the convenience of preparing a meal	5.33	5.55 a	5.54 a	5.30 a	4.94 b	5.24 ab	0.0030
the overworked	5.23	5.34 a	5.28 ab	5.33 a	4.90 b	5.20 ab	0.1103
cooking lovers	4.66	4.50 bc	4.97 a	4.33 c	4.80 ab	4.76 abc	0.0156
bargain hunters	4.53	4.11 c	4.60 ab	4.37 bc	4.81 a	4.78 ab	0.0030

One-Way ANOVA, $p < 0.05$; a, b, c—Means with the same letter are not significantly different in Waller–Duncan test.

Respondents from segment No. 4 ("Particularly demanding in terms of quality") in the least degree compared to the other segments agreed with the opinion that such consumers are people who can be characterized as: doing sports, young people, as well as people with abnormal intestinal motility and people who are particularly health-conscious.

3.3. Impact of Selected Attributes on Methods of Improving Quality of Animal Origin Food

In the next stage of the study, the extent in which consumers would accept 3 methods aimed at increasing the level of food quality was assessed.

The rise in the acceptance of opinion that the content of live bacterial cultures in dairy products should be increased resulted in a 47% increase in the willingness of accepting production ensuring welfare of farm animals (OR: 1.47; 95% CI: 1.14–1.90), while maintaining other model parameters at a constant level. The level of education had an impact on the acceptance of production, ensuring the welfare of farm animals. The higher the level of education, the greater the willingness of this acceptance. This willingness in the case of people with secondary education increased four times compared to people with primary education (OR: 4.06; 95% CI: 1.82–12.93). In the case of higher education, the willingness of acceptance increased more than 10-fold (OR: 10.25; 95% CI: 1.95–22.49) in relation to people with primary education (Table 7).

Table 7. Prediction of the acceptance of production ensuring welfare of farm animals.

Variable	e^β	β	95% Wald CI		p-Value
Intercept		0.142			0.8728
Independent variables (regressors):					
Increasing the content of live bacterial cultures in dairy products	1.47	0.390	1.14	1.90	0.0024
Basic vocational education vs. primary education	2.32	0.839	0.56	9.51	0.2441
Secondary education vs. primary education	4.06	1.400	1.82	12.93	0.0447
Higher education vs. primary education	10.25	2.327	1.95	22.49	0.0250

e^β (OR)—point estimate; β—estimate; 95% Wald CI—95% Wald confidence interval.

The rise (by 1 point) in acceptance of the opinion that the content of minerals in dairy products should be increased resulted in a 21% increase in the willingness of acceptance of adding health-promoting ingredients to livestock feed (OR: 1.21; 95% CI: 1.06–1.37). The rise in importance of the opinion that high-quality yoghurts are bought by people involved in sports resulted in a 27% decrease in willingness of acceptance of adding health-promoting ingredients to livestock feed (OR: 0.73; 95% CI: 0.58–0.91). The increase in the rank that high-quality yoghurts are bought by those looking for novelty foods increased by 19% compared to the willingness of acceptance of adding health-promoting ingredients to livestock feed (OR: 1.19; 95% CI: 1.02–1.40). People declaring that quality is important to them when choosing dairy products showed a 19% lower willingness of accepting the addition of health-promoting ingredients to livestock feed (OR: 0.81; 95% CI: 0.65–0.99) along with increasing the rank of this opinion by 1 level (Table 8).

Table 8. Prediction of acceptance of adding health-promoting ingredients to livestock feed.

Variable	e^β	β	95% Wald CI		p-Value
Intercept		2.248			0.006
Independent variables (regressors):					
Increasing the mineral content in dairy products	1.21	0.187	1.06	1.37	0.004
Purchase of high quality yoghurt by people involved in sport	0.73	−0.315	0.58	0.91	0.004
Purchase of high quality yoghurts by people seeking nutrition novelties	1.19	0.177	1.02	1.40	0.031
Quality is important when choosing dairy products	0.81	−0.212	0.65	0.99	0.044

e^β (OR)—point estimate; β—estimate; 95% Wald CI—95% Wald confidence interval.

The rise in importance of the opinion that the content of cholesterol-lowering ingredients should be increased in dairy products resulted in a 29% increase in the willingness to accept enhancing food products with pro-health ingredients at the processing stage (OR: 1.29; 95% CI: 1.13–1.47). Obviously, while maintaining the remaining model parameters at a constant level. The increase in the rank referring to opinion that high-quality yoghurts are bought by the professionally active individuals resulted in a 30% decrease in the level of acceptance of enhancing food products with pro-health ingredients at the processing stage (OR: 0.70; 95% CI: 0.56–0.86). The increase in the rank of the opinion that high-quality yoghurts are bought by people with abnormal intestinal motility gave a 31% greater willingness of accepting enhancing food products with health-promoting ingredients at the processing stage (OR: 1.31; 95% CI: 1.06–1.60). Similar results were seen in the responses that high-quality yoghurt is bought by those looking for price bargains. In this case, the willingness of accepting enhancing food products with health-promoting ingredients at the processing stage increased by 24% (OR: 1.24; 95%

CI: 1.08–1.41). The willingness to accept enhancing food products with health-promoting ingredients at the processing stage decreased by 26% in the case of persons agreeing with the opinion that quality is important for them when choosing dairy products (OR: 0.74; 95% CI: 0.60–0.90). The rise in the rank of the opinion that *I buy high-quality dairy products for those family members who have health issues* resulted in a 17% increase in enhancing food products with health-promoting ingredients at the processing stage (OR: 1.17; 95% CI: 1.04–1.33) (Table 9).

Table 9. Prediction of acceptance of enhancing food products with health-promoting ingredients at the processing stage.

Variable	e^β	β	95% Wald CI		p-Value
Intercept		0.115			0.8825
Independent variables (regressors):					
Increasing the content of cholesterol-lowering ingredients in dairy products	1.29	0.258	1.13	1.47	0.0001
Purchase of high-quality yoghurt by professionally active people	0.70	−0.355	0.56	0.86	0.0010
Purchase of high-quality yoghurt by people with abnormal intestinal motility	1.31	0.267	1.06	1.60	0.0104
Purchase of high quality yoghurts by people looking for bargains	1.24	0.212	1.08	1.41	0.0021
Quality is important when choosing dairy products	0.74	−0.304	0.60	0.90	0.0032
Purchase of high-quality dairy products for family members who have health issues	1.17	0.162	1.04	1.33	0.0101

e^β (OR)—point estimate; β—estimate; 95% Wald CI—95% Wald confidence interval.

4. Discussion

The study presents the results of a survey on a representative sample of consumers. The analysis of the obtained results indicated that the quality of animal origin food with particular emphasis on dairy products is of great importance to consumers, and they are willing to accept new methods of production and ingredients in dairy products.

4.1. The Food Quality from a Consumer Point of View

Our study revealed that the consumer's perception of food quality differ among segments. Furthermore, the high-quality products of animal origin were perceived by consumers in various ways depending on the segment, so the results are consistent with previous studies stating that understanding the personal and context specific influences on consumer quality perceptions is important in developing products that meet consumer needs [31].

In general, referring to the aspect of food quality, the results showed that taking into account the consumer segments, consumers in segment No. 4 ("Particularly demanding in terms of quality") slightly agreed with the opinion that the group of people buying yoghurts perceived as high-quality yoghurts includes people involved in sports, young people, people with abnormal intestinal motility and those concerned about their health. Analysis of the research findings showed that consumers in segment 2 ("Uninvolved") showed high levels of indications in terms of increasing the level of health-promoting ingredients, which may suggest that despite a relatively indifferent position on food quality compared to other consumer segments, these people were interested in increasing the amount of selected ingredients, and at the same time, it may prove that consumers expect producers and processors to take appropriate action on their behalf to improve their health. This is reflected in the studies by other authors, which emphasized the importance of health as a value influencing the acceptance of specific type of food [32]. Moreover, referring to milk products, the totality of available scientific evidence supports the fact that intake of milk and dairy products contributes to meet the nutrient recommendations and may protect against the most prevalent chronic diseases, whereas, very few adverse effects have been reported [33]. Furthermore, lactose malabsorption is widespread in most parts of the world, with wide variation between different regions and an overall frequency of around two-thirds of the world's population [34].

4.2. The Acceptance of Production Ensuring Welfare of Farm Animals

Our study assessed the level of acceptance of methods used to increase the food quality and selected factors that may affect the level of this acceptance among consumers. In general, animal welfare is the credence quality attribute [7] that is of great interest to consumer. The results showed that the increase in education contributed to the acceptance of production ensuring welfare of farm animals as a way of increasing the level of healthy ingredients in food. The results of other studies indicate that individuals involved in health and/or sustainable eating are more likely to be better educated than those who are not involved [35]. This may be due to a greater awareness of ensuring adequate welfare of farm animals (and/or probably due to the sensitivity of this group of people to animal suffering) [36]. Results also show that the use of appropriate production ensuring welfare of farm animals as a method of increasing food quality is also accepted by consumers who willingly accept increasing the content of live bacterial cultures in dairy products. This can be associated with the positive consumer perception of yoghurt through the aspect referring to health issue, which is confirmed by the studies of other authors [26,37].

The results of other studies indicate that consumers with a higher income and higher education were willing to pay more for farm animal welfare [38]. The results of studies revealed also that referring to animal welfare, the provision of additional information significantly increased the intention to purchase higher than the conventional welfare products. The empathy measures revealed that younger participants, females and those with lower household incomes all had significantly higher AES (Animal Empathy Score). Moreover, this score was associated with the intent to purchase higher welfare products [39]. However, some studies showed that consumers are, in general, unaware about welfare issues at the farming level [40,41]. In addition, an analysis of the results of surveys performed under Euro barometer 2019 [42] indicated that the most important factors for Europeans when buying food are where the food comes from (53%), cost (51%), food safety (50%) and taste (49%). Nutrient content is considered slightly less important (44%), while ethics and beliefs (e.g., considerations of animal welfare, environmental concerns) rank lowest in importance (19%) [42]. However, Vanhonacker and Verbeke [43] noticed that the role of information on animal welfare as well as the type of consumer to whom this information is presented is important when making purchasing decisions. Moreover, in their opinion, the issue involves acknowledging that not everyone has the same level of interest in animal welfare or in purchasing higher welfare products. Furthermore, not all individuals with an interest in higher welfare products share the same motivation. Information sharing should thus be adjusted to specific target segments [43]. On the other hand, the results of more recent research [39] suggest that concern for the welfare of animals farmed for food remains high and continues to grow. Moreover, this research indicates that providing consumers with descriptive signals referring to the welfare condition at the point-of-purchase can boost welfare purchase intentions [39].

4.3. The Acceptance of Adding Health-Promoting Ingredients to Livestock Feed

The second method of increasing the food quality that was accessed in the survey, was adding health-promoting ingredients to livestock feed. It was observed that consumer acceptance of opinions on increasing mineral components at the same time inclines them to accept adding health-promoting ingredients to livestock feed. The results of other studies showed that in the area of animal nutrition, the opportunity for improving quality is by adding health-promoting ingredients to livestock feed containing additives such as vitamins, vitamin-like compounds, minerals including trace elements, fatty acids, probiotics and other bioactive compounds [44,45].

The results also showed that with the increase in acceptance of novelty on the food market, the level of acceptance of the production method which entails adding health-promoting ingredients to livestock feed increases. This may be associated with generally greater openness to changes in the food market and a higher level of acceptance of changes in this market in relation to some consumer groups [46]. When it comes to Polish consumers, in general, the new generation of Poles is relatively more open to new food products due to the wide range of food products available on the free market. Furthermore, the group of well-educated consumers with a higher level of income has increased in size, and this includes people interested in knowledge of a product's nutritional value and its impact on health [47]. Results showed that people for whom quality was important when choosing dairy products and people doing sports did not accept this method of improving quality, which confirms their special interest in health aspects and possible consequences and/or concerns related to the consumption of this category of food.

4.4. The Acceptance of Enhancing Food Products with Health-Promoting Ingredients at The Processing Stage

Among various food choice motives, health is thought to be the highly important factor in consumer opinion [3]. Our results showed that with regard to the third method of increasing the food quality, people seeking the possibility of lowering blood cholesterol levels, people who believe that "high-quality yoghurt is bought by people with abnormal intestinal motility" and those who bought yoghurt for members of their families with health issues expressed their willingness to accept increasing the level of ingredients at the food processing stage.

The results of other studies indicated that respondents with a history of familial diseases were more likely than others to have consumed margarine with plant sterol, fruit juices fortified with vitamin C, and breakfast cereals fortified with vitamins and minerals [48]. It was also found in other studies that consumers who considered health, sensory appeal, natural content, and ethicality to be important factors in their food choices and were concerned about their health, considered yoghurts which were reasonably sour, thick, and genuine in flavor to be more pleasant [3]. Results of similar studies among consumers regarding functional food showed that food benefits were more positively evaluated when attached to a more attractive carrier (e.g., yoghurt). Moreover, benefits of improving the body's natural defense system were most favored by all groups of surveyed consumers while benefits about specific diseases were suitable to tailor for certain groups [13]. Generally, our results are consistent with other studies showing that claims referring to prevention of the diseases are accepted by the consumers [49,50].

It should be emphasized, however, that the results of our study showed that consumers paying special attention to quality were afraid of the above-mentioned methods, declaring their low level of acceptance, which may indicate that they believe there are some concerns related to increasing the level of some ingredients or the fear of using selected methods are not well known to them, which may be associated with so-called food technology neophobia [51]. Research shows that the majority of consumers have relatively little knowledge about the technologies used in food production [52]. However, when it comes to advertising and marketing to consumers about new technologies, campaigns that incorporate convenience, naturalness, taste and benefit for the consumer could have a positive impact on consumer food choices, particularly when the message is concise and from trusted sources [53].

5. Conclusions

The significant role of food quality in decisions taken on the food market, as well as the availability of products with special health benefits, encourages the assessment of consumer behavior in relation to food perceived as the high-quality food and learning about consumer opinions on products that have a positive impact on health.

The results show that in general, food quality is perceived by consumers mainly through the following attributes: freshness, naturalness, production method, as well as appearance, taste and smell, while when it comes to the quality of food of animal origin, convenience connected with availability, nutritional value and health benefits are of primary importance.

The most accepted production method of high-quality food is animal production with respect to the welfare of farm animals. It is a particularly important aspect, because animal welfare is an important element of sustainable development, including food consumption and human diet and can positively contribute to food quality. Results also show that the increase in the level of education of the surveyed people contributed to the acceptance of production, ensuring welfare of farm animals as a method of increasing food quality.

With regard to the acceptance of other methods aimed at increasing the content of health-promoting ingredients, it should be emphasized that consumer openness to new products favored the acceptance of adding health-promoting ingredients to livestock feed. Regarding the assessment of the level of acceptance by consumers of enhancing food with beneficial ingredients at the processing stage, people for whom health aspects were important declared their willingness to accept such a method of increasing food quality.

Moreover, the research findings can be used to develop educational campaigns as well as in marketing communication of enterprises operating on the food market. When it comes to the educational campaigns, there are opportunities to increase the level of consumer awareness referring to animal welfare that is still not considered an issue for many of the Central and Eastern European citizens. Moreover, to strengthen the competitive position of food enterprises, an important point could be the development of the food products labelled with information on animal production, ensuring welfare of farm animals.

In addition, in terms of information on the methods of improving quality communicated to consumers, for some people, the manner in which this information is presented will play an important role, as well as the level of awareness of the recipients to whom it was addressed, and the possible health consequences they perceive. Therefore, the observed impact of the level of education as well as the health benefits of accepting some methods of increasing food quality should be used on the food market. This aspect could be particularly important according to the development of health and nutrition claims.

The present study fills the relevant research gaps regarding enhancing the quality of food from animal origin and explores methods referring to increasing the level of food quality, providing a new perspective to food industry and the scholars. On the one hand, there are some differences regarding the consumer acceptance referring to methods of enhancing food quality. On the other hand, the determinants that impact the level of acceptance are also various. Future research studies should concentrate on investigating and developing the level of consumers' acceptance of new production methods. Nevertheless, it should be noted that our results indicated the main directions regarding the acceptance of the used productions methods as well as designated the possible changes that may be accepted by consumers.

Author Contributions: M.S. developed the concept of the study, supervised the survey, interpreted the data and wrote the manuscript. J.G. analyzed the data and contributed to its interpretation. M.S., J.G., D.G., S.Ż.-B. and K.G. were involved in critically revising the manuscript, and have given their approval to the manuscript submitted. K.G. was responsible for funding acquisition and supervision. All authors have read and agreed to the published version of the manuscript.

Funding: This research was funded by "BIOFOOD—Innovative, Functional Products of Animal Origin" grant number [POIG.01.01.02-014-090/09] that was co-financed by the European Union from the European Regional Development Fund within the Innovative Economy Operational Programme 2007–2013 And The APC was funded by Polish Ministry of Science and Higher Education within funds of Institute of Human Nutrition Sciences and Faculty of Human Nutrition, Warsaw University of Life Sciences (WULS), for scientific research.

Conflicts of Interest: The authors declare no conflict of interest.

References

1. Markovina, J.; Stewart-Knox, B.J.; Rankin, A.; Gibney, M.; de Almeida, M.D.V.; Fischer, A.; Kuznesof, S.A.; Poínhos, R.; Panzone, L.; Frewer, L.J. Food4Me study: Validity and reliability of Food Choice Questionnaire in 9 European countries. *Food Qual. Prefer.* **2015**, *45*, 26–32. [CrossRef]

2. Román, S.; Sánchez-Siles, L.M.; Siegrist, M. The importance of food naturalness for consumers: Results of a systematic review. *Trends Food Sci. Technol.* **2017**, *67*, 44–57. [CrossRef]
3. Pohjanheimo, T.; Sandell, M. Explaining the liking for drinking yoghurt: The role of sensory quality, food choice motives, health concern and product information. *Int. Dairy J.* **2009**, *19*, 459–466. [CrossRef]
4. Saba, A.; Sinesio, F.; Moneta, E.; Dinnella, C.; Laureati, M.; Torri, L.; Peparaio, M.; Saggia Civitelli, E.; Endrizzi, I.; Gasperi, F.; et al. Measuring consumers attitudes towards health and taste and their association with food-related life-styles and preferences. *Food Qual. Prefer.* **2019**, *73*, 25–37. [CrossRef]
5. Albertsen, L.; Wiedmann, K.P.; Schmidt, S. The impact of innovation-related perception on consumer acceptance of food innovations—Development of an integrated framework of the consumer acceptance process. *Food Qual. Prefer.* **2020**, *84*, 103958. [CrossRef]
6. Grunert, K.G. Current issues in the understanding of consumer food choice. *Trends Food Sci. Technol.* **2002**, *13*, 275–285. [CrossRef]
7. Grunert, K.G.; Bech-Larsen, T.; Bredahl, L. Three issues in consumer quality perception and acceptance of dairy products. *Int. Dairy J.* **2000**, *10*, 575–584. [CrossRef]
8. Mascarello, G.; Pinto, A.; Parise, N.; Crovato, S.; Ravarotto, L. The perception of food quality. Profiling Italian consumers. *Appetite* **2015**, *89*, 175–182. [CrossRef]
9. Kraus, A. Development of functional food with the participation of the consumer. Motivators for consumption of functional products. *Int. J. Consum. Stud.* **2015**, *39*, 2–11. [CrossRef]
10. Bryła, P. The perception of EU quality signs for origin and organic food products among Polish consumers. *Qual. Assur. Saf. Crop. Foods* **2017**, *9*, 345–355. [CrossRef]
11. Grunert, K.G. European consumers' acceptance of functional foods. *Ann. N. Y. Acad. Sci.* **2010**, *1190*, 166–173. [CrossRef]
12. Bimbo, F.; Bonanno, A.; Nocella, G.; Viscecchia, R.; Nardone, G.; De Devitiis, B.; Carlucci, D. Consumers' acceptance and preferences for nutrition-modified and functional dairy products: A systematic review. *Appetite* **2017**, *113*, 141–154. [CrossRef] [PubMed]
13. Huang, L.; Bai, L.; Gong, S. The effects of carrier, benefit, and perceived trust in information channel on functional food purchase intention among Chinese consumers. *Food Qual. Prefer.* **2020**, *81*, 103854. [CrossRef]
14. Kaur, N.; Singh, D.P. Deciphering the consumer behaviour facets of functional foods: A literature review. *Appetite* **2017**, *112*, 167–187. [CrossRef] [PubMed]
15. Teratanavat, R.; Hooker, N.H. Consumer valuations and preference heterogeneity for a novel functional food. *J. Food Sci.* **2006**, *71*. [CrossRef]
16. Toma, L.; Stott, A.W.; Revoredo-Giha, C.; Kupiec-Teahan, B. Consumers and animal welfare. A comparison between European Union countries. *Appetite* **2012**, *58*, 597–607. [CrossRef]
17. Eurobarometer. *Special Eurobarometer 442 Report Attitudes of Europeans towards Animal Welfare*; European Commission: Brussels, Belgium, 2016; ISBN 9789279568787.
18. Clark, B.; Stewart, G.B.; Panzone, L.A.; Kyriazakis, I.; Frewer, L.J. Citizens, consumers and farm animal welfare: A meta-analysis of willingness-to-pay studies. *Food Policy* **2017**, *68*, 112–127. [CrossRef]
19. Williams, E.B.; Hooper, B.; Spiro, A.; Stanner, S. The contribution of yogurt to nutrient intakes across the life course. *Nutr. Bull.* **2015**, *40*, 9–32. [CrossRef]
20. Webb, D.; Donovan, S.M.; Meydani, S.N. The role of Yogurt in improving the quality of the American diet and meeting dietary guidelines. *Nutr. Rev.* **2014**, *72*, 180–189. [CrossRef]
21. Sfakianakis, P.; Tzia, C. Conventional and Innovative Processing of Milk for Yogurt Manufacture; Development of Texture and Flavor: A Review. *Foods* **2014**, *3*, 176–193. [CrossRef]
22. Das, K.; Choudhary, R.; Thompson-Witrick, K.A. Effects of new technology on the current manufacturing process of yogurt-to increase the overall marketability of yogurt. *LWT* **2019**, *108*, 69–80. [CrossRef]
23. Rahnama, H.; Rajabpour, S. Factors for consumer choice of dairy products in Iran. *Appetite* **2017**, *111*, 46–55. [CrossRef] [PubMed]
24. Hashemi Gahruie, H.; Eskandari, M.H.; Mesbahi, G.; Hanifpour, M.A. Scientific and technical aspects of yogurt fortification: A review. *Food Sci. Hum. Wellness* **2015**, *4*, 1–8. [CrossRef]
25. Rad, A.H.; Javadi, M.; Kafil, H.S.; Pirouzian, H.R.; Khaleghi, M. The safety perspective of probiotic and non-probiotic yoghurts: A review. *Food Qual. Saf.* **2019**, *3*, 9–14. [CrossRef]

26. Fazilah, N.F.; Ariff, A.B.; Khayat, M.E.; Rios-Solis, L.; Halim, M. Influence of probiotics, prebiotics, synbiotics and bioactive phytochemicals on the formulation of functional yogurt. *J. Funct. Foods* **2018**, *48*, 387–399. [CrossRef]
27. Bauer, J.J. Selection Errors of Random Route Samples. *Sociol. Methods Res.* **2014**, *43*, 519–544. [CrossRef]
28. Kent, R. *Marketing Research in Action. Sampling Cases*; Routledge: London, UK, 1993.
29. Hosmer, D.W.; Lemeshow, S. *Applied Logistic Regression*, 2nd ed.; John Wiley & Sons Inc.: New York, NY, USA, 2000.
30. Field, A. *Discovering Statistics Using IBM SPSS Statistics*, 5th ed.; SAGE Publications Ltd.: London, GB, 2017.
31. Henchion, M.; McCarthy, M.; Resconi, V.C.; Troy, D. Meat consumption: Trends and quality matters. *Meat Sci.* **2014**, *98*, 561–568. [CrossRef]
32. Loebnitz, N.; Grunert, K.G. Impact of self-health awareness and perceived product benefits on purchase intentions for hedonic and utilitarian foods with nutrition claims. *Food Qual. Prefer.* **2018**, *64*, 221–231. [CrossRef]
33. Thorning, T.K.; Raben, A.; Tholstrup, T.; Soedamah-Muthu, S.S.; Givens, I.; Astrup, A. Milk and dairy products: Good or bad for human health? An assessment of the totality of scientific evidence. *Food Nutr. Res.* **2016**, *60*, 32527. [CrossRef]
34. Storhaug, C.L.; Fosse, S.K.; Fadnes, L.T. Country, regional, and global estimates for lactose malabsorption in adults: A systematic review and meta-analysis. *Lancet Gastroenterol Hepatol.* **2017**, *2*, 738–746. [CrossRef]
35. Van Loo, E.J.; Hoefkens, C.; Verbeke, W. Healthy, sustainable and plant-based eating: Perceived (mis)match and involvement-based consumer segments as targets for future policy. *Food Policy* **2017**, *69*, 46–57. [CrossRef]
36. Alonso, M.E.; González-Montaña, J.R.; Lomillos, J.M. Consumers' concerns and perceptions of farm animal welfare. *Animals* **2020**, *10*, 385. [CrossRef] [PubMed]
37. Nowak, A.; Ślizewska, K.; Libudzisz, Z.; Socha, J. Probiotyki—Efekty zdrowotne. *Zywn. Nauk. Technol. Jakosc/Food. Sci. Technol. Qual.* **2010**, *17*, 20–36.
38. Clark, B.; Stewart, G.B.; Panzone, L.A.; Kyriazakis, I.; Frewer, L.J. A Systematic Review of Public Attitudes, Perceptions and Behaviours Towards Production Diseases Associated with Farm Animal Welfare. *J. Agric. Environ. Ethics* **2016**, *29*, 455–478. [CrossRef]
39. Cornish, A.R.; Briley, D.; Wilson, B.J.; Raubenheimer, D.; Schlosberg, D.; McGreevy, P.D. The price of good welfare: Does informing consumers about what on-package labels mean for animal welfare influence their purchase intentions? *Appetite* **2020**, *148*, 104577. [CrossRef] [PubMed]
40. Mceachern, M.G.; Schroder, M.J.A. The Role of Livestock Production Ethics in Consumer Values Towards Meat. *J. Agric. Environ. Ethics* **2002**, *15*, 221–237. [CrossRef]
41. Schröder, M.J.A.; McEachern, M.G. Consumer value conflicts surrounding ethical food purchase decisions: A focus on animal welfare. *Int. J. Consum. Stud.* **2004**, *28*, 168–177. [CrossRef]
42. EFSA. *Special Eurobarometer Wave EB91.3 Food Safety in the EU Report Fieldwork*; European Commission: Brussels, Belgium, 2019; ISBN 9789294990822.
43. Vanhonacker, F.; Verbeke, W. Public and Consumer Policies for Higher Welfare Food Products: Challenges and Opportunities. *J. Agric. Environ. Ethics* **2014**, *27*, 153–171. [CrossRef]
44. Jóźwik, A.; Strzałkowska, N.; Bagnicka, E.; Łagodziński, Z.; Pyzel, B.; Chyliński, W.; Czajkowska, A.; Grzybek, W.; Słoniewska, D.; Krzyzewski, J.; et al. The effect of feeding linseed cake on milk yield and milk fatty acid profile in goats. *Anim. Sci. Pap. Rep.* **2010**, *28*, 245–251.
45. Pinotti, L.; Baldi, A.; Krogdahl, A.; Givens, I.; Knight, C.; Baeten, V.; Van Raamsdonk, L.; Woodgate, S.; Marin, D.P.; Luten, J. The role of animal nutrition in designing optimal foods of animal origin as reviewed by the COST action feed for health (FA0802). *Biotechnol. Agron. Soc. Environ.* **2014**, *18*, 471–479.
46. Henchion, M.; McCarthy, M.; Dillon, E.J.; Greehy, G.; McCarthy, S.N. Big issues for a small technology: Consumer trade-offs in acceptance of nanotechnology in food. *Innov. Food Sci. Emerg. Technol.* **2019**, *58*. [CrossRef]
47. Sajdakowska, M.; Jankowski, P.; Gutkowska, K.; Guzek, D.; Żakowska-Biemans, S.; Ozimek, I. Consumer acceptance of innovations in food: A survey among Polish consumers. *J. Consum. Behav.* **2018**, *17*, 253–267. [CrossRef]
48. Büyükkaragöz, A.; Bas, M.; Sağlam, D.; Cengiz, Ş.E. Consumers' awareness, acceptance and attitudes towards functional foods in Turkey. *Int. J. Consum. Stud.* **2014**, *38*, 628–635. [CrossRef]

49. Van Kleef, E.; Van Trijp, H.C.M.; Luning, P. Functional foods: Health claim-food product compatibility and the impact of health claim framing on consumer evaluation. *Appetite* **2005**, *44*, 299–308. [CrossRef] [PubMed]
50. Williams, P.; Ridges, L.; Batterham, M.; Ripper, B.; Hung, M.C. Australian consumer attitudes to health claim—Food product compatibility for functional foods. *Food Policy* **2008**, *33*, 640–643. [CrossRef]
51. Giordano, S.; Clodoveo, M.L.; De Gennaro, B.; Corbo, F. Factors determining neophobia and neophilia with regard to new technologies applied to the food sector: A systematic review. *Int. J. Gastron. Food Sci.* **2018**, *11*, 1–19. [CrossRef]
52. Bruhn, C.M. Enhancing consumer acceptance of new processing technologies. *Innov. Food Sci. Emerg. Technol.* **2007**, *8*, 555–558. [CrossRef]
53. Rollin, F.; Kennedy, J.; Wills, J. Consumers and new food technologies. *Trends Food Sci. Technol.* **2011**, *22*, 99–111. [CrossRef]

 © 2020 by the authors. Licensee MDPI, Basel, Switzerland. This article is an open access article distributed under the terms and conditions of the Creative Commons Attribution (CC BY) license (http://creativecommons.org/licenses/by/4.0/).

Article

In Vitro and In Vivo Anti-inflammatory Activity of Bovine Milkfat Globule (MFGM)-derived Complex Lipid Fractions

Kate P. Palmano [1], Alastair K. H. MacGibbon [2,*], Caroline A. Gunn [2] and Linda M. Schollum [2]

1. Retired from Fonterra Research & Development Centre, Palmerston North 4442, New Zealand; kate.palmano@xtra.co.nz
2. Fonterra Research & Development Centre, Palmerston North 4442, New Zealand; Caroline.Gunn@fonterra.com (C.A.G.); Linda.schollum@fonterra.com (L.M.S.)
* Correspondence: Alastair.MacGibbon@fonterra.com; Tel.: +64-6-350-4649

Received: 18 June 2020; Accepted: 9 July 2020; Published: 15 July 2020

Abstract: Numerous health related properties have been reported for bovine milk fat globule membrane (MFGM) and its components. Here we present novel data on the in vitro and in vivo anti-inflammatory activity of various MFGM preparations which confirm and extend the concept of MFGM as a dietary anti-inflammatory agent. Cell-based assays were used to test the ability of MFGM preparations to modulate levels of the inflammatory mediators IL-1β, nitric oxide, superoxide anion, cyclo-oxygenase-2, and neutrophil elastase. In rat models of arthritis, using MFGM fractions as dietary interventions, the phospholipid-enriched MFGM isolates were effective in reducing adjuvant-induced paw swelling while there was a tendency for the ganglioside-enriched isolate to reduce carrageenan-induced rat paw oedema. These results indicate that the anti-inflammatory activity of MFGM, rather than residing in a single component, is contributed to by an array of components acting in concert against various inflammatory targets. This confirms the potential of MFGM as a nutritional intervention for the mitigation of chronic and acute inflammatory conditions.

Keywords: MFGM; phospholipids; gangliosides; dairy; inflammation; polar lipids; anti-inflammatory; IL-1β; nitric oxide; superoxide anion; cyclo-oxygenase-2; neutrophil elastase

1. Introduction

Fat droplets in milk are enclosed in a thin triple-layer membrane, called the milk fat globule membrane (MFGM), which stabilises the droplets as a colloidal dispersion throughout the liquid and prevents them from coalescing [1,2]. The MFGM is a complex association of both protein and lipid components and in particular, is a rich source of polar lipids such as phospholipids, sphingolipids and gangliosides [3]. The membrane proteins are distinct from milk serum proteins and comprise only 1–2% of the total milk protein. There are several dairy streams containing higher levels of MFGM than fresh milk and these can be used to produce various extracts with specific enrichments in either the protein or polar lipid components [4–6].

In recent years, the MFGM has garnered much interest as a specialty ingredient, due to the numerous health benefits associated with the bioactivities of both its protein and lipid constituents (for reviews see [1,7]). In animal and human trials, positive health outcomes have been reported following dietary intake of MFGM and in particular its polar lipid-enriched fractions, including enhanced neural and cognitive development [8,9], improved brain function and neuroplasticity [10], enhanced postnatal neuromuscular development [11], anti-infective and anti-febrile activity in infants [12,13] and postprandial cholesterol modulation [14]. Additionally, MFGM isolates or fractions have been shown to have anti-inflammatory properties in vitro [15–17], in animal models [14,18,19], and human clinical trials [14,20,21], including protection against intestinal inflammation in neonatal rat pups [22] and low birth weight mouse pups [23]. Furthermore, safety and tolerability of MFGM for use in infant formula has been confirmed [7].

It is likely that some of these bioactivities can be associated with specific components, or classes of components, which are present in the MFGM [24]. For example, sphingolipids and their various metabolites have been shown to have anti-cancer, anti-infective, anti-cholesterolaemic and gut maturation properties, in particular, by modulating the inflammatory responses associated with the various pathologies (for reviews see [1,3,25–29]). Dietary phospholipids have also been demonstrated to have an array of bioactivities including protection against cardiovascular disease, chemopreventive and chemotherapeutic activities, enhancement of memory and cognition, and anti-inflammatory activity in diseases such as arthritis and inflammatory ulcerative colitis (reviewed in [1,8,25,26,30]). Further, lack of expression of bacteria- derived phospholipids has been demonstrated to increase intestinal inflammation [31].

Clearly MFGM has the potential to confer a spectrum of health benefits, based on the activities of individual components or components acting in concert. However, its anti-inflammatory activity is of interest as inflammation is an underlying factor in many adverse health outcomes and diseases [32], and most chronic inflammatory diseases as well as allergic diseases are strongly influenced by nutrition [33]. There is some indication from the literature that differentially enriched MFGM fractions may deliver different anti-inflammatory outcomes. Park et al., [18] showed that a ganglioside enriched milkfat (MFGM) supplement inhibited release of pro-inflammatory signals in the intestinal mucosa and blood in a rat model of acute gut inflammation. Using in vitro models of acute gout, Dalbeth et al. [15] demonstrated that the MFGM-derived ganglioside-enriched dairy fraction G600, but not MFGM-derived phospholipid enriched fraction PC500, decreased pro-inflammatory cytokine release from THP-1 cell lines. In addition, the G600 concentrate inhibited the influx of inflammatory cells in a murine urate peritonitis model [15]. In a recent dietary intervention trial, however, a beta serum powder (BSP) that was lactose depleted [34], was shown to have no effect on post-prandial pro-inflammatory markers in obese and overweight adults [35]. This suggests that dose and composition could both be important.

In this paper, we present novel in vitro and in vivo data which provide further insights into the anti-inflammatory activity of various MFGM preparations. In an attempt to elucidate the most effective anti-inflammatory components of the MFGM, we tested a range of isolates derived from the beta serum starting material, each with varying concentrations of bioactive components, against a panel of in vitro assays of inflammatory biomarkers, with certain fractions being further selected for screening in well-established animal models of acute and chronic inflammatory arthritis.

2. Materials and Methods

2.1. Analytical Assays

The phospholipid concentration was determined by 31P nuclear magnetic resonance spectroscopy (NMR) as described by MacKenzie et al. [36]. The ganglioside concentration was determined by liquid chromatography high-resolution electrostatic ion-trap mass spectrometric (HPLC-MS) analysis as described by Fong et al. [37].

2.2. MFGM Fractions

MFGM complex lipid products BSP, BPC70, PC500, PC600, PC700, G500 and G600 were provided by Fonterra Co-operative Group Ltd., Auckland, New Zealand. Table 1 gives the gross composition of each of these products and indicates the various enrichments in phospholipid and ganglioside content. The MFGM isolates that were tested were concentrates derived from a beta serum powder (BSP) from anhydrous milkfat production. The manufacturing flow has been described by Gallier et al. [4] and Fontecha et al. [6]. The P) series (PC500, PC600, PC700) are phospholipid concentrates (PC) with varying degrees of residual neutral lipid (milkfat). The G series (G500, G600) are a more polar ganglioside concentrated fraction which includes a greater concentration of the more polar phospholipids. BPC70 is a MFGM protein dominant fraction produced by supercritical extraction using carbon dioxide and dimethylether [34].

Table 1. Composition of Milk Fat Globule Membrane Complex Lipid Fractions (g/100 g).

	BSP	PC600	PC700	PC500	G500	G600	BPC70
Total Lipids	17.3	86	84	89	33	30	7.1
Total Phospholipids	7.3	80	60	32.9	17.6	13.3	4.7
PI	0.6	2.6	2	1	2.8	2.5	0.3
PS	0.9	2.4	2.4	1.2	3.6	3.6	0.4
PC	2.0	26.5	19.1	10.6	3.1	1.8	1.5
PE	2.2	21.8	17	10.3	4.9	3.9	0.9
SM	1.5	24.9	16.6	9.2	2.8	1.5	1.5
Total Ganglioside	0.36	-	-	-	1.1	1.7	0.28
Protein	28.5				5.5	6.0	70.8
Lactose	46	6	6.2	4.1	56	58	10.5
Ash	6.4	11	7.4	4.5	5.0	8.3	5.3
Moisture	1.7	1.3	2.0	4.3	3.2	3.5	4.3
Neutral Lipid	10.0	6	24	56.1	14.3	15.0	2.1
GA/(GA+PL)	0.05	0	0	0	0.06	0.13	0.04

Key: BSP, beta serum powder; BPC70, beta serum protein concentrate; PC500/600/700, phospholipid concentrates; G500/600, ganglioside concentrates; PI, phosphatidylinositol; PS, phosphatidylserine; PC, phosphatidylcholine; PE, phosphatidylethanolamine; SM, sphingomyelin. BSP is the parent from which all fractions are derived. Other fractions represent increasing enrichment of phospholipids (PL), or gangliosides (GA) or protein.

2.3. Animal Procedures

All animal procedures and in vivo experiments were approved by the Wellington School of Medicine Animal Ethics Committee, NZ, according to national guidelines and regulations on animal welfare at the time of the individual studies. Animals were housed in conventional facilities with temperature control (20 ± 2 °C), a 12h light/dark cycle, and ad libitum access to food and water. Rats were euthanised with intraperitoneal ketamine/xylamine immediately prior to blood sampling at the end of each trial, or if necessary, during a trial for humane reasons.

Rat Monocytes were isolated for nitric oxide (NO) and interleukin-1β (IL-1β) assays as described in (Current Protocol in Immunology, John Wiley & Sons, Inc. Hoboken, NJ, USA.) (2001) [38].

Rat Neutrophils were isolated for neutrophil elastase (NE) and superoxide anion (SO) production assays as described [39]. Whole blood taken by cardiac puncture from healthy male Dark Agouti (DA) rats was collected into heparinised tubes and layered over Polymorphprep™ (density 1.113 g/mL; Alere ASA, Oslo, Norway). After centrifugation at $500 \times g$ for 30 min at 20 °C the lower of the two visible leukocyte bands, containing polymorphonuclear cells was harvested, washed twice and resuspended to a concentration of 10^7 cells/mL in Hanks Balanced Salt Solution (HBSS). Cells were maintained on ice and used within 2–4 h of collection. Purity of the cells as assessed by cytocentrifugation and staining was >95%.

2.4. In Vitro Assays

MFGM test samples were initially solubilised in 20% ethanol in HBSS (Hanks Balanced Salt Solution, Gibco Laboratories, Grand Is, NY, USA) to form stock solutions of 10 mg/mL. Samples were further diluted as necessary in 20% ethanol in HBSS. Unless otherwise stated the samples were tested at 1:100, 1:200 and 1:400 dilutions and measured in triplicate.

In vitro assays were principally cell and whole blood-based, and targeted the inflammatory mediators Interleukin-1β (IL-1β), nitric oxide (NO), superoxide anion (SO), cyclo-oxygenase-2 (COX-2), and neutrophil elastase (NE), all of which have key roles in the inflammatory responses underlying numerous chronic health conditions and pathologies [8,40–46]. Data on cyclo-oxygenase-1 (COX-1) activity are also included to indicate COX selectivity. Unlike inducible COX-2, COX-1 is not involved in acute inflammatory responses but is a constitutive enzyme involved in gastric mucosal protection [47].

2.4.1. Measurement of Neutrophil Elastase Activity

Assay of NE in activated rat neutrophils was based on the method of Yoshimura et al. [48]. PMA (phorbol 12-myristate 13-acetate, Sigma Aldrich, St Louis, MO, USA) was used to activate the cells, Alpha-1 proteinase inhibitor (Sigma-Aldrich, St. Louis, MO, USA) was used as a positive control and N-methoxysuccinyl-ala-ala-pro-val p-nitroaniline (N-A) (Sigma-Aldrich, St. Louis, MO, USA) was used as the chromogenic enzyme substrate in the assay.

2.4.2. Measurement of Superoxide Anion Production

Measurement of SO production by PMA-activated neutrophils was based on the colorimetric assay of Tan and Berridge ([49] in which a chromogenic substrate, the tetrazolium salt WST-1 (PreMix WST-1 Cell Proliferation Assay System, Takara Bio Inc., Shiga, Japan), is oxidised by SO produced during the respiratory burst. Aspirin (Sigma-Aldrich, St. Louis, MO, USA) dissolved in HBSS was used as a positive control.

2.4.3. Measurement of Nitric Oxide Production

This assay measured the release of NO following LPS-stimulated monocyte to macrophage conversion (Current Protocols in Immunology, Chapter 14: Oxidative metabolism of murine macrophages Unit 14 15. John Wiley & Sons, Inc., 2001) [50] and was based on the methods described in Yoshimura et al. [48] The NO concentration was measured by the colorimetric Griess reagent procedure using a kit according to the manufacturer's instructions (Cat No. 23479, Sigma, St Louis, MO, USA). LPS (Sigma-Aldrich, St. Louis, MO, USA) was used as the positive control for NO, and L-NMMA (NG-Methyl-L-Arginine, Sigma-Aldrich, St. Louis, MO, USA) was used as an inhibitor for NO.

2.4.4. Measurement of Interleukin -1β Release

This assay measured the release of the cytokine IL-1β following LPS-stimulated monocyte to macrophage conversion, as described in Current Protocols in Immunology, Chapter 14: Oxidative metabolism of murine macrophages, Unit 14 15., John Wiley & Sons, Inc., 2001 [50].

Indomethacin (Sigma-Aldrich, St. Louis, MO, USA) was used as the positive control. IL-1β was measured using an ELISA kit (R & D systems, Cat. No. RLB00) following the manufacturer's instructions.

2.4.5. Measurement of COX-1 and COX-2 Activity

COX-1 and COX-2 activities were determined using either a whole rat blood assay [51,52]) or using human cell line U937 monocytes ([53]. The human monocyte U937 cell line was purchased from American Type Culture Collection (ATCC, Rockville, MD, USA). Production of thromboxane B2 (TXB2,) or prostaglandin E2 (PGE2) following stimulation with LPS, were used as measures for COX-1 and COX-2, respectively. Indomethacin was used as a positive control. Both markers were determined using ELISA kits according to the manufacturer's instructions: TXB2 ELISA kit (R & D Systems, Minneapolis, MI, USA; Cat. No. DE0700), PGE2 ELISA kit (Cat. No. DE0100, R & D Systems, Minneapolis, MI, USA).

Final assay concentrations of MFGM test samples were 5–500 µg/mL. Controls and test compounds were assayed in duplicate.

2.4.6. Measurement of Human Neutrophil Elastase Activity

The effect of MFGM fractions on the in vitro activity of HNE was measured using the release of p-nitroalinine from the chromogenic substrate N-A. This assay was based on the methods described in [48]. p-Nitroaniline release was measured spectrophotometrically at 405 nm using a Bio-Rad ELISA reader.

2.5. In Vivo Experiments

2.5.1. Adjuvant-Induced Rat Model of Rheumatoid Arthritis

MFGM preparations were assessed for their abilities to modulate joint inflammation using the adjuvant-induced model of joint swelling in rats [54,55]. Control and treatment groups each comprised 8 female DA rats, aged 21 to 24 weeks (Trial A), or 6 female DA rats aged 16–19 weeks (Trial B), at the commencement of the study. Rats were fed either normal AIN-93 diet (control groups) or a diet supplemented with MFGM preparations at 38.3mg/g of diet (treatment groups) for two weeks prior to challenge [55]. On the same day as adjuvant injection, a positive control group was established by daily administration of an anti-arthritic drug via oral gavage (0.12 mg meloxicam/kg body weight (Metacam®, Boehringer-Ingelheim, St. Joseph, MO, USA)), until the completion of the trial 17 days after CFA administration.

Rat weights were recorded throughout the trial and food consumption was recorded every three days. From the 10th day after CFA administration, the size of each hind foot and ankle was measured daily with a plethysmometer. The volume measured was used to determine the foot volume change relative to the control at 17 days. Clinical scoring of the inflammation was also carried out, using the American Rheumatism Association guidelines [56,57] and a modification of the clinical severity scoring systems used by Larsson et al. [58] and Kawahito et al. [59]. Scores were assigned as follows: (A) for the three joints of the lateral four fingers or toes, 0 = normal, 1 = swelling; (B) for midfoot, mid-forepaw, ankle and wrist joints, 0 = normal, 1 = mild swelling, 2 = moderate swelling, 3 = severe swelling, 4 = non-weight bearing. The individual rat arthritic score was obtained by summing the scores recorded for each limb to give the foot score. The foot score measures the clinical severity of the inflammation, thus a score of zero is the normal condition. Maximum score per animal was 80.

2.5.2. Carrageenan-Induced Rat Model of Acute Inflammation

MFGM preparations were assessed for their abilities to modulate carrageenan-induced acute inflammation in the hind foot-pads of rats using 100 µL of 2.5% (w/v) carrageenan (Type IV lambda, Sigma Aldrich St Louis, MO, US) [60,61]. Each study group comprised 4 male Lewis (300–325 g) and 4 female Lewis rats (240–260 g). Rats were fed either normal AIN-93 diet (control groups) or diet supplemented with MFGM preparations at 38.3 mg/g of diet (treatment groups) for 15 days prior to the induction of acute inflammation. Food consumption and weight were recorded every three days. A positive control group was established by administering meloxicam by oral gavage on each of the two days prior to carrageenan injection. Paw swelling as determined by the volume displacement of the hind feet of all rats using a plethysmometer, was measured prior to, and 4 h after carrageenan injection. The percentage increase in volume of each foot was calculated.

2.5.3. Statistical Analyses

Statistical analyses were aimed at combining data from different experiments to obtain representative estimates of anti-inflammatory activity of MFGM preparations at different dose levels. In short, sample results were firstly expressed as percentage of their respective control (% Control), results from different trials were then pooled using weighted means, and lastly the significance of the pooled results was evaluated using the student *t*-test.

Results from the original experiments were recorded in a centralised repository as summary data, namely group mean, number of replicates, and standard deviation (SD) or standard error of the mean (SEM). Most of the group means were recorded in the original units but in some cases, data were only available as percentage of the control value (% Control). The detailed procedure, as follows, was applied to both in vitro and in vivo data: (1) mean sample results for each product (and the inhibitor control if included) were expressed as % Control using the mean of the control from the respective experiment. The SD of % Control was calculated from the approximate variance of the sample-to-control ratio using Taylor expansions [62]. (2) Where the same preparation was tested in multiple experiments, % Control results were pooled by calculating weighted means and SDs [63]. (3) The two-sided t-value for the difference between pooled mean and 100% was calculated (using the total number of replicates (n) to calculate the SEM of %Control) and its p-value found from the t distribution for n-1 degrees of freedom [63]. For preparations where only % Control data (mean, SD, and p-value) was originally recorded, results are reported accordingly herein, with or without pooling. Statistical significance for pooled results was evaluated as under (3).

All calculations were performed using Microsoft Excel 2016 and significance is declared if $p < 0.05$.

3. Results

3.1. In Vitro Assays

The effect of MFGM fractions on the production or activity of the various inflammatory mediators are presented below. Values are expressed as % control or as % stimulated control (LPS+, or PMA+) in those assays where cells were activated to produce an inflammatory response. Indomethacin, alpha-1-proteinase inhibitor, L-NMMA and aspirin were all effective inhibitors in their respective target assays, with aspirin inhibiting superoxide production in a dose-dependent manner. Since indomethacin gave similar levels of inhibition in both the cell and whole blood-based COX assays, inhibitor data from both methods of assay were pooled for COX-1 and COX-2, respectively. It must be noted that cell based in vitro assays are inherently variable, batch to batch and day to day, due to changes in the live cell concentrations and activity. To mitigate this the results have been expressed as % of the control activity to make comparison easier. Even so, variation was seen in the % of control response of the positive controls (especially where experiments were pooled). Thus while trends are observed, absolute results were more variable (as seen in the error bars).

3.1.1. Neutrophil Elastase (NE)

Results for NE activity following activation of rat neutrophils were collated from two datasets which are distinguished from each other as Series 1 (PI-1) and Series 2 (PI-2).

All the MFGM products tested appeared to inhibit the activity of NE. The first dataset shows that BSP, G500 and G600 all gave similar levels of inhibition, while the response with PC600 appeared to be more marked.

PC500 and PC700 were also mildly inhibitory, but as alpha-1 proteinase inhibitor appeared to be much less effective in this assay set, levels of inhibition for these two fractions cannot be directly compared with those of the other MFGM fractions (Figure 1).

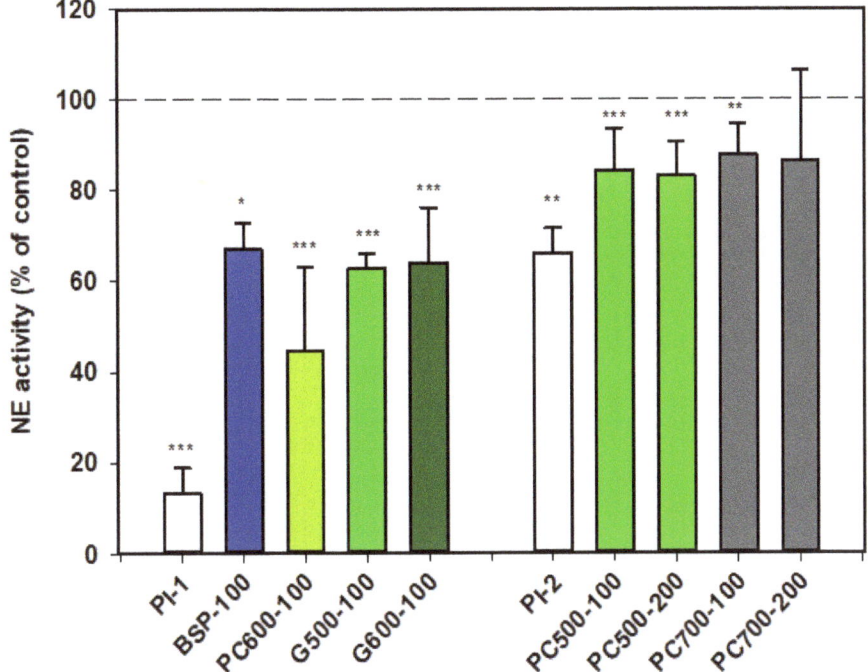

Figure 1. Neutrophil elastase activity relative to the control, two datasets. Values are means and SD (represented by error bars). Different coloured bars refer to different ingredients as indicated on the bottom axis (as described in Methods section and Table 1), with the last set of digits on each label indicating the ingredient dose (µg/mL). The positive control α-1-proteinase inhibitor (PI, 100 µg/mL) is shown prior to the samples in each series (PI-1 and PI-2 referring to series 1 and 2). Values for Series 1 represent the means from 1 experiment with triplicate measurements, or pooled means from 2 independent experiments, each with triplicate measurements. Values for Series 2 represent the pooled means from 2 or 3 independent experiments, each with triplicate measurements. The first series represents data where only % Control (mean, SD, and p-value) was originally recorded. Significance relative to control, * $p < 0.05$; ** $p \leq 0.01$; *** $p \leq 0.001$.

3.1.2. Superoxide Anion (SO)

Results for SO production by activated neutrophils were collated from two dose-response datasets, and as for NE, one set is distinguished from the other (Figure 2). Aspirin inhibitor values were consistent between datasets and the overall dose response shown in the first dataset (Figure 2) is representative of both sets. The MFGM total isolate BSP had no significant effect on SO production by neutrophils while the higher protein BPC70 extract gave significant, although similar, levels of inhibition at all three doses tested. The patterns of response of SO production to phospholipid enriched extracts were very similar, with generally mild inhibition across the dose range but no well-defined dose response. PC600 was slightly more inhibitory than either PC500 or PC700. The two ganglioside enriched fractions were also effective at inhibiting SO production. A weak dose response was observed with G500 with inhibition being significant at all doses. While a stronger dose response was observed with G600, the results were significant only at the highest dose (400 µg/mL). The level of inhibition corresponded to that produced by aspirin at the same dose (Figure 2).

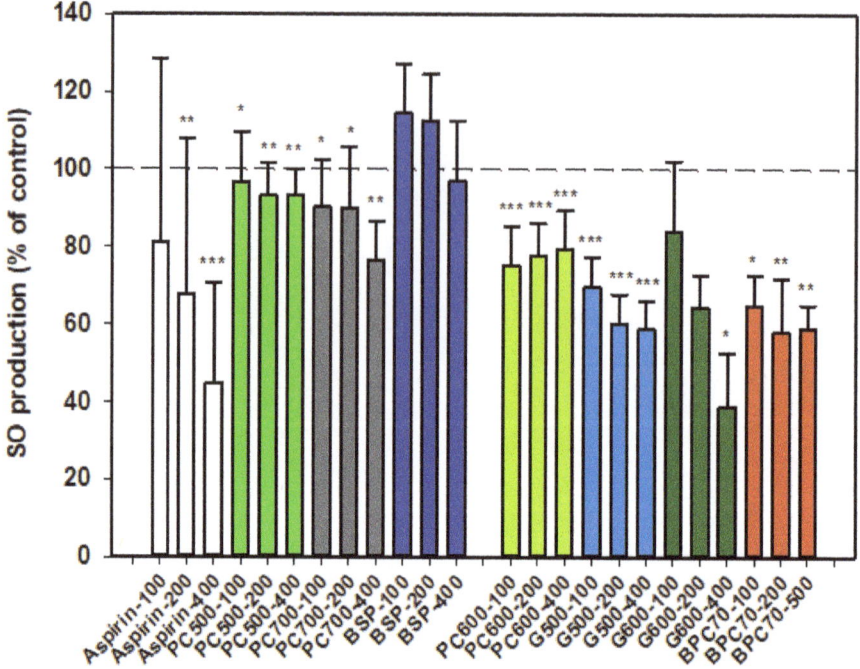

Figure 2. Superoxide anion production relative to the control. Two dose-response data sets. Values are means and SD (represented by error bars). Different bar colours refer to different ingredients as indicated on the bottom axis (as described in Methods section and Table 1), with the last set of digits on each label indicating the ingredient dose (µg/mL). The first data set includes aspirin as a comparison. The second set represents data where only % Control (mean, SD, and *p*-value) was originally recorded. Values for the first data set represent either the means from one experiment with triplicate measurement, or the pooled means from 3–4 independent experiments with triplicate measurement. Values for the second dataset represent either the means from one experiment with triplicate measurement, or the pooled means from 2 independent experiments with triplicate measurement. Significance relative to control, * $p < 0.05$; ** $p \leq 0.01$; *** $p \leq 0.001$.

3.1.3. Nitric Oxide (NO)

At all three doses tested, BSP inhibited NO production by LPS-stimulated monocytes, although the strongest inhibition was observed at the lowest dose (100 µg/mL) (Figure 3). On the other hand, PC700 had no effect at the lower doses while being somewhat stimulatory at the highest dose of 400 µg/mL. BPC70 also appeared to have a slight stimulatory effect, albeit only a single dose was tested. PC500 was mildly inhibitory at 100 µg/mL while PC600 was inhibitory at 100 µg/mL but significantly stimulatory at 400 µg/mL. The ganglioside fractions G500 and G600 both gave similar patterns of response, with the lower doses (100 and 200 µg/mL) being inhibitory (although not significant for G500 at 200 µg/mL), with no significant effect at the highest dose (400 µg/mL). Response at the lowest doses (100 µg/mL) approximated to the positive control, L-NMMA (52.8% ± 13.75 for G500, 49.2% ± 14.29 for G600 vs. 33.2% ± 6.28 for L-NMMA). Notably, this tendency toward an inverse dose response was observed with all fractions where more than one dose was tested (Figure 3).

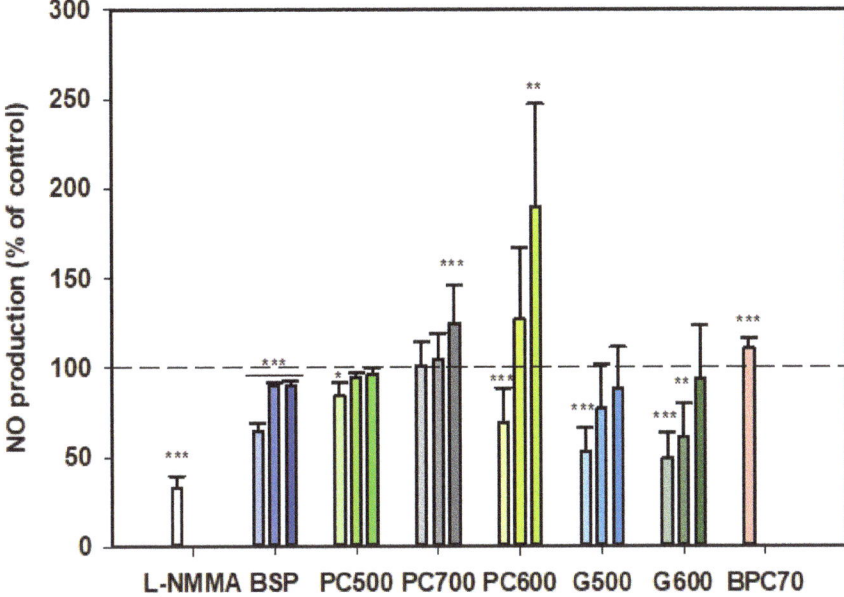

Figure 3. Nitric oxide production dose responses for samples relative to the control. Values are means and SD (represented by error bars). Different bar colours refer to different ingredients as indicated on the bottom axis (as described in Methods section and Table 1). Doses shown as filled bars (light, medium and dark) for 100, 200, 400 µg/mL, respectively. The BPC70 dose is 100 µg/mL. Positive control was L-NMMA (1 mM final concentration). Values represent either the means from one experiment with triplicate measurement, or the pooled means from 2–3 independent experiments with triplicate measurement. Data for L-NMMA represents the pooled mean from 3 independent experiments. Significance relative to control, * $p < 0.05$; ** $p \leq 0.01$; *** $p \leq 0.001$.

3.1.4. Interleukin-1ß (IL-1ß)

The protein enriched fraction BPC70 had no effect on production of IL-1β (Figure 4) from activated rat monocytes, while the phospholipid fractions PC500 and PC700 significantly inhibited the production of the cytokine in an apparently dose-dependent manner. The level of inhibition at the higher dose of 200 µg/mL was similar to that of indomethacin. PC600, the fraction with the highest phospholipid content, was potently inhibitory of IL-1β, while of the two ganglioside concentrates, G500 and G600, only G500 was inhibitory, producing a response similar to that of indomethacin (Figure 4).

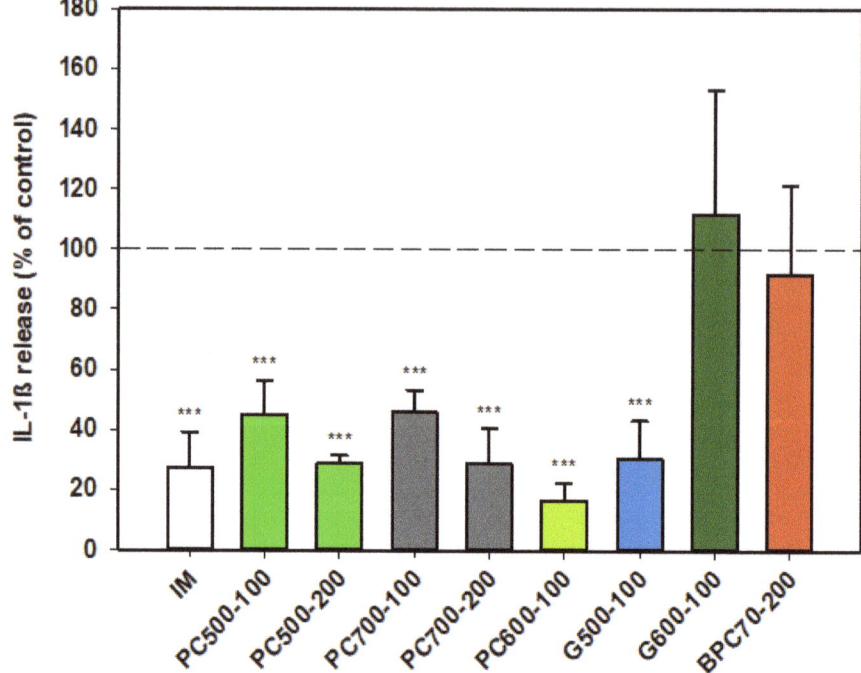

Figure 4. Interleukin-1ß responses relative to the control. Values are means and SD (represented by error bars). Different bar colours represent different ingredients as indicated on the bottom axis (as described in Methods section and Table 1), with the last set of digits on each label indicating the ingredient dose (µg/mL). The positive control is indomethacin (IM, 0.1 mM final concentration). Values represent either the means from one experiment with triplicate measurement, or the pooled means from 2 independent experiments with triplicate measurement. Data for indomethacin represents the mean from one experiment with triplicate measurement. Significance relative to control, *** $p \leq 0.001$.

3.1.5. COX-1/COX-2

Although indomethacin is known to be a non-selective COX inhibitor with a higher activity against the COX-1 isoform [51,52] in the assays reported here (Figure 5) indomethacin was significantly more inhibitory of COX-2 (than COX-1), $p = 0.002$. As expected, activation of monocytes by LPS (in both cell and whole blood assays) had no effect on COX-1 activity while stimulating COX-2 activity 2.5-fold (not shown). The MFGM isolate BSP was inhibitory of COX-2 at all doses tested over the range 5–500 µg/mL, with a slight inverse dose responsiveness. A stronger inverse dose-response was observed with COX-1, the MFGM isolate inhibiting the enzyme at the lower doses but significantly elevating activity at the highest dose (500 µg/mL). At the lower doses (5, 50 µg/mL) the effect of BSP on both COX isoforms was similar to that of indomethacin. BPC70, the membrane protein-enriched MFGM fraction had no meaningful effect on either of the COX enzymes. The phospholipid-enriched fractions PC500 and PC700, at a dose of 100 µg/mL, were both inhibitory of COX-2 while having no effect on COX-1. PC700 at the higher dose of 500 µg/mL inhibited COX-2, but also significantly increased COX-1 activity. The effect of G600, the ganglioside-enriched MFGM extract, on both COX isoforms was very similar to that of PC700 at the same dose (Figure 5).

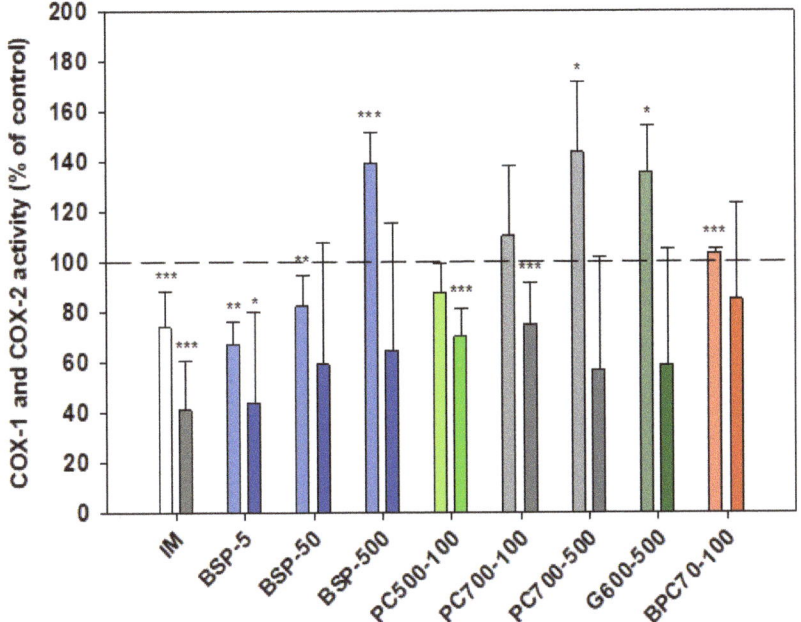

Figure 5. Cox-1 (lighter bars) and Cox 2 (darker bars) responses presented relative to control. Values are means and SD (represented by error bars). Cox-2 response is in the presence of LPS. Different bar colours refer to different ingredients as indicated on the bottom axis (as described in Methods section and Table 1), with the last set of digits on each label indicating the ingredient dose (µg/mL). The positive control is indomethacin (IM, 1 mM final concentration). Values represent either the means from one experiment with triplicate measurement, or the pooled means from 2–4 independent experiments with triplicate measurement. Data for indomethacin represents the mean from 3 independent experiments with triplicate measurement. Significance relative to control, * $p < 0.05$; ** $p \leq 0.01$; *** $p \leq 0.001$.

3.1.6. Human Neutrophil Elastase (HNE)

Neither of the ganglioside fractions G500 and G600 had any effect on the activity of the isolated HNE enzyme (Figure 6). The MFGM parent extract, BSP, was not significantly inhibitory, but the protein-enriched BPC70 product demonstrated dose response inhibition between 200–1000 µg/mL, with enzyme activity being inhibited up to 70% at the highest doses. At lower doses (50 and 100 µg/mL), there was some stimulation of the enzyme, but the overall trend was a decrease in enzyme activity with increasing dose (Figure 6).

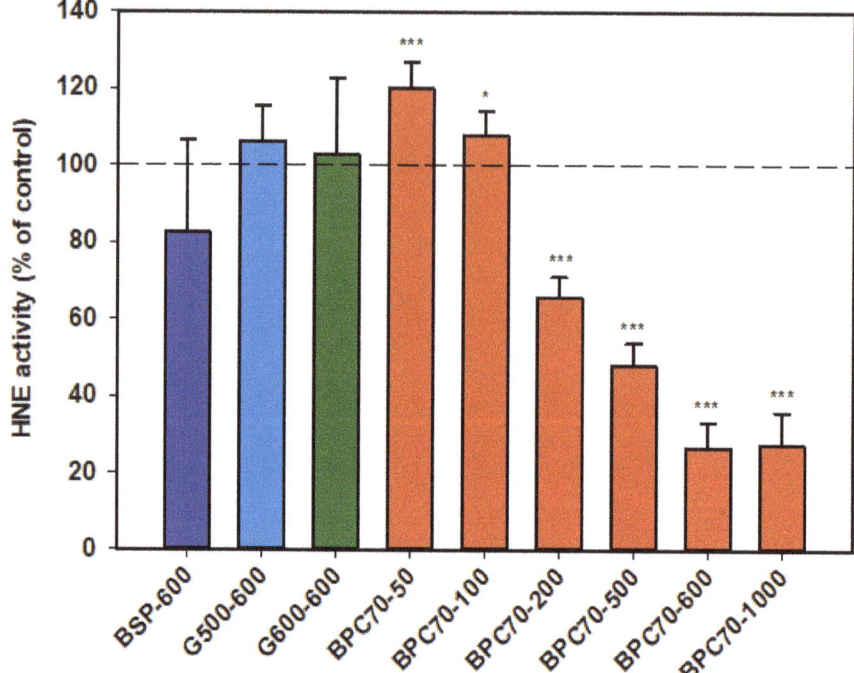

Figure 6. Human neutrophil elastase responses relative to the control. Values are means and SD (represented by error bars). Different bar colours refer to different ingredients as indicated on the bottom axis (as described in Methods section and Table 1), with the last set of digits on each label indicating the ingredient dose (µg/mL). Values represent either the means from one experiment with triplicate measurement, or the pooled means from 2 independent experiments with triplicate measurement. Significance relative to control, * $p < 0.05$; *** $p \leq 0.001$.

3.2. In Vivo Experiments

3.2.1. Adjuvant-Induced Arthritis

Three of the eight rats in the control group (Trial A) were lost to the trial. In accordance with reported disease trajectories for this animal model [56,64–66], clinical symptoms of arthritis were observed 10 days following adjuvant administration and increased rapidly for the next 4 days after which a plateau was reached at around 17 days, as seen in foot scores and foot volumes (Figure 7a,b Trial A), at which time the animals were euthanised. In the group administered meloxicam, inflammatory score and foot volume changes were ameliorated from Day 12 onwards. This was also the case for foot scores for the groups supplemented with MFGM fractions PC500 and PC700 (Figure 7), where a decline in the rate of swelling was observed at Day 12. As also expected with this model, rats lost weight following adjuvant administration with the greatest weight loss occurring between Days 10 and 17, corresponding to onset of clinical manifestations (not shown). Overall mean weight loss was 16–18% for all groups except the meloxicam controls, where mean weight loss was 14.5%. There was no significant difference in food intake between groups over the period of the trial (ANOVA, Tukey HSD post-hoc analysis) and appetite was not affected by weight loss.

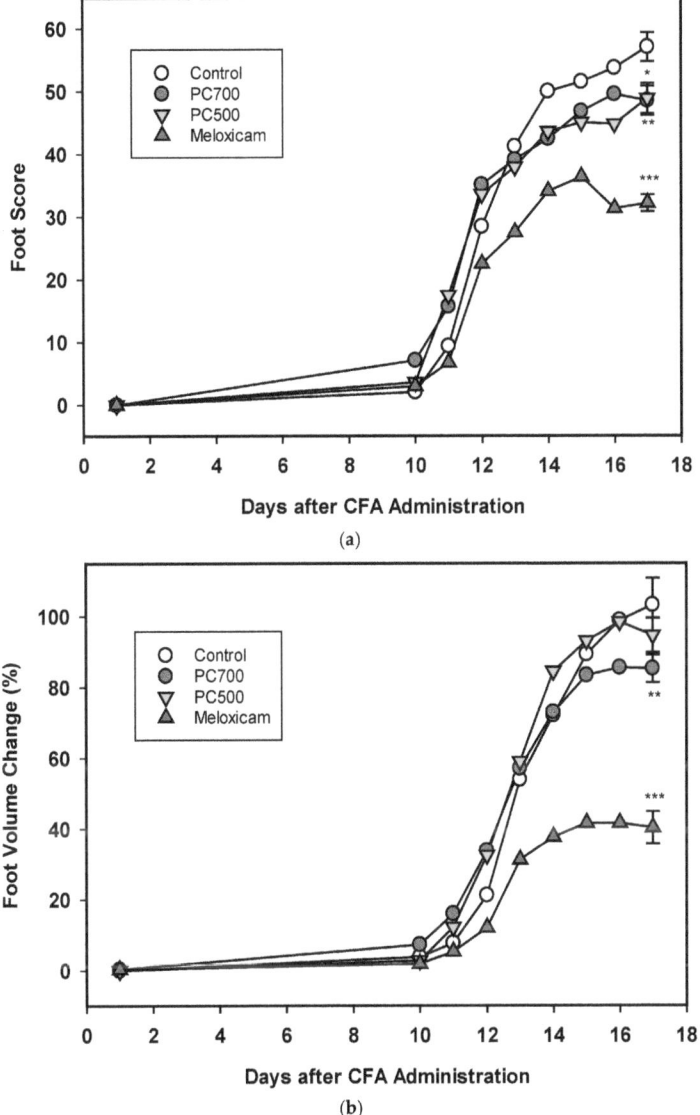

Figure 7. (**a**) The effects of PC700 and PC500 (as described in Methods section and Table 1) in the diet of rats on joint inflammation after the administration of complete Freund's' adjuvant (CFA) were compared with a control and the anti-arthritic drug meloxicam, based on the American Rheumatism Association (ARA) Scoring System (foot score). Values are means, n = 8 for all groups except control (n = 5). Error bars are SEM and probabilities * $p < 0.05$, ** $p < 0.01$, *** $p < 0.001$. (**b**) The effects of PC700 and PC500 in the diet of rats on joint inflammation after the administration of complete Freund's adjuvant (CFA) were compared with a control and the anti-arthritic drug meloxicam, based on the change on foot volume (relative to the control at Day 17). Values are means, $n = 8$ for all groups except control ($n = 5$). Error bars are SEM and probabilities * $p < 0.05$, ** $p < 0.01$, *** $p < 0.001$.

The degree of inflammation of the joints at Day 17, the clinical end-point of the trial, was assessed both by foot score (all four feet) and foot volume (volume displacement of both hind feet). The positive control meloxicam effectively ameliorated joint swelling as determined by reduction in foot score (Figure 7a), which is a measure of the clinical severity of the inflammation with zero being the normal state. PC700 and to a lesser extent PC500 produced significant anti-inflammatory effects (Figure 7a) (82.2%control ± 0.89 (mean ± SEM) and 86.8% control ± 1.27, respectively).

With regard to the foot volume change (Figure 7b), which is a measure of the extent of swelling of the foot, the positive control meloxicam was very effective in ameliorating joint swelling. PC700 showed a significant, although milder, anti-inflammatory effect while there was no significant effect with PC500 (85.1% control ± 0.93 (mean ± SEM) and 91.9% control ± 1.78, respectively). Neither PC600 nor G600 were effective in this model. Although there were some tendency towards a reduction in foot score, results did not reach significance (Trial B; foot scores 90.9% control ± 1.83 (mean ± SEM), and 94.5% control ± 2.10 respectively; foot volumes 95.5% control ± 1.99 and 99.2% control ± 1.78, respectively).

3.2.2. Carrageenan-induced Paw Oedema

There was no significant difference in food intake between the treatment groups over the period of the trial (ANOVA, Tukey HSD post-hoc analysis). Carrageenan injection into the hind feet of rats induced a rapid inflammatory response, as observed by an 80% increase in foot volume in the control group (79.8% ± 0.71 (mean ± SEM), $n = 16$) 4h post-injection. Meloxicam administration two days prior to carrageenan significantly reduced the inflammation as measured by a 31% reduction in foot volume compared to the control (68.7% control ±1.46 (mean ± SEM, $n = 16$, $p < 0.001$)), although the response to meloxicam in this acute model was not as marked as in the longer-induction adjuvant model (63% reduction, Figure 7b). There was a 10% reduction in the foot volume of rats fed the G600-supplemented diet, compared to the control; however, the result did not quite reach significance (90.5% control ± 1.25 (mean ± SEM), $n = 16$, $p = 0.072$). The PC600 and PC700 supplemented diets had no effect on foot volume changes induced by carrageenan (data not shown).

4. Discussion

Ascribing an anti-inflammatory effect to any one component or class of component within a complex food substrate is not straightforward, as there may be bioactive synergies or even antagonism at play within the milieu. However, by using selectively enriched fractions it is at least possible to determine which group, or groups of components may contribute to an effect, and further, whether they might have potential as targeted nutritional interventions to modulate inappropriate inflammatory responses. The inflammatory response is multi-layered and involves recruitment of an array of inflammatory cytokines and other mediators in response to infection or injury. However, if these factors are not down-regulated appropriately following resolution of the insult or injury, or if the response is sustained, the inflammation can be become chronic and damaging [25]. In respect of the five inflammatory markers described in this report, IL-1β, a key inflammatory cytokine [45], is implicated in osteoarthritis (OA) [46], and also plays a pivotal role in the pathogenesis of numerous other acute and chronic inflammatory diseases including atherosclerosis, type 2 diabetes, and neurodegenerative disease [67]. The serine proteinase NE, secreted by neutrophils and macrophages during inflammation, has been implicated in lung disease and progression [44,68], while the inducible inflammatory mediator COX-2, a primary target for non-steroidal anti-inflammatory drugs (NSAIDs), has been implicated in arthritis, atherosclerosis, cancer and the neuronal cell injury involved in Alzheimer's disease [53,69]. In OA, excessive amounts of NO produced largely by inducible nitric oxide synthase (iNOS) inhibit matrix synthesis and promote its degradation, while reaction of NO with reactive oxygen species such as SO promotes cellular injury and cytokine-induced apoptosis [70]. Similarly, SO produced at high levels during the respiratory burst phase of the inflammatory response, can be toxic to tissues and result in endothelial damage [41].

The hierarchy of the immune response is such that the expression of one pro-inflammatory mediator is generally regulated through another [45]. However, any of the individual mediators in an inflammatory pathway may act as targets for therapeutic intervention. In this paper, we have presented a collation of data on the effects of various MFGM preparations on the in vitro and in vivo activity of the above key inflammatory mediators with the aim of comparing the patterns of response and gaining further insights into the potential of these fractions as nutritional anti-inflammatory interventions.

From the available data a number of the MFGFM fractions demonstrated a broad range of anti-inflammatory activity rather than being selective towards expression or activity of one particular inflammatory mediator, suggesting the potential for use in a wide range of anti-inflammatory indications. However, level of inhibition was generally more marked against IL-1β. IL-1β is implicated in inflammatory processes at multiple levels [45], and is a pivotal cytokine, affecting the expression of other pro-inflammatory mediators such as iNOS, COX-2 and reactive oxygen species (ROS) (e.g., SO) in diseases such as OA [71]. Although it is possible that IL-1β expression might have been a primary target of inhibition in some fractions, the patterns of inhibition were, however, complex and in the case of NO, inhibition was more marked at lower doses with an inverse dose-response observed for all MFGM substrates tested. In the case of the phospholipid fractions PC600 and PC700, there was in fact a significant stimulation of NO production at the higher doses. Given that some of the MFGM fractions have quite varied compositions, the reason for this consistent pattern of activity is not clear. It is possible, however, that all preparations have various other factors in common which when presented at a threshold dose counteract and override inhibitory activity. In the assays described herein, monocytes and neutrophils were activated with LPS and PMA respectively, to stimulate upregulation of pro-inflammatory markers. However, MFGM fractions were pre-incubated with the cells (or whole blood) prior to addition of the inflammatory agent, so it is possible that the inflammatory response may have been modified at the level of cell activation or enzyme expression, or as in the case of NE, release from neutrophil granulocytes. The results from the assay of isolated enzyme human neutrophil elastase, although limited, provide an insight. The parent MFGM preparation, BSP which contains both protein and lipid components, was effective at inhibiting NE in the activated neutrophil assay, while the protein-enriched fraction BPC70 was strongly and dose-dependently inhibitory of HNE in the isolated enzyme assay. The ganglioside-enriched phospholipid fractions, G500 and G600, were ineffective in the latter assay at comparatively high doses, and yet both were inhibitory of NE in the neutrophil assay, as were MFGM phospholipid fractions in general This suggests that there may have been dual mechanisms of inhibition at play, with the protein component of MFGM directly affecting enzyme activity, and the lipid component acting at an earlier step in the pathway of NE release from neutrophils. While it is difficult to draw conclusions in the absence of a full spectrum of data, it appeared that of the MFGM fractions studied, the protein fraction BPC70 had the least effect on inflammatory mediators expressed in activated monocytes and neutrophils.

Comparing the profiles of activity between different MFGM preparations, some overall trends can be observed. None of the phospholipid or ganglioside-enriched fractions were inhibitory towards the constitutive enzyme COX-1 while being inhibitory, or tending to be inhibitory, of COX-2. The parent MFGM preparation BSP, containing both protein and lipid constituents, tended to be inhibitory of COX-2 across the dose range, although inhibitory of COX-1 only at the lower doses tested. At the single dose tested, the BPC70 fraction, containing predominantly MFGM membrane protein with lesser amounts of phospholipid and ganglioside, had no significant effect on either COX-1 or COX-2. This suggests that the COX-2 inhibitory activity of MFGM resides largely in the phospholipid fraction, and furthermore the null effect of these fractions on COX-1 suggests that there is either a relative selectivity towards the COX-2 isoform, or COX-2 enzyme activity is attenuated via another mechanism.

Within the phospholipid-enriched MFGM groups PC500, PC600 and PC700, the increasing order of phospholipid enrichment is PC500 < PC700 < PC600. While PC500 and PC700 demonstrated broadly similar inhibitory activities across the range of inflammatory mediators tested, levels of inhibition seen with PC600, for those mediators assayed in common, were substantially higher. This supports the notion that the major anti-inflammatory activity lies largely within the phospholipid portion of the MFGM. This is perhaps not surprising since phospholipids administered in the diet have been reported to have anti-inflammatory effects [1,8,25,30,72] and phosphatidylserine (PS), a constituent of the MFGM phospholipid fraction, has been shown to inhibit the production of inflammatory mediators IL-6, IL-8 and PGE2 in vitro, as well as alleviate carrageenan-induced inflammation in rats when co-administered with the irritant [73].

In general, the ganglioside-enriched MFGM fractions G500 and G600 demonstrated similar patterns of inhibitory behaviour towards inflammatory markers, and the levels of inhibition were like those observed with PC600. Although both G500 and G600 products contain some phospholipid, and in fact a higher content of PS than the other phospholipid fractions (Table 1), the overall phospholipid content including SM is significantly less than that for PC600, indicating that the gangliosides themselves may contribute to anti-inflammatory activity. This is in line with observations that dietary gangliosides can interrupt pro-inflammatory signalling via IL-1β in the intestinal mucosa, specifically suppressing production of IL-1β, NO, PGE2, hydrogen peroxide, IL-6 and IL-8, and subsequently alleviating symptoms of intestinal disease [18,74]. Interestingly, in the assays reported here, G500 was highly inhibitory of IL-1β whereas G600 had no effect on this marker. Given that G600 and G600 both inhibited production of NO, SO and NE to a similar degree, this seems an anomaly. However, G600 has a higher content of GD3 (the predominant bovine milk ganglioside, [75] while G500 has a higher content of GM3 (the predominant human milk ganglioside, produced in bovine MFGM fractions by desialylation of GD3) [24]. This suggests that the mono-sialylated ganglioside (GM3), but not the di-sialylated ganglioside (GD3) might somehow downregulate IL-1β expression. A final point worth mentioning in respect of the in vitro data is that the parent MFGM fraction BSP had no effect on neutrophil SO production, while the protein extract BPC70 was inhibitory although not dose-responsively. This supports the notion that there may be anti-inflammatory factors within the protein as well as the lipid components of the MFGM. It has been reported that MFG-E8 (milk fat globule-epidermal growth factor 8), also known as PA6/7 and a major protein component of the MFGM [16], attenuated intestinal inflammation in murine experimental colitis by significantly down-regulating LPS-induced proinflammatory cytokines [76].

Adjuvant-induced inflammation is a well-established and robust model for chronic inflammatory disease and is an important tool for screening and testing of new therapeutics for rheumatoid arthritis. Moreover, it has a proven track record for predictability of efficacy in humans [64,65]. Equally, the carrageenan-induced paw oedema is a useful and reproducible model for testing the efficacy of potential anti-inflammatory agents in acute inflammation [54,60,61]. Both models were used to screen MFGM fractions, selected based on results from primary in vitro screens for their potential as nutritional interventions in the amelioration of inflammatory conditions. Of the four MFGM complex lipid fractions screened as dietary supplements through the adjuvant model (PC500, PC700, PC600, G600) the two phospholipid fractions, P500 and PC700, were effective in ameliorating the delayed-onset foot and ankle swelling which is characteristic of this model. In contrast, of the three complex lipid fractions screened as dietary supplements through the carrageenan model of acute paw oedema (PC500, PC600, G600), only G600 showed a tendency towards anti-inflammatory activity. It is of interest to note here that in a proof of concept clinical trial, G600 as a dietary supplement in combination with a dairy macropeptide, was able to reduce the number of gout flares in gout patients with a history of chronic flares [20]. This suggests that the G600 may well have utility in mitigating acute joint inflammation.

The in vivo behaviours of MFGM fractions observed in this study did not precisely follow the patterns of anti-inflammatory shown in vitro, i.e., in vitro, PC600 was generally more effective against all pro-inflammatory mediators than PC500 or PC700, and ganglioside fractions G500 and G600 were as effective as all phospholipid fractions. However, in vitro activity is not necessarily predictive of in vivo efficacy and the outcome may be dose-dependent. Moreover, since the two rat models are inherently different in terms of the temporal recruitment of inflammatory mediators, the pro-inflammatory targets for the MFGM fractions or components thereof, may differ between models or be influenced differently. It is likely that the anti-inflammatory activity of the dietary MFGM fractions, as observed in the adjuvant model of arthritis (PC500, PC700) and to some extent in the carrageenan-induced paw oedema model (G600), is mediated through the gut mucosa and thence via the systemic immune system.

In this context, it was recently reported that dietary SM (0.1% w/w) attenuated hepatic steatosis and adipose tissue inflammation, and strongly attenuated serum inflammatory cytokines and chemokines in diet-induced obese mice [27] while dietary input of the phospholipids PC and PE, in conjunction with N-acyl PEs reduced the acute inflammatory response in a mice model of carrageenan-induced pleurisy [77]. In addition, it has been reported that dietary supplementation with MFGM protected against intestinal permeability in LPS-challenged mice by controlling the inflammatory response, causing the gut barrier to remain less permeable [78] and similar results had been observed in vitro [79]. In the MFGM supplemented animals, significant decreases in serum levels of the proinflammatory cytokines were also observed. Thus, immune regulation at the level of the gut provides a plausible mechanism through which anti-inflammatory dietary supplements may exert their effects remotely.

5. Conclusions

The novel in vitro data and in vivo results presented in the paper confirm and extend observations on the anti-inflammatory effects of MFGM and its derivative membrane protein, phospholipid and ganglioside enriched fractions, and further demonstrate the potential of MFGM as a nutritional intervention to mitigate chronic and acute inflammatory conditions, especially those relevant to joint disease. Additionally, due to the ability of MFGM to influence the activity and recruitment of inflammatory mediators, as observed from our in vitro data and from in vivo data elsewhere, MFGM supplementation may well have application as a moderator of heightened immune responses. In this study, no single class of components appeared to be responsible for all the observed anti-inflammatory effects, which suggests that there is synergy between the various components involved in the MFGM response. Though statistically significant anti-inflammatory activities have been observed in these primary in vitro and in vivo activity screens, clinical impact of MFGM fractions on inflammatory indications will require further study in proof of concept human trials.

Author Contributions: Conceptualization, K.P.P. and A.K.H.M.; validation, K.P.P., A.K.H.M., L.M.S. and C.A.G.; formal analysis, B.K.S.; data curation, A.K.H.M.; writing—original draft preparation, K.P.P.; writing—review and editing, K.P.P., A.K.H.M., L.M.S. and C.A.G.; visualization, K.P.P., A.K.H.M., L.M.S. and C.A.G. All authors have read and agreed to the published version of the manuscript.

Funding: This work was supported by a grant from the FOUNDATION FOR RESEARCH, SCIENCE, AND TECHNOLOGY, NEW ZEALAND (LPHM0201).

Acknowledgments: Thanks to Paul Davis of Trinity Bioactives and Geoff Krissansen of Auckland Medical School for providing assay services and Barbara Kuhn Sherlock for the formal statistical analysis.

Conflicts of Interest: A.K.H.M., L.M.S. and C.A.G. declare that they are employees of Fonterra Cooperative Ltd.

References

1. Conway, V.; Couture, P.; Gauthier, S.; Pouliot, Y.; Lamarche, B. Effect of buttermilk consumption on blood pressure in moderately hypercholesterolemic men and women. *Nutrition* **2014**, *30*, 116–119. [CrossRef] [PubMed]
2. MacGibbon, A.K.H.; Taylor, M.W. Composition and structure of bovine milk lipids. In *Advanced Dairy Chemistry: Lipids*, 3rd ed.; Fox, M., Ed.; Springer: Berlin/Heidelberg, Germany, 2006; Volume 2, pp. 1–43.
3. El-Loly, M. Composition, Properties and Nutritional Aspects of milk fat globule membrane—A Review. *Pol. J. Food Nutr. Sci.* **2011**, *61*, 7–32. [CrossRef]
4. Gallier, S.; MacGibbon, A.K.H.; McJarrow, P. Milk fat globule membrane supplementation and cognition. *Agro Food Ind. Hi Tech* **2018**, *29*, 14–16.
5. Huang, Z.; Zheng, H.; Brennan, C.; Mohan, M.; Stipkovits, L.; Li, L.; Kulasiri, D. Production of Milk Phospholipid-Enriched Dairy Ingredients. *Foods* **2020**, *9*, 263. [CrossRef] [PubMed]
6. Fontecha, J.B.; Brink, L.; Wu, S.; Pouliot, Y.; Visioli, F.; Jimenez-Flores, R. Sources, production, and clinical treatments of milk fat globule membrane for infant nutrition and well-being. *Nutrients* **2020**, *12*, 1607. [CrossRef]

7. Hernell, O.; Timby, N.; Domellof, M.; Lonnerdal, B. Clinical benefits of milk fat globule membranes for infants and children. *J. Pediatrics* **2016**, *173*, S60–S65. [CrossRef] [PubMed]
8. Kullenberg, D.; Taylor, L.A.; Schneider, M.; Massing, U. Health effects of dietary phospholipids. *Lipids Health Dis.* **2012**, *11*, 3. [CrossRef]
9. Vickers, M.H.; Guan, J.; Gustavsson, M.; Krageloh, C.U.; Breier, B.H.; Davison, M.; Fong, B.; Norris, C.; McJarrow, P.; Hodgkinson, S.C. Supplementation with a mixture of complex lipids derived from milk to growing rats results in improvements in parameters related to growth and cognition. *Nutr. Res.* **2009**, *29*, 426–435. [CrossRef]
10. Guillermo, R.B.; Yang, P.Z.; Vickers, M.H.; McJarrow, P.; Guan, J. Supplementation with complex milk lipids during brain development promotes neuroplasticity without altering myelination or vascular density. *Food Nutr. Res.* **2015**, *59*, 25765. [CrossRef]
11. Markworth, J.F.; Durainayagam, B.; Figueiredo, V.C.; Liu, K.; Guan, J.; MacGibbon, A.K.H.; Fong, B.Y.; Fanning, A.C.; Rowan, A.; McJarrow, P.; et al. Dietary supplementation with bovine-derived milk fat globule membrane lipids promotes neuromuscular development in growing rats. *Nutr. Metab.* **2017**, *14*, 9. [CrossRef]
12. Zavaleta, N.; Kvistgaard, A.S.; Graverholt, G.; Respicio, G.; Guija, H.; Valencia, N.; Lonnerdal, B. Efficacy of an MFGM-enriched complementary food in diarrhea, anemia, and micronutrient status in infants. *J. Pediatric Gastroenterol. Nutr.* **2011**, *53*, 561–568. [CrossRef]
13. Timby, N.; Hernell, O.; Vaarala, O.; Melin, M.; Lonnerdal, B.; Domellof, M. Infections in infants fed formula supplemented with bovine milk fat globule membranes. *J. Pediatric Gastroenterol. Nutr.* **2015**, *60*, 384–389. [CrossRef] [PubMed]
14. Demmer, E.; Van Loan, M.D.; Rivera, N.; Rogers, T.S.; Gertz, E.R.; German, J.B.; Smilowitz, J.T.; Zivkovic, A.M. Addition of a dairy fraction rich in milk fat globule membrane to a high-saturated fat meal reduces the postprandial insulinaemic and inflammatory response in overweight and obese adults. *J. Nutr. Sci.* **2016**, *5*, e14. [CrossRef] [PubMed]
15. Dalbeth, N.; Gracey, E.; Pool, B.; Callon, K.; McQueen, F.M.; Cornish, J.; MacGibbon, A.; Palmano, K. Identification of dairy fractions with anti-inflammatory properties in models of acute gout. *Ann. Rheum. Dis.* **2010**, *69*, 766–769. [CrossRef]
16. Snow, D.R.; Ward, R.E.; Olsen, A.; Jimenez-Flores, R.; Hintze, K.J. Membrane-rich milk fat diet provides protection against gastrointestinal leakiness in mice treated with lipopolysaccharide. *J. Dairy Sci.* **2011**, *94*, 2201–2212. [CrossRef] [PubMed]
17. Zanabria, R.; Tellez, A.M.; Griffiths, M.; Sharif, S.; Corredig, M. Modulation of immune function by milk fat globule membrane isolates. *J. Dairy Sci.* **2014**, *97*, 2017–2026. [CrossRef] [PubMed]
18. Park, E.J.; Suh, M.; Thomson, B.; Ma, D.W.; Ramanujam, K.; Thomson, A.B.; Clandinin, M.T. Dietary ganglioside inhibits acute inflammatory signals in intestinal mucosa and blood induced by systemic inflammation of Escherichia coli lipopolysaccharide. *Shock* **2007**, *28*, 112–117. [CrossRef]
19. Zhou, A.L.; Ward, R.E. Milk polar lipids modulate lipid metabolism, gut permeability, and systemic inflammation in high-fat-fed C57BL/6J ob/ob mice, a model of severe obesity. *J. Dairy Sci.* **2019**, *102*, 4816–4831. [CrossRef]
20. Dalbeth, N.; Ames, R.; Gamble, G.D.; Horne, A.; Wong, S.; Kuhn-Sherlock, B.; MacGibbon, A.; McQueen, F.M.; Reid, I.R.; Palmano, K. Effects of skim milk powder enriched with glycomacropeptide and G600 milk fat extract on frequency of gout flares: A proof-of-concept randomised controlled trial. *Ann. Rheum. Dis.* **2012**, *71*, 929–934. [CrossRef] [PubMed]
21. Beals, E.; Kamita, S.G.; Sacchi, R.; Demmer, E.; Rivera, N.; Rogers-Soeder, T.S.; Gertz, E.R.; Van Loan, M.D.; German, J.B.; Hammock, B.D.; et al. Addition of milk fat globule membrane-enriched supplement to a high-fat meal attenuates insulin secretion and induction of soluble epoxide hydrolase gene expression in the postprandial state in overweight and obese subjects. *J. Nutr. Sci.* **2019**, *8*, e16. [CrossRef]
22. Bhinder, G.; Allaire, J.M.; Garcia, C.; Lau, J.T.; Chan, J.M.; Ryz, N.R.; Bosman, E.S.; Graef, F.A.; Crowley, S.M.; Celiberto, L.S.; et al. Milk fat globule membrane supplementation in formula modulates the neonatal gut microbiome and normalizes intestinal development. *Sci. Rep.* **2017**, *7*, 45274. [CrossRef] [PubMed]
23. Huang, S.; Wu, Z.; Liu, C.; Han, D.; Feng, C.; Wang, S.; Wang, J. Milk fat globule membrane supplementation promotes neonatal growth and alleviates inflammation in low-birth-weight mice treated with lipopolysaccharide. *BioMed. Res. Int.* **2019**, *2019*, 4876078. [CrossRef] [PubMed]

24. Palmano, K.; Rowan, A.; Guillermo, R.; Guan, J.; McJarrow, P. The role of gangliosides in neurodevelopment. *Nutrients* **2015**, *7*, 3891–3913. [CrossRef] [PubMed]
25. Lordan, R.; Zabetakis, I. Invited review: The anti-inflammatory properties of dairy lipids. *J. Dairy Sci.* **2017**, *100*, 4197–4212. [CrossRef] [PubMed]
26. Schverer, M.; O'Mahony, S.; O'Riordan, K.; Donoso, F.; Roy, B.; Stanton, C.; Dinan, T.; Schellekens, H.; Cryan, J. Dietary phospholipids: Role in cognitive processes across the lifespan. *Neurosci. Biobehav. Rev.* **2020**, *111*, 183–193. [CrossRef] [PubMed]
27. Norris, G.H.; Blesso, C.N. Dietary and endogenous sphingolipid metabolism in chronic inflammation. *Nutrients* **2017**, *9*, 1180. [CrossRef]
28. Norris, G.H.; Milard, M.; Michalski, M.C.; Blesso, C.N. Protective properties of milk sphingomyelin against dysfunctional lipid metabolism, gut dysbiosis, and inflammation. *J. Nutr. Biochem.* **2019**, *73*, 108224. [CrossRef]
29. Anto, L.; Warykas, S.W.; Torres-Gonzalez, M.; Blesso, C.N. Milk Polar Lipids: Underappreciated Lipids with Emerging Health Benefits. *Nutrients* **2020**, *12*, 1001. [CrossRef]
30. Contarini, G.; Povolo, M. Phospholipids in milk fat: Composition, biological and technological significance, and analytical strategies. *Int. J. Mol. Sci.* **2013**, *14*, 2808–2831. [CrossRef]
31. Brown, E.M.; Ke, X.; Hitchcock, D.; Jeanfavre, S.; Avila-Pacheco, J.; Nakata, T.; Arthur, T.D.; Fornelos, N.; Heim, C.; Franzosa, E.A.; et al. Bacteroides-Derived Sphingolipids Are Critical for Maintaining Intestinal Homeostasis and Symbiosis. *Cell Host Microbe* **2019**, *25*, 668–680.e667. [CrossRef]
32. Libby, P. Inflammatory mechanisms: The molecular basis of inflammation and disease. *Nutr. Rev.* **2007**, *65*, S140–S146. [CrossRef] [PubMed]
33. Bordoni, A.; Danesi, F.; Dardevet, D.; Dupont, D.; Fernandez, A.S.; Gille, D.; Nunes Dos Santos, C.; Pinto, P.; Re, R.; Remond, D.; et al. Dairy products and inflammation: A review of the clinical evidence. *Crit. Rev. Food Sci. Nutr.* **2017**, *57*, 2497–2525. [CrossRef]
34. Fletcher, K.; Catchpole, O.; Grey, J.B.; Pritchard, M. Beta-serum Dairy Products, Neutral Lipid-Depleted and/or Polar Lipid-Enriched Dairy Products, and Processes for Their Production. U.S. Patent 8471002, 25 June 2013.
35. Rogers, T.S.; Demmer, E.; Rivera, N.; Gertz, E.R.; German, J.B.; Smilowitz, J.T.; Zivkovic, A.M.; Van Loan, M.D. The role of a dairy fraction rich in milk fat globule membrane in the suppression of postprandial inflammatory markers and bone turnover in obese and overweight adults: An exploratory study. *Nutr. Metab.* **2017**, *14*. [CrossRef] [PubMed]
36. MacKenzie, A.V.; Vyssotski, M.; Nekrasov, E. Quantitative Analysis of Dairy Phospholipids by 31P NMR. *J. Am. Oil Chem. Soc.* **2009**, *86*, 757–763. [CrossRef]
37. Fong, B.; Norris, C.; McJarrow, P. Liquid chromatography–high-resolution electrostatic ion-trap mass spectrometric analysis of GD3 ganglioside in dairy products. *Int. Dairy J.* **2011**, *21*, 42–47. [CrossRef]
38. Wahl, L.M.; Smith, P.D. Isolation of Monocyte/Macrophage Populations. In *Current Protocols in Immunology*; John Wiley & Sons, Inc.: Hoboken, NJ, USA, 2001.
39. Clark, R.A.; Nauseef, W.M. Isolation and functional analysis of neutrophils. In *Current Protocols in Immunology*; John Wiley & Sons, Inc.: Hoboken, NJ, USA, 2001.
40. Guilak, F.; Fermor, B.; Keefe, F.J.; Kraus, V.B.; Olson, S.A.; Pisetsky, D.S.; Setton, L.A.; Weinberg, J.B. The role of biomechanics and inflammation in cartilage injury and repair. *Clin. Orthop. Relat. Res.* **2004**, *423*, 17–26. [CrossRef]
41. Guzik, T.J.; Korbut, R.; Adamek-Guzik, T. Nitric oxide and superoxide in inflammation and immune regulation. *J. Physiol. Pharmacol. Off. J. Pol. Physiol. Soc.* **2003**, *54*, 469–487.
42. Minghetti, L. Cyclooxygenase-2 (COX-2) in inflammatory and degenerative brain diseases. *J. Neuropathol. Exp. Neurol.* **2004**, *63*, 901–910. [CrossRef]
43. Catterall, J.B.; Stabler, T.V.; Flannery, C.R.; Kraus, V.B. Changes in serum and synovial fluid biomarkers after acute injury (NCT00332254). *Arthritis Res. Ther.* **2010**, *12*, R229. [CrossRef]
44. Sandhaus, R.A.; Turino, G. Neutrophil elastase-mediated lung disease. *Copd* **2013**, *10* (Suppl. 1), 60–63. [CrossRef]
45. Turner, M.D.; Nedjai, B.; Hurst, T.; Pennington, D.J. Cytokines and chemokines: At the crossroads of cell signalling and inflammatory disease. *Bba-Mol. Cell Res.* **2014**, *1843*, 2563–2582. [CrossRef]
46. Lieberthal, J.; Sambamurthy, N.; Scanzello, C.R. Inflammation in joint injury and post-traumatic osteoarthritis. *Osteoarthr. Cartil.* **2015**, *23*, 1825–1834. [CrossRef] [PubMed]

47. Jackson, L.M.; Wu, K.C.; Mahida, Y.R.; Jenkins, D.; Hawkey, C.J. Cyclooxygenase (COX) 1 and 2 in normal, inflamed, and ulcerated human gastric mucosa. *Gut* **2000**, *47*, 762–770. [CrossRef] [PubMed]
48. Yoshimura, K.; Nakagawa, S.; Koyama, S.; Kobayashi, T.; Homma, T. Roles of neutrophil elastase and superoxide anion in leukotriene B4-induced lung injury in rabbit. *J. Appl. Physiol.* **1994**, *76*, 91–96. [CrossRef] [PubMed]
49. Tan, A.S.; Berridge, M.V. Superoxide produced by activated neutrophils efficiently reduces the tetrazolium salt, WST-1 to produce a soluble formazan: A simple colorimetric assay for measuring respiratory burst activation and for screening anti-inflammatory agents. *J. Immunol. Methods* **2000**, *238*, 59–68. [CrossRef]
50. Green, S.J.; Aniagolu, J.; Raney, J.J. Oxidative metabolism of murine macrophages. In *Current Protocols in Immunology*; John Wiley & Sons Inc.: Hoboken, NJ, USA, 2001.
51. Brideau, C.; Kargman, S.; Liu, S.; Dallob, A.L.; Ehrich, E.W.; Rodger, I.W.; Chan, C.C. A human whole blood assay for clinical evaluation of biochemical efficacy of cyclooxygenase inhibitors. *Inflamm. Res.* **1996**, *45*, 68–74. [CrossRef]
52. Riendeau, D.; Percival, M.D.; Boyce, S.; Brideau, C.; Charleson, S.; Cromlish, W.; Ethier, D.; Evans, J.; Falgueyret, J.P.; FordHutchinson, A.W.; et al. Biochemical and pharmacological profile of a tetrasubstituted furanone as a highly selective COX-2 inhibitor. *Brit. J. Pharmacol.* **1997**, *121*, 105–117. [CrossRef]
53. Barbieri, S.S.; Eligini, S.; Brambilla, M.; Tremoli, E.; Colli, S. Reactive oxygen species mediate cyclooxygenase-2 induction during monocyte to macrophage differentiation: Critical role of NADPH oxidase. *Cardiovasc. Res.* **2003**, *60*, 187–197. [CrossRef]
54. Fehrenbacher, J.C.; Vasko, M.R.; Duarte, D.B. Models of inflammation: Carrageenan- or complete Freund's Adjuvant (CFA)-induced edema and hypersensitivity in the rat. *Curr. Protoc. Pharmacol.* **2012**. [CrossRef]
55. Zhang, Z.Y.; Lee, C.S.; Lider, O.; Weiner, H.L. Suppression of adjuvant arthritis in Lewis rats by oral administration of type II collagen. *J. Immunol.* **1990**, *145*, 2489–2493.
56. Cremer, M.A.; Rosloniec, E.F.; Kang, A.H. The cartilage collagens: A review of their structure, organization, and role in the pathogenesis of experimental arthritis in animals and in human rheumatic disease. *J. Mol. Med.* **1998**, *76*, 275–288. [CrossRef]
57. Trentham, D.E.; Townes, A.S.; Kang, A.H. Autoimmunity to type II collagen an experimental model of arthritis. *J. Exp. Med.* **1977**, *146*, 857–868. [CrossRef] [PubMed]
58. Larsson, P.; Kleinau, S.; Holmdahl, R.; Klareskog, L. Characterisation of the disease and demonstration of clinically distinct forms of arthritis in two strains of rats after immunization with the same collagen preparation. *Arth. Rheum.* **1990**, *33*, 693–701. [CrossRef]
59. Kawahito, Y.; Cannon, G.W.; Gulko, P.S.; Remmers, E.F.; Longman, R.E.; Reese, V.R.; Wang, J.; Griffiths, M.M.; Wilder, R.L. Localization of quantitative trait loci regulating adjuvant-induced arthritis in rats: Evidence for genetic factors common to multiple autoimmune diseases. *J. Immunol.* **1998**, *161*, 4411–4419. [PubMed]
60. Otterness, I.G.; Moore, P.F. Carrageenan foot edema test. *Methods Enzymol.* **1988**, *162*, 320–327. [CrossRef] [PubMed]
61. Annamalai, P.; Thangam, E.B. Local and Systemic Profiles of Inflammatory Cytokines in Carrageenan-induced Paw Inflammation in Rats. *Immunol. Investig.* **2017**, *46*, 274–283. [CrossRef]
62. Seltman, H. Approximations for Mean and Variance of a Ratio. Available online: http://www.stat.cmu.edu//~{}hseltman/files/ratio.pdf (accessed on 6 November 2017).
63. Cochran, W.G.; Cox, G.M. *Experimental Designs*; John Wiley & Sons Inc: Hoboken, NJ, USA, 1957.
64. Bendele, A. Animal models of rheumatoid arthritis. *J. Musculoskelet. Neuronal Interact.* **2001**, *1*, 377–385.
65. Mossiat, C.; Laroche, D.; Prati, C.; Pozzo, T.; Demougeot, C.; Marie, C. Association between arthritis score at the onset of the disease and long-term locomotor outcome in adjuvant-induced arthritis in rats. *Arthritis Res. Ther.* **2015**, *17*, 184. [CrossRef]
66. Tuncel, J.; Haag, S.; Hoffmann, M.H.; Yau, A.C.; Hultqvist, M.; Olofsson, P.; Backlund, J.; Nandakumar, K.S.; Weidner, D.; Fischer, A.; et al. Animal models of rheumatoid arthritis (I): Pristane-induced arthritis in the rat. *PLoS ONE* **2016**, *11*, e0155936. [CrossRef]
67. Lukens, J.R.; Kanneganti, T.D. Beyond canonical inflammasomes: Emerging pathways in IL-1-mediated autoinflammatory disease. *Semin. Immunopathol.* **2014**, *36*, 595–609. [CrossRef]
68. Korkmaz, B.; Horwitz, M.S.; Jenne, D.E.; Gauthier, F. Neutrophil Elastase, Proteinase 3, and Cathepsin G as Therapeutic Targets in Human Diseases. *Pharmacol. Rev.* **2010**, *62*, 726–759. [CrossRef]
69. Adelizzi, R.A. COX-1 and COX-2 in health and disease. *J. Am. Osteopath. Assoc.* **1999**, *99*, S7–S12. [CrossRef]

70. Abramson, S.B.; Attur, M.; Amin, A.R.; Clancy, R. Nitric oxide and inflammatory mediators in the perpetuation of osteoarthritis. *Curr. Rheumatol. Rep.* **2001**, *3*, 535–541. [CrossRef]
71. Wojdasiewicz, P.; Poniatowski, L.A.; Szukiewicz, D. The role of inflammatory and anti-inflammatory cytokines in the pathogenesis of osteoarthritis. *Mediat. Inflamm.* **2014**, *2014*, 561459. [CrossRef] [PubMed]
72. Norris, G.; Porter, C.; Jiang, C.; Blesso, C. Dietary milk sphingomyelin reduces systemic inflammation in diet-induced obese mice and inhibits lps activity in macrophages. *Beverages* **2017**, *3*, 37. [CrossRef]
73. Yeom, M.; Hahm, D.H.; Sur, B.J.; Han, J.J.; Lee, H.J.; Yang, H.I.; Kim, K.S. Phosphatidylserine inhibits inflammatory responses in interleukin-1beta-stimulated fibroblast-like synoviocytes and alleviates carrageenan-induced arthritis in rat. *Nutr. Res.* **2013**, *33*, 242–250. [CrossRef] [PubMed]
74. Schnabl, K.L.; Larsen, B.; Van Aerde, J.E.; Lees, G.; Evans, M.; Belosevic, M.; Field, C.; Thomson, A.B.; Clandinin, M.T. Gangliosides protect bowel in an infant model of necrotizing enterocolitis by suppressing proinflammatory signals. *J. Pediatric Gastroenterol. Nutr.* **2009**, *49*, 382–392. [CrossRef] [PubMed]
75. Laegreid, A.; Otnaess, A.B.; Fuglesang, J. Human and bovine milk: Comparison of ganglioside composition and enterotoxin-inhibitory activity. *Pediatric Res.* **1986**, *20*, 416–421. [CrossRef]
76. Aziz, M.M.; Ishihara, S.; Mishima, Y.; Oshima, N.; Moriyama, I.; Yuki, T.; Kadowaki, Y.; Rumi, M.A.; Amano, Y.; Kinoshita, Y. MFG-E8 attenuates intestinal inflammation in murine experimental colitis by modulating osteopontin-dependent alphavbeta3 integrin signaling. *J. Immunol.* **2009**, *182*, 7222–7232. [CrossRef]
77. Eros, G.; Varga, G.; Varadi, R.; Czobel, M.; Kaszaki, J.; Ghyczy, M.; Boros, M. Anti-inflammatory action of a phosphatidylcholine, phosphatidylethanolamine and N-acylphosphatidylethanolamine-enriched diet in carrageenan-induced pleurisy. *Eur. Surg. Res. Eur. Chir. Forschung. Rech. Chir. Eur.* **2009**, *42*, 40–48. [CrossRef]
78. Hintze, K.; Snow, D.; Burtenshaw, I.; Ward, R. Nutraceutical Properties of Milk Fat Globular Membrane. In *Biotechnology of Biopolymers*; Elnashar, P.M., Ed.; Intech: Ruijeka, Croatia, 2011; pp. 321–342. [CrossRef]
79. Anderson, R.C.; MacGibbon, A.K.H.; Haggarty, N.; Armstrong, K.M.; Roy, N.C. Bovine dairy complex lipids improve in vitro measures of small intestinal epithelial barrier integrity. *PLoS ONE* **2018**, *13*, e0190839. [CrossRef] [PubMed]

© 2020 by the authors. Licensee MDPI, Basel, Switzerland. This article is an open access article distributed under the terms and conditions of the Creative Commons Attribution (CC BY) license (http://creativecommons.org/licenses/by/4.0/).

Article

Association between Dairy Intake and Linear Growth in Chinese Pre-School Children

Yifan Duan, Xuehong Pang, Zhenyu Yang *, Jie Wang, Shan Jiang, Ye Bi, Shuxia Wang, Huanmei Zhang and Jianqiang Lai

Key Laboratory of Trace Element Nutrition of National Health and Family Planning Commission, National Institute for Nutrition and Health, Chinese Center for Disease Control and Prevention, No. 29 Nanwei Road, Xicheng District, Beijing 100050, China; duanyf@ninh.chinacdc.cn (Y.D.); pangxh@ninh.chinacdc.cn (X.P.); wangjie@ninh.chinacdc.cn (J.W.); jiangshan@ninh.chinacdc.cn (S.J.); biye@ninh.chinacdc.cn (Y.B.); wangsx@ninh.chinacdc.cn (S.W.); zhanghm@ninh.chinacdc.cn (H.Z.); jq_lai@126.com (J.L.)
* Correspondence: yangzy@ninh.chinacdc.cn; Tel.: +86-10-6623-7198

Received: 30 June 2020; Accepted: 21 August 2020; Published: 25 August 2020

Abstract: Stunting remains a major public health issue for pre-school children globally. Dairy product consumption is suboptimal in China. The aim of this study was to investigate the relationship between dairy intake and linear growth in Chinese pre-school children. A national representative survey (Chinese Nutrition and Health Surveillance) of children aged under 6 years was done in 2013. Stratified multistage cluster sampling was used to select study participants. A food frequency questionnaire was used to collect dietary information. We calculated height-for-age Z-scores (HAZs) and estimated stunting using the 2006 WHO growth standard. In total, 12,153 children aged two to four years old (24 to <60 months) were studied from 55 counties in 30 provinces in China. Approximately 39.2% (4759/12,153) of those children consumed dairy at least once per day, 11.9% (1450/12,153) consumed dairy at least once in the last week, and nearly half (48.9%, 5944/12,153) did not have any dairy in the last week. The HAZ was −0.15 ± 1.22 and the prevalence of stunting was 6.5% (785/12,153). The HAZ for children who consumed dairy at least once per day or per week was 0.11 points or 0.13 points higher than the children without dairy intake. The risk of stunting for children who consumed dairy at least once per day was 28% lower than the children without dairy intake in the last week, and the risk was similar between weekly dairy consumption and no dairy consumption (AOR: 1.03, 95% CI: 0.74–1.42) after adjusting for potential confounders, including socioeconomic characteristics, lifestyle, health status, and the intake frequency of other foods. Dairy intake was significantly associated with a higher HAZ and a lower risk of stunting for children aged 2–4 years old in China. The proportion of dairy intake was still low in Chinese pre-school children. The promotion of dairy consumption might be an effective and feasible measurement for improving linear growth in Chinese pre-school children.

Keywords: HAZ; stunting; dairy; pre-school children; cross-sectional study; China

1. Introduction

Linear growth is the best overall indicator of children's well-being and should be promoted for assessing nutritional status, designing programs, and assessing impacts [1–3]. Stunting is the most prevalent form of child malnutrition worldwide. Stunting is associated with increased morbidity and mortality, loss of physical growth potential, neurodevelopmental and cognitive function retardation, and an elevated risk of chronic diseases in adulthood [4]. Despite the slow reduction in recent years, stunting still affected an estimated 21.3% or 144 million children under five years old globally in 2019 [5]. One of the six global nutrition targets is to reduce the number of stunted children under 5 years of age by 40% before 2025 [6]. At the current rate of progress, it will be a challenge to achieve the global goal of decreasing the number of stunted children to 100 million in 2025 [4]. Asia is the continent with the most stunted children globally, which has more than half of all stunted children under five years old, with an estimated 78.2 million in 2019 [5].

As an optimal source of nutrients and bioactive factors, milk and dairy products play an important role in childhood growth and development [7]. Several studies have investigated the association between dairy intake and linear growth since the 1920s [8,9]. However, the conclusions are inconsistent in both the observational and interventional studies. In one of the first studies in 1928, Orr et al. estimated an increase in height of 20% for Scottish children aged 5–14 years who consumed milk in addition to their normal diet for seven months in comparison to children who did not [8]. An early study in 1978 found significant increases in height for children consuming milk compared to a control group in a school milk intervention in New Guinea [10]. A recent randomized controlled trial in Vietnam showed that the height-for-age Z-score (HAZ) significantly improved over 6 months of milk intervention in 454 children aged 7–8 years and stunting dropped by 10% [11], while another randomized controlled trial found no significant change in HAZs in Kenyan school children [12].

Dairy consumption is generally low in the Chinese population. Although dairy consumption increased from 1982 to 2002 in the Chinese population and reached a peak in 2002, there was a pronounced decline in the following decade [13]. In 2010, the amount of dairy consumption by Chinese people was 24.9 g per day, most of which was liquid cow milk. Children aged 2–3 years old took part in higher dairy consumption ($\approx$80 g per day) than 4–6-year-old children ($\approx$45 g per day) did. Even for 2–3-year-old children, only 4.3% of them achieved the daily dairy consumption recommendation and this proportion decreased to 1.1% in 4–6-year-old children [13]. Cow's milk allergy and lactose intolerance may contribute to low dairy consumption to some extent in the Chinese population. Cow's milk allergy is one of the most common food allergies, with an estimated prevalence in developed countries ranging from 0.5–3% at 1 year of age [14]. Gupta et al. estimated the overall prevalence of cow's milk allergy was 1.7% (95% confidence interval: 1.5–1.8%) and peaked in children aged 0–5 years at 2.0% [15]. Approximately 20% of Hispanic, Asian, and Black children younger than 5 years of age display evidence of lactase deficiency and lactose malabsorption [16]. In China, a study showed that the incidences of lactase deficiency and lactose intolerance were 38.5% and 12.2% in 3–5-year-old children, respectively [17].

The prevalence of stunting was high in Chinese children under 6 years old, especially in poor rural areas [18]. Among other reasons for stunting, inadequate animal-sourced food intake is a critical risk factor for the stunting of children under 5 years old [19], of which dairy consumption was associated with higher HAZs of children aged 6–23 months [20]. The first 1000 days is considered the crucial period for correcting child stunting. Few studies have focused on the effects of dairy consumption on the linear growth of pre-school children in developing countries. In China, low dairy intake and the high prevalence of stunting in pre-school children co-existed. This study aimed to investigate the relationship between dairy intake and linear growth in Chinese pre-school children.

2. Materials and Methods

2.1. Study Design

The study was a secondary data analysis from a nationally representative survey (Chinese Nutrition and Health Surveillance (CNHS)) in 2013 and a subpopulation was selected from the original study. The detailed methods of the CNHS were described previously [18]. Briefly, this was a cross-sectional survey among children under 6 years of age and lactating mothers from 30 provinces, autonomous regions, and municipalities in mainland China (the Tibet Autonomous Region was not included in the survey). Multi-stage stratified cluster random sampling was used in the study. In total, 2865 districts/counties in China were categorized into four strata (large cities, medium and small cities, non-poor rural areas, and poor rural areas) based on the population size and the definition of urban or rural from the National Bureau of Statistics of the People's Republic of China. A city with a population size of more than 1,000,000 was defined as a large city, and other cities belonged to medium and small cities category. Poor rural areas were the key counties for poverty alleviation and development identified by the Framework for Poverty Alleviation and Development in Chinese Rural Areas, while other counties belonged to the non-poor rural areas. Then, 55 counties (12 metropolises, 15 medium and small cities, 18 non-poor rural areas, and 10 poor rural areas) were chosen in the study. In each selected county, three communities/townships were systematically sampled. In each selected township, three neighborhoods/villages were systematically selected. Finally, 10 children from each age group were randomly selected in each selected village. The total sample size was 34,650 for children under 6 years old, of which, 14,850 were children aged two to four years old (24 to <60 months). The sample size calculation was based on an estimation of the prevalence of anemia in children under 6 years old after taking the complex sampling design into account.

2.2. Subjects

Children were selected according to three age groups (24–35 months, 36–47 months, and 48–59 months). Their caregivers were asked to finish a face-to-face interview with well-trained staff. The Ethics Review Board of the National Institute for Nutrition and Health, Chinese Center for Disease Control and Prevention, approved the protocol (No. 2013–018). All caregivers gave their informed consent in writing to participate before starting the interview.

2.3. Data Collection

The socioeconomic, family care, dietary intake, lifestyle, and health-related information were collected using the questionnaires during the face-to-face interviews mentioned above. The caregivers of children older than 2 years old were asked to finish a food frequency questionnaire (FFQ) in order to collect the dietary information regarding the past week before the survey, which was modified from a Chinese food frequency questionnaire established by Zhao et al. [21], who also examined the validity and reliability. The FFQ questionnaire, which was adopted in this survey, consisted of 44 food and beverage items, and was categorized into 10 food groups (1. cereal grains, roots, and tubers; 2. legumes and legume products; 3. dairy products from cows, goats, buffalo, etc. (whole milk, skim milk, milk powder, infant formula, yogurt, cheese); 4. flesh foods (meat, fish, poultry, and liver/organ meats); 5. eggs; 6. vegetables; 7. fruits; 8. snacks; 9. beverages; 10. nuts). The frequency of dairy product consumption was divided into three levels for this study. Eating at least once per day in the week before the survey was defined as "daily" consumption. "Weekly" consumption means the food was eaten once or more in the last week but less than once per day. If there was no consumption of the food or the frequency was less than once in the last week, the frequency was named "none."

2.4. Primary Outcomes

The standing height of children was measured by well-trained staff using a stadiometer with an accuracy of 0.1 cm. The height measurement was asked to be without the children having braided hair or shoes on. The primary outcome measures were the height-for-age Z-scores (HAZs) and the prevalence of stunting. Z-scores were calculated using the WHO Anthro software (WHO Anthro for personal computers, version 3.2.2, World Health Organization, Geneva, Switzerland.) Values were expressed as SD scores (Z-scores) using the reference population of the 2006 WHO growth standard [22]. The conversion of anthropometric variables to sex- and age-specific Z-scores were performed using the WHO standard. Stunting was defined as a height-for-age Z-score less than −2 SD of the median height of the WHO reference population.

2.5. Statistical Analysis

Data were entered via a standardized data management platform and were cleaned for all variables. All the data were analyzed using SAS 9.4 software (SAS Institute Inc., Cary, NC, USA). The frequency of consumption of each food group, including dairy, was tested using correlation analysis, which showed that each of the other food groups was uncorrelated with dairy consumption. Therefore, the frequency of consumption of each food group was viewed as an independent variable in the model. HAZs were expressed as mean ± SD and the differences were compared using ANOVA tests. The multivariate linear regression analysis was used to assess the relationship between HAZs and the frequency of dairy consumption after controlling for the potential confounders (e.g., residential area, children's age group, ethnicity, parental education level, parental age and occupation, birth weight and length, major caretaker, duration of daytime outdoors, regular growth monitoring, and the frequency of egg consumption). Categorical variables were expressed as a percentage (%). Chi-square tests were used for categorical variable comparisons. The logistic procedure was used to assess the relationship between stunting prevalence and the frequency of dairy consumption after controlling for the potential confounders (e.g., residential area, ethnicity, maternal occupation and migrant status, birth weight, major caretaker, sleep duration, regular growth monitoring, incidence of respiratory system disease in the last two weeks, and frequency of egg and fruit consumption). First, a bivariate analysis was conducted between each linear growth indicator and each potential confounder. The variables marginally associated with outcome variables in the bivariate analysis were selected for multivariate analyses ($p < 0.20$). Then, a multivariate analysis was conducted for each linear growth indicator with potential confounders. In the final model, only variables significantly associated with outcome variables were retained ($p < 0.05$). The adjusted β value and standard error (SE) were reported in the final linear regression model while retaining all significant variables, and the adjusted odds ratio (OR) and 95% confidence intervals (CIs) were reported in the final logistic regression model while retaining all significant variables.

3. Results

In total, 12,153 children aged two to four years old (24 to <60 months) were included in the study, where 51.5% (6261/12,153) of the children were boys and 48.9% of the children lived in an urban area. Approximately 39.2% (4759/12,153) of the children consumed dairy at least once per day, 11.9% (1450/12,153) of them consumed dairy at least once in the past week, and nearly a half (48.9%, 5944/12,153) of them did not have any dairy in the past week. The average height-for-age Z-score (HAZ) was −0.15 ± 1.22 and the prevalence of stunting was 6.5% (785/12,153).

The residential area, ethnicity, parental education status, parental age, occupation, migrant status, and household income was significantly associated with the HAZ and stunting in the bivariate analysis ($p < 0.001$) (Table 1). The birth weight and length, incidence of respiratory system disease in the last two weeks, major caretaker, regular growth monitoring, duration of daytime outdoors, and frequency of egg or fruit consumption was significantly associated with the HAZ and stunting in the bivariate analysis ($p < 0.001$) (Table 2).

Table 1. The relationship between socioeconomic status and height-for-age Z-score (HAZ) and the prevalence of stunting.

Variables	% (n/N)	HAZ (Mean ± SD)	p-Value [1]	Prevalence of Stunting (%)	p-Value [2]
Residential area					
Urban—metropolis	21.1% (2561/12,153)	0.28 ± 1.12	<0.001	2.0	<0.001
Urban—middle or small cities	27.9% (3386/12,153)	0.01 ± 1.15		4.0	
Rural—non-poor areas	32.9% (3997/12,153)	−0.20 ± 1.16		6.0	
Rural—poor areas	18.2% (2209/12,153)	−0.82 ± 1.23		16.3	
Age group (years)					
2~	31.8% (3858/12,153)	−0.11 ± 1.27	0.016	7.0	0.076
3~	33.5% (4073/12,153)	−0.18 ± 1.22		6.7	
4~	34.7% (4222/12,153)	−0.17 ± 1.17		5.8	
Gender					
Male	51.5% (6261/12,153)	−0.14 ± 1.24	0.083	6.8	0.148
Female	48.5% (5892/12,153)	−0.17 ± 1.19		6.1	
Ethnicity					
Han	85.1% (10,346/12,153)	−0.06 ± 1.19	<0.001	4.9	<0.001
Minority	14.9% (1807/12,153)	−0.71 ± 1.26		15.2	
Maternal education					
Primary or below	15.3% (1783/11,695)	−0.63 ± 1.19	<0.001	11.8	<0.001
Junior high	49.2% (5750/11,695)	−0.27 ± 1.20		7.4	
Senior high or above	35.6% (4162/11,695)	0.23 ± 1.15		2.7	
Paternal education					
Primary or below	11.1% (1274/11,501)	−0.71 ± 1.21	<0.001	13.8	<0.001
Junior high	49.3% (5672/11,501)	−0.30 ± 1.19		7.5	
Senior high or above	39.6% (4555/11,501)	0.19 ± 1.15		3.1	
Maternal age group (years)					
≤26	20.7% (2421/11,708)	−0.35 ± 1.23	<0.001	9.0	<0.001
27–30	31.2% (3650/11,708)	−0.14 ± 1.22		6.3	
31–34	25.6% (2991/11,708)	−0.04 ± 1.19		5.2	
≥35	22.6% (2646/11,708)	−0.10 ± 1.21		5.6	
Paternal age group (years)					
≤28	23.0% (2643/11,516)	−0.23 ± 1.23	<0.001	8.0	0.003
29–32	28.1% (3233/11,516)	−0.12 ± 1.22		5.9	
33–36	23.5% (2706/11,516)	−0.11 ± 1.20		5.9	
≥37	25.5% (2934/11,516)	−0.15 ± 1.21		6.1	
Maternal occupation					
Unemployed	34.8% (4072/11,689)	−0.20 ± 1.19	<0.001	6.3	<0.001
Farmer	17.1% (1999/11,689)	−0.63 ± 1.24		13.6	
Others	48.1% (5618/11,689)	0.06 ± 1.18		4.0	
Paternal occupation					
Unemployed	6.1% (701/11,526)	−0.43 ± 1.19	<0.001	9.4	<0.001
Farmer	23.1% (2660/11,526)	−0.56 ± 1.25		12.5	
Others	70.8% (8165/11,526)	0.00 ± 1.17		4.3	
Annual household income (per capita CNY *)					
≥15,000	35.3% (4282/12,147)	0.03 ± 1.18	<0.001	4.4	<0.001
10,000–14,999	17.7% (2155/12,147)	−0.16 ± 1.19		6.0	
5000–9999	19.9% (2420/12,147)	−0.31 ± 1.25		8.8	
<5000	16.8% (2035/12,147)	−0.47 ± 1.25		10.3	
refuse	10.3% (1255/12,147)	0.03 ± 1.15		3.6	
Maternal migrant status					
Migrant mother	15.0% (1824/12,153)	−0.45 ± 1.22	<0.001	10.8	<0.001
Mother living at home	85.0% (10,329/12,153)	−0.10 ± 1.21		5.7	
Paternal migrant status					
Migrant father	24.0% (2919/12,153)	−0.36 ± 1.22	<0.001	8.7	<0.001
Father living at home	76.0% (9234/12,153)	−0.09 ± 1.21		5.7	

[1] The result of the variance analysis between the variable and the HAZ. [2] The result of the chi-square test between the variable and the prevalence of stunting. * CNY: Chinese Yuan.

Table 2. The relationship between the health status and lifestyles and the HAZ and the prevalence of stunting.

Variables	% (n/N)	HAZ (Mean ± SD)	p-Value [1]	Prevalence of Stunting (%)	p-Value [2]
Birth weight (g)					
<2500	3.6% (436/12,142)	−0.64 ± 1.27	<0.001	12.6	<0.001
2500–3200	41.3% (5013/12,142)	−0.33 ± 1.19		8.0	
3201–3999	40.4% (4909/12,142)	0.06 ± 1.18		4.2	
≥4000	7.5% (915/12,142)	0.29 ± 1.18		3.4	
Unknown	7.2% (869/12,142)	−0.54 ± 1.25		10.6	
Birth length (cm)					
<50	12.8% (1549/12,133)	−0.23 ± 1.17	<0.001	6.7	<0.001
=50	36.4% (4410/12,133)	−0.01 ± 1.15		4.5	
>50	19.5% (2365/12,133)	0.18 ± 1.23		4.1	
Unknown	31.4% (3809/12,133)	−0.50 ± 1.22		10.1	
Premature					
Yes	9.9% (1179/11,958)	−0.21 ± 1.16	0.055	6.0	0.656
No	90.1% (10,779/11,958)	−0.14 ± 1.22		6.4	
Incidence of respiratory system disease in the last two weeks					
Yes	24.5% (2973/12,109)	−0.06 ± 1.15	<0.001	4.3	<0.001
No	75.5% (9136/12,109)	−0.18 ± 1.24		7.2	
Incidence of diarrhea in the last two weeks					
Yes	4.9% (597/12,122)	−0.15 ± 1.10	0.997	4.7	0.073
No	95.1% (11,525/12,122)	−0.15 ± 1.22		6.5	
Major caretaker					
Grandmothers	18.0% (2187/12,153)	−0.30 ± 1.20	<0.001	8.4	<0.001
Mother and father	42.9% (5219/12,153)	−0.07 ± 1.23		5.7	
Mother	36.8% (4470/12,153)	−0.16 ± 1.21		6.1	
Father	1.6% (198/12,153)	−0.58 ± 1.23		10.6	
Others	0.7% (79/12,153)	−0.57 ± 1.24		15.2	
Regular growth monitoring					
Yes	72.6% (8819/12,141)	−0.01 ± 1.18	<0.001	4.8	<0.001
No	27.4% (3322/12,141)	−0.52 ± 1.23		10.8	
Sleep duration (hours)					
<10	13.8% (1678/12,141)	−0.23 ± 1.26	0.010	7.6	<0.001
10 to <10.5	33.8% (4104/12,141)	−0.15 ± 1.19		5.4	
10.5 to <12	25.5% (3098/12,141)	−0.11 ± 1.19		6.2	
≥12	26.9% (3261/12,141)	−0.16 ± 1.25		7.5	
Duration of daytime outdoors (minutes)					
≤90	25.1% (3049/12,134)	0.03 ± 1.24	<0.001	5.1	<0.001
91–150	26.8% (3257/12,134)	−0.06 ± 1.20		5.7	
151–240	34.7% (4212/12,134)	−0.27 ± 1.21		7.3	
>240	13.3% (1616/12,134)	−0.37 ± 1.17		8.6	
Have been breastfed in the last 24 h					
Yes	1.4% (164/12,136)	−0.24 ± 1.30	0.355	9.8	0.083
No	98.7% (11,972/12,136)	−0.15 ± 1.22		6.4	
Cow's milk allergy					
Yes	0.8% (93/12,081)	−0.27 ± 1.27	0.344	8.6	0.388
No	99.2% (11,988/12,081)	−0.15 ± 1.22		6.4	
Frequency of egg consumption					
Daily	36.0% (4369/12,153)	0.11 ± 1.17	<0.001	3.8	<0.001
Weekly	48.5% (5898/12,153)	−0.23 ±1.20		6.7	
None	15.5% (1886/12,153)	−0.52 ± 1.23		11.8	
Frequency of fruit consumption					
Daily	56.1% (6812/12,153)	−0.01 ± 1.18	<0.001	4.6	<0.001
Weekly	38.3% (4657/12,153)	−0.33 ± 1.23		8.4	
None	5.6% (684/12,153)	−0.43 ± 1.34		11.8	

[1] The result of the variance analysis between the variable and the HAZ. [2] The result of the chi-square test between the variable and the prevalence of stunting.

According to the results of the bivariate analysis, which are listed in Tables 1 and 2, the variables marginally associated with the HAZ or the prevalence of stunting in the bivariate analysis were selected for multivariate analyses ($p < 0.20$). In the final model, only variables significantly associated with outcome variables were retained ($p < 0.05$).

After adjusting for residential area, children's age group, ethnicity, parental education level, parental age and occupation, birth weight and length, major caretaker, duration of daytime outdoors, regular growth monitoring, and the frequency of egg consumption, the HAZ was significantly associated with the frequency of dairy intake. The HAZ was 0.11 points or 0.13 points greater for children who consumed dairy at least once per day or per week, respectively, than the children without dairy intake in the past week (Table 3).

Table 3. Association between the HAZ and the frequency of dairy consumption.

Frequency of Dairy Consumption	HAZ (Mean ± SD)	β [#]	SE	t	p-Value
Daily	0.13 ± 1.14	0.11	0.03	4.23	<0.001
Weekly	−0.01 ± 1.16	0.13	0.04	3.64	<0.001
None	−0.42 ± 1.23	Ref.	-	-	-

Ref. means the reference group. [#] Adjusted by residential area, children's age group, ethnicity, parental education level, parental age and occupation, birth weight and length, major caretaker, duration of daytime outdoors, regular growth monitoring, and the frequency of egg consumption.

After adjusting for residential area, ethnicity, maternal occupation and migrant status, birth weight, major caretaker, sleeping duration, regular growth monitoring, incidence of respiratory system disease in the last two weeks, and frequency of egg and fruit consumption, the children with daily dairy consumption had a 28% lower risk of stunting than the children without dairy intake in the past week. Meanwhile, the risk of stunting was similar between weekly dairy consumption and without dairy consumption in the past week (Table 4).

Table 4. Association between the prevalence of stunting and the frequency of dairy consumption.

Frequency of Dairy Consumption	Prevalence of Stunting (%)	Crude OR (95% CI)	Adjusted OR [#] (95% CI)	p-Value
Daily	3.2	0.32 (0.26, 0.38)	0.72 (0.58, 0.90)	0.003
Weekly	4.1	0.78 (0.57, 1.05)	1.03 (0.74, 1.42)	0.875
None	9.6	1.00	1.00	-

[#] Adjusted by residential area, ethnicity, maternal occupation and migrant status, birth weight, major caretaker, sleeping duration, regular growth monitoring, incidence of respiratory system disease in the last two weeks, and the frequency of egg and fruit consumption.

4. Discussion

Dairy consumption was quite low in Chinese pre-school children, where nearly half of the children did not consume dairy during the past week. Dairy intake was significantly associated with greater height-for-age Z-scores and a lower risk of stunting for children aged 2–4 years old in China.

Linear growth failure was common for 2–4-year-old Chinese children, especially in poor rural areas. Although the etiology of stunting is poorly understood, prenatal and postnatal nutritional deficits could contribute to the stunting of children under 5 years of age [23]. Inadequate intake of one or more nutrients, including energy, protein, and micronutrients, such as zinc, vitamin A, and phosphorus, may result in the growth retardation of children. In addition, repeated infections worsen nutrient deficiency and impair the absorption of nutrients [24].

Dairy consumption in childhood has long been assumed to be beneficial for growth, which was proposed as the "milk hypothesis" by Bogin [25]. The specific stimulating effect of milk on linear growth may be related to several components of milk, such as high-quality protein, bioactive peptides, amino acids, insulin-like growth factor-1 (IGF-1), or minerals, including calcium. Dairy products provide high-quality protein with peptides and bioactive factors that could have specific effects on growth. A prospective cohort found that dairy protein intake was a significant predictor of peak height velocity and adult height, while animal or vegetable protein was not [26]. Approximately 80% of the protein in cow's milk is casein, and the remaining 20% is whey [7]. Whey, as a soluble milk protein, may have some insulinotropic components [27]. Another study conducted on eight-year-old boys suggested that casein might have a stronger IGF-1 stimulating effect than whey does [28].

Apart from proteins, milk IGF-1 is another major relevant factor for children's development and growth. Milk IGF-1 is structurally identical to human IGF-1 and the milk IGF-1 concentration is approximately 30 ng/mL [29]. IGF-1 is a potential growth factor in bone and mediates the effects of the pituitary growth hormone [30]. IGF-1 facilitates bone growth by increasing the uptake of amino acids, which are then integrated into new proteins in bone tissue [31]. It is the most abundant growth factor in bone and has a strong anabolic effect on growing bone tissue since it stimulates the chondrocytes in the epiphyseal plate [32]. Serum levels of IGF-1 rise after milk consumption, although it is not clear whether this is due to the IGF-1 in milk or whether milk consumption stimulates endogenous IGF-1 production [33]. In studies of children, milk consumption, circulating IGF-1, and height were positively correlated [7,34,35], and milk supplementation resulted in an elevation of IGF-1 levels [7,36,37]. Thus, milk may promote linear growth through an IGF-1-mediated process, perhaps in concert with calcium or other milk constituents [30].

With respect to linear growth, the effect of calcium was unclear. Some calcium supplementation or calcium fortification studies did not appear to positively influence height in children [38–42]. However, a study with a sample of 1002 children aged 24–59 months from the National Health and Nutrition Examination Survey (NHANES) showed that total calcium intake was positively associated with height and it might mediate the relationship between milk intake and height. The author hypothesized that calcium appears to play a role in the increased height of young children, but it may act synergistically with other components of milk [30].

The lactase persistence (LP) phenotype was studied as a proxy for dairy intake in some studies in recent years. So far, the relationship between the LP phenotype and milk consumption is inconclusive. Some studies have found that milk or total dairy consumption is associated with the LP genotype [43–47]. Other studies have shown that LP genetics has no influence over whether a participant is a cow milk consumer [48–50]. Most of those studies were conducted in populations with high frequencies of the LP phenotype and high amounts of milk consumption. A study showed that the incidence of lactase deficiency was 38.5% and the proportion of lactose intolerance was 12.2% in Chinese 3–5-year-old children [17].

The relationship between dairy intake and height is inconclusive for children. A systematic review and meta-analysis of controlled trials assessed the effects of supplementing a usual diet with dairy products on physical growth, including twelve studies conducted in Europe, USA, China, Northern Vietnam, Kenya, Indonesia, and India between the 1920s and 2000s. Only one of these studies was conducted in preschool children in Beijing suburbs; the others were all conducted in school-aged children from 7–13 years old. The meta-analysis with a random effects model yielded a pooled estimate of 0.59 cm. This additional growth was the result of giving a daily milk supplement of 245 mL for 12 months on average. In addition, the results of the sensitivity analysis suggested an effect size of 0.4 cm could be considered a conservative estimate [51]. The protective effect against stunting was also found in a study involving 68 low- and middle-income countries, which suggested that milk consumption is associated with a reduced probability of being stunted of 1.9 percentage points for children aged 6–59 months. This study showed that a child aged 24 to 59 months that consumed milk had a 0.14 points greater HAZ ($p < 0.001$) than those that did not consume milk [52], which was in accordance with our study. Despite these results, the relationship may be different in the populations who consumed different quantities of dairy. Wiley found that children in the highest quartile of milk intake were taller (1.1–1.2 cm, $p < 0.01$) than those in the middle quartiles. Interestingly, this difference was not found between the group in the highest quartile of milk drinkers and the lowest quartile.

Furthermore, this study also evaluated the association between milk consumption and height among preschool-age children in the USA, 89% of whom reported daily dairy consumption. Results showed that children who drank milk daily were 1.0 cm taller ($p < 0.02$) than those with a less frequent intake [30].

As mentioned above, intervention studies that focus on preschool children are scarce. Although growth was relatively stable during the preschool years, the growth velocity is still high in this period. This period should be viewed as a sensitive life stage for intervention. Interventions beyond 24 months that prove successful in enhancing adult stature (especially in girls) may offer additional opportunities to improve nutritional status and would likely foster advantages throughout the mothers' entire reproductive life and benefit future generations [53].

This study has some limitations. First, the CNHS is a cross-sectional survey; as such, causality cannot be attributed to dairy in this study design. Second, the amount of dairy intake was not included in our study. However, the FFQ is valid and with good reliability for assessing preschool children's food intake [54]. Furthermore, the frequency of dairy intake can be recalled by subjects more easily and accurately than the amount of dairy intake and the results can be modifiable for education and intervention design. Third, since the frequency was recalled over the past week, it was only a snapshot of the dairy consumption of children. This snapshot may or may not be reflective of the overall patterns of dairy consumption.

5. Conclusions

The amount of dairy intake remains low in Chinese pre-school children. The dairy intake was positively associated with linear growth in pre-school children in China. A more frequent dairy intake might promote a greater HAZ and a lower prevalence of stunting in Chinese pre-school children. The promotion of dairy consumption might be an effective and feasible measurement for improving linear growth in pre-school children. Further intervention studies are warranted to test the relationship.

Author Contributions: Conceptualization, Z.Y.; data curation, Y.D., X.P., and J.W.; formal analysis, Y.D., X.P., and Z.Y.; investigation, Y.D., X.P., Z.Y., J.W., S.J., Y.B., S.W., and H.Z.; methodology, Z.Y.; project administration, J.L.; supervision, Z.Y. and J.L.; writing—review and editing, Y.D. and Z.Y. All authors have read and agreed to the published version of the manuscript.

Funding: This research was funded by National Special Program for Science & Technology Basic Resources Investigation of China (Grant Number: 2017FY101100 and 2017FY101103) and the Major Program for Healthcare Reform from the Chinese National Health and Family Planning Commission.

Acknowledgments: We thank all the participants in our study and all the staff working for the Chinese Nutrition and Health Surveillance 2013 (CNHS 2013) study.

Conflicts of Interest: We declare that we have no conflict of interest.

References

1. Black, R.E.; Allen, L.H.; Bhutta, Z.A.; Caulfield, L.E.; De Onis, M.; Ezzati, M.; Mathers, C.; Rivera, J. Maternal and child undernutrition: Global and regional exposures and health consequences. *Lancet* **2008**, *371*, 243–260. [CrossRef]
2. Victora, C.G.; Adair, L.; Fall, C.H.; Hallal, P.C.; Martorell, R.; Richter, L.; Sachdev, H.P.S. Maternal and child undernutrition: Consequences for adult health and human capital. *Lancet* **2008**, *371*, 340–357. [CrossRef]
3. Bhutta, Z.A.; Ahmed, T.; Black, R.E.; Cousens, S.; Dewey, K.; Giugliani, E.; Haider, B.A.; Kirkwood, B.; Morris, S.S.; Maternal and Child Undernutrition Study Group; et al. What works? Interventions for maternal and child undernutrition and survival. *Lancet* **2008**, *371*, 417–440. [CrossRef]
4. De Onis, M.; Branca, F. Childhood stunting: A global perspective. *Matern. Child Nutr.* **2016**, *12* (Suppl. S1), 12–26. [CrossRef]
5. UNICEF; World Health Organisation; World Bank. *Levels and Trends in Child Malnutrition*; UNICEF: New York, NY, USA, 2020.
6. Global Nutrition Targets 2025. Available online: https://www.who.int/nutrition/global-target-2025/en/ (accessed on 22 June 2020).

7. Hoppe, C.; Mølgaard, C.; Michaelsen, K.F. Cow's Milk and Linear Growth in Industrialized and Developing Countries. *Annu. Rev. Nutr.* **2006**, *26*, 131–173. [CrossRef]
8. Orr, J.B. Influence of Amount of Milk Consumption on the Rate of Growth of School Children. *Br. Med. J.* **1928**, *1*, 140–141. [CrossRef]
9. Leighton, G.; Clark, M.L. Milk Consumption and the Growth of School Children: Second Preliminary Report on Tests to the Scottish Board of Health. *Br. Med. J.* **1929**, *1*, 23–25. [CrossRef]
10. Lampl, M.; Johnston, F.E.; Malcolm, L.A. The effects of protein supplementation on the growth and skeletal maturation of New Guinean school children. *Ann. Hum. Biol.* **1978**, *5*, 219–227. [CrossRef]
11. Lien, D.T.K.; Nhung, B.T.; Khan, N.C.; Hop, L.T.; Nga, N.T.Q.; Hung, N.T.; Kiers, J.; Shigeru, Y.; Biesebeke, R.T. Impact of milk consumption on performance and health of primary school children in rural Vietnam. *Asia Pac. J. Clin. Nutr.* **2009**, *18*, 326–344.
12. Grillenberger, M.; Neumann, C.G.; Murphy, S.P.; Bwibo, N.O.; Van't Veer, P.; Hautvast, J.G.A.J.; West, C.E. Food Supplements Have a Positive Impact on Weight Gain and the Addition of Animal Source Foods Increases Lean Body Mass of Kenyan Schoolchildren. *J. Nutr.* **2003**, *133*, 3957S–3964S. [CrossRef]
13. Zhao, L.Y.; He, Y.N. *The Series Report of Chinese National Nutrition and Health Survey: The Status of Dietary and Nutrients Intake in 2010–2013*; People's Medical Publishing House: Beijing, China, 2018. (In Chinese)
14. Flom, J.D.; Sicherer, S.H. Epidemiology of Cow's Milk Allergy. *Nutrients* **2019**, *11*, 1051. [CrossRef]
15. Gupta, R.S.; Springston, E.E.; Warrier, M.R.; Smith, B.; Kumar, R.; Pongracic, J.; Holl, J.L. The Prevalence, Severity, and Distribution of Childhood Food Allergy in the United States. *Pediatrics* **2011**, *128*, e9–e17. [CrossRef] [PubMed]
16. Heyman, M.B.; Committee on Nutrition. Lactose Intolerance in Infants, Children, and Adolescents. *Pediatrics* **2006**, *118*, 1279–1286. [CrossRef] [PubMed]
17. Yang, Y.; He, M.; Cui, H.; Bian, L. Study on the incidence of lactose intolerance of children in China. *J. Hyg. Res.* **1999**, *28*, 44–46.
18. Yu, D.M.; Zhao, L.Y.; Yang, Z.Y.; Chang, S.Y.; Yu, W.T.; Fang, H.Y.; Wang, X.; Yu, D.; Guo, Q.Y.; Xu, X.L.; et al. Comparison of Undernutrition Prevalence of Children under 5 Years in China between 2002 and 2013. *Biomed. Environ. Sci.* **2016**, *29*, 165–176. [CrossRef] [PubMed]
19. Sari, M.; De Pee, S.; Bloem, M.W.; Sun, K.; Thorne-Lyman, A.L.; Moench-Pfanner, R.; Akhter, N.; Kraemer, K.; Semba, R.D. Higher Household Expenditure on Animal-Source and Nongrain Foods Lowers the Risk of Stunting among Children 0–59 Months Old in Indonesia: Implications of Rising Food Prices. *J. Nutr.* **2009**, *140*, 195S–200S. [CrossRef]
20. Choudhury, S.; Headey, D.D. Household dairy production and child growth: Evidence from Bangladesh. *Econ. Hum. Biol.* **2018**, *30*, 150–161. [CrossRef]
21. Zhao, W.-H.; Huang, Z.-P.; Zhang, X.; He, L.; Willett, W.; Wang, J.; Hasegawa, K.; Chen, J.-S. Reproducibility and Validity of a Chinese Food Frequency Questionnaire. *Biomed. Environ. Sci.* **2010**, *23*, 1–38. [CrossRef]
22. World Health Organization. *WHO Child Growth Standards: Length/Height-For-Age, Weight-For-Age, Weight-For-Length, Weight-For-Height and Body Mass Index-For-Age: Methods and Development*; World Health Organization: Geneva, Switzerland, 2006.
23. Black, R.E.; Heidkamp, R. Causes of Stunting and Preventive Dietary Interventions in Pregnancy and Early Childhood. *Nestle Nutr. Inst. Workshop Ser.* **2018**, *89*, 105–113. [CrossRef]
24. Branca, F.; Ferrari, M. Impact of micronutrient deficiencies on growth: The stunting syndrome. *Ann. Nutr. Metab.* **2002**, *46* (Suppl. S1), 8–17. [CrossRef]
25. Bogin, B. Milk and human development: An essay on the milk hypothesis. *Antropol. Port.* **1998**, *15*, 23–36.
26. Berkey, C.S.; Colditz, G.A.; Rockett, H.R.; Frazier, A.L.; Willett, W.C. Dairy consumption and female height growth: Prospective cohort study. *Cancer Epidemiol. Biomark. Prev.* **2009**, *18*, 1881–1887. [CrossRef] [PubMed]
27. Nilsson, M.; Stenberg, M.; Frid, A.H.; Holst, J.J.; Bjorck, I.M. Glycemia and insulinemia in healthy subjects after lactose-equivalent meals of milk and other food proteins: The role of plasma amino acids and incretins. *Am. J. Clin. Nutr.* **2004**, *80*, 1246–1253. [CrossRef] [PubMed]
28. Hoppe, C.; Mølgaard, C.; Vaag, A.; Michaelsen, K.F. The effect of seven-day supplementation with milk protein fractions and milk minerals on IGFs and glucose-insulin metabolism. *Scand. J. Food Nutr.* **2006**, *50*, 46.
29. Outwater, J.L.; Nicholson, A.; Barnard, N. Dairy products and breast cancer: The IGF-I, estrogen, and bGH hypothesis. *Med. Hypotheses* **1997**, *48*, 453–461. [CrossRef]

30. Wiley, A.S. Consumption of milk, but not other dairy products, is associated with height among US preschool children in NHANES 1999–2002. *Ann. Hum. Biol.* **2009**, *36*, 125–138. [CrossRef]
31. Cameron, N. *Human Growth and Development*; Academic Press: New York, NY, USA, 2002.
32. Kelly, O.; Cusack, S.; Cashman, K.D. The effect of bovine whey protein on ectopic bone formation in young growing rats. *Br. J. Nutr.* **2003**, *90*, 557–564. [CrossRef]
33. Holmes, M.D.; Pollak, M.N.; Willett, W.C.; Hankinson, S.E. Dietary correlates of plasma insulin-like growth factor I and insulin-like growth factor binding protein 3 concentrations. *Cancer Epidemiol. Biomark. Prev.* **2002**, *11*, 852–861.
34. Garnett, S.P.; Cowell, C.; Bradford, D.; Lee, J.; Tao, C.; Petrauskas, V.; Fay, R.; Baur, L.A. Effects of gender, body composition and birth size on IGF-I in 7- and 8-year-old children. *Horm. Res.* **1999**, *52*, 221–229. [CrossRef]
35. Rogers, I.; Emmett, P.; Gunnell, D.; Dunger, D.; Holly, J. Milk as a food for growth? The insulin-like growth factors link. *Public Health Nutr.* **2006**, *9*, 359–368. [CrossRef]
36. Cadogan, J.; Eastell, R.; Jones, N.; Barker, M.E. Milk intake and bone mineral acquisition in adolescent girls: Randomised, controlled intervention trial. *BMJ* **1997**, *315*, 1255–1260. [CrossRef] [PubMed]
37. Zhu, K.; Greenfield, H.; Zhang, Q.; Ma, G.; Zhang, Z.; Hu, X.; Fraser, D.R. Bone mineral accretion and growth in Chinese adolescent girls following the withdrawal of school milk intervention: Preliminary results after two years. *Asia Pac. J. Clin. Nutr.* **2004**, *13*, S83.
38. Dibba, B.; Prentice, A.; Ceesay, M.; Stirling, D.M.; Cole, T.J.; Poskitt, E.M. Effect of calcium supplementation on bone mineral accretion in Gambian children accustomed to a low-calcium diet. *Am. J. Clin. Nutr.* **2000**, *71*, 544–549. [CrossRef] [PubMed]
39. Bonjour, J.P.; Chevalley, T.; Ammann, P.; Slosman, D.; Rizzoli, R. Gain in bone mineral mass in prepubertal girls 3–5 years after discontinuation of calcium supplementation: A follow-up study. *Lancet* **2001**, *358*, 1208–1212. [CrossRef]
40. Cameron, M.A.; Paton, L.M.; Nowson, C.A.; Margerison, C.; Frame, M.; Wark, J.D. The Effect of Calcium Supplementation on Bone Density in Premenarcheal Females: A Co-Twin Approach. *J. Clin. Endocrinol. Metab.* **2004**, *89*, 4916–4922. [CrossRef]
41. Gibbons, M.J.; Gilchrist, N.L.; Frampton, C.; Maguire, P.; Reilly, P.H.; March, R.L.; Wall, C.R. The effects of a high calcium dairy food on bone health in pre-pubertal children in New Zealand. *Asia Pac. J. Clin. Nutr.* **2004**, *13*, 341–347.
42. Winzenberg, T.; Shaw, K.; Fryer, J.; Jones, G. Calcium Supplements in Healthy Children Do Not Affect Weight Gain, Height, or Body Composition. *Obesity* **2007**, *15*, 1789–1798. [CrossRef]
43. Alharbi, O.; El-Sohemy, A. Lactose Intolerance (LCT-13910C> T) Genotype Is Associated with Plasma 25-Hydroxyvitamin D Concentrations in Caucasians: A Mendelian Randomization Study. *J. Nutr.* **2017**, *147*, 1063–1069. [CrossRef]
44. Travis, R.C.; Appleby, P.N.; Siddiq, A.; Allen, N.E.; Kaaks, R.; Canzian, F.; Feller, S.; Tjønneland, A.; Johnsen, N.F.; Overvad, K.; et al. Genetic variation in thelactasegene, dairy product intake and risk for prostate cancer in the European prospective investigation into cancer and nutrition. *Int. J. Cancer* **2013**, *132*, 1901–1910. [CrossRef]
45. Bergholdt, H.K.; Nordestgaard, B.G.; Ellervik, C. Milk intake is not associated with low risk of diabetes or overweight-obesity: A Mendelian randomization study in 97,811 Danish individuals. *Am. J. Clin. Nutr.* **2015**, *102*, 487–496. [CrossRef]
46. Lamri, A.; Poli, A.; Emery, N.; Bellili, N.; Velho, G.; Lantieri, O.; Balkau, B.; Marre, M.; Fumeron, F. The lactase persistence genotype is associated with body mass index and dairy consumption in the D.E.S.I.R. study. *Metabolism* **2013**, *62*, 1323–1329. [CrossRef] [PubMed]
47. Yang, Q.; Lin, S.L.; Au Yeung, S.L.; Kwok, M.K.; Xu, L.; Leung, G.M.; Schooling, C.M. Genetically predicted milk consumption and bone health, ischemic heart disease and type 2 diabetes: A Mendelian randomization study. *Eur. J. Clin. Nutr.* **2017**, *71*, 1008–1012. [CrossRef] [PubMed]
48. Sacerdote, C.; Guarrera, S.; Smith, G.D.; Grioni, S.; Krogh, V.; Masala, G.; Mattiello, A.; Palli, D.; Panico, S.; Tumino, R.; et al. Lactase Persistence and Bitter Taste Response: Instrumental Variables and Mendelian Randomization in Epidemiologic Studies of Dietary Factors and Cancer Risk. *Am. J. Epidemiol.* **2007**, *166*, 576–581. [CrossRef] [PubMed]
49. Enattah, N.S.; Sulkava, R.; Halonen, P.; Kontula, K.; Järvelä, I. Genetic Variant of Lactase-Persistent C/T-13910 Is Associated with Bone Fractures in Very Old Age. *J. Am. Geriatr. Soc.* **2005**, *53*, 79–82. [CrossRef]

50. Chin, E.L.; Huang, L.; Bouzid, Y.Y.; Kirschke, C.P.; Durbin-Johnson, B.; Baldiviez, L.M.; Bonnel, E.L.; Keim, N.L.; Korf, I.F.; Stephensen, C.B.; et al. Association of Lactase Persistence Genotypes (rs4988235) and Ethnicity with Dairy Intake in a Healthy U.S. Population. *Nutrients* **2019**, *11*, 1860. [CrossRef]
51. De Beer, H. Dairy products and physical stature: A systematic review and meta-analysis of controlled trials. *Econ. Hum. Biol.* **2012**, *10*, 299–309. [CrossRef]
52. Herber, C.; Bogler, L.; Subramanian, S.V.; Vollmer, S. Association between milk consumption and child growth for children aged 6–59 months. *Sci. Rep.* **2020**, *10*, 6730. [CrossRef]
53. Prentice, A.M.; Ward, K.A.; Goldberg, G.R.; Jarjou, L.M.; Moore, S.E.; Fulford, A.J.; Prentice, A. Critical windows for nutritional interventions against stunting. *Am. J. Clin. Nutr.* **2013**, *97*, 911–918. [CrossRef]
54. Noor Hafizah, Y.; Ang, L.C.; Yap, F.; Nurul Najwa, W.; Cheah, W.L.; Ruzita, A.T.; Jumuddin, F.A.; Koh, D.; Lee, J.A.C.; Essau, C.A.; et al. Validity and Reliability of a Food Frequency Questionnaire (FFQ) to Assess Dietary Intake of Preschool Children. *Int. J. Environ. Res. Public Health* **2019**, *16*, 4722. [CrossRef]

© 2020 by the authors. Licensee MDPI, Basel, Switzerland. This article is an open access article distributed under the terms and conditions of the Creative Commons Attribution (CC BY) license (http://creativecommons.org/licenses/by/4.0/).

Review

Branched-Chain Fatty Acids—An Underexplored Class of Dairy-Derived Fatty Acids

Victoria M. Taormina [1,†], Allison L. Unger [2,†], Morgan R. Schiksnis [2], Moises Torres-Gonzalez [3,*] and Jana Kraft [2,4]

1. Department of Nutrition and Food Sciences, The University of Vermont, Burlington, VT 05405, USA; Victoria.Taormina@uvm.edu
2. Department of Animal and Veterinary Sciences, The University of Vermont, Burlington, VT 05405, USA; Allison.Unger@uvm.edu (A.L.U.); Morgan.Schiksnis@uvm.edu (M.R.S.); Jana.Kraft@uvm.edu (J.K.)
3. National Dairy Council, Rosemont, IL 60018, USA
4. Department of Medicine, Division of Endocrinology, Metabolism and Diabetes, The University of Vermont, Colchester, VT 05446, USA
* Correspondence: Moises.Torres-Gonzalez@dairy.org; Tel.: +1-847-627-3275
† These authors contributed equally to this work.

Received: 21 August 2020; Accepted: 17 September 2020; Published: 20 September 2020

Abstract: Dairy fat and its fatty acids (FAs) have been shown to possess pro-health properties that can support health maintenance and disease prevention. In particular, branched-chain FAs (BCFAs), comprising approximately 2% of dairy fat, have recently been proposed as bioactive molecules contributing to the positive health effects associated with the consumption of full-fat dairy products. This narrative review evaluates human trials assessing the relationship between BCFAs and metabolic risk factors, while potential underlying biological mechanisms of BCFAs are explored through discussion of studies in animals and cell lines. In addition, this review details the biosynthetic pathway of BCFAs as well as the content and composition of BCFAs in common retail dairy products. Research performed with in vitro models demonstrates the potent, structure-specific properties of BCFAs to protect against inflammation, cancers, and metabolic disorders. Yet, human trials assessing the effect of BCFAs on disease risk are surprisingly scarce, and to our knowledge, no research has investigated the specific role of dietary BCFAs. Thus, our review highlights the critical need for scientific inquiry regarding dairy-derived BCFAs, and the influence of this overlooked FA class on human health.

Keywords: *anteiso*; branched-chain amino acids; cancer; diabetes; inflammation; *iso*; metabolic diseases; milk; phytanic acid

1. Introduction

Milk and dairy products have been a staple of the human diet for thousands of years [1] and represent one of the most important agricultural commodities in regard to human nutrition. Dairy products have been instituted in most dietary guidelines around the world as they provide a large variety of essential nutrients and several key shortfall nutrients. Moreover, milk and dairy products hold a unique position among all foods as they are considered the largest single source of natural bioactive components [2]. In particular, dairy fat is the most complex and diverse dietary fat source in nature comprised of an impressive fatty acid (FA) repertoire (>400 different FAs and FA derivatives [3]) that accounts for the myriad of nutritional, organoleptic, and technological characteristics of milk and dairy products. Specifically, dairy fat contains a unique variety of bioactive FAs that are synthesized by rumen microbes (i.e., bacteria and protozoa) and the mammary gland. These FAs consist of short- and medium-chain FAs (4 to 13 carbons), positional and geometric isomers of octadecanoate

(18:1), conjugated linoleic acids, odd-chain FAs (15:0 and 17:0), and branched-chain FAs (BCFAs). Of the repertoire of FAs in dairy fat, about 14% of them are these unique dairy-derived FAs [4] and several are known to impact normal mammalian physiology by functioning as bioactive molecules, exerting beneficial properties that support health and well-being [5–7]. To date, most research efforts in the arena of dairy FAs have been centered on the biological effects of conjugated linoleic acids, specifically rumenic acid (18:2 c9,t11), which was originally driven by the discovery of its demonstrated anticarcinogenic activity [8]. However, far less work has focused on other bioactive dairy-derived FAs.

BCFAs are an emerging group of bioactive FAs sparking growing research interest within the scientific community due to their biological effects and potential pro-health benefits. Because BCFAs are principally derived from rumen bacteria, milk and dairy products pose unique dietary sources of BCFAs. The objectives of this paper are to review and summarize the current knowledge on BCFAs and specifically to evaluate the possible role of dairy-derived BCFAs in human health.

2. Structure and Origin of BCFAs in Ruminants

BCFAs are commonly saturated FAs substituted with one (mono-) or more (di-/poly-) methyl branch(es) on the carbon chain. Typically, BCFAs possess either an *iso* structure where the FA has the branch point on the penultimate carbon atom (i.e., one from the end) or an *anteiso* structure where the branch point is located on the antepenultimate carbon atom (i.e., two from the end; Figure 1). More than 50 BCFAs have been identified in ruminant-derived fats [9,10] but *iso*- and *anteiso*-mono-methyl BCFAs with chain lengths from 14 to 17 carbon atoms are quantitatively the most abundant BCFAs in milk fat [4,11–16].

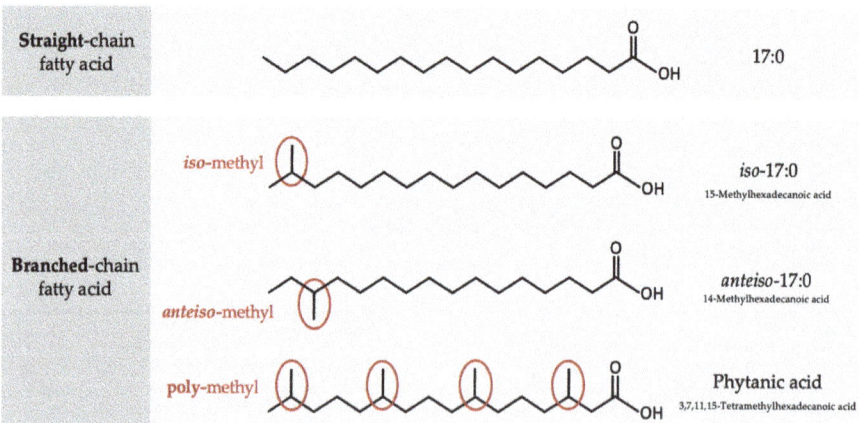

Figure 1. Structural differences between straight chain fatty acids and *iso-*, *anteiso-*, and multimethyl branched-chain fatty acids.

In ruminants, monomethyl BCFAs are synthesized by the microorganisms that reside within the rumen, specifically bacteria and protozoa [17], which utilize dietary branched-chain amino acids (BCAAs), i.e., valine, leucine, and isoleucine, to form BCFAs (Figure 2) [18]. Through a common biosynthetic pathway, BCAAs are first transformed into branched-chain α-ketoacids through the removal of the amino group by a BCAA transferase enzyme [19]. These α-ketoacid products, α-ketoisovalerate, α-keto-β-methylvalerate, and α-ketoisocaproate, are subsequently decarboxylated by branched-chain-α-ketoacid dehydrogenase producing the respective branched short-chain carboxylic acids isobutyral-CoA, isovaleryl-CoA, and 2-methylbutyral-CoA [20]. Finally, branched short-chain carboxylic acids are elongated by BCFA synthetase, with malonyl-CoA as the chain extender, to form *iso-* and *anteiso-*BCFAs [18]. The products of the BCFA biosynthetic pathway are *iso-*14:0 and *iso-*16:0, derived from valine, *iso-*15:0 and *iso-*17:0 from leucine, and *anteiso-*15:0 and *anteiso-*17:0 from isoleucine.

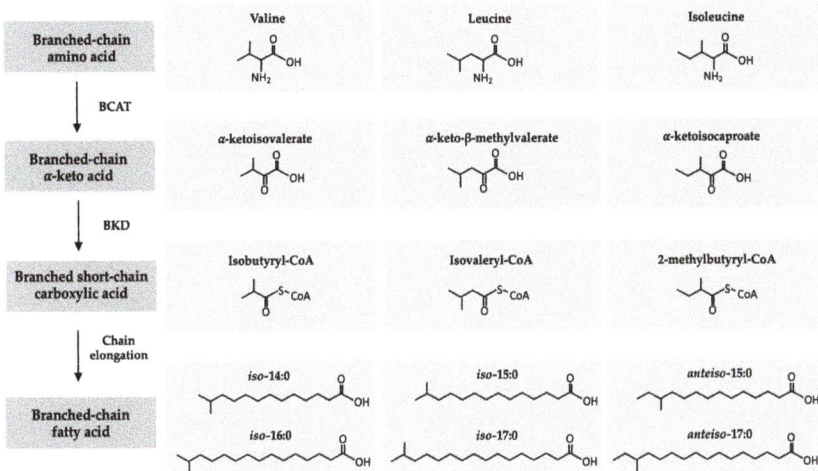

Figure 2. Biosynthetic pathway of branched-chain fatty acids from branched-chain amino acids. BCAT: branched-chain amino acid transferase (BCAT) enzyme. BKD: branched-chain-α-ketoacid dehydrogenase.

Phytanic acid (3,7,11,15-tetramethylhexadecanoic acid) and its metabolite pristanic acid (2,6,10,14-tetramethylpentadecanoic acid), two multimethyl BCFAs, are also predominantly synthesized in the rumen (Figure 3) [21,22]. Phytanic acid is derived from the phytol component of chlorophyll found in forages [21]. While mammals are unable to cleave the ring structure from the phytol moiety of chlorophyll, bacteria release phytol in the rumen [23]. The primary synthesis pathway of phytanic acid in ruminants is initiated by the biohydrogenation of phytol to produce dihydrophytol [21,24]. Dihydrophytol is subsequently converted to phytanal before it is synthesized into phytanic acid [21,24]. Notably, a potential secondary synthesis pathway has been described in non-ruminant tissues and marine bacteria [22,24,25] and may be applicable to ruminant animals as well, although there is only limited evidence [26]. In this case, phytol is first converted to phytenal by an alcohol dehydrogenase [22]. Fatty aldehyde dehydrogenase is used to transform phytenal to phytenic acid before it is modified into phytenoyl-CoA by an acyl-CoA synthetase. Subsequently, phytenoyl-CoA is converted by an enoyl-CoA reductase to form phytanoyl-CoA [24,25]. The final synthesis step of phytanoyl-CoA to phytanic acid is thought to occur through α-oxidation [25] due to the enzymatic activity of thioesterase, a protein involved in α-oxidation [24]. In order to form pristanic acid, phytanic acid is activated by phytanoyl-CoA synthetase to create phytanoyl-CoA, which can then undergo α-oxidation to produce pristanic acid.

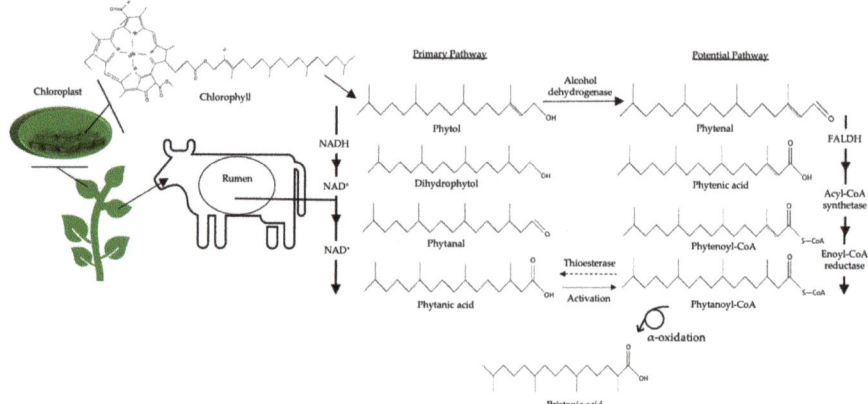

Figure 3. Synthesis of phytanic and pristanic acid via rumen microorganisms derived from chlorophyll within forages NAD: Nicotinamide adenine dinucleotide. NAD$^+$: the oxidized form of NAD. NADH: the reduced form of NAD (protonated with a hydrogen).

BCFAs are key structural lipid constituents of the bacterial membranes [17] where they play an important regulatory role in membrane fluidity and permeability [27]. Importantly, bacterial membrane lipids make an important contribution to ruminant milk fat, as bacterial cells leaving the rumen pass to the duodenum where their membrane FAs are absorbed and subsequently incorporated into milk fat and other tissues [21].

3. Occurrence and Metabolism of BCFAs in Humans

BCFAs are common constituents of microbial lipids present in abundant quantities but have also been found in many other organisms (e.g., *C. elegans*, [28]) including mammals [29–31], although in much lower amounts. In humans, BCFAs have been detected in various tissues and fluids including vernix caseosa, the biofilm covering the skin of the fetus [32,33], colostrum and mature breast milk [34–36], adipose tissue [37], and serum [38]. Early work by Nicolaides and co-authors [39–41] established that the meibomian and sebaceous glands of the human skin produce BCFAs secreted into meibum and sebum, respectively. Subsequent research in vitro and in vivo using mouse models confirmed the endogenous synthesis of BCFAs from their respective BCAA precursors (i.e., valine, leucine, isoleucine) as a result of the BCAA catabolic pathway [30,31,42,43]. BCAA degradation in humans is comparable to that in bacteria (Figure 2). Despite the endogenous synthesis, dietary intake of BCFAs is presumably the principal source of BCFAs in the human body.

The intake of BCFAs depends on the type of food and the fat content of the food consumed. While ruminant-derived foods (i.e., milk and meat and their respective products) are the chief source of BCFAs, fish and non-dairy fermented foods (e.g., kimchi, sauerkraut, miso, tempeh) may also supply BCFAs, although only at very small quantities. The mean daily intake of BCFAs in the United States has been estimated to range between 220 mg/day (for milk only [44]) and 500 mg/day (from dairy and beef products combined [11]). The contribution of total dairy products to the BCFA intake has been calculated at 317 mg/day [11]. There is no requirement for BCFAs per se but it is conceivable that the daily dietary intake of BCFAs needs to be higher than the current estimates to achieve potential health benefits.

Very little is known about the fate and metabolism of dietary BCFAs in humans, thus representing an important area for future research. While information is scarce, it can be presumed that digestion, absorption, and transport of BCFAs in humans is comparable to that of other long-chain FAs. Our knowledge of BCFA uptake by cells is based largely on work in cell lines, specifically from the intestine (Caco-2), fetal small intestine (H4), and breast cancer (SKBR-3 and MCF-7), following exposure to one or a mixture of BCFAs [45–49]. From this limited perspective, it is thought that the cellular uptake of BCFAs is dependent on BCFA length and configuration. Using Caco-2 cells, Yan et al. [45] demonstrated that BCFAs can be taken up and further metabolized into their elongation or chain-shortened products. Other work indicates that in fetal intestinal cells incubated with identical concentrations of BCFAs (i.e., *iso*-16:0, *anteiso*-17:0, *iso*-18:0, and *iso*-20:0), *anteiso*-17:0 exhibited the greatest uptake efficiency, followed by *iso*-16:0, *iso*-18:0, then *iso*-20:0, with significant differences, and a specific hierarchy, between these BCFAs [47]. In contrast, MCF-7 human breast cancer cells accumulated greater amounts of *iso*-15:0 and *iso*-17:0 compared to their *anteiso* analogs [49]. Dietary supplementation of BCFAs to neonatal rat pups demonstrated that BCFAs incorporate into ileal phospholipids as well as liver tissue and serum [50]. Research conflicts as to whether BCFAs are preferentially incorporated into cellular triacylglycerols or phospholipids [47,48], and whether the FA structure plays a determinant role requires further investigation.

4. Occurrence of BCFAs in Dairy Products

4.1. Content and Composition of BCFAs in Milk

Within dairy foods, research to date has mainly focused on the content and composition of BCFAs derived from cow's milk. From this work, studies established that cow's milk is a significant source of BCFAs, typically comprising 1.7–3.4% BCFAs of total FAs (Table 1). When considered on a per serving basis, three servings of whole milk (3.25% milk fat) per day can thus provide 367–763 mg of BCFAs (Table 1). In milk fat, monomethyl BCFAs are the principal BCFAs present, with a chain length of 14–17 carbons and the methyl group occurring in either an *iso* or *anteiso* configuration. In general, total *iso*- and *anteiso*-BCFAs isomers occur in an approximate ratio of 1:1 (Table 1). Short-chain (4:0–6:0) and medium-chain (7:0–13:0) *iso*- and *anteiso*-BCFA isomers are also present in cow's milk but in rather a minor abundance, accounting for approximately 0.01 and 0.12% of total BCFAs, respectively [51].

Table 1. Branched-chain fatty acid composition and content of cow's milk.

FA [1]	Proportion (% of Total FAs) [2]	Content (mg/Three Daily Servings) [3]
iso-13:0	0.02–0.03 [4,13–15] [4]	5–7 [4,13–15] [4]
anteiso-13:0	0.07–0.09 [4,13] [4]	15–19 [4,13] [4]
iso-14:0	0.08–0.22 [4,11–15] [4]	18–48 [4,11–15] [4]
iso-15:0	0.13–0.44 [4,11–15] [4]	29–97 [4,11–15] [4]
anteiso-15:0	0.37–0.93 [4,11–15] [4]	81–206 [4,11–15] [4]
iso-16:0	0.17–0.45 [4,11–15] [4]	38–100 [4,11–15] [4]
iso-17:0	0.26–0.56 [4,11–15] [4]	58–123 [4,11–15] [4]
anteiso-17:0	0.11–0.76 [4,11–15] [4]	24–169 [4,11–15] [4]
iso-18:0	0.01–0.09 [4,11–15] [4]	2–20 [4,11–15] [4]
Σ *iso*-BCFAs	0.87–1.75 [4,11–15] [4]	193–387 [4,11–15] [4]
Σ *anteiso*-BCFAs	0.67–1.69 [4,11–15] [4]	149–375 [4,11–15] [4]
Σ BCFAs [5]	1.66–3.44 [4,11–15] [4]	367–763 [4,11–15] [4]

[1] FA = fatty acid. [2] Averages were estimated by calculating the median of reported FA values when appropriate. [3] Content of FAs per serving was calculated from FA values listed (% of total FAs) as described by Bainbridge et al. [13], assuming 3.25% fat per serving whole milk, then multiplied by three. [4] Raw data obtained from previously published work [4]. [5] BCFAs = branched-chain FAs; sum of *iso* and *anteiso* BCFA isomers (13:0–18:0).

Multimethyl BCFAs, namely phytanic and pristanic acid, are consistent constituents within milk fat, but are often overlooked or ignored, likely because of their very low content and analytical limitations. In cow's milk, the content of phytanic acid typically ranges between 0.10 and 0.50% of total FAs, equivalent to 7–37 mg/serving whole milk [12,23,52–56]. The presence of phytanic acid in milk is specifically dependent on the feed type of the dairy cow, with higher amounts associated with the intake of grass-based rations rich in chlorophyll [23,52,53,55]. Pristanic acid, a metabolite of phytanic acid, occurs in very low amounts in milk fat, at approximately 0.04–0.06% (3–4 mg/serving whole milk) [23].

4.2. BCFAs in Milk across Ruminant Species

A survey of the recent literature regarding the BCFA composition of ruminant milk shows that milk from sheep and goats comprises approximately 1.8–3.1% and 1.2–2.4% BCFAs of total FAs, respectively, and is thus comparable to cow's milk (1.7–3.4% BCFAs of total FAs; Table 1). Yet, the content of total and individual BCFAs in milk varies considerably based on the ruminant species of origin (Table 2). In particular, on a per serving basis, milk from sheep contains considerably more BCFAs per serving compared to milk from cows (204–502 mg/serving versus 123–254 mg/serving). In the U.S., cow's milk is standardized to 3.25% milk fat (i.e., whole milk), however, this practice is less common for specialty milk such as goat and sheep milk. Differences in the BCFA content of ruminant-derived milks therefore appear to be chiefly reflective of milk fat standardization practices in retail milk, rather than species differences in BCFA occurrence.

4.3. Comparison of BCFAs among Dairy Products

The content of BCFAs in dairy also differs depending upon the type of dairy product (Table 2). For example, Ran Ressler et al. [11] found that yogurt contained an equivalent of 152 mg BCFAs/serving, while butter contained an equivalent of 195 mg/serving. Limited work assessing the BCFA content in cheese demonstrates that cheese type appears to be an important consideration as well (82–322 mg/serving depending upon cheese variety). One notable source of variation in the content of BCFAs per serving among cheese types is the differing fat content. For example, one serving (28.35 g) of low-moisture Mozzarella (22.1% fat) contains 82 mg BCFAs, whereas Cheddar (33.0% fat) contains 148 mg BCFAs [11]. One serving of butter (14.2 g), another dairy-derived product high in fat (<80%), can contain more than 200 mg of BCFAs (137–204 mg/serving; Table 2).

5. Human Trials Assessing Health Effects of BCFAs

To date, epidemiological research indicates that dairy fat consumption is neutral or protective against cardiometabolic diseases. However, the specific role of BCFAs is less understood, as these studies largely rely upon either food frequency questionnaires and/or the measurement of specific dairy FA biomarkers (i.e., 15:0, 17:0, and 16:1 t9) in blood or tissues [57–60], but not BCFAs.

5.1. Obesity

Epidemiological studies have noted that the intake of full-fat dairy products can be beneficial for long-term weight maintenance [61,62]. Other research which specifically focused on the consumption of BCFAs suggests that BCFAs have an important role in energy homeostasis. For example, one study found that total BCFAs in serum were higher in non-obese women than obese women and that *iso*-BCFAs were inversely associated with body mass index [38]. Similar results have been reported in a recent study by Pakiet et al. [63]. Another study found that the adipose tissue of lean subjects had higher proportions of total BCFAs and individual BCFAs than obese individuals [37].

One niche of scientific interest is the relationship between tissue concentrations of BCFAs and weight status in individuals post-gastric bypass surgery. In a longitudinal study, the BCFAs of adipose tissue in obese subjects was assessed at baseline and one year following Roux-en-Y gastric bypass surgery [37]. This study demonstrated that, after one year, the proportion of total BCFAs increased after surgery-induced weight loss. A follow-up study found similar results, showing that serum BCFAs increased in individuals after one-anastomosis gastric bypass surgery [63]. Of note, a similar study found no changes in serum BCFAs two weeks after one-anastomosis gastric bypass surgery, suggesting that a minimum time period may be needed to observe changes [64].

Table 2. Branched-chain fatty acid content (mg/serving) of common dairy products [1].

FA [2]	Milk [3]	Cheese [4] Soft/Semi-Soft [9]	Cheese [4] Semi-Hard/Hard [10]	Cheese [4] Unknown [11]	Yogurt [5]	Butter [6]	Sheep Milk [7]	Goat Milk [8]
iso-13:0	2 [4,13–15] [12]					2 [65]	8 [66]	1 [67]
anteiso-13:0	5–6 [4,13] [12]					9 [65]	5 [66]	1–7 [67,68]
iso-14:0	6–16 [4,11–15] [12]	5–14 [11,69]	6–20 [11,12,69]	56–13 [11,69]	9 [11]	8–18 [11,65,70]	8–24 [66,71,72]	5–10 [67,68,73,74]
iso-15:0	10–32 [4,11–15] [12]	6–33 [11,69]	8–41 [11,12,69]	12–33 [11,69]	11 [11]	1–22 [11,65,70]	21–71 [66,71,72,75]	13–36 [67,68,73,74,76]
anteiso-15:0	27–69 [4,11–15] [12]	26–68 [11,69]	33–87 [11,12,69]	35–62 [11,69]	47 [11]	36–67 [11,65,70]	45–122 [66,71,72,75]	25–47 [67,68,73,74,76]
iso-16:0	13–33 [4,11–15] [12]	12–30 [11,69]	9–42 [11,12,69]	11–26 [11,69]	22 [11]	20–36 [11,65,70]	25–69 [66,71,72,75]	13–22 [67,68,73,74]
iso-17:0	19–41 [4,11–15] [12]	8–41 [11,69]	8–52 [11,12,69]	14–38 [11,69]	19 [11]	22–34 [11,65,70]	34–115 [66,71,72,75]	20–58 [67,68,73,74,76]
anteiso-17:0	8–56 [4,11–15] [12]	23–71 [11,69]	17–71 [11,12,69]	25–61 [11,69]	45 [11]	34–42 [11,65,70]	48–124 [66,71,72,75]	25–57 [67,68,73,74,76]
iso-18:0	1–7 [4,11–15] [12]	<1–8 [11,69]	<1–8 [11,12,69]	<1–7 [11,69]	3 [11]	<1–7 [11,65,70]	17–20 [66,72]	5 [67]
Σ iso-BCFAs	64–129 [4,11–15] [12]	32–127 [11,69]	32–164 [11,12,69]	53–116 [11,69]	64 [11]	68–106 [11,65,70]	108–255 [66,71,72,75]	53–96 [67,68,73,74,76]
Σ anteiso-BCFAs	50–125 [4,11–15] [12]	51–139 [11,69]	50–158 [11,12,69]	68–123 [11,69]	91 [11]	69–107 [11,65,70]	96–247 [66,71,72,75]	50–104 [67,68,73,74,76]
Σ BCFAs [13]	123–254 [4,11–15] [12]	83–264 [11,69]	82–322 [11,12,69]	123–239 [11,69]	152 [11]	137–204 [11,65,70]	204–502 [66,71,72,75]	109–180 [67,68,73,74,76]

[1] Averages were estimated by calculating the median of reported FA values when appropriate; content of FAs per serving was calculated from reported FA values (% of total FAs), assuming 93.3% of FAs in milk fat (correction for glycerol) [77]. [2] FA = fatty acid. [3] Cow-derived; based on the assumption of 3.25% fat, 244 g per serving [13]. [4] Cow-derived; based on reported % fat and serving sizes (g) listed by FoodData Central for each cheese variety [78–89] with the exception of Ricotta cheese (62 g/serving). [5] Cow-derived; based on reported % fat and serving size (245 g) listed by FoodData Central [90]. [6] Cow-derived; based on reported % fat (80%) and serving size (14.2 g) listed by FoodData Central [91]; when FA values were reported as μg/g butter [70], FAs per serving were calculated by assuming 14.2 g per serving [91] without correction factor for glycerol. [7] Based on reported % fat (median of values calculated when appropriate) or % fat listed by FoodData Central [92] if not reported; based on serving size (245 g) listed by FoodData Central [92]. [8] Based on reported % fat (median of values calculated when appropriate) or % fat listed by FoodData Central [93] if not reported; based on serving size (244 g) listed by FoodData Central [93]. [9] Includes Ricotta, Romadur, Cottage cheese, Camembert, Brie, Limburger, Feta, and Bavaria Blue. [10] Includes Provolone, low-moisture Mozzarella, Emmental, Cheddar, Montasio, Gouda, and Butter cheese (Butterkäse). [11] Includes Swiss, American, Alpine cheese, curd cheese, and Mozzarella (varieties without sufficient information reported to categorize). [12] Raw data underlying previously published work [4]. [13] BCFAs = branched-chain FAs; sum of iso and anteiso BCFA isomers (13:0–18:0).

5.2. Insulin Sensitivity

Serum collected from fasted individuals showed a modest inverse correlation between serum BCFA concentrations and an index of insulin resistance (homeostatic model of assessment of insulin resistance), suggesting that BCFAs may promote insulin sensitivity [63]. Similarly, total BCFAs in adipose tissue have been associated with measurements of insulin sensitivity (i.e., insulin-stimulated glucose rate of disappearance) [37]. Moreover, Mika et al. [38] observed that two BCFAs, *anteiso*-15:0 and *iso*-17:0, were negatively correlated with fasted serum insulin levels but not with the homeostatic model assessment of insulin resistance.

5.3. Limitations of Current Evidence Available in Humans

Research suggests that BCFAs promote weight maintenance and metabolic health [37,38,63], however, more studies are needed to ascertain the specific impact of dietary BCFAs, and to what extent dietary BCFAs contribute to these outcome measures. Research assessing the metabolic effects of dairy-derived BCFAs is still lacking. Therefore, for this review, we considered studies evaluating the role of BCFAs on health, regardless of the source. While dietary patterns are the plausible source of variation in tissue BCFA occurrence in humans, BCFAs derived from an individual's gut microbiota or metabolism from BCAAs cannot be discounted. Future work evaluating the relationship between diet and disease risk should consider the inclusion of BCFAs as biomarkers of dairy fat consumption.

6. Role of Dairy-Derived BCFAs in Health: Potential Mechanisms

6.1. Inflammation

Inflammation is an important underlying factor contributing to the pathogenesis of many metabolic diseases [94–96]. In particular, gastrointestinal health is critical for proper immune function and regulation of inflammation [97,98]. Research utilizing gastrointestinal cell lines has demonstrated the potent anti-inflammatory potential of dietary BCFAs. For example, exposure of Caco-2 cells to BCFAs reduced the lipopolysaccharide-induced gene expression of important proinflammatory mediators (i.e., IL-8, TLR-4, and NF-κB) [45]. In addition, certain shorter chain BCFAs (*i.e.*, *iso*-14:0, *iso*-16:0, *anteiso*-13:0) improved the lipopolysaccharide-induced decrease in cell viability. Similar results were reported in a follow-up study [46] when Caco-2 cells were treated with monoacylglycerols or free FAs comprised of ~30% BCFAs isolated from the vernix.

Animal work is limited, but also supports that BCFAs attenuate inflammation and related diseases. Ran-Ressler et al. [50] examined the effect of a diet containing a mixture of BCFAs (*iso*-14:0, *anteiso*-15:0, *iso*-16:0, *anteiso*-17:0, *iso*-18:0, and *iso*-20:0) on necrotizing enterocolitis in neonatal Sprague–Dawley pups. BCFA-fed pups had a lower incidence of necrotizing enterocolitis and an enhanced gene expression of IL-10, an anti-inflammatory cytokine. Moreover, the cecal samples of pups fed the BCFA-enriched rat formula had a greater abundance of *Bacillus subtilis* and *Pseudomonas aeruginosa* compared to those fed standard rat formula, which is notable as both are known to contain BCFAs in their membranes [18,50,99]. Of note, an inverse association has been found in humans for serum *iso*-BCFAs (i.e., *iso*-15:0, *iso*-16:0, *iso*-17:0, and *anteiso*-15:0) and serum C-reactive protein [38], an important inflammatory marker for type 2 diabetes risk.

6.2. Anticarcinogenic Properties

One of the first beneficial effects attributed to BCFAs were their anticancer properties discovered 20 years ago. Specifically, *iso*-15:0 is known to inhibit the growth of T-cell non-Hodgkin lymphoma (Jurkat, Hut78, and EL4 cell lines [100]) and various types of carcinoma cell lines including, breast (MCF-7 and SKBR-3), prostate (DU145), lung (NCI-H1688), pancreas (BxPC-3), liver (SNU-423), bladder (T24, 5637, and UM-UC-3), leukemia (K-562), and gastric (NCI-SNU-1) and colorectal carcinoma (HCT 116) in a dose- and/or time-dependent manner by inhibiting proliferation and inducing apoptosis [48,49,101,102]. In a comparison of eight *iso*-BCFA species of varying carbon chain lengths (*iso*-12:0 through *iso*-20:0), Wongtangtintharn et al. [103] demonstrated that the anticarcinogenic activity of BCFAs are dependent on the chain length, with *iso*-16:0 exerting the highest cytotoxic activity. Furthermore, a more recent study showed that BCFAs with an *iso* structure (i.e., *iso*-15:0 and *iso*-17:0) were more potent than their *anteiso* counterparts (i.e., *anteiso*-15:0 and *anteiso*-17:0) [49]. Similarly, in vivo experiments have shown that BCFAs inhibit tumor growth in both mouse xenograft tumor models in which cancer cells were co-grafted [100,101] and in the orthotopic VX2 squamous cell carcinoma model in rabbits [104].

BCFAs may also induce beneficial, location-specific effects on angiogenesis. Treatment with *iso*-15:0 promoted angiogenesis post cerebral ischemia/reperfusion injury [105], while *anteiso*-15:0 suppressed angiogenesis to aid corneal recovery [106]. These effects appear to be structure specific, but studies in this area are still very limited, hence more research is needed. Moreover, the biological significance of the role of BCFAs on angiogenesis in the context of cancer growth is not yet established.

6.3. Energy and Glucose Homeostasis

Previous research has suggested that the body content, and specifically the liver content, of BCFAs reflects the ingestion of their respective BCAA precursors [30,107]. Furthermore, Brooks et al. [107] and Garcia-Caraballo et al. [30] observed that the body or hepatic content of BCFAs were inversely correlated with the body triacylglycerol content, as shown in *C. elegans*, or the liver triacylglycerol content, as shown in mice, respectively. Likewise, a recent in vitro study, using a human fatty liver cell line model (i.e., L02 cells), concluded that individual BCFAs (*iso*-15:0 and *iso*-18:0) reduced the cellular triacylglycerol content and were associated with an upregulation of multiple genes involved in lipid catabolism [108]. These observations are in accordance with human trials which showed that, in serum, total BCFAs [63] and specific *iso*-BCFAs [38] had an inverse relationship with triacylglycerols. Thus, the results from these studies indicate that BCFA may be directly or indirectly involved in the regulation of fat storage in the body.

As described above, BCFAs may favorably influence insulin sensitivity in humans, however, studies assessing potential mechanisms involved are scarce. One study performed with rat insulinoma INS-1 β-cells demonstrated that *iso*-17:0 may beneficially modulate β-cell function via the upregulation of Pdx1 and PPAR-γ transcription factors [109]. Yet, more in vivo research is necessary to confirm these results.

6.4. Biological Functions of Polymethyl BCFAs: Phytanic Acid

The health effects of dietary phytanic acid, both positive and negative, have been extensively examined in a recent review [110]. In particular, Roca-Saavedra et al. [110] summarize the biological complications that arise due to an excess of phytanic acid within the body, such as neurological injury, oxidative stress, and cancers. To that end, recent in vitro work continues to show that phytanic acid, when provided at concentrations exceeding normal physiological ranges, is neurotoxic [111,112]. Importantly, the nutritional relevance of such studies is called into question for the general population, as the accumulation of phytanic acid due to impaired lipid metabolism (e.g., Refsum disease) is thought to be rare in terms of prevalence [113]. Nakanishi et al. [114] demonstrated that when cells are treated with concentrations of phytanic acid comparable to that found in the plasma of healthy individuals, phytanic acid exerted beneficial immunomodulatory effects in a PPAR-α-dependent manner. Furthermore, there is no clinical evidence that the consumption of phytanic acid in naturally occurring amounts, e.g., phytanic acid derived from milk and dairy products, detrimentally impacts health in individuals with normal lipid metabolism.

Roca-Saavedra et al. [110] also describe favorable effects of phytanic acid on glucose and lipid metabolism, energy expenditure, as well as immune and anticancer functions. Indeed, efforts to ameliorate or prevent obesity have been gaining momentum, and phytanic acid has been attracting scientific interest as an active compound for potential novel therapeutic drugs. Emerging research identified phytanic acid as a potent agonist of PPAR-α [115], a transcription factor expressed in metabolically active tissues controlling cellular FA oxidation and regulating energy homeostasis [116]. An et al. [115] showed that phytol treatment in a high-fat diet-induced mouse model of obesity resulted in greater contents of phytanic acid in liver and brown adipose tissue as well as the activation of PPAR-α and its target genes. Additionally, a recent study reported that phytanic acid promotes beige adipogenesis in murine 3T3-L1 (white) adipocytes (also called browning) in a PPAR-α ligand-dependent manner [117]. Taken together, these novel findings add support to phytanic acid possibly playing a role in the regulation of energy homeostasis.

7. Conclusions

BCFAs are a class of saturated FAs that comprise a significant portion of total FAs in milk and dairy products. While humans can derive a limited amount of BCFAs from endogenous synthesis from BCAAs, dietary BCFAs remain the most important source of BCFAs within the body. Recent human trials indicate that BCFAs in tissues have a beneficial influence on metabolism. Mechanistic research validates these studies by demonstrating that BCFAs possess a wide range of structure-specific functions that may favorably influence health at the cellular and systemic level. Despite the promise of BCFAs as pro-health dietary constituents, our narrative review reveals a critical scarcity in the scientific understanding of the relationship between diet-derived BCFAs and disease risk. Of note, research has been slow to utilize these FAs as markers of dairy fat intake. Here we propose that BCFAs may be a useful and complimentary biomarker of dairy fat intake for future studies. It is clear that more research is needed to clarify the diverse biological roles of BCFAs in vivo, particularly through clinical and epidemiolocal studies that evaluate the relationship between BCFA consumption, food source, BCFA tissue concentrations, and disease.

Author Contributions: Conceptualization, J.K and M.T.-G.; Data curation, V.M.T., A.L.U., M.R.S., and J.K.; Writing—original draft preparation, V.M.T., A.L.U., M.R.S., and J.K.; Writing—review and editing, M.T.-G. and J.K.; Visualization, M.R.S. and J.K.; Supervision, J.K.; Project administration, M.T.-G. and J.K. All authors have read and agreed to the published version of the manuscript.

Funding: This research received no external funding.

Conflicts of Interest: M.T.-G. is employee of National Dairy Council. J.K. has received research funding from National Dairy Council.

References

1. Cordain, L.; Eaton, S.B.; Sebastian, A.; Mann, N.; Lindeberg, S.; Watkins, B.A.; O'Keefe, J.H.; Brand-Miller, J. Origins and evolution of the Western diet: Health implications for the 21st century. *Am. J. Clin. Nutr.* **2005**, *81*, 341–354. [CrossRef]
2. German, J.B.; Dillard, C.J.; Ward, R.E. Bioactive components in milk. *Curr. Opin. Clin. Nutr. Metab. Care* **2002**, *5*, 653–658. [CrossRef]
3. Jensen, R.G. The composition of bovine milk lipids: January 1995 to December 2000. *J. Dairy Sci.* **2002**, *85*, 295–350. [CrossRef]
4. Unger, A.L.; Bourne, D.E.; Walsh, H.; Kraft, J. Fatty acid content of retail cow's milk in the northeastern United States—What's in it for the consumer? *J. Agric. Food Chem.* **2020**, *68*, 4268–4276. [CrossRef]
5. Fuke, G.; Nornberg, J.L. Systematic evaluation on the effectiveness of conjugated linoleic acid in human health. *Crit. Rev. Food Sci. Nutr.* **2017**, *57*, 1–7. [CrossRef]
6. Shokryazdan, P.; Rajion, M.A.; Meng, G.Y.; Boo, L.J.; Ebrahimi, M.; Royan, M.; Sahebi, M.; Azizi, P.; Abiri, R.; Jahromi, M.F. Conjugated linoleic acid: A potent fatty acid linked to animal and human health. *Crit. Rev. Food Sci. Nutr.* **2017**, *57*, 2737–2748. [CrossRef] [PubMed]
7. Den Hartigh, L.J. Conjugated linoleic acid effects on cancer, obesity, and atherosclerosis: A review of pre-clinical and human trials with current perspectives. *Nutrients* **2019**, *11*, 370. [CrossRef] [PubMed]
8. Pariza, M.W.; Ha, Y.L. Conjugated dienoic derivatives of linoleic acid: A new class of anticarcinogens. *Med. Oncol. Tumor Pharmacother.* **1990**, *7*, 169–171. [CrossRef] [PubMed]
9. Kim Ha, J.; Lindsay, R.C. Method for the quantitative analysis of volatile free and total branched-chain fatty acids in cheese and milk fat. *J. Dairy Sci.* **1990**, *73*, 1988–1999. [CrossRef]

10. Alonso, L.; Fontecha, J.; Lozada, L.; Fraga, M.J.; Juárez, M. Fatty acid composition of caprine milk: Major, branched-chain, and trans fatty acids. *J. Dairy Sci.* **1999**, *82*, 878–884. [CrossRef]
11. Ran-Ressler, R.R.; Bae, S.; Lawrence, P.; Wang, D.H.; Thomas Brenna, J. Branched-chain fatty acid content of foods and estimated intake in the USA. *Br. J. Nutr.* **2014**, *112*, 565–572. [CrossRef] [PubMed]
12. Corazzin, M.; Romanzin, A.; Sepulcri, A.; Pinosa, M.; Piasentier, E.; Bovolenta, S. Fatty acid profiles of cow's milk and cheese as affected by mountain pasture type and concentrate supplementation. *Animals* **2019**, *9*, 68. [CrossRef] [PubMed]
13. Bainbridge, M.L.; Cersosimo, L.M.; Wright, A.-D.G.; Kraft, J. Content and composition of branched-chain fatty acids in bovine milk are affected by lactation stage and breed of dairy cow. *PLoS ONE* **2016**, *11*, e0150386. [CrossRef] [PubMed]
14. Schwendel, B.H.; Wester, T.J.; Morel, P.C.H.; Fong, B.; Tavendale, M.H.; Deadman, C.; Shadbolt, N.M.; Otter, D.E. Pasture feeding conventional cows removes differences between organic and conventionally produced milk. *Food Chem.* **2017**, *229*, 805–813. [CrossRef] [PubMed]
15. Schwendel, B.H.; Morel, P.C.H.; Wester, T.J.; Tavendale, M.H.; Deadman, C.; Fong, B.; Shadbolt, N.M.; Thatcher, A.; Otter, D.E. Fatty acid profile differs between organic and conventionally produced cow milk independent of season or milking time. *J. Dairy Sci.* **2015**, *98*, 1411–1425. [CrossRef] [PubMed]
16. Yan, Y.; Wang, Z.; Wang, X.; Wang, Y.; Xiang, J.; Kothapalli, K.S.D.; Thomas Brenna, J. Branched chain fatty acids positional distribution in human milk fat and common human food fats and uptake in human intestinal cells. *J. Funct. Foods* **2017**, *29*, 172–177. [CrossRef]
17. Or-Rashid, M.M.; Odongo, N.E.; McBride, B.W. Fatty acid composition of ruminal bacteria and protozoa, with emphasis on conjugated linoleic acid, vaccenic acid, and odd-chain and branched-chain fatty acids. *J. Anim. Sci.* **2007**, *85*, 1228–1234. [CrossRef]
18. Kaneda, T. Iso-and anteiso-fatty acids in bacteria: Biosynthesis, function, and taxonomic significance. *Microbiol. Rev.* **1991**, *55*, 288–302. [CrossRef]
19. Vlaeminck, B.; Fievez, V.; Cabrita, A.R.J.; Fonseca, A.J.M.; Dewhurst, R.J. Factors affecting odd- and branched-chain fatty acids in milk: A review. *Anim. Feed Sci. Technol.* **2006**, *131*, 389–417. [CrossRef]
20. Harper, A.E.; Miller, R.H.; Block, K.P. Branched-chain amino acid metabolism. *Ann. Rev. Nutr.* **1984**, *4*, 409–454. [CrossRef]
21. Harfoot, C.G. Lipid metabolism in ruminant animals. In *Lipid Metabolism in Ruminant Animals*; Christie, W.W., Ed.; Pergamon Press Ltd: Oxford, UK, 1981; pp. 1–19.
22. Jansen, G.A.; Wanders, R.J.A. Alpha-oxidation. *Biochim. Biophys. Acta* **2006**, *1763*, 1403–1412. [CrossRef] [PubMed]
23. Vetter, W.; Schröder, M. Concentrations of phytanic acid and pristanic acid are higher in organic than in conventional dairy products from the German market. *Food Chem.* **2010**, *119*, 746–752. [CrossRef]
24. Van Veldhoven, P.P. Biochemistry and genetics of inherited disorders of peroxisomal fatty acid metabolism. *J. Lipid Res.* **2010**, *51*, 2863–2895. [CrossRef]
25. Gloerich, J.; Van Den Brink, D.M.; Ruiter, J.P.N.; Van Vlies, N.; Vaz, F.M.; Wanders, R.J.A.; Ferdinandusse, S. Metabolism of phytol to phytanic acid in the mouse, and the role of PPARa in its regulation. *J. Lipid Res.* **2007**, *48*, 77–85. [CrossRef] [PubMed]
26. Lough, A.K. The phytanic acid content of the lipids of bovine tissues and milk. *Lipids* **1977**, *12*, 115–119. [CrossRef] [PubMed]
27. Harfoot, C.; Hazelwood, G. Lipid metabolism in the rumen. In *The Rumen Microbial Ecosystem*; Hobson, P., Stewart, C., Eds.; Blackie Academic & Professional: London, UK, 1997; pp. 382–426.
28. Kniazeva, M.; Crawford, Q.T.; Seiber, M.; Wang, C.Y.; Han, M. Monomethyl branched-chain fatty acids play an essential role in Caenorhabditis elegans development. *PLoS Biol.* **2004**, *2*, e257. [CrossRef]
29. Garcia-Caraballo, S.C.; Comhair, T.M.; Verheyen, F.; Gaemers, I.; Schaap, F.G.; Houten, S.M.; Hakvoort, T.B.M.; Dejong, H.C.; Lamers, W.H.; Koehler, S.E. Prevention and reversal of hepatic steatosis with a high-protein diet in mice. *Biochim. Biophys. Acta* **2013**, *1832*, 685–695. [CrossRef] [PubMed]
30. Garcia-Caraballo, S.C.; Comhair, T.M.; Houten, S.M.; Dejong, H.C.; Lamers, W.H.; Koehler, S.E. High-protein diets prevent steatosis and induce hepatic accumulation of monomethyl branched-chain fatty acids. *J. Nutr. Biochem.* **2014**, *25*, 1263–1274. [CrossRef]

31. Hirosuke, O.; Noriyasu, Y.; Junichi, N.; Isao, C. Precursor role of branched-chain amino acids in the biosynthesis of iso and anteiso fatty acids in rat skin. *Biochim. Biophys. Acta (BBA)/Lipids Lipid Metab.* **1994**, *1214*, 279–287. [CrossRef]
32. Nicolaides, N. The structures of the branched fatty acids in the wax esters of vernix caseosa. *Lipids* **1971**, *6*, 901–905. [CrossRef]
33. Ran-Ressler, R.; Devapatla, S.; Lawrence, P.; Brenna, J.T. Comparison of BCFA types in vernix and meconium of healthy neonates. *FASEB J.* **2008**, *22*. [CrossRef]
34. Egge, H.; Murawski, U.; Ryhage, R.; György, P.; Chatranon, W.; Zilliken, F. Minor constitutents of human milk IV: Analysis of the branched chain fatty acids. *Chem. Phys. Lipids* **1972**, *8*, 42–55. [CrossRef]
35. Gibson, R.A.; Kneebone, G.M. Fatty acid composition of human colostrum and mature breast milk. *Am. J. Clin. Nutr.* **1981**, *43*, 252–257. [CrossRef] [PubMed]
36. Dingess, K.A.; Valentine, C.J.; Ollberding, N.J.; Davidson, B.S.; Woo, J.G.; Summer, S.; Peng, Y.M.; Guerrero, M.L.; Ruiz-Palacios, G.M.; Ran-Ressler, R.R.; et al. Branched-chain fatty acid composition of human milk and the impact of maternal diet: The global exploration of human milk (GEHM) study. *Am. J. Clin. Nutr.* **2017**, *105*, 177–184. [CrossRef]
37. Su, X.; Magkos, F.; Zhou, D.; Eagon, J.C.; Fabbrini, E.; Okunade, A.L.; Klein, S. Adipose tissue monomethyl branched-chain fatty acids and insulin sensitivity: Effects of obesity and weight loss. *Obesity* **2015**, *23*, 329–334. [CrossRef] [PubMed]
38. Mika, A.; Stepnowski, P.; Kaska, L.; Proczko, M.; Wisniewski, P.; Sledzinski, M.; Sledzinski, T. A comprehensive study of serum odd- and branched-chain fatty acids in patients with excess weight. *Obesity* **2016**, *24*, 1669–1676. [CrossRef]
39. Nicolaides, N.; Fu, H.C.; Ansari, M.N.A.; Rice, G.R. The fatty acids of wax esters and sterol esters from vernix caseosa and from human skin surface lipid. *Lipids* **1972**, *7*, 506–517. [CrossRef]
40. Nicolaides, N.; Apon, J.M.B. The saturated methyl branched fatty acids of adult human skin surface lipid. *Biol. Mass Spectrom.* **1977**, *4*, 337–347. [CrossRef]
41. Nicolaides, N.; Kaitaranta, J.K.; Rawdah, T.N.; Macy, J.I.; Boswell, F.M.; Smith, R.E. Meibomian gland studies: Comparison of steer and human lipids. *Investig. Ophthalmol. Vis. Sci.* **1981**, *20*, 522–536.
42. Horning, M.G.; Martin, D.B.; Karmen, A.; Vagelos, P.R. Fatty Acid Synthesis in Adipose Tissue. II. Enzymatic Synthesis of Branched Chain and Odd-Numbered Fatty Acids. *J. Biol. Chem.* **1961**, *236*, 669–672.
43. Crown, S.B.; Marze, N.; Antoniewicz, M.R. Catabolism of branched chain amino acids contributes significantly to synthesis of odd-chain and even-chain fatty acids in 3T3-L1 adipocytes. *PLoS ONE* **2015**, *10*, e0145850. [CrossRef] [PubMed]
44. Ran-Ressler, R.R.; Sim, D.; O'Donnell-Megaro, A.M.; Bauman, D.E.; Barbano, D.M.; Brenna, J.T. Branched chain fatty acid content of United States retail cow's milk and implications for dietary intake. *Lipids* **2011**, *46*, 569–576. [CrossRef] [PubMed]
45. Yan, Y.; Wang, Z.; Greenwald, J.; Kothapalli, K.S.D.; Park, H.G.; Liu, R.; Mendralla, E.; Lawrence, P.; Wang, X.; Brenna, J.T. BCFA suppresses LPS induced IL-8 mRNA expression in human intestinal epithelial cells. *Prostaglandins Leukot. Essent. Fat. Acids* **2017**, *116*, 27–31. [CrossRef] [PubMed]
46. Yan, Y.; Wang, Z.; Wang, D.; Lawrence, P.; Wang, X.; Kothapalli, K.S.D.; Greenwald, J.; Liu, R.; Park, H.G.; Brenna, J.T. BCFA-enriched vernix-monoacylglycerol reduces LPS-induced inflammatory markers in human enterocytes in vitro. *Pediatric Res.* **2018**, *83*, 874–879. [CrossRef]
47. Liu, L.; Wang, Z.; Park, H.G.; Xu, C.; Lawrence, P.; Su, X.; Wijendran, V.; Walker, W.A.; Kothapalli, K.S.D.; Brenna, J.T. Human fetal intestinal epithelial cells metabolize and incorporate branched chain fatty acids in a structure specific manner. *Prostaglandins Leukot. Essent. Fat. Acids* **2017**, *116*, 32–39. [CrossRef]
48. Wongtangtintharn, S.; Oku, H.; Iwasaki, H.; Inafuku, M.; Toda, T.; Yanagita, T. Incorporation of branched-chain fatty acid into cellular lipids and caspase-independent apoptosis in human breast cancer cell line, SKBR-3. *Lipids Health Dis.* **2005**, *4*, 29. [CrossRef]
49. Vahmani, P.; Salazar, V.; Rolland, D.C.; Gzyl, K.E.; Dugan, M.E.R. Iso- but not anteiso-branched chain fatty acids exert growth-inhibiting and apoptosis-inducing effects in MCF-7 cells. *J. Agric. Food Chem.* **2019**, *67*, 10042–10047. [CrossRef]
50. Ran-Ressler, R.R.; Khailova, L.; Arganbright, K.M.; Adkins-Rieck, C.K.; Jouni, Z.E.; Koren, O.; Ley, R.E.; Brenna, J.T.; Dvorak, B. Branched chain fatty acids reduce the incidence of necrotizing enterocolitis and alter gastrointestinal microbial ecology in a neonatal rat model. *PLoS ONE* **2011**, *6*, e29032. [CrossRef]

51. Shingfield, K.J.; Chilliard, Y.; Toivonen, V.; Kairenius, P.; Givens, D.I. Trans fatty acids and bioactive lipids in ruminant milk. In *Bioactive Components in Milk*; Bösze, Z., Ed.; Springer: New York, NY, USA, 2008; pp. 3–66.
52. Leiber, F.; Kreuzer, M.; Nigg, D.; Wettstein, H.R.; Scheeder, M.R.L. A study on the causes for the elevated n-3 fatty acids in cows' milk of alpine origin. *Lipids* **2005**, *40*, 191–202. [CrossRef]
53. Schröder, M.; Larissa Lutz, N.; Chick Tangwan, E.; Hajazimi, E.; Vetter, W. Phytanic acid concentrations and diastereomer ratios in milk fat during changes in the cow's feed from concentrate to hay and back. *Eur. Food Res. Technol.* **2012**, *234*, 955–962. [CrossRef]
54. Che, B.N.; Kristensen, T.; Nebel, C.; Dalsgaard, T.K.; Hellgren, L.I.; Young, J.F.; Larsen, M.K. Content and distribution of phytanic acid diastereomers in organic milk as affected by feed composition. *J. Agric. Food Chem.* **2013**, *61*, 225–230. [CrossRef] [PubMed]
55. Baars, T.; Schröder, M.; Kusche, D.; Vetter, W. Phytanic acid content and SRR/RRR diastereomer ratio in milk from organic and conventional farms at low and high level of fodder input. *Org. Agric.* **2012**, *2*, 13–21. [CrossRef]
56. Vetter, W.; Schröder, M. Phytanic acid—A tetramethyl-branched fatty acid in food. *Lipid Technol.* **2011**, *23*, 175–178. [CrossRef]
57. Chen, M.; Li, Y.; Sun, Q.; Pan, A.; Manson, J.E.; Rexrode, K.M.; Willett, W.C.; Rimm, E.B.; Hu, F.B. Dairy fat and risk of cardiovascular disease in 3 cohorts of US adults. *Am. J. Clin. Nutr.* **2016**, *104*, 1209–1217. [CrossRef] [PubMed]
58. Yakoob, M.Y.; Shi, P.; Willett, W.C.; Rexrode, K.M.; Campos, H.; Orav, E.J.; Hu, F.B.; Mozaffarian, D. Circulating biomarkers of dairy fat and risk of incident diabetes mellitus among men and women in the United States in two large prospective cohorts. *Circulation* **2016**, *133*, 1645–1654. [CrossRef]
59. Santaren, I.D.; Watkins, S.M.; Liese, A.D.; Wagenknecht, L.E.; Rewers, M.J.; Haffner, S.M.; Lorenzo, C.; Hanley, A.J. Serum pentadecanoic acid (15:0), a short-term marker of dairy food intake, is inversely associated with incident type 2 diabetes and its underlying disorders. *Am. J. Clin. Nutr.* **2014**, *100*, 1532–1540. [CrossRef]
60. Mozaffarian, D.; De Oliveira Otto, M.C.; Lemaitre, R.N.; Fretts, A.M.; Hotamisligil, G.; Tsai, M.Y.; Siscovick, D.S.; Nettleton, J.A. Trans-Palmitoleic acid, other dairy fat biomarkers, and incident diabetes: The multi-ethnic study of atherosclerosis (MESA). *Am. J. Clin. Nutr.* **2013**, *97*, 854–861. [CrossRef]
61. Rautiainen, S.; Wang, L.; Lee, I.M.; Manson, J.E.; Buring, J.E.; Sesso, H.D. Dairy consumption in association with weight change and risk of becoming overweight or obese in middle-aged and older women: A prospective cohort study. *Am. J. Clin. Nutr.* **2016**, *103*, 979–988. [CrossRef]
62. Mozaffarian, D.; Hao, T.; Rimm, E.B.; Willett, W.C.; Hu, F.B. Changes in diet and lifestyle and long-term weight gain in women and men. *N. Engl. J. Med.* **2011**, *364*, 2392–2404. [CrossRef]
63. Pakiet, A.; Wilczynski, M.; Rostkowska, O.; Korczynska, J.; Jabłonska, P.; Kaska, L.; Proczko-Stepaniak, M.; Sobczak, E.; Stepnowski, P.; Magkos, F.; et al. The effect of one anastomosis gastric bypass on branched-chain fatty acid and branched-chain amino acid metabolism in subjects with morbid obesity. *Obes. Surg.* **2020**, *30*, 304–312. [CrossRef]
64. Mika, A.; Wilczynski, M.; Pakiet, A.; Kaska, L.; Proczko-stepaniak, M.; Stankiewicz, M.; Stepnowski, P.; Sledzinski, T. Short-term effect of one-anastomosis gastric bypass on essential fatty acids in the serum of obese patients. *Nutrients* **2020**, *12*, 187. [CrossRef] [PubMed]
65. Unger, A.L.; Eckstrom, K.; Jetton, T.L.; Kraft, J. Colonic bacterial composition is sex-specific in aged CD-1 mice fed diets varying in fat quality. *PLoS ONE* **2019**, *14*, e0226635. [CrossRef] [PubMed]
66. Pellattiero, E.; Cecchinato, A.; Tagliapietra, F.; Schiavon, S.; Bittante, G. The use of 2-dimensional gas chromatography to investigate the effect of rumen-protected conjugated linoleic acid, breed, and lactation stage on the fatty acid profile of sheep milk. *J. Dairy Sci.* **2015**, *98*, 2088–2102. [CrossRef]
67. Gómez-Cortés, P.; Cívico, A.; De La Fuente, M.A.; Núñez Sánchez, N.; Peña Blanco, F.; Martinez Marin, A.L. Effects of dietary concentrate composition and linseed oil supplementation on the milk fatty acid profile of goats. *Animal* **2018**, *12*, 2310–2317. [CrossRef]
68. Cossignani, L.; Giua, L.; Urbani, E.; Simonetti, M.S.; Blasi, F. Fatty acid composition and CLA content in goat milk and cheese samples from Umbrian market. *Eur. Food Res. Technol.* **2014**, *239*, 905–911. [CrossRef]
69. Hauff, S.; Vetter, W. Quantification of branched chain fatty acids in polar and neutral lipids of cheese and fish samples. *J. Agric. Food Chem.* **2010**, *58*, 707–712. [CrossRef] [PubMed]

70. Wang, D.H.; Wang, Z.; Brenna, J.T. Gas chromatography chemical ionization mass spectrometry and tandem mass spectrometry for identification and straightforward quantification of branched chain fatty acids in foods. *J. Agric. Food Chem.* **2020**, *68*, 4973–4980. [CrossRef]
71. Goudjil, H.; Fontecha, J.; Luna, P.; De La Fuente, M.A.; Alonso, L.; Juárez, M. Quantitative characterization of unsaturated and trans fatty acids in ewe's milk fat. *Lait* **2004**, *84*, 473–482. [CrossRef]
72. Tzamaloukas, O.; Orford, M.; Miltiadou, D.; Papachristoforou, C. Partial suckling of lambs reduced the linoleic and conjugated linoleic acid contents of marketable milk in Chios ewes. *J. Dairy Sci.* **2015**, *98*, 1739–1749. [CrossRef]
73. Lopez, A.; Vasconi, M.; Moretti, V.M.; Bellagamba, F. Fatty acid profile in goat milk from high- and low-input conventional and organic systems. *Animal* **2019**, *9*, 452. [CrossRef]
74. Serment, A.; Schmidely, P.; Giger-Reverdin, S.; Chapoutot, P.; Sauvant, D. Effects of the percentage of concentrate on rumen fermentation, nutrient digestibility, plasma metabolites, and milk composition in mid-lactation goats. *J. Dairy Sci.* **2011**, *94*, 3960–3972. [CrossRef] [PubMed]
75. Valenti, B.; Luciano, G.; Morbidini, L.; Rossetti, U.; Codini, M.; Avondo, M.; Priolo, A.; Bella, M.; Natalello, A.; Pauselli, M. Dietary pomegranate pulp: Effect on ewe milk quality during late lactation. *Animals* **2019**, *9*, 283. [CrossRef] [PubMed]
76. Currò, S.; Manuelian, C.; De Marchi, M.; Claps, S.; Rufrano, D.; Neglia, G. Effects of breed and stage of lactation on milk fatty acid composition of Italian goat breeds. *Animals* **2019**, *9*, 764. [CrossRef] [PubMed]
77. Glasser, F.; Doreau, M.; Ferlay, A.; Chilliard, Y. Technical note: Estimation of milk fatty acid yield from milk fat data. *J. Dairy Sci.* **2007**, *90*, 2302–2304. [CrossRef] [PubMed]
78. FoodData Central—Butterkase. Available online: https://fdc.nal.usda.gov/fdc-app.html#/food-details/522677/nutrients (accessed on 11 May 2020).
79. FoodData Central—Cheese, Brie. Available online: https://fdc.nal.usda.gov/fdc-app.html#/food-details/172177/nutrients (accessed on 11 May 2020).
80. FoodData Central—Cheese, Swiss. Available online: https://fdc.nal.usda.gov/fdc-app.html#/food-details/171251/nutrients (accessed on 7 May 2020).
81. FoodData Central—Ricotta Cheese, Ricotta. Available online: https://fdc.nal.usda.gov/fdc-app.html#/food-details/678748/nutrients (accessed on 14 May 2020).
82. FoodData Central—Cheese, Camembert. Available online: https://fdc.nal.usda.gov/fdc-app.html#/food-details/172178/nutrients (accessed on 11 May 2020).
83. FoodData Central—Cheese, Cottage, Creamed, Large or Small Curd. Available online: https://fdc.nal.usda.gov/fdc-app.html#/food-details/172179/nutrients (accessed on 7 May 2020).
84. FoodData Central—Cheese, Gouda. Available online: https://fdc.nal.usda.gov/fdc-app.html#/food-details/171241/nutrients (accessed on 11 May 2020).
85. FoodData Central—Cheese, Limburger. Available online: https://fdc.nal.usda.gov/fdc-app.html#/food-details/172175/nutrients (accessed on 11 May 2020).
86. FoodData Central—Cheese, Mozzarella, Whole Milk. Available online: https://fdc.nal.usda.gov/fdc-app.html#/food-details/170845/nutrients (accessed on 7 May 2020).
87. FoodData Central—Cheese, Pasteurized Process, American, Fortified with Vitamin D. Available online: https://fdc.nal.usda.gov/fdc-app.html#/food-details/170853/nutrients (accessed on 7 May 2020).
88. FoodData Central—Cheese, Provolone. Available online: https://fdc.nal.usda.gov/fdc-app.html#/food-details/170850/nutrients (accessed on 7 May 2020).
89. FoodData Central—Cheese, Ricotta, Whole Milk. Available online: https://fdc.nal.usda.gov/fdc-app.html#/food-details/170851/nutrients (accessed on 7 May 2020).
90. FoodData Central—Yogurt, Plain, Whole Milk. Available online: https://fdc.nal.usda.gov/fdc-app.html#/food-details/171284/nutrients (accessed on 4 May 2020).
91. FoodData Central—Butter, Salted. Available online: https://fdc.nal.usda.gov/fdc-app.html#/food-details/173410/nutrients (accessed on 4 May 2020).
92. FoodData Central—Milk, Sheep, Fluid. Available online: https://fdc.nal.usda.gov/fdc-app.html#/food-details/170882/nutrients (accessed on 5 May 2020).
93. FoodData Central—Milk, Goat, Fluid, with Added Vitamin D. Available online: https://fdc.nal.usda.gov/fdc-app.html#/food-details/171278/nutrients (accessed on 5 May 2020).

94. Hotamisligil, G.S.; Shargill, N.S.; Spiegelman, B.M. Adipose expression of tumor necrosis factor-α: Direct role in obesity-linked insulin resistance. *Science* **1993**, *259*, 87–91. [CrossRef] [PubMed]
95. Blackburn, P.; Després, J.P.; Lamarche, B.; Tremblay, A.; Bergeron, J.; Lemieux, I.; Couillard, C. Postprandial variations of plasma inflammatory markers in abdominally obese men. *Obesity* **2006**, *14*, 1747–1754. [CrossRef] [PubMed]
96. Skinner, A.C.; Steiner, M.J.; Henderson, F.W.; Perrin, E.M. Multiple markers of inflammation and weight status: Cross-sectional analyses throughout childhood. *Pediatrics* **2010**, *125*, e801–e809. [CrossRef] [PubMed]
97. Liu, Y.; Chen, F.; Odle, J.; Lin, X.; Jacobi, S.K.; Zhu, H.; Wu, Z.; Hou, Y. Fish oil enhances intestinal integrity and inhibits TLR4 and NOD2 signaling pathways in weaned pigs after LPS challenge. *J. Nutr.* **2012**, *142*, 2017–2024. [CrossRef]
98. Lee, S.I.; Kang, K.S. Function of capric acid in cyclophosphamide-induced intestinal inflammation, oxidative stress, and barrier function in pigs. *Sci. Rep.* **2017**, *7*, 16530. [CrossRef]
99. Chao, J.; Wolfaardt, G.M.; Arts, M.T. Characterization of pseudomonas aeruginosa fatty acid profiles in biofilms and batch planktonic cultures. *Can. J. Microbiol.* **2010**, *56*, 1028–1039. [CrossRef]
100. Cai, Q.; Huang, H.; Qian, D.; Chen, K.; Luo, J.; Tian, Y.; Lin, T.; Lin, T. 13-methyltetradecanoic acid exhibits anti-tumor activity on T-cell lymphomas in vitro and in vivo by down-regulating p-AKT and activating caspase-3. *PLoS ONE* **2013**, *8*, e65308. [CrossRef] [PubMed]
101. Yang, Z.; Liu, S.; Chen, X.; Chen, H.; Huang, M.; Zheng, J. Induction of apoptotic cell death and in vivo growth inhibition of human cancer cells by a saturated branched-chain fatty acid, 13-methyltetradecanoic acid. *Cancer Res.* **2000**, *60*, 505–509. [PubMed]
102. Lin, T.; Yin, X.B.; Cai, Q.; Fan, X.; Xu, K.; Huang, L.; Luo, J.; Zheng, J.; Huang, J. 13-Methyltetradecanoic acid induces mitochondrial-mediated apoptosis in human bladder cancer cells. *Urol. Oncol. Semin. Orig. Investig.* **2012**, *30*, 339–345. [CrossRef]
103. Wongtangtintharn, S.; Oku, H.; Iwasaki, H.; Toda, T. Effect of branched-chain fatty acids on fatty acid biosynthesis of human breast cancer cells. *J. Nutr. Sci. Vitaminol.* **2004**, *50*, 137–143. [CrossRef] [PubMed]
104. Wright, K.C.; Yang, P.; Van Pelt, C.S.; Hicks, M.E.; Collin, P.; Newman, R.A. Evaluation of targeted arterial delivery of the branched chain fatty acid 12-methyltetradecanoic acid as a novel therapy for solid tumors. *J. Exp. Ther. Oncol.* **2005**, *5*, 55–68. [PubMed]
105. Yu, J.; Yang, L.N.; Wu, Y.Y.; Li, B.H.; Weng, S.M.; Hu, C.L.; Han, Y.L. 13-Methyltetradecanoic acid mitigates cerebral ischemia/reperfusion injury. *Neural Regen. Res.* **2016**, *11*, 1431–1437. [CrossRef]
106. Cole, N.; Hume, E.B.; Jalbert, I.; Vijay, A.K.; Krishnan, R.; Willcox, M.D. Effects of topical administration of 12-methyl tetradecanoic acid (12-MTA) on the development of corneal angiogenesis. *Angiogenesis* **2007**, *10*, 47–54. [CrossRef]
107. Brooks, K.K.; Liang, B.; Watts, J.L. The influence of bacterial diet on fat storage in *C. elegans*. *PLoS ONE* **2009**, *4*, e7545. [CrossRef]
108. Liu, L.; Xiao, D.; Lei, H.; Peng, T.; Li, J.; Cheng, T.; He, J. Branched-chain fatty acids lower triglyceride levels in a fatty liver model in vitro. *FASEB J.* **2017**, *31*. [CrossRef]
109. Kraft, J.; Jetton, T.; Satish, B.; Gupta, D. Dairy-derived bioactive fatty acids improve pancreatic ß-cell function. *FASEB J.* **2015**, *29*. [CrossRef]
110. Roca-Saavedra, P.; Mariño-Lorenzo, P.; Miranda, J.M.; Porto-Arias, J.J.; Lamas, A.; Vazquez, B.I.; Franco, C.M.; Cepeda, A. Phytanic acid consumption and human health, risks, benefits and future trends: A review. *Food Chem.* **2017**, *221*, 237–247. [CrossRef] [PubMed]
111. Chaudhary, S.; Sahu, U.; Kar, S.; Parvez, S. Phytanic acid-induced neurotoxicological manifestations and apoptosis ameliorated by mitochondria-mediated actions of melatonin. *Mol. Neurobiol.* **2017**, *54*, 6960–6969. [CrossRef]
112. Chaudhary, S.; Parvez, S. Phytanic acid induced neurological alterations in rat brain synaptosomes and its attenuation by melatonin. *Biomed. Pharmacother.* **2017**, *95*, 37–46. [CrossRef] [PubMed]
113. Refsum Disease—Genetics Home Reference—NIH. Available online: https://ghr.nlm.nih.gov/condition/refsum-disease#statistics (accessed on 30 June 2020).
114. Nakanishi, T.; Motoba, I.; Anraku, M.; Suzuki, R.; Yamaguchi, Y.; Erickson, L.; Eto, N.; Sugamoto, K.; Matsushita, Y.; Kawahara, S. Naturally occurring 3RS, 7R, 11R-phytanic acid suppresses in vitro T-cell production of interferon-gamma. *Lipids Health Dis.* **2018**, *17*, 147. [CrossRef]

115. An, J.-Y.; Jheng, H.-F.; Nagai, H.; Sanada, K.; Takahashi, H.; Iwase, M.; Watanabe, N.; Kim, Y.-I.; Teraminami, A.; Takahashi, N.; et al. A Phytol-Enriched Diet Activates PPAR-α in the Liver and Brown Adipose Tissue to Ameliorate Obesity-Induced Metabolic Abnormalities. *Mol. Nutr. Food Res.* **2018**, *62*, 1700688. [CrossRef]
116. Lamichane, S.; Lamichane, B.D.; Kwon, S.M. Pivotal roles of peroxisome proliferator-activated receptors (PPARs) and their signal cascade for cellular and whole-body energy homeostasis. *Int. J. Mol. Sci.* **2018**, *19*, 914. [CrossRef] [PubMed]
117. Wang, H.; Mao, X.; Du, M. Phytanic acid activates PPARα to promote beige adipogenic differentiation of preadipocytes. *J. Nutr. Biochem.* **2019**, *67*, 201–211. [CrossRef]

© 2020 by the authors. Licensee MDPI, Basel, Switzerland. This article is an open access article distributed under the terms and conditions of the Creative Commons Attribution (CC BY) license (http://creativecommons.org/licenses/by/4.0/).

Review

Dairy Consumption and Metabolic Health

Claire M. Timon [1], Aileen O'Connor [2,3], Nupur Bhargava [2,3], Eileen R. Gibney [2,3,*] and Emma L. Feeney [2,3]

1. School of Nursing, Psychotherapy and Community Health, Dublin City University, Glasnevin, 9 Dublin, Ireland; claire.timon@dcu.ie
2. UCD Institute of Food and Health, University College Dublin, Belfield, 4 Dublin, Ireland; aileen.oconnor@ucd.ie (A.O.); nupur.bhargava@ucd.ie (N.B.); emma.feeney@ucd.ie (E.L.F.)
3. School of Agriculture and Food Science, University College Dublin, Belfield, 4 Dublin, Ireland
* Correspondence: Eileen.gibney@ucd.ie

Received: 26 August 2020; Accepted: 1 October 2020; Published: 3 October 2020

Abstract: Milk and dairy foods are naturally rich sources of a wide range of nutrients, and when consumed according to recommended intakes, contribute essential nutrients across all stages of the life cycle. Seminal studies recommendations with respect to intake of saturated fat have been consistent and clear: limit total fat intake to 30% or less of total dietary energy, with a specific recommendation for intake of saturated fat to less than 10% of total dietary energy. However, recent work has re-opened the debate on intake of saturated fat in particular, with suggestions that recommended intakes be considered not at a total fat intake within the diet, but at a food-specific level. A large body of evidence exists examining the impact of dairy consumption on markers of metabolic health, both at a total-dairy-intake level and also at a food-item level, with mixed findings to date. However the evidence suggests that the impact of saturated fat intake on health differs both across food groups and even between foods within the same food group such as dairy. The range of nutrients and bioactive components in milk and dairy foods are found in different levels and are housed within very different food structures. The interaction of the overall food structure and the nutrients describes the concept of the 'food matrix effect' which has been well-documented for dairy foods. Studies show that nutrients from different dairy food sources can have different effects on health and for this reason, they should be considered individually rather than grouped as a single food category in epidemiological research. This narrative review examines the current evidence, mainly from randomised controlled trials and meta-analyses, with respect to dairy, milk, yoghurt and cheese on aspects of metabolic health, and summarises some of the potential mechanisms for these findings.

Keywords: dairy; health; matrix; metabolism; nutrient; composition; saturated fats

1. Contribution of Dairy to a Balanced Diet

Milk and dairy foods are naturally rich sources of a wide range of nutrients such as proteins, fats, oligosaccharides and micronutrients including vitamins A, D, E and K and Ca, Mg, P and Zn [1], Figure 1. Milk proteins are of high biological value, not only because they contain essential amino acids but also because of their high digestibility and bioavailability. Approximately 80% of milk protein is casein and the remaining 20% is serum, or whey protein for cow's milk [2]. Fat, mainly in the form of triacylglycerols (98%), is present in milk as globules which are surrounded by a membrane (or milk fat globule membrane (MFGM)). This component of milk fat has been suggested to elicit favourable lipid and low-density lipoprotein (LDL) cholesterol response to dairy consumption [3], and will be discussed later. With respect to micronutrients, milk is considered a major source of calcium in the diet [4]. As well as being a rich source of this nutrient, the bioavailability of calcium from dairy sources has also been shown to be higher compared to other dietary sources [5,6]. Furthermore, modelling of dietary intake data has indicated that, without consuming dairy products, less than half of the dietary calcium requirements would be met [7]. The authors of that study also noted that nutrients from dairy foods are difficult to replace and modelled removal and replacement with available alternatives, which resulted in lower amounts of several nutrients including protein, phosphorus, riboflavin, zinc and vitamin B12 [7]. While there is some concern that avoidance of dairy may have implications for some nutrients, this does not suggest that adequate nutrient intake from a low-dairy or dairy-free diet is unattainable, but rather indicates that dairy can significantly contribute towards a healthy diet.

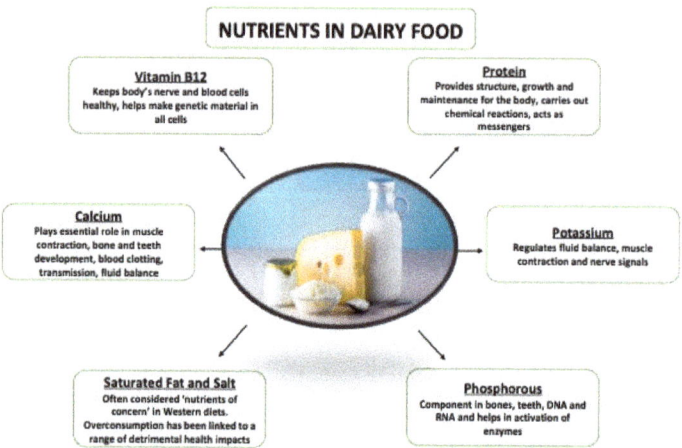

Figure 1. Nutrient content and associated health benefits of dairy consumption.

When consumed according to recommended intakes of national guidelines, milk and dairy products contribute essential nutrients across all stages of the life cycle [1]. For example, milk and dairy products are an important part of a young child's diet as they are a good source of energy and protein and contain a wide range of vitamins and minerals, especially calcium, that young children need for healthy bones and teeth [8]. In Europe, it is reported that milk contributes proportionally more to the diets of young children than to adults [9,10]. Data from cross-sectional studies [11] and intervention studies [12] have reported the positive effect of milk consumption in childhood and adolescence on bone mineral content and bone mineral density. In addition, some research studies have indicated that the consumption of milk and milk products during adolescence is associated with neutral or reduced risk of adiposity [13,14]. During pregnancy, dairy products can be an important means of providing adequate calcium and other key nutrients in the diet [15]. Evidence from prospective cohort studies suggests that moderate milk consumption compared to none or low intakes during pregnancy is positively associated with foetal growth and infant birth weight in healthy, Western populations [16]. Finally, several studies point to the benefits of milk and dairy products in diets of the elderly and highlight that, in combination with physical activity, milk and dairy products can improve muscle mass and function resulting in a lower risk of sarcopenia and vertebral fractures [16], although another recent review of the area did not find strong evidence for a benefit of milk on muscle health in older adults [17].

Current consumption patterns of dairy are in a period of considerable change, with reported decreases in dairy consumption in countries who have traditionally consumed large quantities, potentially due to increased intake of 'dairy-free' alternative products [18] and reported increases in some global regions, where dairy has not been commonly consumed [19]. These changes are reported to be due to several factors, but are predominantly driven by consumer perception of the health effects of dairy consumption and the environmental impact of dairy production [20]. Particular nutrients of concern, when considering intake of dairy, are sodium and fat. Whilst dairy has been shown to contribute beneficially to the diet, dairy foods also contribute significantly to sodium and saturated fat intakes (Figure 1). High dietary salt intake is a prominent factor for the development of hypertension, a strong risk factor for cardiovascular disease [21,22]. In the US and UK, dairy products are significant contributors to dietary salt intakes, providing approximately 11% and 8% of overall intakes, respectively [23,24]. Dietary efforts to reduce hypertension include the well-established DASH diet (Dietary Approaches to Stop Hypertension), where low-fat dairy products are important characteristics [25]. In relation to dairy fat, the 2011–2014 National Health and Nutrition Examination Survey (NHANES) demonstrated that dairy foods contributed 26% of saturated fat and 14.2% of total fat to the diets of US adults [26].

Similarly, Feeney et al. observed that, within the Irish diet, dairy foods contributed 12.8% of total fat to the diet, and 19.8% of the saturated fat [27]. For this reason, many healthy eating guidelines recommend 3–5 portions of dairy daily, with consumption of "low/reduced-fat dairy" when possible [28,29]. However, research investigating the importance of the food source of saturated fatty acids (SFA) suggests that although SFAs from meat and processed-meat are associated with detrimental health effects [30], SFA intake from dairy sources may be associated with either neutral [31] or beneficial effects on cardiovascular health markers [32,33]. Further, the individual dairy sources may have different impacts. Much work in this area is underway, which is summarised below.

2. Dairy Fat and the Link to Health

Since the seminal Seven Countries Study and other subsequent studies [34,35], recommendations with respect to intake of saturated fat have been consistent and clear: limit total fat intake to 30% or less of total dietary energy, with a specific recommendation for intake of saturated fat to less than 10% of total dietary energy [36–38]. However, recent work has re-opened the debate on intake of saturated fat in particular [39]. In a large review and meta-analysis, de Souza et al. examined associations between intake of total fat, saturated fat and trans-unsaturated fat with all-cause mortality and differing morbidities. The authors concluded that, contrary to previous evidence, saturated fat intake was not associated with all-cause mortality, cardiovascular disease (CVD) mortality, total coronary heart disease (CHD), ischemic stroke or type 2 diabetes [39]. Drouin-Chartier et al. in 2016 also examined the impact of dairy consumption and dairy fat on cardiometabolic disease risk factors, and also reported that the purported detrimental effects of SFAs on cardiometabolic health may in fact be nullified when they are consumed as part of complex food matrices such as those in cheese and other dairy foods [40]. Similarly, Alexander et al. (2016) completed a meta-analysis of prospective studies and the intake of dairy products and CVD risk. These authors more cautiously concluded that although for some individual dairy products risk estimates below 1.0 were observed, additional data are needed to more comprehensively examine potential dose–response patterns [41].

2.1. Dietary Guidelines—Nutrient-Based vs. Food-Based

Despite these more recent findings, the most recent review of published literature by the UK Scientific Advisory Committee on Nutrition (SACN) concluded that there is a significant body of evidence demonstrating a relationship between intake of saturated fats and CVD and CHD events, but not CVD and CHD mortality [28], and noted that, irrespective of the lack of evidence for an effect on mortality, non-fatal CVD and CHD events have a serious adverse impact on health and quality of life, and that existing public health recommendations for saturated fat to be <10% of total dietary energy intake were still valid [42]. This mirrors recommendations in other countries and regions of the world [36–38]. However, some criticisms on the continued support for such recommendations note that such policies are based on evidence from total dietary saturated fat intake, and may have not considered the source of fat, whereby the food source, or matrix may in fact have a differing influence on metabolism and subsequently, health [43]. In addition, different food sources contain different types and amounts of SFA, and the continued promotion of an overall <10% of total dietary energy recommendation may be perceived to overlook this [43]. Further, by focusing on SFA content alone, foods that are nutrient-dense but also high in SFA may be excluded from the diet and inadvertently result in a reduced intake of important micronutrients [43]. For this reason, many are advocating for food-based guidelines rather than nutrient-based advice. As such, evidence may need to be considered at a food-item level within the dairy food group, rather than together. In light of a reported shift in recent dairy consumption [19], where instances of decreased dairy consumption may be potentially explained by an increase in dairy-free alternatives due to the perceived impact of dairy on health [18], it is important to consider some of the recent evidence at both a total-dairy level and for individual dairy products cheese, milk and yoghurt, and then further discuss potential mechanisms influencing the metabolic response to consumption.

2.2. Total Dairy vs. Specific Products

Dehghan et al. (2018) specifically examined the associations between total dairy and specific types of dairy products with mortality and major cardiovascular disease. Dietary intakes of dairy products for 136,384 individuals were recorded using country-specific validated food frequency questionnaires and associations with mortality or major cardiovascular events were examined [44]. The authors reported that a higher intake of total dairy (>2 servings per day compared with no intake) was associated with a lower risk of the composite of mortality or major cardiovascular events, total mortality, non-cardiovascular mortality, cardiovascular mortality, major cardiovascular disease and stroke. No significant association with myocardial infarction was observed. Higher intake (>1 serving vs. no intake) of milk and yoghurt was associated with lower risk of the composite outcome, whereas cheese intake was not. Butter intake was low and was not significantly associated with clinical outcomes. The authors concluded that dairy consumption was associated with lower risk of mortality and major cardiovascular disease events in a diverse multinational cohort [44].

Fontecha and colleagues [45] specifically examined evidence regarding the influence of dairy product consumption on the risk of major cardiovascular-related outcomes and how various doses of different dairy products affected such responses. In this overview of 12 meta-analyses involving randomised controlled trials (RCTs), as well as the updated meta-analyses of RCTs, increasing consumption of dairy products did not result in significant changes of known risk biomarkers such as systolic and diastolic blood pressure and total cholesterol and LDL cholesterol. They concluded that consumption of total dairy products (either regular or low-fat content), did not adversely affect the risk of CVD [45].

Considering total dairy consumption initially, characteristics of fifteen published RCTs that studied the effects of overall dairy consumption on markers of metabolic health and CVD risk on variable age groups are included in Table 1. RCTs included in the table were either parallel or crossover trials and the participants' ages ranged from 20 to 75 years. Duration of interventions varied widely from 4 to 8 weeks in some to 12 to 24 weeks in the others, with a washout period included in some of the trials. Most of the interventions were based on consumption of different dairy products such as milk, cheese or butter. Zemel et al. (2010) compared the effects of consumption of soy-based smoothie to a dairy-based smoothie and observed suppressed inflammatory and stress markers for the latter [46]. Groups that consumed a low-fat or non-fat dairy diet showed decreased levels of total cholesterol (TC) and low-density cholesterol (LDL) as compared to the diet containing conventional levels of fat. Most recently, Vasilopoulou et al. (2020) examined the impact of modified dairy fat consumption, through the provision of products with modified monounsaturated fatty acid (MUFA) content in adults at moderate CVD risk—the RESET study (controlled REplacement of SaturatEd fat in dairy on Total cholesterol). The authors concluded that consumption of a high-fat diet containing modified dairy products with reduced saturated fatty acids, and enriched monounsaturated fatty acids showed beneficial effects on fasting LDL cholesterol and endothelial function compared with conventional dairy products [47]. Finally, although not an intervention study, Drouin-Chartier (2019), recently examined dairy intake and risk of diabetes, and reported positive impacts for yoghurt and reduced-fat milk, but a negative association for cheese [48].

Table 1. Randomised controlled trials (RCT) demonstrating an effect of overall dairy consumption on markers of metabolic health and cardiovascular disease (CVD) risk.

Author	Country	Study Design	Population	Age (Years)	Intervention	Duration	Main Findings
Vasilopoulou et al. 2020 [47]	UK	Crossover	n = 54 (31 male; 23 female), with risk of CVD	25–70	2 arm: (A) monounsaturated fatty acid (MUFA)-modified dairy—340 g UHT milk, 45 g cheese, 25.1 g butter. (B) Control—340 g ultrahigh temperature (UHT) pasteurised milk, 45 g cheese, 25.1 g butter	Two 12-week periods separated by an 8-week washout period	No significant change from baseline in serum total cholesterol (TC) between diets. Group A had a significant beneficial effect in terms of attenuation of the rise of the low-density lipoprotein (LDL) cholesterol. No changes in high-density lipoprotein (HDL) cholesterol between diets. The LDL:HDL ratio decreased significantly after group A, and increased after the control. No significant differences were observed for indexes of insulin sensitivity/resistance. Fasting plasma nitrite concentrations increased after the modified diet, yet decreased after the control.
Markey et al. 2017 [49]	UK	Crossover	n = 54 (31 male; 23 female), with risk of CVD	25–70	2 arm: (A) MUFA-modified dairy—340 g UHT milk, 45 g cheese, 25.1 g butter. (B) Control—340 g UHT milk, 45 g cheese, 25.1 g butter.	Two 12-week periods, separated by an 8-week washout period	Group A showed a smaller increase in saturated fatty acids (SFA) and greater increase in MUFA intake when compared with the control.
Rosqvist et al. 2015 [3]	Sweden	Parallel	n = 57 (gender split not stated), overweight or obese	20–70	2 arm: (A) milk-fat globule membrane (MFGM) group—100 mL whipping cream (40%fat)/d, 100 mL fat-free milk (0.1% fat)/d and 1 scone/d (baked with wheat flour, water, sodium chloride and baking powder). (B) Fat-free milk, 100 mL, (0.1% fat)/d and 1 scone/d (baked with wheat flour, water, butter oil (98.7% fat), sodium chloride, baking powder and milk protein isolate).	8 weeks	Control diet increased TC, LDL, apolipoprotein B:apolipoprotein A-I ratio and non-HDL plasma lipids, whereas the MFGM diet did not. HDL, triglyceride, sitosterol, lathosterol, campesterol and proprotein convertase subtilisin/kexin type 9 concentrations and fatty acid compositions did not differ between groups.
Benatar et al. 2013 [50]	New Zealand	Parallel	n = 180 (54 male; 126 female), healthy volunteers	>18	3 arm: (A) increased dairy—an extra two-to-three servings per day, and to change to high-fat milk and dairy solids. (B) Habitual dairy intake remains unchanged. (C) Decreased dairy were asked to eliminate all possible sources of dairy.	1 month	No significant change in LDL or HDL, triglycerides, systolic or diastolic BP, C-reactive protein, glucose or insulin across groups. There was a small increase in weight in group A.
Nestel et al. 2013 [51]	Australia	Crossover	n = 12 (gender not specified) overweight or obese	40–70	3 arm: (A) low-fat dairy, 1% fat milk (400 mL/d) and 1% fat yoghurt (200 g/d). (B) Full-fat dairy (fermented), cheddar cheese (85 g/d) and full-cream yoghurt (three servings, 600 g/d). (C) Full-fat dairy (non-fermented), butter (30 g/d) and cream (70 mL/d) and small amounts of ice-cream.	Two 3-week periods, for group B + C (full fat diets). Group A diet (low fat) was consumed twice—between and at the end of the full-fat dairy dietary periods, for a duration of 2 weeks.	Lowest LDL and HDL concentrations were observed in group A, but plasma Triacylglycerol (TAG) concentrations did not differ significantly across the groups. Concentrations of plasma sphingomyelin and IL-6 were significantly higher after the non-fermented dairy diet (group C) than the low-fat dairy diet

Table 1. *Cont.*

Author	Country	Study Design	Population	Age (Years)	Intervention	Duration	Main Findings
Crichton et al. 2012 [52]	Australia	Crossover	n = 61 (18 male; 43 female), overweight or obese	18–75	2 arm: **(A)** 4 servings of reduced-fat dairy/d, 1 serving = 250 mL milk, 175–200 g yoghurt and 190 g custard. **(B)** Control—1 serving of dairy/d, reflecting habitual intake.	Two 6-month periods, no washout period	No significant changes in resting metabolic rate or total energy expenditure, systolic and diastolic BP, fasting blood glucose, TC, HDL or LDL, triglycerides or *hs*-CRP (high sensitivity C-reactive protein). Additionally no differences between groups for waist circumference (WC), body weight and fat mass.
Palacios et al. 2011 [53]	Puerto Rico	Parallel	n = 25 (5 male; 20 female), obese	22–50	3 arm: **(A)** 4 servings of dairy/d (low-fat milk, low-fat cheese and low-fat yoghurt), with a dairy-calcium intake goal of 1200–1300 mg/d. **(B)** calcium supplement (600 mg/d, calcium carbonate). **(C)** Control—habitual diet with placebo tablet.	21 weeks	No significant group effects were observed for anthropometric measurements or serum lipids such as TC, HDL, LDL and TAG levels. Although TAG levels decreased by 18% in group **A** (high dairy).
Stancliffe et al. 2011 [54]	USA	Parallel	n = 40 (19 male; 21 female), overweight or obese with metabolic syndrome	37.0 ± 9.9	2 arm: **(A)** low-dairy diet— 0.5 servings/d and provided with 3 servings/d of non-dairy foods that is low-sodium luncheon meats, soy-based luncheon meat substitutes, packaged fruit cups, granola bars and peanut butter crackers. **(B)** adequate dairy diet—3.5 servings/d, of which 2/3 servings were milk and/or yoghurt.	12 weeks	Group **A** decreased malondialdehyde and oxidised LDL. Inflammatory markers were suppressed with intake of AD, with decreases in TNF-α (Tumor Necrosis Factor-α); decreases in IL-6 (Interleukin-6) and monocyte chemoattractant protein 1 and an increase in adiponectin. Group **B** exerted no effect on oxidative or inflammatory markers. Group **A** significantly reduced waist circumference and trunk fat but group **B** exerted no effects.
van Meijl and Mensink, 2010 [55]	Netherlands	Crossover	n = 35 (10 male; 25 female), overweight or obese	18–70	2 arm: **(A)** 500 mL low-fat milk and 150 g low-fat yoghurt per day. **(B)** Control—600 mL fruit juice and 43 g fruit biscuits per day	Two 8-week periods separated by a 2-week washout period	Plasma concentrations of TNF-α decreased, and soluble TNF-α receptor-1 increased after low-fat dairy consumption compared to the control. s-TNFR-2 also increased. Low-fat dairy consumption had no effect on IL-6, monocyte chemoattractant protein-1, intracellular adhesion molecule-1 and vascular cell adhesion molecule-1 concentrations. Lipid profiles were not analysed.
Zemel et al. 2010 [46]	USA	Crossover	n = 20 (14 male; 6 female), overweight or obese	Average— 31 ± 10.3	2 arm: **(A)** dairy smoothie, 3 times/d, with non-fat dry milk as the protein source, and containing 350 mg calcium per smoothie. **(B)** Soy smoothie, 3 times/d, with soy protein isolate as the protein source and 50 mg calcium per smoothie.	Two 4-week periods separated by a 4-week washout period	Group **A** resulted in significant suppression of oxidative stress and lower inflammatory markers; tumour necrosis factor-α; IL-6; monocyte chemoattractant protein-1 and increased adiponectin. Group **B** exerted no significant effects. Lipid profiles were not analysed.

Table 1. Cont.

Author	Country	Study Design	Population	Age (Years)	Intervention	Duration	Main Findings
Wennersberg et al. 2009 [56]	Norway	Parallel	n = 121 (41 male; 80 female)	30–65	2 arm: (A) milk group, 3–5 portions of dairy/d. Portion = 200 g milk, 200–250 g yoghurt or sour milk, 75 g cream or crème fraiche, 15–40 g cheese, 3–10 g butter or butter-containing spreads, 50 mL cottage cheese, and ice-cream occasionally. (B) Control—habitual daily diet.	6 months	No significant differences between changes in body weight or body composition, BP, markers of inflammation, endothelial function, adiponectin or oxidative stress in group A and B. There was a modest unfavourable increase in serum TC concentrations in the group A.
van Meijl and Mensink, 2009 [57]	Netherlands	Crossover	n = 35 (10 males; 25 females), overweight or obese	18–70	2 arm: (A) 500 mL low-fat milk and 150 g low-fat yoghurt per day. (B) Control—600 mL fruit juice and 43 g fruit biscuits per day	Two 8-week periods separated by a 2-week washout period	In group A, systolic BP significantly decreased compared with the control, but diastolic BP did not reach significance. Decreases in HDL and apo A-1 concentrations were also observed in group A. Serum TC, LDL, apo B, TAG, non-esterified fatty acids, glucose, insulin, C-reactive protein and plasminogen activator inhibitor-1 remained unchanged.
Tricon et al. 2006 [58]	UK	Crossover	n = 32 males, healthy volunteers	34–60	2 arm: (A) 500 mL UHT full-fat milk, 12.5 g butter and 36.3 g cheese per day, naturally enriched with CLA (conjugated linoleic acid). (B) Control, 500 mL UHT full-fat milk, 12.5 g butter and 28 g cheese per day.	Two 6 week periods, separated by a 7 week washout period	Diet A did not significantly affect body weight, inflammatory markers, insulin, glucose, TAG, or TC, LDL and HDL cholesterol but resulted in a small increase in the LDL:HDL ratio. The modified dairy products changed LDL fatty acid composition but had no significant effect on LDL particle size or the susceptibility of LDL to oxidation.
Zemel et al. 2005 [59]	America	Parallel	Study 1. n = 34 (11 male; 23 female), Study 2. n = 29 (4 male; 25 female), obese	26–55	Study 1. (A) Dairy group—3 servings of dairy/d, at least one serving to be milk (B) Control—low dairy, 0–1 servings of low-fat dairy/d. Study 2. (A) Dairy group—3 servings of dairy/d, at least one serving to be milk and a 500-kcal/d deficit (B) Control—low dairy, 0–1 servings of low-fat dairy/d and 500 kcal/d deficit diet.	24 weeks (both studies)	Study 1. Body weight remained stable for both groups. Group A resulted in decreases in total body fat, trunk fat, insulin and BP and an increase in lean mass and no significant changes in the control group. Study 2. Both diets produced significant weight and fat loss. Weight and fat loss within group A were 2-fold higher, and loss of lean body mass significantly reduced compared with the control. There were no effects on circulating lipids in either group.
Tholstrup et al. 2004 [60]	Denmark	Crossover	n = 14 males, healthy volunteers	20–31	3 arms: (A) 1.5 L of whole milk/10 MJ (54 g of fat and 1779 mg calcium per 10 MJ). (B) Butter 64 g/10 MJ (54 g of fat and 10 mg calcium per 10 MJ) (C) 205 g of hard cheese, "Samsø", 45% fat of dry weight, i.e., 26% fat/10 MJ (1989 mg calcium/10 MJ).	Three 3-week periods separated by a 4-week washout period	Fasting LDL concentrations were significantly higher after butter than cheese diet, with a borderline significant difference in TC after the experimental periods. Postprandial glucose showed a higher response after cheese diet compared to milk diet. No differences were found between groups for HDL, Very Low-Density Lipoprotein (VLDL), apo A-1 and apo B concentrations.

2.3. Cheese

Focusing specifically on cheese, several published RCTs have demonstrated a beneficial effect of cheese consumption on markers of metabolic health and CVD risk, summarised in Table 2. Brassard et al. (2017) compared the impact of consuming equal amounts of SFAs from cheese and butter on cardiometabolic risk factors [61]. In this multicentre, crossover, randomised controlled trial, participants were assigned to a randomised sequence of five isoenergetic diets of 4-week duration (separated by 4-week washout periods). The diets were rich in SFAs from either cheese or butter, or a monounsaturated fatty acid (MUFA)–rich diet, a polyunsaturated fatty acid (PUFA)–rich diet and a low-fat, high-carbohydrate diet. The authors reported that serum HDL-cholesterol concentrations were similar after the cheese and butter diets but were significantly higher in comparison to response after the carbohydrate diet. Comparing cheese and butter, LDL-cholesterol concentrations after the cheese diet were lower than after the butter diet but were higher than after all of the other diets. Some variation in response was noted. Work conducted by this research group previously has both supported and added to the existing evidence [32]. A 6-week randomised parallel intervention involving 164 volunteers who received ~40 g of dairy fat/d, in 1 of 4 treatments: 120 g full-fat Irish cheddar cheese (group A), 120 g reduced-fat Irish cheddar cheese + butter (21 g) (group B); butter (49 g), calcium caseinate powder (30 g) and Ca supplement ($CaCO_3$) (500 mg) (group C) or 120 g full-fat Irish cheddar cheese, for 6 weeks following completion of a 6-week "run-in" period, where this group excluded all dietary cheese before commencing the intervention (group D). This study found that a stepwise-matrix effect was observed between the groups for total cholesterol (TC) ($P = 0.033$) and LDL cholesterol ($P = 0.026$), with significantly lower post-intervention TC and LDL cholesterol when all of the fat was contained within the cheese matrix (Group A), compared with Group C when it was not. These findings suggest that dairy fat, when eaten in the form of cheese, appears to differently affect blood lipids compared with the same constituents eaten in different matrices, with significantly lower total cholesterol observed when all nutrients are consumed within a cheese matrix. This 'dairy matrix' concept, whereby the nutrients within a dairy food interact with the overall structure providing different health effects, is becoming increasingly studied, [62], and is discussed further in Section 3, below.

Table 2. Randomised controlled trials (RCT) demonstrating an effect of cheese consumption on markers of metabolic health and CVD risk.

Author	Country	Study Design	Population	Age (Years)	Intervention	Duration	Main Findings
Feeney et al. 2018 [32]	Ireland	Parallel	$n = 164$ (75 male; 89 female), BMI > 25 kg/m^2	>50	4 arm: **(A)** 120 g full-fat cheddar cheese (FFCC). **(B)** Reduced-fat Irish cheddar cheese, 120 g, +butter (21 g) (RFC + B). **(C)** Butter (49 g), calcium caseinate powder (30 g), Ca supplement (BCC). **(D)** Full-fat Irish cheddar cheese, 120 g, (as per "a" but with 6-week run-in period).	6 weeks	There was a significant difference in total cholesterol (TC) and LDL between groups. Group **A** had significantly lower TC and LDL compared with other groups. No differences were observed for HDL cholesterol, anthropometry, fasting glucose or insulin.
Limongi et al. 2018 [63]	Italy	Crossover	$n = 58$ (16 male; 42 female), healthy volunteers	>60	2 arm: **(A)** 90 g/d of CLA-enriched Pecorino cheese. **(B)** Control, 90 g/d of Pecorino cheese.	Two 2-month periods, separated by 1-month washout period	No significant differences found in relation to LDL between diet **A** + **B**. Participants consuming enriched cheese had a lower increase in glycaemia compared to control but did not display an increase in lipid levels.
Brassard et al. 2017 [61]	Canada	Crossover	$n = 92$ (43 male; 49 female), abdominally obese	18–65	5 arm: **(A)** 90 g/2500 kcal cheese (type not specified). **(B)** Butter 49 g/2500 kcal. **(C)** MUFA-rich diet **(D)** PUFA-rich diet. **(E)** High-cholesterol, low-fat diet. The SFA content was matched in diets **A** + **B**.	Five 4-week periods, separated by 4-week washout periods	No changes were evident in HDL after cheese consumption. LDL was lower in group **A** compared with group **B**, but higher than groups **C**, **D** and **E**. No significant differences were found in inflammation markers, blood pressure and insulin-glucose homeostasis.
Raziani et al. 2016 [64]	Denmark	Parallel	$n = 139$ (47 male; 92 female), risk of metabolic syndrome	18–70	3 arm: **(A)** 40 g equal parts regular-fat Danbo and cheddar cheese (REG). **(B)** Reduced-fat Danbo and cheddar cheese, 40 g, (RED). **(C)** Noncheese, carbohydrate control (CHO40 g/d)—90 g bread, 25 g jam.	12 weeks	No differences were evident on lipid profile between groups. In addition, insulin, glucose, and triacylglycerol concentrations as well as blood pressure and waist circumference did not differ.
Thorning et al. 2015 [65]	Denmark	Crossover	$n = 14$ female, overweight, post-menopausal	45–68	3 arm: **(A)** 96–120 g of equal parts Dando and cheddar cheese per 8–10 MJ diet. **(B)** High fat meat—164 g per 10-MJ diet. **(C)** Non-dairy, low-fat control (CHO)—fruit (84 g), white bread, pasta and rice (58 g), marmalade (20 g), and cake, sweetened biscuits and chocolate (13 g) per 10-MJ diet.	Three 2-week periods, separated by 2-week washout periods	Group **A** caused higher levels of circulating HDL levels and apo A-I concentrations, and a lower apoB:apo A-I ratio compared to group **C**. Faecal fat excretion was also higher in group **A**. TC and LDL was similar across all groups.
Nilsen et al. 2015 [66]	Norway	Parallel	$n = 153$ (73 male; 80 female), normotensive and hypertensive	>18	3 arm: **(A)** 50 g/d Gamalost. **(B)** Norvegia 80 g/d. **(C)** Control—limited intake of Gamalost and Norvegia.	8 weeks	There were no changes in MetS (metabolic equivalents) factors between the intervention groups and control. Significant reductions were noted for TC in those with MetS in group **B**. Those in group **A** with high TC also showed significant decreases compared with control.

Table 2. Cont.

Author	Country	Study Design	Population	Age (Years)	Intervention	Duration	Main Findings
Soerensen et al. 2014 [67]	Denmark	Crossover	n = 15 males, healthy volunteers	18–50	3 arm: **(A)** 500 mg Ca/d-non-dairy control. **(B)** Semi-skimmed milk, 670 mL per 10 MJ. **(C)** Semi hard cheese, 120 g per 10 MJ (45% fat).	Three 2-week periods, separated by 2-week washout period	Significantly lower increases in TC and LDL were found in the group **B** and **C** compared with control. Faecal fat excretion also increased group **B** and **C** compared with control. No changes were found in blood pressure, high-density lipoprotein cholesterol, triglycerides and lipid ratios.
Hjerpsted et al. 2011 [68]	Denmark	Crossover	n = 49 (28 male; 21 female), healthy volunteers	22–69	Test-food amounts were dependent on participants' energy levels. 2 arm: **(A)** 143 g/d (based on medium energy level) hard cheese "Samsø" (27 g fat/100 g). **(B)** Salted butter, 47 g/d (based on medium energy level).	Two 6-week periods, separated by 2-week washout period	Group **A** had significantly lower serum total, LDL and HDL cholesterol, and increased glucose concentrations compared with group **B**. Faecal fat excretion did not differ between groups.
Intorre et al. 2011 [69]	Italy	Crossover	n = 30 (11 male; 19 female), healthy volunteers	20–40	2 arm: **(A)** 150 g of hard cheese per week (milk from cows fed a grass and maize silage-based diet with 5% of linseed oil added). **(B)** Control, 150 g of hard cheese per week (from normal cow milk).	Two 4-week periods, separated by a 4-week washout	The blood lipid profile did not change after diet **A**. Although it led to higher levels of vitamin C and E and stearic acid in blood, while myristic acid and oxidised LDL concentrations were significantly lower.
Pintus et al. 2013 [70]	Italy	Crossover	n = 42 (19 male; 23 female), mildly hypercholesterolaemic	30–60	2 arm: **(A)** control, 90 g/d sheep cheese. **(B)** Sheep cheese, 90 g/d, naturally enriched with CLA	Two 3-week periods, separated by a 6-week washout	The findings confirmed an association between anandamide and adiposity. Diet **B** significantly increased the plasma levels of fatty acid hydrocyperoxidases and LDL decreased. However, no changes were detected in levels of inflammatory markers.
Sofi et al. 2010 [71]	Italy	Crossover	n = 10 (4 male; 6 female) healthy volunteers	30–65	2 arm: **(A)** 200 g/week (3 times a week) of pecorino cheese, naturally rich in CLA. **(B)** Placebo cheese—control, 200 g/week (3 times a week).	Two 10-week periods, separated by a 10-week washout period	Consumption of cheese naturally rich in CLA determined a significant reduction in some inflammatory parameters as well as some haemorheological, appearing to cause favourable biochemical changes of atherosclerotic markers, albeit limited. No significant effects on lipid profile were evident.
Nestel et al. 2005 [72]	Australia	Crossover	n = 19 (14 male; 5 female), overweight and mildly hypercholesterolaemic	Average— 56.3 ± 7.8	2 arm: **(A)** 120 g/d mature cheddar (40 g fat). **(B)** Similar amount of butter fat to group **A**, from preweighed portions of butter and one butter-rich muffin.	Two 4-week periods, separated by 2-week washout period	Lipid values did not differ significantly between the group **A** and run-in periods, but TC and LDL were significantly higher with group **B**. Group **B** (butter) also raised total and LDL cholesterol significantly. This was not evident for cheese.

Table 2. Cont.

Author	Country	Study Design	Population	Age (Years)	Intervention	Duration	Main Findings
Biong et al. 2004 [73]	Norway	Crossover	n = 22 (9 male; 13 female), healthy volunteers	21–54	3 arm: (A) 150 g/d (per 8 MJ diet) Jarlsberg 'Swiss-type' cheese. (B) Butter, 52 g/d (per 8 MJ diet) + casein (as calcium caseinate). (C) Butter, 52 g/d (per 8 MJ diet) + egg white.	Three 3-week periods, separated by a 1-week washout periods	TC was significantly lower after diet A compared to diet B. While LDL was lower after diet A, this was not statistically significant. There were also no significant differences in HDL-cholesterol, triacylglycerols, apo A-1, apo B or lipoprotein (a), haemostatic variables and homocysteine between groups.
Karvonen et al. 2002 [74]	Finland	Crossover	n = 31 (17 male; 14 female), hyperlipidaemic	25–65	2 arm: (A) 65 g/d low-fat rapeseed oil-based cheese (11 g fat, of which 1 g was SFA). (B) Hard cheese, 65 g/d (15 g fat, of which 10 g was SFA).	Two 4-week periods, washout period not specified	Serum TC and LDL concentration was lower in group A, 2 and 4 weeks after use of rapeseed oil-based cheese compared to group B (control).

Other groups have also looked at postprandial response to cheese consumption, as the postprandial response to lipid consumption is considered an independent indicator of CVD risk [75]. Drouin-Chartier (2017) reported minor differences in postcirculating TAG concentrations, in a postprandial RCT where participants ingested 33 g fat from a firm cheese (young cheddar), a soft cream cheese (cream cheese) or butter (control) incorporated into standardised macronutrient-matched meals. They conclude that the study demonstrates that the cheese matrix modulates the impact of dairy fat on postprandial lipemia in healthy subjects [75].

Hansson et al. 2019, examining postprandial response to sour cream, whipped cream, cheese and butter, noted that sour cream resulted in a larger postprandial triacylglycerol (TAG) area under the curve (AUC), compared to whipped cream, butter and cheese (P = 0.05). Intake of sour cream also induced a larger HDL cholesterol AUC compared to cheese. Intake of cheese induced a 124% larger insulin AUC compared to butter. Hansson et al. concluded that high-fat meals containing similar amount of fat from different dairy products induce different postprandial effects on serum TAGs, HDL cholesterol and insulin in healthy adults [76].

2.4. Milk

Characteristics of seven RCTs that studied the effects of milk consumption on markers of metabolic health and CVD risk are described in Table 3 [77–83]. RCTs included in the table were either parallel or crossover trials and the participants' ages ranged from 20–85 years. Duration of interventions ranged from 4 to 16 weeks with a washout period included in some of the studies. Gardner et al. (2007) compared the effects of soy-based drinks to dairy milk consumption on metabolic health markers, and found that LDL was significantly lower after consuming soy milk compared to dairy milk, but no significant differences between groups were observed for HDL, triacylglycerols, insulin or glucose [77]. In 2016, Lee et al. examined the impact of milk consumption (400 mL per day) compared to habitual intake on markers of metabolic health, and found no significant differences in body mass index, blood pressure or lipid profile [78]. Hidaka et al. compared intakes of full-fat vs. non-fat milk intakes and showed lower plasma triglyceride and phospholipid levels in the no-fat group [79].

2.5. Yoghurt

Table 4 presents a summary of RCTs that specifically studied the effects of yoghurt consumption on markers of metabolic health and CVD risk. Most of these studies focused specifically on the microbial content of yoghurts, where consumption of probiotic yoghurt or modified-bacterial-strain-containing yoghurts or comparisons between commonly available varieties of yoghurts such as non-fat yoghurt, natural yoghurt and heated yoghurt were examined. In general, when compared to participants on a controlled diet (diet with zero consumption of fermented products), the participants who consumed probiotic or conventional yoghurt showed significant decrease in TC and LDL cholesterol [84–92].

Table 3. Randomised controlled trials (RCT) demonstrating an effect of milk consumption on markers of metabolic health and cardiovascular disease (CVD) risk.

Author	Country	Study Design	Population	Age (Years)	Intervention	Duration	Main Findings
Lee et al. 2016 [78]	Korea	Parallel	n = 58 (29 male; 29 female), overweight and obese with metabolic syndrome	35–65	2 arm: (A) 400 mL per day (200 mL twice daily) of low-fat milk. (B) Control—maintain habitual diet.	6 weeks	No significant differences in body mass index, blood pressure, lipid profile and adiponectin levels, as well as levels of inflammatory markers, oxidative stress markers and atherogenic markers were found between groups.
Hidaka et al. 2012 [79]	Japan	Parallel	n = 14 (8 male; 6 female), healthy volunteers	Mean: 28.6 ± 6.0 S.D	2 arm: (A) 500 mL/d whole milk. (B) non-fat milk, 500 mL/d.	4 weeks	Group B showed lowering of plasma triglyceride (TG), phospholipid levels, TG level in HDL and increased plasma apolipoprotein (apo) C-III level. TG/cholesterol ratios in HDL and LDL also significantly decreased in group B. Whole milk consumption showed increases in plasma levels of apoC-III and apoE. C.
Rosado et al. 2011 [80]	Mexico	Parallel	n = 139 females, obese	25–45	3 arm: (A) 250 mL of low-fat milk, 3 times per day, and an energy-restricted diet (500 kcal/day). (B) Low-fat milk, 250 mL, with added micronutrients, 3 times per day, an energy-restricted diet (500 kcal/day). (C) Control—an energy-restricted diet (500 kcal/day) with no intake of milk.	16 weeks	Group B lost significantly more weight compared with group A + C. BMI and body fat changes were also significantly greater in the group B compared with group A + C. No differences were found between groups in glucose level, blood lipid profile, C-reactive protein level or blood pressure.
Venkatramanan et al. 2010 [81]	Canada	Crossover	n = 18 (11 male; 7 female) moderately overweight and borderline hyperlipidaemic	30–60	3 arm: (A) 1000 mL/d milk naturally enriched CLA. (B) Milk enriched with synthetic CLA, 1000 mL/d. (C) Control—1000 mL/d untreated milk.	Three 8-week periods separated by 4-week washout periods	Group A + B failed to alter plasma TC, LDL, HDL or triacylglycerol concentrations; body weight; or fat composition compared with the control group. CLA consumption did not significantly affect plasma ALT (alanine transaminase), TBIL (total bilirubin in plasma), CRP (C-reactive protein) or TNF-α (tumor necrosis factor) concentrations.
Faghih et al. 2009 [82]	Iran	Parallel	n = 100 females, premenopausal and overweight or obese	20–50	4 arm: (A) control—500 kcal/d deficit (500–600 mg/d dietary calcium). (B) Calcium-supplemented diet identical to control diet (800 mg/d of calcium carbonate). (C) Servings of low-fat milk, 220 mL, (1.5%) and 500 kcal/d deficit. (D) Three servings of calcium-fortified soy milk and 500 kcal/d deficit.	8 weeks	Body weight, BMI, waist circumference (WC), waist-to-hip ratio (WHR), body fat mass and percent body fat decreased significantly across all groups. The changes in WC and WHR were significantly higher in groups C and D compared to controls. Reductions in weight and BMI were significantly greater in the group C compared to controls. Lipid profiles were not analysed.

Table 3. *Cont.*

Author	Country	Study Design	Population	Age (Years)	Intervention	Duration	Main Findings
Gardner et al. 2007 [77]	USA	Crossover	n = 28 (6 male; 22 female), hypercholesteraemic	30–65	3 arm: **(A)** 32 oz/d whole soy bean drink. **(B)** Soy protein isolate drink, 28 oz/d. **(C)** Dairy milk, 18.5 oz/d, (all volumes were standardised to yield 25 g protein/d.	Three 4-week periods, separated by 4-week washout periods	LDL was significantly lower after consuming soy milk in groups **A + B** compared to dairy milk (group C). No significant differences between groups were observed for HDL, triacylglycerols, insulin or glucose.
Barr et al. 2000 [83]	USA	Parallel	n = 200 (70 male; 130 female), healthy volunteers	55–85	2 arm: **(A)** three 8 oz/d of skimmed 1% milk. **(B)** Maintain habitual diet and consuming <1.5 dairy servings per day.	12 weeks	Similar decreases in blood pressure were apparent across both groups. TC, LDL and the ratio of TC:HDL remained unchanged. Triglyceride levels increased within the normal range in group A.

Table 4. Randomised controlled trials (RCT) demonstrating an effect of yoghurt consumption on markers of metabolic health and CVD risk.

Author	Country	Study Design	Population	Age (Years)	Intervention	Duration	Main Findings
El Khoury et al. 2014 [84]	Canada	Crossover	n = 20 males, BMI 20–24.9 kg/m²	20–30	5 arm: **(A)** 250 g non-fat yoghurt—plain. **(B)** Non-fat yoghurt with honey—plain, 250 g. **(C)** Non-fat yoghurt, strawberry flavoured, 250 g. **(D)** Skimmed milk, 250 g. **(E)** Orange juice, 250 g.	Postprandial	Pre-meal glucose responses were dose-dependent to increasing protein and decreasing sugars in dairy. Protein:carbohydrate ratio correlated negatively with pre-meal glucose due to improved efficacy of insulin action. Compared with treatment E, blood glucose was lower after dairy snack treatments, contribution of dairy products to post-meal glucose was independent of their protein:carbohydrate ratio
Shab-Bidar et al. 2011 [85]	Iran	Parallel	n = 100 (43 male; 57 female), type 2 diabetes	29–67	2 arm: **(A)** 250 mL vitamin D3 fortified yoghurt drink, twice a day **(B)** 250 mL plain yoghurt fortified drink, twice a day.	12 weeks	Diet **A** showed significant improvement in fasting glucose, glycated haemoglobin (HbA1c), TAG, HDL cholesterol, endothelin-1, E-selectin and MMP-9 compared with the control (diet B).
Sadrzadeh-Yeganeh et al. 2010 [86]	Iran	Parallel	n = 90 females, healthy volunteers	19–49	3 arm: **(A)** 300 g/d probiotic yoghurt. **(B)** Conventional yoghurt, 300 g/d. **(C)** Control, no consumption of fermented products	6 weeks	No significant difference in lipid profile within any group. No difference in TAG and LDL across the groups. There was a decrease in cholesterol in both group A + B compared with the control as well as a decrease in TC:HDL ratio. HDL increased in group A compared with the control.

Table 4. *Cont.*

Author	Country	Study Design	Population	Age (Years)	Intervention	Duration	Main Findings
Ejtahed et al. 2010 [87]	Iran	Parallel	$n = 60$ (23 male; 37 female), type 2 diabetes	30–60	2 arm: **(A)** 300 g/d probiotic yoghurt. **(B)** Control, 300 g/d conventional yoghurt	6 weeks	Diet A caused a decrease in TC and LDL compared with the control. No significant changes from baseline were shown in TAG and HDL in diet A. The TC:HDL ratio and LDL:HDL ratio as atherogenic indices significantly decreased in after diet A compared with the control.
Ataie-Jafari et al. 2009 [88]	Iran	Crossover	$n = 14$ (4 male; 10 female), mild to moderate hypercholesteraemic	40–64	2 arm: **(A)** 3 × 100 g/d probiotic yoghurt, *Lactobacillus acidophilus* and *Bifidobacterium lactis* **(B)** Three × 100 g/day control yoghurt. Both yoghurts contain 2.5% fat.	two 6-week periods, separated by a 4-week washout period	Consumption of diet A caused a significant decrease in serum TC compared with the control. No differences were reported for remaining blood lipids examined between the two diets.
Kiessling et al. 2002 [89]	Germany	Crossover	$n = 29$ females, normo- and hypercholesterolaemic	19–56	2 arm: **(A)** 300 g/d control yoghurt *streptococcus thermophilus* and *L. lactis*. **(B)** Probiotic yoghurt, 300 g/d, enriched with *L. acidophilus* 145, *B. longum* 913 and 1% oligofructose.	6 weeks (all on control diet), followed by two 6-week periods, separated by a 9-day washout periods.	Serum TC and LDL concentrations were not influenced by diet A. The HDL concentration increased significantly after diet A. The ratio of LDL:HDL cholesterol decreased. Long-term consumption of 300 g yoghurt increased HDL and lead to desired improvement of LDL:HDL ratio in both diets.
Rizkalla et al. 2000 [90]	France	Crossover	$n = 24$ males, healthy volunteers	20–60	2 arm: **(A)** 500 g/d fresh yoghurt w/ live bacterial cultures. **(B)** Heated yoghurt, 500 g/d	Two 15-day periods, separated by a 15-day washout period	No changes detected in fasting plasma glucose, insulin, fatty acid, TAG or cholesterol concentrations in both groups. Plasma butyrate was higher and plasma propionate tended to be higher in subjects without lactose malabsorption after diet A than B. Subjects with lactose malabsorption increased propionate production after fresh yoghurt consumption compared with baseline measures.
Agerholm-Larson et al. 2000 [91]	Denmark	Parallel	$n = 70$ (20 male; 50 female), healthy volunteers	18–55	5 arm: **(A)** 450 mL/d yoghurt fermented with two strains of *Streptococcus thermophilus* and two strains of *Lactobacillus acidophilus*. **(B)** Placebo yoghurt, 450 mL/d, fermented with delta-acid-lactone. **(C)** Yoghurt, 450 mL/d, fermented with two strains of *Streptococcus thermophilus* and one strain of *Lactobacillus rhamnosus*. **(D)** Yoghurt, 450 mL/d, fermented with one strain of *Enterococcus faecium* and two strains of *Streptococcus thermophilus*. **(E)** Two placebo pills per day.	8 weeks	Comparing all 5 groups, no statistical effects on LDL were observed after consumption of diet D. After adjusting for small changes in body weight, LDL decreased by 8.4% and fibrinogen increased. This was significantly different from the control groups, **B** + **E**. After diets **A** + **D**, systolic blood pressure reduced significantly more compared to diet C.

Table 4. *Cont.*

Author	Country	Study Design	Population	Age (Years)	Intervention	Duration	Main Findings
Anderson and Gilliland, 1999 [92]	USA	Study 1: single-blind, parallel. Study 2: double-blind, crossover.	Study 1: $n = 29$ (9 male; 20 female). Study 2: $n = 40$ (18 male; 22 female), all hypercholesterolaemic	49–55	**Study 1:** 2 arm: **(A1)** 200 g/d fermented milk yoghurt containing human *L. acidophilus* L1. **(B1)** Fermented milk yoghurt, 200 g/d, containing swine *L. acidophilus* strain ATCC 43121 **Study 2:** 2 arm: **(A2)** 200 g/d fermented milk yoghurt containing *L. acidophilus* L1 strain. **(B)** Fermented milk yoghurt, 200 g/d, without these active bacteria.	Study 1: 3 weeks Study 2: Two 4-week periods with a 2-week washout period	Study 1: Diet A1 showed a significant 2.4% reduction in TC. LDL was lower after both treatment groups although not significant. HDL decreased significantly in both groups. Study 2: diet **A2** reduced TC in the 1st treatment period but not in the 2nd. A combined analysis of the two treatment study interventions demonstrated a significant reduction in TC.

3. The 'Dairy Matrix'

While dairy products are often considered together as a food category in nutritional epidemiology, they vary considerably in terms of their content and structure and how these interact with other food components, which describes the 'dairy matrix' concept [93]. Values from the Composition of Foods Integrated Dataset (CoFID) were used to compare nutrient composition across the range of commonly available dairy products (summarised in Figure 2). While they represent a wide range of products, only plain, unflavoured versions with no added sugar were included in this analysis. Figure 1 demonstrates how nutrient composition across the range of products varies greatly, as does their overall matrix or structure, depending on the product type. For example, whole milk contains 3.6% fat in a liquid oil-in-water emulsion with lactose and protein (both casein and whey) while cheeses mainly consist of casein proteins and fat, in a solid matrix, with only trace levels of lactose and whey [62]. Butter is an emulsion of water-in oil, and contains mostly fat and water, with no protein or carbohydrate, while (liquid) cream is also a water-in-oil emulsion, and contains low levels of protein (approx. 2%) [62] and lactose (approx. 3%) (in 35%-fat cream). The various processing steps that different products undergo from raw milk to final foodstuff impact the level of the different nutrients and the overall macro and microstructure of these foods. The nature of these differences may result in the different health outcomes associated with their consumption.

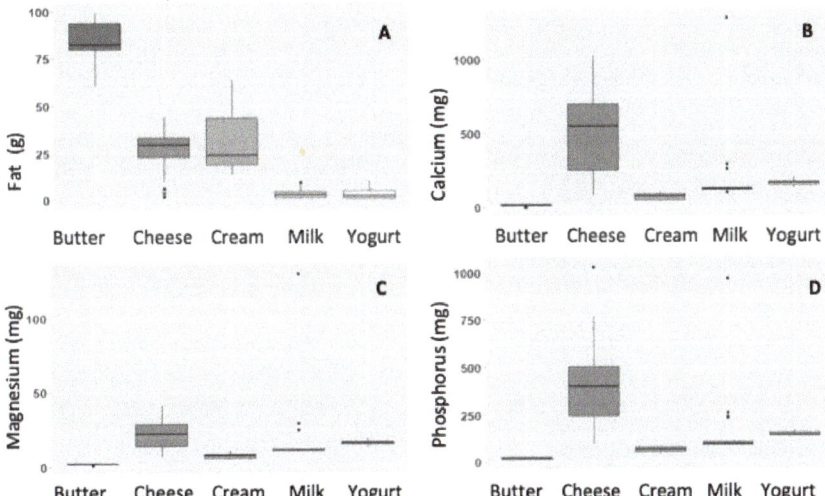

Figure 2. Boxplot showing average fat and mineral content in dairy products per 100 g. Values from the Composition of Foods Integrated Dataset (CoFID). Only plain, unflavoured products with no added sugar were included in this analysis. Nutrients shown are: (**A**) total fat (g), (**B**) calcium (mg), (**C**) magnesium (mg) and (**D**) phosphorus (mg), calculated from $n = 4$ butter, $n = 43$ cheese, $n = 8$ cream, $n = 20$ milk and $n = 4$ yoghurt samples.

Cheese in particular is associated with lower levels of blood cholesterol than other dairy products and especially when compared to butter (see Table 2 for an overview of studies in this area). Cheese structures contain aggregated casein micelles [94] which may impact the ability of lipases to break down the fat contained within the matrix, compared to the same fat contained within other food matrices (e.g., milk and butter). The structure of cheese, including the degree of hardness and cohesiveness, can result in it being more physically resistant to digestion than other matrices [94] which affects the degree to which the fat can be digested and absorbed. There may be additional effects from the calcium contained within this matrix, reacting with the fatty acids to form insoluble calcium soaps [95], as well as the separate textural effects from calcium that increase the cohesiveness [94] which may result in enhanced digestive resistance. This mechanism also appears to be supported by a higher faecal fat excretion following cheese consumption compared to other sources, in postprandial studies [67] although is not fully confirmed [68]. In addition to the calcium content of cheese, the phosphorus content is also implicated in the reduction of fat digestibility, by affecting the ability of cheese constituents to form insoluble soaps during digestion (calcium phosphate). This is thought to further increase fat excretion via the adsorption of bile acids to the surface, and has been implicated in the reduction of LDL-c [62,95]. Both phosphorus and calcium are particularly concentrated in cheese compared to other dairy products [96].

Cheese is a fermented dairy product, and the fermentation process may also be one of the contributors to the cardiometabolic protective nature of cheese via a number of mechanisms. Lactic acid bacteria found in fermented dairy products can result in platelet-activating factor (PAF)-inhibitory lipid production [97,98]. Further, as cheeses ripen and age, shorter peptides are produced and in some cases there is a release of latent bioactives, as some of these peptides have specific bioactivity that is not apparent in the intact 'parent' protein [99,100]. In cheeses, antihypertensive peptides are produced during fermentation, including V-P-P and I-P-P, which are tripeptides that exert their effects via inhibition of the angiotension converting enzyme (ACE) pathway [101,102]. Some cheeses have also been found to have bioactivity related to glycaemic control [103], which may also contribute to the cardio-metabolic benefits from cheese consumption. The starter culture used in the cheesemaking process can have an additional impact on the inherent bioactivity produced during fermentation [104,105]. Finally, the form in which fat is contained in cheese compared to butter may also result in some of the differences observed between these two products. The polar lipids found in dairy products in general appear to have anti-inflammatory properties compared to other oxidised dietary lipids [106–108] and they are mostly contained in a bioactive envelope surrounding the fat, known as the milk fat globule membrane (MFGM) [109]. Cheese contains particularly high levels of polar sphingolipids that are not present in the same levels in butter, since the membrane is disrupted during the churning process [62,110]. Studies suggest that polar lipids can impact blood lipid levels in the acute postprandial period, with lower lipaema (and insulin) observed following a liquid meal of palm fat when an MGFM-rich dairy fraction was added, compared to the same meal without this addition [111].

With research strongly suggestive of postprandial hyperlipidema as an independent risk factor for CVD [112], this could be a further explanation for the growing list of studies showing a protective effect of cheese consumption on CVD risk [32,60,61,68,72,113–115] despite the relatively high SFA content.

4. Conclusions/Future Directions

This paper summarises and discusses the evidence that examines the link between dairy intake and CVD risk. Whilst the evidence is mixed for some dairy foods (milk), it is more consistent for others (cheese/yoghurt), and supports the concept that the source of saturated fat intake has an important impact on cardiometabolic response to consumption. This is not a new concept, but whilst this evidence is growing, more research is needed before any significant change in public health recommendations are implemented.

While the link between dairy foods and metabolic health has been well-studied, research gaps still remain, and must be considered in future work. Many of the RCTs completed are short-term studies with single products, and there is a need to consider combinations of foods in a dietary pattern, considered by only a few studies to date [116,117]. The manner of the food consumption also needs to be considered, as dairy is often consumed in many forms (heated, melted) and as part of meals or recipes or in sweetened beverages. In addition, the amount of dairy product given in many of the randomised controlled trial studies to date are largely outside of recommended portion size intakes, as such, caution is required in interpreting and generalising findings.

Looking to the future, to date many of the studies have focused on traditional markers of cardiometabolic risk, particularly circulating lipid levels, which may not show subtle changes between products and/or further elucidate the cause/mechanisms for difference in response following consumption of specific foods. It has been suggested that the use of such traditional markers alone may limit the ability to predict health outcomes from the fat in dairy products, since many other components in dairy may have impact on CVD risk [118]. For this reason, more recent studies are also examining novel biomarkers that include vascular function (arterial stiffness, flow-mediated vasodilation (FMD) [47] and LDL-c particle size distribution [119]. It is important that we more fully understand the mechanisms underlying the variance in response to consumption, and the use of such novel markers will help develop this knowledge.

In conclusion, dairy foods are diverse in their structure and their nutrient content, resulting in differing biological responses and associated health outcomes. Thus, continuing to treat them as a single food category in food-based healthy eating guidelines may obscure the individual effects of these foods. As diets transition, there is an urgent need to understand the impact of different dairy foods, their preparation methods and how they are consumed, within the overall patterns of dietary intake in different cultural groups.

Author Contributions: Conceptualization, C.M.T., E.R.G. and E.L.F.; investigation, writing—original draft preparation, writing—review and editing, C.M.T., N.B., A.O., E.R.G. and E.L.F.; supervision, project administration, and funding acquisition, E.R.G. and E.L.F. All authors have read and agreed to the published version of the manuscript.

Funding: This research was supported by the Food for Health Ireland project, funded by Enterprise Ireland, grant number TC20180025.

Conflicts of Interest: E.R.G. and E.L.F. have previously received travel expenses and speaking honoraria from the National Dairy Council. E.R.G. and E.L.F. have received research funding through the Food for Health Ireland project, funded by Enterprise Ireland, grant number TC20180025. The other authors are currently, or have been in the past, employed on the Food for Health Ireland project, funded by Enterprise Ireland, grant number TC20180025. The funders had no role in the design of the study; in the collection, analyses or interpretation of data; in the writing of the manuscript or in the decision to publish the results.

References

1. Haug, A.; Høstmark, A.T.; Harstad, O.M. Bovine milk in human nutrition—A review. *Lipids Health Dis.* **2007**, *6*, 25. [CrossRef] [PubMed]
2. Marangoni, F.; Pellegrino, L.; Verduci, E.; Ghiselli, A.; Bernabei, R.; Calvani, R.; Cetin, I.; Giampietro, M.; Perticone, F.; Piretta, L. Cow's milk consumption and health: A health professional's guide. *J. Am. Coll. Nutr.* **2019**, *38*, 197–208. [CrossRef] [PubMed]
3. Rosqvist, F.; Smedman, A.; Lindmark-Månsson, H.; Paulsson, M.; Petrus, P.; Straniero, S.; Rudling, M.; Dahlman, I.; Risérus, U. Potential role of milk fat globule membrane in modulating plasma lipoproteins, gene expression, and cholesterol metabolism in humans: A randomized study. *Am. J. Clin. Nutr.* **2015**, *102*, 20–30. [CrossRef] [PubMed]
4. Institute of Medicine Committee to Review Dietary Reference Intakes for Calcium and Vitamin D. The National Academies Collection: Reports funded by National Institutes of Health. In *Dietary Reference Intakes for Calcium and Vitamin D*; Ross, A.C., Taylor, C.L., Yaktine, A.L., Del Valle, H.B., Eds.; National Academies Press: Washington, DC, USA, 2011.
5. Gueguen, L.; Pointillart, A. The bioavailability of dietary calcium. *J. Am. Coll. Nutr.* **2000**, *19*, 119S–136S. [CrossRef]
6. Zhao, Y.; Martin, B.R.; Weaver, C.M. Calcium bioavailability of calcium carbonate fortified soymilk is equivalent to cow's milk in young women. *J. Nutr.* **2005**, *135*, 2379–2382. [CrossRef] [PubMed]
7. Fulgoni, V.L., III; Keast, D.R.; Auestad, N.; Quann, E.E. Nutrients from dairy foods are difficult to replace in diets of Americans: Food pattern modeling and an analyses of the National Health and Nutrition Examination Survey. *Nutr. Res.* **2011**, *31*, 759–765. [CrossRef] [PubMed]

8. Dror, D.K.; Allen, L.H. Dairy product intake in children and adolescents in developed countries: Trends, nutritional contribution, and a review of association with health outcomes. *Nutr. Rev.* **2014**, *72*, 68–81. [CrossRef]
9. Vissers, P.A.; Streppel, M.T.; Feskens, E.J.; de Groot, L.C. Contribution of dairy products to micronutrient intake in The Netherlands. *J. Am. Coll. Nutr.* **2011**, *30*, 415S–421S. [CrossRef]
10. Coudray, B. Contribution of dairy products to micronutrient intake in France. *J. Am. Coll. Nutr.* **2011**, *30*, 410S–414S. [CrossRef]
11. Kalkwarf, H.J.; Khoury, J.C.; Lanphear, B.P. Milk intake during childhood and adolescence, adult bone density, and osteoporotic fractures in US women. *Am. J. Clin. Nutr.* **2003**, *77*, 257–265. [CrossRef]
12. Ma, D.F.; Zheng, W.; Ding, M.; Zhang, Y.M.; Wang, P.Y. Milk intake increases bone mineral content through inhibiting bone resorption: Meta-analysis of randomized controlled trials. *e-SPEN J.* **2013**, *8*, E1–E7. [CrossRef]
13. Dror, D.K. Dairy consumption and pre-school, school-age and adolescent obesity in developed countries: A systematic review and meta-analysis. *Obes. Rev.* **2014**, *15*, 516–527. [CrossRef] [PubMed]
14. Spence, L.A.; Cifelli, C.J.; Miller, G.D. The Role of Dairy Products in Healthy Weight and Body Composition in Children and Adolescents. *Curr. Nutr. Food. Sci.* **2011**, *7*, 40–49. [CrossRef] [PubMed]
15. Brantsæter, A.L.; Olafsdottir, A.S.; Forsum, E.; Olsen, S.F.; Thorsdottir, I. Does milk and dairy consumption during pregnancy influence fetal growth and infant birthweight? A systematic literature review. *Food Nutr. Res.* **2012**, *56*. [CrossRef] [PubMed]
16. Geiker, N.R.W.; Mølgaard, C.; Iuliano, S.; Rizzoli, R.; Manios, Y.; van Loon, L.J.C.; Lecerf, J.M.; Moschonis, G.; Reginster, J.Y.; Givens, I.; et al. Impact of whole dairy matrix on musculoskeletal health and aging–current knowledge and research gaps. *Osteoporos. Int.* **2020**, *31*, 601–615. [CrossRef] [PubMed]
17. Granic, A.; Hurst, C.; Dismore, L.; Aspray, T.; Stevenson, E.; Witham, M.D.; Sayer, A.A.; Robinson, S. Milk for Skeletal Muscle Health and Sarcopenia in Older Adults: A Narrative Review. *Clin. Interv. Aging* **2020**, *15*, 695–714. [CrossRef] [PubMed]
18. Thorning, T.K.; Raben, A.; Tholstrup, T.; Soedamah-Muthu, S.S.; Givens, I.; Astrup, A. Milk and dairy products: Good or bad for human health? An assessment of the totality of scientific evidence. *J. Food Nutr. Res.* **2016**, *60*, 32527. [CrossRef]
19. Food and Agricultural Organization of the United Nations. Dairy Development in Asia. Available online: http://www.fao.3/i0588e/I0588E02.htm (accessed on 24 September 2020).
20. Cifelli, C.J.; Houchins, J.A.; Demmer, E.; Fulgoni, V.L. Increasing Plant Based Foods or Dairy Foods Differentially Affects Nutrient Intakes: Dietary Scenarios Using NHANES. *Nutrients* **2016**, *8*, 422. [CrossRef]
21. Weinberger, M.H. Salt Sensitivity of Blood Pressure in Humans. *Hypertension* **1996**, *27*, 481–490. [CrossRef]
22. Strazzullo, P.; D'Elia, L.; Kandala, N.-B.; Cappuccio, F.P. Salt intake, stroke, and cardiovascular disease: Meta-analysis of prospective studies. *BMJ* **2009**, *339*, b4567. [CrossRef]
23. Ni Mhurchu, C.; Capelin, C.; Dunford, E.K.; Webster, J.L.; Neal, B.C.; Jebb, S.A. Sodium content of processed foods in the United Kingdom: Analysis of 44,000 foods purchased by 21,000 households. *Am. J. Clin. Nutr.* **2010**, *93*, 594–600. [CrossRef] [PubMed]
24. Anderson, C.A.; Appel, L.J.; Okuda, N.; Brown, I.J.; Chan, Q.; Zhao, L.; Ueshima, H.; Kesteloot, H.; Miura, K.; Curb, J.D.; et al. Dietary sources of sodium in China, Japan, the United Kingdom, and the United States, women and men aged 40 to 59 years: The INTERMAP study. *J. Am. Diet. Assoc.* **2010**, *110*, 736–745. [CrossRef]
25. Sacks, F.M.; Svetkey, L.P.; Vollmer, W.M.; Appel, L.J.; Bray, G.A.; Harsha, D.; Obarzanek, E.; Conlin, P.R.; Miller, E.R.; Simons-Morton, D.G.; et al. Effects on Blood Pressure of Reduced Dietary Sodium and the Dietary Approaches to Stop Hypertension (DASH) Diet. *N. Engl. J. Med.* **2001**, *344*, 3–10. [CrossRef] [PubMed]
26. Huth, P.J.; Fulgoni, V.L.; Keast, D.R.; Park, K.; Auestad, N. Major food sources of calories, added sugars, and saturated fat and their contribution to essential nutrient intakes in the U.S. diet: Data from the National Health and Nutrition Examination Survey (2003–2006). *Nutr. J.* **2013**, *8*. [CrossRef] [PubMed]
27. Feeney, E.L.; Nugent, A.P.; Mc Nulty, B.; Walton, J.; Flynn, A.; Gibney, E.R. An overview of the contribution of dairy and cheese intakes to nutrient intakes in the Irish diet: Results from the National Adult Nutrition Survey. *Br. J. Nutr.* **2016**, *115*, 709–717. [CrossRef]

28. Scientific Advisory Committee on Nutrition. Saturated Fats and Health. Available online: https://assets.publishing.service.gov.uk/government/uploads/system/uploads/attachment_data/file/814995/SACN_report_on_saturated_fat_and_health.pdf (accessed on 24 September 2020).
29. Department of Health, Ireland. Your Guide to Healthy Eating Using the Food Pyramid. 2016. Available online: https://www.hse.ie/eng/about/who/healthwellbeing/our-priority-programmes/heal/food-pyramid-images/food-pyramid-simple-version.pdf (accessed on 24 September 2020).
30. De Oliveira, M.C.O.; Mozaffarian, D.; Kromhout, D.; Bertoni, A.G.; Sibley, C.T.; Jacobs, D.R., Jr.; Nettleton, J.A. Dietary intake of saturated fat by food source and incident cardiovascular disease: The Multi-Ethnic Study of Atherosclerosis. *Am. J. Clin. Nutr.* **2012**, *96*, 397–404. [CrossRef]
31. O'Sullivan, T.A.; Hafekost, K.; Mitrou, F.; Lawrence, D. Food sources of saturated fat and the association with mortality: A meta-analysis. *Am. J. Public Health* **2013**, *103*, e31–e42. [CrossRef]
32. Feeney, E.L.; Barron, R.; Dible, V.; Hamilton, Z.; Power, Y.; Tanner, L.; Flynn, C.; Bouchier, P.; Beresford, T.; Noronha, N. Dairy matrix effects: Response to consumption of dairy fat differs when eaten within the cheese matrix—A randomized controlled trial. *Am. J. Clin. Nutr.* **2018**, *108*, 667–674. [CrossRef]
33. Chen, M.; Pan, A.; Malik, V.S.; Hu, F.B. Effects of dairy intake on body weight and fat: A meta-analysis of randomized controlled trials. *Am. J. Clin. Nutr.* **2012**, *96*, 735–747. [CrossRef]
34. Menotti, A.; Keys, A.; Aravanis, C.; Blackburn, H.; Dontas, A.; Fidanza, F.; Karvonen, M.J.; Kromhout, D.; Nedeljkovic, S.; Nissinen, A.; et al. Seven Countries Study. First 20-Year Mortality Data in 12 Cohorts of Six Countries. *Ann. Med.* **1989**, *21*, 175–179. [CrossRef]
35. Kromhout, D.; Bloemberg, B.; Feskens, E.; Menotti, A.; Nissinen, A.; Seven Countries Study Group. Saturated fat, vitamin C and smoking predict long-term population all-cause mortality rates in the Seven Countries Study. *Int. J. Epidemiol.* **2000**, *29*, 260–265. [CrossRef] [PubMed]
36. Jacobson, T.A.; Maki, K.C.; Orringer, C.E.; Jones, P.H.; Kris-Etherton, P.; Sikand, G.; La Forge, R.; Daniels, S.R.; Wilson, D.P.; Morris, P.B. National Lipid Association recommendations for patient-centered management of dyslipidemia: Part 2. *J. Clin. Lipidol.* **2015**, *9*. [CrossRef] [PubMed]
37. Eckel, R.H.; Jakicic, J.M.; Ard, J.D.; de Jesus, J.M.; Miller, N.H.; Hubbard, V.S.; Lee, I.-M.; Lichtenstein, A.H.; Loria, C.M.; Millen, B.E. 2013 AHA/ACC guideline on lifestyle management to reduce cardiovascular risk: A report of the American College of Cardiology/American Heart Association Task Force on Practice Guidelines. *J. Am. Coll. Cardiol.* **2014**, *63*, 2960–2984. [CrossRef] [PubMed]
38. USDA. Scientific Report of the 2015 Dietary Guidelines Advisory Committee: Advisory Report to the Secretary of Health and Human Services and the Secretary of Agriculture. Available online: https://health.gov/sites/default/files/2019-09/Scientific-Report-of-the-2015-Dietary-Guidelines-Advisory-Committee.pdf (accessed on 24 September 2020).
39. De Souza, R.J.; Mente, A.; Maroleanu, A.; Cozma, A.I.; Ha, V.; Kishibe, T.; Uleryk, E.; Budylowski, P.; Schünemann, H.; Beyene, J. Intake of saturated and trans unsaturated fatty acids and risk of all cause mortality, cardiovascular disease, and type 2 diabetes: Systematic review and meta-analysis of observational studies. *BMJ* **2015**, *351*, h3978. [CrossRef] [PubMed]
40. Drouin-Chartier, J.-P.; Brassard, D.; Tessier-Grenier, M.; Côté, J.A.; Labonté, M.-È.; Desroches, S.; Couture, P.; Lamarche, B. Systematic review of the association between dairy product consumption and risk of cardiovascular-related clinical outcomes. *Adv. Nutr.* **2016**, *7*, 1026–1040. [CrossRef] [PubMed]
41. Alexander, D.D.; Bylsma, L.C.; Vargas, A.J.; Cohen, S.S.; Doucette, A.; Mohamed, M.; Irvin, S.R.; Miller, P.E.; Watson, H.; Fryzek, J.P. Dairy consumption and CVD: A systematic review and meta-analysis. *Br. J. Nutr.* **2016**, *115*, 737–750. [CrossRef]
42. Antoni, R.; Griffin, B. Draft reports from the UK's Scientific Advisory Committee on Nutrition and World Health Organization concur in endorsing the dietary guideline to restrict intake of saturated fat. *Nutr. Bull.* **2018**, *43*, 206–211. [CrossRef]
43. Astrup, A.; Bertram, H.C.; Bonjour, J.-P.; De Groot, L.C.; de Oliveira Otto, M.C.; Feeney, E.L.; Garg, M.L.; Givens, I.; Kok, F.J.; Krauss, R.M. WHO draft guidelines on dietary saturated and trans fatty acids: Time for a new approach? *BMJ* **2019**, *366*, l4137. [CrossRef]
44. Dehghan, M.; Mente, A.; Rangarajan, S.; Sheridan, P.; Mohan, V.; Iqbal, R.; Gupta, R.; Lear, S.; Wentzel-Viljoen, E.; Avezum, A.; et al. Association of dairy intake with cardiovascular disease and mortality in 21 countries from five continents (PURE): A prospective cohort study. *Lancet* **2018**, *392*, 2288–2297. [CrossRef]

45. Fontecha, J.; Calvo, M.V.; Juarez, M.; Gil, A.; Martínez-Vizcaino, V. Milk and dairy product consumption and cardiovascular diseases: An overview of systematic reviews and meta-analyses. *Adv. Nutr.* **2019**, *10*, S164–S189. [CrossRef]
46. Zemel, M.B.; Sun, X.; Sobhani, T.; Wilson, B. Effects of dairy compared with soy on oxidative and inflammatory stress in overweight and obese subjects. *Am. J. Clin. Nutr.* **2010**, *91*, 16–22. [CrossRef] [PubMed]
47. Vasilopoulou, D.; Markey, O.; Kliem, K.E.; Fagan, C.C.; Grandison, A.S.; Humphries, D.J.; Todd, S.; Jackson, K.G.; Givens, D.I.; Lovegrove, J.A. Reformulation initiative for partial replacement of saturated with unsaturated fats in dairy foods attenuates the increase in LDL cholesterol and improves flow-mediated dilatation compared with conventional dairy: The randomized, controlled REplacement of SaturatEd fat in dairy on Total cholesterol (RESET) study. *Am. J. Clin. Nutr.* **2020**, *111*, 739–748. [CrossRef] [PubMed]
48. Drouin-Chartier, J.-P.; Li, Y.; Ardisson Korat, A.V.; Ding, M.; Lamarche, B.; Manson, J.E.; Rimm, E.B.; Willett, W.C.; Hu, F.B. Changes in dairy product consumption and risk of type 2 diabetes: Results from 3 large prospective cohorts of US men and women. *Am. J. Clin. Nutr.* **2019**, *110*, 1201–1212. [CrossRef] [PubMed]
49. Markey, O.; Vasilopoulou, D.; Kliem, K.E.; Koulman, A.; Fagan, C.C.; Summerhill, K.; Wang, L.Y.; Grandison, A.S.; Humphries, D.J.; Todd, S. Plasma phospholipid fatty acid profile confirms compliance to a novel saturated fat-reduced, monounsaturated fat-enriched dairy product intervention in adults at moderate cardiovascular risk: A randomized controlled trial. *Nutr. J.* **2017**, *16*, 33. [CrossRef] [PubMed]
50. Benatar, J.R.; Jones, E.; White, H.; Stewart, R.A. A randomized trial evaluating the effects of change in dairy food consumption on cardio-metabolic risk factors. *Eur. J. Prev. Cardiol.* **2014**, *21*, 1376–1386. [CrossRef]
51. Nestel, P.J.; Mellett, N.; Pally, S.; Wong, G.; Barlow, C.K.; Croft, K.; Mori, T.A.; Meikle, P.J. Effects of low-fat or full-fat fermented and non-fermented dairy foods on selected cardiovascular biomarkers in overweight adults. *Br. J. Nutr.* **2013**, *110*, 2242–2249. [CrossRef]
52. Crichton, G.E.; Howe, P.R.; Buckley, J.D.; Coates, A.M.; Murphy, K.J. Dairy consumption and cardiometabolic health: Outcomes of a 12-month crossover trial. *Nutr. Metab.* **2012**, *9*, 19. [CrossRef]
53. Palacios, C.; Bertrán, J.J.; Ríos, R.E.; Soltero, S. No effects of low and high consumption of dairy products and calcium supplements on body composition and serum lipids in Puerto Rican obese adults. *Nutrition* **2011**, *27*, 520–525. [CrossRef]
54. Stancliffe, R.A.; Thorpe, T.; Zemel, M.B. Dairy attenuates oxidative and inflammatory stress in metabolic syndrome. *Am. J. Clin. Nutr.* **2011**, *94*, 422–430. [CrossRef]
55. Van Meijl, L.E.; Mensink, R.P. Effects of low-fat dairy consumption on markers of low-grade systemic inflammation and endothelial function in overweight and obese subjects: An intervention study. *Br. J. Nutr.* **2010**, *104*, 1523–1527. [CrossRef]
56. Wennersberg, M.H.; Smedman, A.; Turpeinen, A.M.; Retterstøl, K.; Tengblad, S.; Lipre, E.; Aro, A.; Mutanen, P.; Seljeflot, I.; Basu, S. Dairy products and metabolic effects in overweight men and women: Results from a 6-mo intervention study. *Am. J. Clin. Nutr.* **2009**, *90*, 960–968. [CrossRef] [PubMed]
57. Van Meijl, L.E.; Mensink, R.P. Low-fat dairy consumption reduces systolic blood pressure, but does not improve other metabolic risk parameters in overweight and obese subjects. *Nutr. Metab. Cardiovasc. Dis.* **2010**, *21*, 355–361. [CrossRef]
58. Tricon, S.; Burdge, G.C.; Jones, E.L.; Russell, J.J.; El-Khazen, S.; Moretti, E.; Hall, W.L.; Gerry, A.B.; Leake, D.S.; Grimble, R.F. Effects of dairy products naturally enriched with cis-9, trans-11 conjugated linoleic acid on the blood lipid profile in healthy middle-aged men. *Am. J. Clin. Nutr.* **2006**, *83*, 744–753. [CrossRef] [PubMed]
59. Zemel, M.B.; Richards, J.; Milstead, A.; Campbell, P. Effects of calcium and dairy on body composition and weight loss in African-American adults. *Obes. Res.* **2005**, *13*, 1218–1225. [CrossRef] [PubMed]
60. Tholstrup, T.; Høy, C.-E.; Andersen, L.N.; Christensen, R.D.; Sandström, B. Does fat in milk, butter and cheese affect blood lipids and cholesterol differently? *J. Am. Coll. Nutr.* **2004**, *23*, 169–176. [CrossRef] [PubMed]
61. Brassard, D.; Tessier-Grenier, M.; Allaire, J.; Rajendiran, E.; She, Y.; Ramprasath, V.; Gigleux, I.; Talbot, D.; Levy, E.; Tremblay, A. Comparison of the impact of SFAs from cheese and butter on cardiometabolic risk factors: A randomized controlled trial. *Am. J. Clin. Nutr.* **2017**, *105*, 800–809. [CrossRef]
62. Thorning, T.K.; Bertram, H.C.; Bonjour, J.-P.; De Groot, L.; Dupont, D.; Feeney, E.; Ipsen, R.; Lecerf, J.M.; Mackie, A.; McKinley, M.C. Whole dairy matrix or single nutrients in assessment of health effects: Current evidence and knowledge gaps. *Am. J. Clin. Nutr.* **2017**, *105*, 1033–1045. [CrossRef]

63. Limongi, F.; Noale, M.; Marseglia, A.; Gesmundo, A.; Mele, M.; Banni, S.; Crepaldi, G.; Maggi, S. Impact of cheese rich in Conjugated Linoleic Acid on low density lipoproteins cholesterol: Dietary Intervention in Older People (CLADIS Study). *J. Food. Nutr. Res.* **2018**, *6*, 1–7. [CrossRef]
64. Raziani, F.; Tholstrup, T.; Kristensen, M.D.; Svanegaard, M.L.; Ritz, C.; Astrup, A.; Raben, A. High intake of regular-fat cheese compared with reduced-fat cheese does not affect LDL cholesterol or risk markers of the metabolic syndrome: A randomized controlled trial. *Am. J. Clin. Nutr.* **2016**, *104*, 973–981. [CrossRef]
65. Thorning, T.K.; Raziani, F.; Bendsen, N.T.; Astrup, A.; Tholstrup, T.; Raben, A. Diets with high-fat cheese, high-fat meat, or carbohydrate on cardiovascular risk markers in overweight postmenopausal women: A randomized crossover trial. *Am. J. Clin. Nutr.* **2015**, *102*, 573–581. [CrossRef]
66. Nilsen, R.; Høstmark, A.T.; Haug, A.; Skeie, S. Effect of a high intake of cheese on cholesterol and metabolic syndrome: Results of a randomized trial. *Food. Nutr. Res.* **2015**, *59*, 27651. [CrossRef] [PubMed]
67. Soerensen, K.V.; Thorning, T.K.; Astrup, A.; Kristensen, M.; Lorenzen, J.K. Effect of dairy calcium from cheese and milk on fecal fat excretion, blood lipids, and appetite in young men. *Am. J. Clin. Nutr.* **2014**, *99*, 984–991. [CrossRef] [PubMed]
68. Hjerpsted, J.; Leedo, E.; Tholstrup, T. Cheese intake in large amounts lowers LDL-cholesterol concentrations compared with butter intake of equal fat content. *Am. J. Clin. Nutr.* **2011**, *94*, 1479–1484. [CrossRef]
69. Intorre, F.; Foddai, M.S.; Azzini, E.; Martin, B.; Montel, M.-C.; Catasta, G.; Toti, E.; Finotti, E.; Palomba, L.; Venneria, E. Differential effect of cheese fatty acid composition on blood lipid profile and redox status in normolipidemic volunteers: A pilot study. *Int. J. Food. Sci. Nutr.* **2011**, *62*, 660–669. [CrossRef] [PubMed]
70. Pintus, S.; Murru, E.; Carta, G.; Cordeddu, L.; Batetta, B.; Accossu, S.; Pistis, D.; Uda, S.; Ghiani, M.E.; Mele, M. Sheep cheese naturally enriched in α-linolenic, conjugated linoleic and vaccenic acids improves the lipid profile and reduces anandamide in the plasma of hypercholesterolaemic subjects. *Br. J. Nutr.* **2013**, *109*, 1453–1462. [CrossRef]
71. Sofi, F.; Buccioni, A.; Cesari, F.; Gori, A.M.; Minieri, S.; Mannini, L.; Casini, A.; Gensini, G.F.; Abbate, R.; Antongiovanni, M. Effects of a dairy product (pecorino cheese) naturally rich in cis-9, trans-11 conjugated linoleic acid on lipid, inflammatory and haemorheological variables: A dietary intervention study. *Nutr. Metab. Cardiovasc. Dis.* **2010**, *20*, 117–124. [CrossRef]
72. Nestel, P.; Chronopulos, A.; Cehun, M. Dairy fat in cheese raises LDL cholesterol less than that in butter in mildly hypercholesterolaemic subjects. *Eur. J. Clin. Nutr.* **2005**, *59*, 1059–1063. [CrossRef]
73. Biong, A.S.; Müller, H.; Seljeflot, I.; Veierød, M.B.; Pedersen, J.I. A comparison of the effects of cheese and butter on serum lipids, haemostatic variables and homocysteine. *Br. J. Nutr.* **2004**, *92*, 791–797. [CrossRef]
74. Karvonen, H.; Tapola, N.; Uusitupa, M.; Sarkkinen, E. The effect of vegetable oil-based cheese on serum total and lipoprotein lipids. *Eur. J. Clin. Nutr.* **2002**, *56*, 1094–1101. [CrossRef]
75. Drouin-Chartier, J.-P.; Tremblay, A.J.; Maltais-Giguère, J.; Charest, A.; Guinot, L.; Rioux, L.-E.; Labrie, S.; Britten, M.; Lamarche, B.; Turgeon, S.L.; et al. Differential impact of the cheese matrix on the postprandial lipid response: A randomized, crossover, controlled trial. *Am. J. Clin. Nutr.* **2017**, *106*, 1358–1365. [CrossRef]
76. Hansson, P.; Holven, K.B.; Øyri, L.K.; Brekke, H.K.; Biong, A.S.; Gjevestad, G.O.; Raza, G.S.; Herzig, K.-H.; Thoresen, M.; Ulven, S.M. Meals with similar fat content from different dairy products induce different postprandial triglyceride responses in healthy adults: A randomized controlled cross-over trial. *J. Nutr.* **2019**, *149*, 422–431. [CrossRef] [PubMed]
77. Gardner, C.D.; Messina, M.; Kiazand, A.; Morris, J.L.; Franke, A.A. Effect of Two Types of Soy Milk and Dairy Milk on Plasma Lipids in Hypercholesterolemic Adults: A Randomized Trial. *J. Am. Coll. Nutr.* **2007**, *26*, 669–677. [CrossRef] [PubMed]
78. Lee, Y.J.; Seo, J.A.; Yoon, T.; Seo, I.; Lee, J.H.; Im, D.; Lee, J.H.; Bahn, K.-N.; Ham, H.S.; Jeong, S.A.; et al. Effects of low-fat milk consumption on metabolic and atherogenic biomarkers in Korean adults with the metabolic syndrome: A randomised controlled trial. *J. Hum. Nutr Diet.* **2016**, *29*, 477–486. [CrossRef]
79. Hidaka, H.; Takiwaki, M.; Yamashita, M.; Kawasaki, K.; Sugano, M.; Honda, T. Consumption of nonfat milk results in a less atherogenic lipoprotein profile: A pilot study. *Ann. Nutr. Metab.* **2012**, *61*, 111–116. [CrossRef]
80. Rosado, J.L.; Garcia, O.P.; Ronquillo, D.; Hervert-Hernández, D.; Caamaño, M.D.C.; Martínez, G.; Gutiérrez, J.; García, S. Intake of milk with added micronutrients increases the effectiveness of an energy-restricted diet to reduce body weight: A randomized controlled clinical trial in Mexican women. *J. Am. Diet. Assoc.* **2011**, *111*, 1507–1516. [CrossRef] [PubMed]

81. Venkatramanan, S.; Joseph, S.V.; Chouinard, P.Y.; Jacques, H.; Farnworth, E.R.; Jones, P.J. Milk enriched with conjugated linoleic acid fails to alter blood lipids or body composition in moderately overweight, borderline hyperlipidemic individuals. *J. Am. Coll. Nutr.* **2010**, *29*, 152–159. [CrossRef]
82. Faghih, S.H.; Abadi, A.R.; Hedayati, M.; Kimiagar, S.M. Comparison of the effects of cows' milk, fortified soy milk, and calcium supplement on weight and fat loss in premenopausal overweight and obese women. *Nutr. Metab. Cardiovasc. Dis.* **2009**, *21*, 499–503. [CrossRef] [PubMed]
83. Barr, S.I.; McCarron, D.A.; Heaney, R.P.; Dawson-Hughes, B.; Berga, S.L.; Stern, J.S.; Oparil, S. Effects of increased consumption of fluid milk on energy and nutrient intake, body weight, and cardiovascular risk factors in healthy older adults. *J. Am. Diet. Assoc.* **2000**, *100*, 810–817. [CrossRef]
84. El Khoury, D.; Brown, P.; Smith, G.; Berengut, S.; Panahi, S.; Kubant, R.; Anderson, G.H. Increasing the protein to carbohydrate ratio in yogurts consumed as a snack reduces post-consumption glycemia independent of insulin. *Clin. Nutr.* **2014**, *33*, 29–38. [CrossRef] [PubMed]
85. Shab-Bidar, S.; Neyestani, T.R.; Djazayery, A.; Eshraghian, M.-R.; Houshiarrad, A.; Gharavi, A.; Kalayi, A.; Shariatzadeh, N.; Zahedirad, M.; Khalaji, N.; et al. Regular consumption of vitamin D-fortified yogurt drink (Doogh) improved endothelial biomarkers in subjects with type 2 diabetes: A randomized double-blind clinical trial. *BMC Med.* **2011**, *9*, 125. [CrossRef]
86. Sadrzadeh-Yeganeh, H.; Elmadfa, I.; Djazayery, A.; Jalali, M.; Heshmat, R.; Chamary, M. The effects of probiotic and conventional yoghurt on lipid profile in women. *Br. J. Nutr.* **2010**, *103*, 1778–1783. [CrossRef] [PubMed]
87. Ejtahed, H.; Mohtadi-Nia, J.; Homayouni-Rad, A.; Niafar, M.; Asghari-Jafarabadi, M.; Mofid, V.; Akbarian-Moghari, A. Effect of probiotic yogurt containing Lactobacillus acidophilus and Bifidobacterium lactis on lipid profile in individuals with type 2 diabetes mellitus. *J. Dairy Sci.* **2011**, *94*, 3288–3294. [CrossRef] [PubMed]
88. Ataie-Jafari, A.; Larijani, B.; Majd, H.A.; Tahbaz, F. Cholesterol-lowering effect of probiotic yogurt in comparison with ordinary yogurt in mildly to moderately hypercholesterolemic subjects. *Ann. Nutr. Metab.* **2009**, *54*, 22–27. [CrossRef] [PubMed]
89. Kiessling, G.; Schneider, J.; Jahreis, G. Long-term consumption of fermented dairy products over 6 months increases HDL cholesterol. *Eur. J. Clin. Nutr.* **2002**, *56*, 843–849. [CrossRef]
90. Rizkalla, S.W.; Luo, J.; Kabir, M.; Chevalier, A.; Pacher, N.; Slama, G. Chronic consumption of fresh but not heated yogurt improves breath-hydrogen status and short-chain fatty acid profiles: A controlled study in healthy men with or without lactose maldigestion. *Am. J. Clin. Nutr.* **2000**, *72*, 1474–1479. [CrossRef]
91. Agerholm-Larsen, L.; Raben, A.; Haulrik, N.; Hansen, A.; Manders, M.; Astrup, A. Effect of 8 week intake of probiotic milk products on risk factors for cardiovascular diseases. *Eur. J. Clin. Nutr.* **2000**, *54*, 288–297. [CrossRef]
92. Anderson, J.W.; Gilliland, S.E. Effect of fermented milk (yogurt) containing Lactobacillus acidophilus L1 on serum cholesterol in hypercholesterolemic humans. *J. Am. Coll. Nutr.* **1999**, *18*, 43–50. [CrossRef]
93. Feeney, E.L.; McKinley, M.C. The dairy food matrix: What it is and what it does. In *Milk and Dairy Foods: Their Functionality in Human Health and Disease*; Academic Press: Cambridge, MA, USA, 2020; pp. 205–225. [CrossRef]
94. Ayala-Bribiesca, E.; Lussier, M.; Chabot, D.; Turgeon, S.L.; Britten, M. Effect of calcium enrichment of Cheddar cheese on its structure, in vitro digestion and lipid bioaccessibility. *Int. Dairy J.* **2016**, *53*, 1–9. [CrossRef]
95. Lorenzen, J.K.; Nielsen, S.; Holst, J.J.; Tetens, I.; Rehfeld, J.F.; Astrup, A. Effect of dairy calcium or supplementary calcium intake on postprandial fat metabolism, appetite, and subsequent energy intake. *Am. J. Clin. Nutr.* **2007**, *85*, 678–687. [CrossRef]
96. Ditscheid, B.; Keller, S.; Jahreis, G. Cholesterol metabolism is affected by calcium phosphate supplementation in humans. *J. Nutr.* **2005**, *135*, 1678–1682. [CrossRef]
97. Antonopoulou, S.; Semidalas, C.E.; Koussissis, S.; Demopoulos, C.A. Platelet-activating factor (PAF) antagonists in foods: A study of lipids with PAF or anti-PAF-like activity in cow's milk and yogurt. *J. Agric. Food Chem.* **1996**, *44*, 3047–3051. [CrossRef]
98. Poutzalis, S.; Anastasiadou, A.; Nasopoulou, C.; Megalemou, K.; Sioriki, E.; Zabetakis, I. Evaluation of the in vitro anti-atherogenic activities of goat milk and goat dairy products. *Dairy Sci. Technol.* **2016**, *96*, 317–327. [CrossRef]

99. Meisel, H. Bioactive peptides from milk proteins: A perspective for consumers and producers. *Aust. J. Dairy Technol.* **2001**, *56*, 83.
100. Gobbetti, M.; Stepaniak, L.; De Angelis, M.; Corsetti, A.; Di Cagno, R. Latent bioactive peptides in milk proteins: Proteolytic activation and significance in dairy processing. *Crit. Rev. Food Sci. Nutr.* **2002**, *42*, 223–239. [CrossRef] [PubMed]
101. FitzGerald, R.J.; Meisel, H. Milk protein-derived peptide inhibitors of angiotensin-I-converting enzyme. *Br. J. Nutr.* **2000**, *84*, 33–37. [CrossRef] [PubMed]
102. Hirota, T.; Ohki, K.; Kawagishi, R.; Kajimoto, Y.; Mizuno, S.; Nakamura, Y.; Kitakaze, M. Casein hydrolysate containing the Antihypertensive Tripeptides Val-Pro-Pro and Ile-Pro-Pro improves vascular endothelial function independent of Blood Pressure–Lowering Effects: Contribution of the inhibitory action of Angiotensin-Converting enzyme. *Hypertens. Res.* **2007**, *30*, 489–496. [CrossRef]
103. Uenishi, H.; Kabuki, T.; Seto, Y.; Serizawa, A.; Nakajima, H. Isolation and identification of casein-derived dipeptidyl-peptidase 4 (DPP-4)-inhibitory peptide LPQNIPPL from gouda-type cheese and its effect on plasma glucose in rats. *Int. Dairy J.* **2012**, *22*, 24–30. [CrossRef]
104. Gupta, A.; Mann, B.; Kumar, R.; Sangwan, R.B. ACE-inhibitory activity of cheddar cheeses made with adjunct cultures at different stages of ripening. *Adv. Dairy. Res.* **2013**, *1*. [CrossRef]
105. Gómez-Ruiz, J.Á.; Ramos, M.; Recio, I. Angiotensin-converting enzyme-inhibitory peptides in Manchego cheeses manufactured with different starter cultures. *Int. Dairy J.* **2002**, *12*, 697–706. [CrossRef]
106. Lordan, R.; Zabetakis, I. Invited review: The anti-inflammatory properties of dairy lipids. *J. Dairy Sci.* **2017**, *100*, 4197–4212. [CrossRef]
107. Norris, G.H.; Milard, M.; Michalski, M.-C.; Blesso, C.N. Protective properties of milk sphingomyelin against dysfunctional lipid metabolism, gut dysbiosis, and inflammation. *J. Nutr. Biochem.* **2019**, *73*, 108224. [CrossRef]
108. Millar, C.L.; Jiang, C.; Norris, G.H.; Garcia, C.; Seibel, S.; Anto, L.; Lee, J.-Y.; Blesso, C.N. Cow's milk polar lipids reduce atherogenic lipoprotein cholesterol, modulate gut microbiota and attenuate atherosclerosis development in LDL-receptor knockout mice fed a Western-type diet. *J. Nutr. Biochem.* **2020**, *79*, 108351. [CrossRef] [PubMed]
109. Singh, H. The milk fat globule membrane—A biophysical system for food applications. *Curr. Opin. Colloid Interface Sci.* **2006**, *11*, 154–163. [CrossRef]
110. Vanderghem, C.; Bodson, P.; Danthine, S.; Paquot, M.; Deroanne, C.; Blecker, C. Milk fat globule membrane and buttermilks: From composition to valorization. *Biotechnol. Agron. Soc. Environ.* **2010**, *14*, 485–500.
111. Demmer, E.; Van Loan, M.D.; Rivera, N.; Rogers, T.S.; Gertz, E.R.; German, J.B.; Smilowitz, J.T.; Zivkovic, A.M. Addition of a dairy fraction rich in milk fat globule membrane to a high-saturated fat meal reduces the postprandial insulinaemic and inflammatory response in overweight and obese adults. *J. Nutr. Sci.* **2016**, *5*. [CrossRef] [PubMed]
112. Burdge, G.C.; Calder, P.C. Plasma cytokine response during the postprandial period: A potential causal process in vascular disease? *Br. J. Nutr.* **2005**, *93*, 3–9. [CrossRef] [PubMed]
113. De Goede, J.; Geleijnse, J.M.; Ding, E.L.; Soedamah-Muthu, S.S. Effect of cheese consumption on blood lipids: A systematic review and meta-analysis of randomized controlled trials. *Nutr. Rev.* **2015**, *73*, 259–275. [CrossRef]
114. Hjerpsted, J.B. *Cheese and Cardiovascular Health: Evidence from Observational, Intervention and Explorative Studies*; University of Copenhagen: Copenhagen, Denmark, 2013.
115. Hjerpsted, J.; Tholstrup, T. Cheese and cardiovascular disease risk: A review of the evidence and discussion of possible mechanisms. *Crit. Rev. Food. Sci. Nutr.* **2016**, *56*, 1389–1403. [CrossRef]
116. Feeney, E.L.; O'Sullivan, A.; Nugent, A.P.; McNulty, B.; Walton, J.; Flynn, A.; Gibney, E.R. Patterns of dairy food intake, body composition and markers of metabolic health in Ireland: Results from the National Adult Nutrition Survey. *Nutr. Diabetes* **2017**, *7*, e243. [CrossRef]
117. Rebholz, C.M.; Appel, L.J. Health effects of dietary patterns: Critically important but vastly understudied. *Am. J. Clin. Nutr.* **2018**, *108*, 207–208. [CrossRef]

118. Markey, O.; Vasilopoulou, D.; Givens, D.I.; Lovegrove, J.A. Dairy and cardiovascular health: Friend or foe? *Nutr. Bull.* **2014**, *39*, 161–171. [CrossRef] [PubMed]
119. Raziani, F.; Ebrahimi, P.; Engelsen, S.B.; Astrup, A.; Raben, A.; Tholstrup, T. Consumption of regular-fat vs reduced-fat cheese reveals gender-specific changes in LDL particle size—A randomized controlled trial. *Nutr. Metab.* **2018**, *15*, 61. [CrossRef] [PubMed]

© 2020 by the authors. Licensee MDPI, Basel, Switzerland. This article is an open access article distributed under the terms and conditions of the Creative Commons Attribution (CC BY) license (http://creativecommons.org/licenses/by/4.0/).

Article

Socio-Demographic Characteristics, Dietary, and Nutritional Intakes of French Elderly Community Dwellers According to Their Dairy Product Consumption: Data from the Three-City Cohort

Hermine Pellay [1,2], Corinne Marmonier [2], Cécilia Samieri [1] and Catherine Féart [1,*]

1 INSERM, University Bordeaux, BPH, U1219, F-33000 Bordeaux, France; hermine.pellay@u-bordeaux.fr (H.P.); cecilia.samieri@u-bordeaux.fr (C.S.)
2 CNIEL, Service Recherche Nutrition-Santé, F-75009 Paris, France; cmarmonier@cniel.com
* Correspondence: catherine.feart-couret@u-bordeaux.fr

Received: 22 September 2020; Accepted: 4 November 2020; Published: 7 November 2020

Abstract: Few data are available regarding dietary habits of the elderly, especially about dairy products (DPs) (total DP and milk, fresh DP, and cheese), whereas these are part of healthy habits. The aim was to describe the socio-demographic characteristics, food, and nutritional intakes of elderly DP consumers. The sample consisted of 1584 participants from the Three-City-Bordeaux cohort (France), who answered a food frequency questionnaire and a 24-h dietary recall. Socio-demographic characteristics, practice of physical activity, Body Mass Index, and polymedication were registered. The sample was 76.2 years (SD 5.0 years) on average, 35% were in line with the French dietary guidelines for DP (3 or 4 servings of DP/day), while 49% were below, and 16% above. Women were significantly more likely to declare the highest total DP (≥4 times/day), milk (>1 time/day), and fresh DP (>1.5 times/day) frequency consumption. The highest cheese frequency consumers (>1.5 times/day) were more likely men, married, and ex-smokers. The highest frequency of fresh DP intake was significantly associated with the lowest energy and lipid intakes, and that of cheese with the highest consumption of charcuteries, meat, and alcohol. This cross-sectional analysis confirmed that the socio-demographics and dietary characteristics varied across DP sub-types consumed, which encourages individual consideration of these confounders in further analyses.

Keywords: dairy products; energy intake; food intakes; nutrient intakes; aging; population-based cohort

1. Introduction

Longevity has remarkably increased over the past decades, notably in developed countries. In France, healthy life expectancy was 63.9 years on average in 2018 and life expectancy at birth is expected to increase by 5 years between 2018 and 2050 for both genders. Moreover, it is estimated that more than one person out of four will be 65 years old in 2050 [1]. This increased proportion of older adults will result in increasing demands of healthcare and medical services. Therefore, maintaining healthy aging represents a tremendous social and economic challenge across the world [2].

Eating a well-balanced diet coupled with regular physical activity are well-known lifestyle factors to promote health; this holds to all age groups but is specifically crucial for healthy aging, which depends on lowering the risk of non-communicable diseases and on maintaining physical and mental capacities in the elderly [3]. Because of age-related physical, physiological, and psychosocial changes, meeting the nutritional needs of older adults through diet can be challenging. Dietary guidelines recommend a well-balanced diet including major food groups for appropriate intake of essential macro- and micro-nutrients [4,5]. As several nutrients (including vitamins D, B1, and B2; calcium; magnesium; and selenium) have been identified at risk of inadequate intake among older adults, it suggests that attention should be paid to the consumption of their main providers [6]. Therefore, dairy products (DPs), which provide proteins of high quality, and numerous nutrients, vitamins, and minerals [6–9], are part of most food-based dietary guidelines that promote a healthy diet [10–12]. Note that DP as a whole are a heterogeneous food group, which encompass milk, fresh DP (yogurt/cottage cheese/petit suisse), and cheese, and their nutrient contents vary according to the sub-type [13] and that lactose intolerance or allergies might reduce their consumption.

Regarding health, a higher DP consumption has been associated with several age-related benefits, such as a lower risk of death, hypertension, type 2 diabetes, and metabolic syndrome and improved bone health [14–19]. The type of DP appears as a key component of such associations [20,21]. For instance, in a meta-analysis on 938,415 participants and 93,518 mortality cases, Guo et al. reported a lack of association between total dairy (high- or low-fat) and milk with the risk of death, while an inverse association between total fermented dairy (including sour milk products, yogurt, or cheese; +20 g/day) and a significant 2% reduced risk of all-cause mortality and cardiovascular diseases [22]. Moreover, the foods consumed in combination with DP (i.e., the food matrix) [23], the dairy structure, and the SFA contents of these DPs appear also as key factors of potential DP-related health outcomes [24]. Although the DP fats content is mostly saturated (65%), it does not seem to adversely affect cardiovascular risk, while debate still remains regarding the SFA recommendations that should be applied, particularly among older adults [5,25–28].

To our knowledge, few studies so far have assessed the contribution of DP consumption on nutritional status (limited to vitamin and nutrient status) in older adults; these few existing studies have highlighted that DP consumption significantly contributed to the protein, SFA, B-, and D vitamins status depending on the DP sub-type among this vulnerable population [29,30]. No study has yet characterized, as a whole, the sociodemographic criteria, dietary patterns (i.e., describing the food group intakes), and nutrient intakes of elderly dairy consumers. Several reports have nevertheless pointed out the need for carefully considered gender, socio-demographic, socio-economic status, and lifestyle characteristics, which might improve the efficiency of targeted public health messages among the oldest old [31–33]. Therefore, the present study aimed to describe the socio-demographic characteristics, dietary habits, and nutrient intakes according to the frequency of consumption of total DP and DP sub-types of French older adults.

2. Methods

2.1. Study Overview

The Three-City Study (3C) is an ongoing population-based study conducted in three French cities (Bordeaux, Dijon, Montpellier, France). This cohort was initiated in 1999–2000 to study the vascular risk factors of dementia [34]. Its protocol was approved by the Consultative Committee for the Protection of Persons participating in Biomedical Research at Kremlin-Bicêtre and all participants gave written informed consent. Participants were randomly sampled from electoral rolls. To be eligible, participants had to be 65 years and older at the time of recruitment and not institutionalized. Among the 9294 participants, 2104 were from the Bordeaux center where the initial data collection was completed in 2001–2002 (wave 1) with a comprehensive dietary survey among 1755 participants.

2.2. Assessment of Food and Nutrient Intakes

2.2.1. Dairy Products

A team of trained dieticians visited all participants at home between 2001 and 2002. Two types of dietary surveys were administered during face-to-face interviews to assess dietary habits. First, a Food Frequency Questionnaire (FFQ) allowed assessment of the daily frequency consumption of 148 foods and beverages (with frequencies assessed in 11 classes, from "never or less than once a month" to "7 times per week") during each of the six meals/snacks of the day, as previously detailed [35]. Regarding DP, the following items were considered: consumption of "coffee with milk", "tea with milk", "chocolate", "chicory", "natural milk or with cereal", and "milk" were considered by adding each response in a single variable called "milk"; those of "yogurt and cottage cheese" were considered as the "fresh DP category" while those of "cheese" were classified as the "cheese" category.

In addition to the FFQ, a 24-h dietary recall was administered at home [36]. Briefly, it allowed estimation of the total amount of all foods and beverages spontaneously ingested the day before the interview, and during and between meals; the 24-h recall was complementary to the FFQ, as it provided greater detail in the food items evaluated along with the quantities consumed daily. No weekend day was recorded. Photographs were used to precisely assess quantities [36]. Therefore, the total amount of DP and of each DP sub-type can account for servings (i.e., amount) and then be compared with the French recommended dietary allowances (RDAs) applyied in 2001 and still in progress today [4].

Using the 24-h dietary recall, 673 foods and beverages were spontaneously reported and we identified 7 items that could be attributed to the "milk" category (expressed in mL); 19 items that could be attributed to the "fresh DP category", including cottage cheese and petit-suisse (expressed in g); and 47 items that could be attributed to the "cheese" category (expressed in g). For each DP subclass, a typical serving was defined as follows: 150 mL of milk (category of milk); 15 g of concentrated milk/skimmed and semi skimmed milk powder (category of milk); 18 g of whole milk powder (category of milk); 125 g of yogurt (category of fresh DP); 100 g of cottage cheese/petit-suisse (category of fresh DP); and 30 g of cheese.

Data about food intakes from both dietary surveys were significantly correlated in an independent sub-sample of the 3C study [37].

2.2.2. Other Food Groups Intake

From the FFQ, we also considered the daily frequency consumption of 19 predetermined food groups, as follows: cereals/bread, pulses, pasta, potatoes, rice, biscuits/cakes, sweets/chocolate/soda, pizza/sandwich, raw vegetables/salad, cooked vegetables, fruits, charcuterie, fish/seafood, eggs, meat, poultry, coffee, tea, and alcohol [35]. As for DPs, all items were again recorded in 11 classes for each of the 3 main meals and 3 between-meal snacks.

2.2.3. Energy and Nutrient Intakes

From the 24-h dietary recall, as previously described, we used the BILNUT® software (SCDA Nutrisoft, Cerelles, France) to determine the total daily energy intake (without considering the energy provided by the alcohol intake), the daily macronutrients intake (i.e., carbohydrates, fatty acids (SFA, mono-unsaturated fatty acids (MUFAs) and polyunsaturated fatty acids (omega-3 and omega-6 PUFAs), proteins (from animal and vegetable origins)) and the daily micronutrients intake (including those relevant to the DP intake) [36]. We also identified participants consuming ≥1 g of proteins/kg of body weight/day and those consuming ≥1200 mg of calcium per day as participants in line with the current RDA for older adults, respectively [11,38,39].

2.3. Socio-Demographic and Lifestyle Characteristics

From the 3C database, we retained the following socio-demographic and lifestyle data: sex; age; education (in three categories: no education or primary school, secondary or high school, university); marital status (in four classes: married; divorced or separated; widowed; single); monthly income (in five classes: very low (less than 750€); low (750€ to 1500€); average (1500€ to 2250€); high (more than 2250€); refused to answer, including those who did not know their monthly income); polymedication, as the number of drugs/day ≥ 6; social isolation, combining living alone and feeling lonely "often enough" or "frequently"; smoking status (in three classes: never smoker; ex-smoker;

current smoker); stoutness according to measured BMI and using the most relevant thresholds for identifying malnutrition among older adults [40] (in three classes: thinness (if BMI < 20 kg/m^2 and age < 70 years) OR (if BMI < 22 kg/m^2 and age ≥ 70 years); normal (if BMI (20–27) kg/m^2 and age < 70 years) OR (if BMI (22–27) kg/m^2 and age ≥ 70 years); overweight/obesity if BMI > 27 kg/m^2); and practice of physical activity (in three classes: yes, no, no answer) [36,39].

2.4. Statistical Analyses

The SAS statistical software program (version 9.3; SAS Institute Inc., Cary, NC, USA) was used for statistical analyses.

We chose to divide the studied sample according to the usual frequency of consumption of (i) total DPs and (ii) milk, fresh DPs, and cheese, both evaluated by the FFQ: 3 categories per DP intakes were built, based on the quartile distribution of consumptions (low frequency: first quartile; moderate frequency: quartiles 2 and 3; high frequency: fourth quartile). This categorization ensured the identification of the most infrequent and frequent consumers. The FFQ database was preferred to define the main exposure, since a single 24-h dietary recall was available.

Then, socio-demographic characteristics, lifestyle, and dietary data (i.e., mean daily energy, macro- and micro-nutrient intakes from the 24-h recall, DPs, and all other food group consumptions from the FFQ) were described according to the 3 categories of frequency of consumption of total DPs and of DP subtypes.

Chi-Squared and ANOVA tests were used as appropriate. The Tukey–Kramer post hoc test was used to compare each mean between them (if ANOVA provided significant results). Statistical significance of different tests was accepted at *p*-value < 0.05.

3. Results

Among 1755 participants enrolled in the 3C Bordeaux cohort and followed at wave 1, 1606 answered the FFQ and 1658 answered the 24-h dietary recall, leading to a studied sample of 1584 participants with no missing data on the main exposure (i.e., total DP, milk, fresh DP, and cheese consumption) for the present analysis. The studied sample was 76.2 years old (SD 5.0 years) on average (ranging from 67.7 to 94.9 years), and 62.0% were women.

3.1. Total Dairy Products

Based on FFQ data, we stratified the sample as low daily frequency consumers of total DPs, such as those who consumed ≤ 2 times DPs per day (*n* = 394, 24.9% of the sample), moderate consumers who consumed 2–4 times DPs per day (*n* = 820, 51.8%), and high consumers who consumed ≥ 4 times DP per day (*n* = 370, 23.3%) (Table 1). Regarding the socio-demographic characteristics and lifestyle data, participants with the highest daily DP frequency intake were significantly more likely to be women (68.1% for the highest DP intake tertile, 56.6% for the lowest one), never smokers (68.4% for the highest DP intake tertile, 53.0% for the lowest one), and less often physically inactive (49.7% for the highest DP intake tertile, 59.7% for the lowest one) (Table 1).

Table 1. Socio-demographic and lifestyle characteristics across increasing daily frequency consumption of dairy products among elderly community dwellers from the 3C study, Bordeaux (France), 2001–2002, n = 1584.

	Total Dairy Products (Time/Day)				Milk (Time/Day)				Fresh Dairy Products (Time/Day)					Cheese (Time/Day)			
	≤2 n = 394	2–4 n = 820	≥4 n = 370	p	0 n = 456	0–1 n = 766	≥1 n = 362	p	<0.5 n = 428	0.5–1.5 n = 770	>1.5 n = 386	p	≤0.5 n = 317	0.5–1.5 n = 831	>1.5 n = 436	p
Sex, women	223 (56.6)	507 (61.8)	252 (68.1)	0.005	279 (61.2)	439 (57.3)	264 (72.9)	<0.0001	201 (47.0)	503 (65.3)	278 (72.0)	<0.0001	225 (70.9)	531 (63.9)	226 (51.8)	<0.0001
Age (years) (m (SD))	75.7 (4.9)	76.4 (5.0)	76.2 (4.9)	0.08	75.8 (4.7) ‡	76.1 (5.0)	76.7 (5.2)	0.03	75.9 (5.0)	76.3 (4.9)	76.3 (5.1)	0.38	75.9 (5.4)	76.3 (4.9)	76.1 (4.7)	0.21
Education				0.32				0.28				0.63				0.67
No/primary	123 (31.2)	284 (34.7)	119 (32.3)		139 (30.5)	267 (34.9)	120 (33.2)		140 (32.8)	267 (34.7)	119 (30.9)		95 (30.1)	284 (34.2)	147 (33.8)	
Secondary or High	191 (48.5)	384 (46.9)	192 (52.0)		223 (48.9)	360 (47.1)	184 (51.0)		207 (48.5)	361 (46.9)	199 (51.7)		163 (51.6)	392 (47.2)	212 (48.7)	
University	80 (20.3)	151 (18.4)	58 (15.7)		94 (20.6)	138 (18.0)	57 (15.8)		80 (18.7)	142 (18.4)	67 (17.4)		58 (18.3)	155 (18.6)	76 (17.5)	
Marital status				0.43				0.58				<0.0001				0.03
Married	222 (56.3)	454 (55.4)	181 (48.9)		247 (54.2)	424 (55.3)	186 (51.4)		261 (61.0)	425 (55.2)	171 (44.3)		160 (50.5)	436 (52.5)	261 (59.9)	
Divorced/separated	26 (6.6)	60 (7.3)	34 (9.2)		42 (9.2)	51 (6.7)	27 (7.5)		29 (6.8)	53 (6.9)	38 (9.8)		23 (7.3)	62 (7.4)	35 (8.0)	
Widowed	122 (31.0)	254 (31.0)	129 (34.9)		141 (30.9)	242 (31.6)	122 (33.7)		117 (27.3)	232 (30.1)	156 (40.5)		118 (37.2)	275 (33.1)	112 (25.7)	
Single	24 (6.1)	52 (6.3)	26 (7.0)		26 (5.7)	49 (6.4)	27 (7.5)		21 (4.9)	60 (7.8)	21 (5.4)		16 (5.0)	58 (7.0)	28 (6.4)	
Monthly income				0.25				0.29				<0.0001				0.12
Very low	25 (6.3)	56 (6.8)	30 (8.1)		34 (7.4)	49 (6.4)	28 (7.7)		18 (4.2)	63 (8.2)	30 (7.8)		19 (6.0)	61 (7.3)	31 (7.1)	
Low	108 (27.4)	245 (29.9)	122 (33.0)		122 (26.8)	232 (30.3)	121 (33.4)		123 (28.7)	216 (28.0)	136 (35.2)		91 (28.7)	261 (31.4)	123 (28.2)	
Average	104 (26.4)	211 (25.7)	83 (22.4)		113 (24.8)	207 (27.0)	78 (21.5)		118 (27.6)	204 (26.5)	76 (19.7)		83 (26.2)	199 (24.0)	116 (26.6)	
High	124 (31.5)	257 (31.4)	100 (27.0)		147 (32.2)	228 (29.8)	106 (29.3)		142 (33.2)	243 (31.6)	96 (24.9)		89 (28.1)	262 (31.5)	130 (29.8)	
Refused answer	33 (8.4)	51 (6.2)	35 (9.5)		40 (8.8)	50 (6.5)	29 (8.0)		27 (6.3)	44 (5.7)	48 (12.4)		35 (11.0)	48 (5.8)	36 (8.3)	
Drugs/day ≥6	148 (37.6)	308 (37.6)	163 (44.0)	0.08	168 (36.8)	305 (39.8)	146 (40.3)	0.50	153 (35.7)	306 (39.7)	160 (41.4)	0.22	122 (38.5)	322 (38.7)	175 (40.1)	0.86
Social isolation	38 (9.7)	57 (7.0)	32 (8.8)	0.25	41 (9.0)	49 (6.5)	37 (10.4)	0.06	42 (9.9)	44 (5.8)	41 (10.8)	0.004	32 (10.2)	71 (8.7)	24 (5.5)	0.049
Smoking status				<0.0001				0.0005				<0.0001				0.02

Table 1. Cont.

	Total Dairy Products (Time/Day)				Milk (Time/Day)				Fresh Dairy Products (Time/Day)				Cheese (Time/Day)			
	≤2 n = 394	2–4 n = 820	≥4 n = 370	p	0 n = 456	0–1 n = 766	>1 n = 362	p	<0.5 n = 428	0.5–1.5 n = 770	>1.5 n = 386	p	≤0.5 n = 317	0.5–1.5 n = 831	>1.5 n = 436	p
Never smoker	209 (53.0)	546 (66.6)	253 (68.4)		265 (58.1)	482 (62.9)	261 (72.1)		219 (51.2)	521 (67.7)	268 (69.4)		213 (67.2)	542 (65.2)	253 (58.0)	
Ex-smoker	152 (38.6)	236 (28.8)	105 (28.4)		160 (35.1)	241 (31.5)	92 (25.4)		175 (40.9)	215 (27.9)	103 (26.7)		86 (27.1)	244 (29.4)	163 (37.4)	
Current smoker	33 (8.4)	38 (4.6)	12 (3.2)		31 (6.8)	43 (5.6)	9 (2.5)		34 (7.9)	34 (4.4)	15 (3.9)		18 (5.7)	45 (5.4)	20 (4.6)	
Stoutness [1]				0.51				0.98				0.91				0.23
Thinness	45 (11.7)	90 (11.3)	45 (12.5)		52 (11.8)	84 (11.3)	44 (12.5)		50 (12.0)	87 (11.5)	43 (11.6)		35 (11.4)	93 (11.5)	52 (12.3)	
Normal	174 (45.3)	403 (50.5)	174 (48.5)		213 (48.3)	367 (49.1)	171 (48.4)		198 (47.5)	377 (50.0)	176 (47.6)		135 (43.8)	398 (49.2)	218 (51.4)	
Overweight/obesity	165 (43.0)	305 (38.2)	140 (39.0)		176 (39.9)	296 (39.6)	138 (39.1)		169 (40.5)	290 (38.5)	151 (40.8)		138 (44.8)	318 (39.3)	154 (36.3)	
Physical activity				0.02				0.18				0.19				0.29
Yes	103 (26.1)	224 (27.3)	102 (27.6)		127 (27.9)	196 (25.6)	106 (29.3)		111 (25.9)	214 (27.8)	104 (26.9)		89 (28.1)	221 (26.6)	119 (27.3)	
No	235 (59.7)	458 (55.9)	184 (49.7)		251 (55.0)	444 (58.0)	182 (50.3)		256 (59.8)	412 (53.5)	209 (54.2)		174 (54.9)	476 (57.3)	227 (52.1)	
Missing	56 (14.2)	138 (16.8)	84 (22.7)		78 (17.1)	126 (16.4)	74 (20.4)		61 (14.3)	144 (18.7)	73 (18.9)		54 (17.0)	134 (16.1)	90 (20.6)	

Values are n (%) except where mentioned [1] Stoutness was based on Body Mass Index (kg/m^2) and on "Global Leadership Initiative on Malnutrition" criteria: thinness (if BMI < 20 and if < 70 years) OR (if BMI < 22 AND if ≥ 70 years)/normal (if BMI (20–27) AND if < 70 years) or (if BMI (22–27) AND if ≥ 70 years)/overweight-obesity if BMI > 27 <1% missing values for social isolation and BMI, (1–5%) missing values for education ‡ mean value of low category was significantly different from high category (pairwise comparisons Tukey–Kramer test). BMI, Body Mass Index; SD, standard deviation.

Participants who declared the highest daily DP frequency intake also significantly reported a higher total energy intake (around +200 kcal/day for the highest DP intake tertile compared with the lowest one), and higher consumption of all macronutrients (including SFAs among total fatty acids and proteins from animal sources among total proteins) compared with others. Consistently, we observed that micronutrient intakes, such as calcium, phosphorus, zinc, and vitamins B1, B2, and B12, were significantly higher among participants with the highest frequency consumption of total DP, compared with others (Table 2).

In the study of all food groups recorded in the FFQ database, when the frequency of consumption of total DP was highest, the frequency of consumption of biscuits, sweets, and cooked vegetables was highest, while the frequency of consumption of charcuterie, meat, coffee, and alcohol was lowest (Table 3).

The consumed amounts of milk, fresh DPs, and cheese were significantly higher when the daily frequency consumption of total DP was the highest (Table 4). Participants with the highest frequency of total DP per day consumed 187 mL (SD 185 mL) of milk, 123 g (SD 111 g) of fresh DP, and 53 g (SD 45 g) of cheese per day on average.

3.2. Sub-Type of Dairy Products Consumed (Milk, Fresh DP, Cheese)

Based on the FFQ data, we stratified the studied sample as low daily frequency consumers of milk, fresh DP, and cheese when participants reported consuming 0 time/day for milk, and <0.5 time/day for fresh DP or cheese, respectively. The high frequency was respectively defined for consumptions of >1 time/day of milk, and >1.5 time/day of fresh DP or cheese (Table 1).

3.2.1. Milk

Regarding the socio-demographic characteristics of milk consumers, we observed that the mean age of participants and proportions of women and never smokers were significantly higher with the highest frequency consumption of milk (i.e., 76.7 years for the highest milk intake tertile vs. 75.8 years for the lowest ones, 72.9% women for the highest milk intake tertile vs. 61.2% for the lowest ones, 72.1% never smokers for the highest milk intake tertile vs. 58.1% for the lowest ones) (Table 1).

With regard to the daily frequency of milk consumption, marginal but significantly lower energy intake was observed among non-consumers of milk with 100 kcal/day less than other consumers (Table 2).

Mean intakes of carbohydrates, SFAs (+1.7 g/day between the highest milk intake tertile and the lowest ones), and proteins (+3.3 g/day between the highest milk intake tertile and the lowest ones) from animal sources were significantly higher with the higher frequency of milk consumption, while the total PUFAs, in particular the omega-6 PUFAs, intake was lower with a higher frequency of milk consumption (all p-value global < 0.05). The proportion of participants in line with the RDA for proteins significantly increased with the frequency of milk intake. Calcium, phosphorus, and vitamin B2 were the only micronutrients provided by DP whose intakes were higher with the higher frequency of milk consumption. The proportion of participants in line with the RDA for calcium significantly increased with the frequency of milk intake. Moreover, the frequency consumption of milk was not significantly associated with the frequency consumption of cheese, but the higher the frequency of milk consumption, the higher the frequency of fresh DP, biscuit, and sweet intakes (Table 3). On the other hand, a higher frequency of milk intake was significantly associated with a lower frequency intake of charcuterie, meat, coffee, and alcohol. A U-shaped relationship was observed between milk and tea intake (Table 3). The frequency consumption of all other food groups was not significantly associated with that of milk. Finally, the frequency of milk intake was not significantly associated with the amount of cheese consumed but was significantly associated with higher amounts of milk and fresh DP intake (Table 4).

Table 2. Daily energy, macro- and micro-nutrient intakes across increasing daily frequency consumption of dairy products among elderly community dwellers from the 3C study, Bordeaux (France), 2001–2002, $n = 1584$.

	Total Dairy Products (Time/Day)				Milk (Time/Day)				Fresh Dairy Products (Time/Day)					Cheese (Time/Day)			
	≤2 $n=394$	2–4 $n=820$	≥4 $n=370$	p	0 $n=456$	0–1 $n=766$	>1 $n=362$	p	<0.5 $n=428$	0.5–1.5 $n=770$	>1.5 $n=386$	p	≤0.5 $n=317$	0.5–1.5 $n=831$	>1.5 $n=436$	p
Energy (Kcal)	1624 (545)‡	1697 (528)§	1830 (549)	<0.0001	1645 (550)†	1745 (539)	1716 (532)	0.007	1755 (536)	1693 (539)	1692 (554)	0.049	1509 (483)†‡	1698 (535)§	1878 (542)	<0.0001
Macronutrients																
Carbohydrates (g)	182.9 (70.4)†‡	193.8 (67.2)§	211.6 (72.4)	<0.0001	183.0 (70.0)†‡	198.8 (70.1)	203.3 (67.5)	<0.0001	191.5 (69.8)	194.6 (68.1)	200.7 (73.4)	0.18	183.4 (67.0)†‡	195.1 (69.8)	204.2 (71.2)	0.0001
Total Fatty Acids (g)	56.8 (27.8)‡	59.2 (27.2)§	64.6 (28.3)	0.0001	58.2 (27.8)	61.1 (27.5)	59.3 (28.2)	0.11	63.2 (28.3)†‡	58.9 (27.3)	58.2 (27.8)	0.005	48.4 (21.9)†‡	60.1 (28.3)§	67.7 (27.7)	<0.0001
SFA (g)	23.4 (12.8)†‡	25.4 (12.7)§	29.1 (13.6)	<0.0001	24.6 (13.3)	26.1 (12.6)	26.3 (13.7)	0.03	26.8 (13.2)	25.4 (12.8)	25.2 (13.5)	0.07	19.7 (10.5)†‡	25.6 (12.9)§	30.4 (13.4)	<0.0001
MUFA (g)	20.6 (10.9)	21.3 (11.1)	22.3 (11.1)	0.06	20.8 (11.0)	21.9 (11.2)	20.9 (11.0)	0.16	22.9 (11.0)†‡	21.0 (11.0)	20.5 (11.2)	0.0002	17.6 (9.0)†‡	21.6 (11.5)§	23.7 (11.0)	<0.0001
PUFA (g)	8.7 (6.7)	8.2 (5.7)	8.6 (5.9)	0.51	8.4 (5.9)	8.7 (6.2)	7.9 (5.8)	0.01	9.1 (6.8)†	8.2 (5.7)	8.3 (5.5)	0.04	7.5 (5.2)†‡	8.6 (6.4)	8.9 (5.7)	<0.0001
Omega-3 PUFA (g)	1.35 (1.68)	1.18 (1.29)	1.26 (1.35)	0.28	1.29 (1.53)	1.27 (1.46)	1.12 (1.11)	0.22	1.42 (1.63)†	1.15 (1.33)	1.21 (1.28)	0.0002	1.07 (1.21)	1.27 (1.53)	1.30 (1.30)	<0.0001
Omega-6 PUFA (g)	6.7 (5.8)	6.4 (4.9)	6.6 (5.2)	0.79	6.5 (5.2)	6.7 (5.3)	6.1 (5.1)	0.01	7.0 (5.9)	6.3 (5.0)	6.3 (4.8)	0.13	5.7 (4.7)†‡	6.6 (5.5)	6.9 (5.0)	<0.0001
Proteins (g)	71.2 (26.8)‡	74.6 (26.0)§	81.6 (28.0)	<0.0001	73.3 (28.3)	76.1 (26.3)	76.5 (26.3)	0.053	76.3 (26.3)	74.7 (26.9)	75.8 (27.6)	0.39	67.5 (23.6)†‡	74.2 (26.6)§	83.4 (27.7)	<0.0001
≥1 g total protein/kg, n (%)	183 (46.8)	447 (55.2)	251 (68.6)	<0.0001	231 (51.2)	426 (56.3)	224 (62.4)	0.006	240 (56.6)	418 (54.6)	223 (59.0)	0.37	138 (43.8)	442 (54.0)	301 (69.5)	<0.0001
Animal sources (g)	49.9 (23.6)‡	53.1 (23.7)§	59.8 (24.8)	<0.0001	52.1 (25.2)	54.2 (23.8)	55.4 (23.7)	0.04	53.9 (23.2)	53.2 (24.5)	55.0 (24.7)	0.41	48.1 (21.8)†‡	52.6 (23.9)§	60.4 (25.1)	<0.0001
Vegetable sources (g)	21.2 (9.4)	21.5 (8.9)	21.9 (9.2)	0.56	21.2 (9.6)	21.9 (9.0)	21.2 (8.4)	0.13	22.4 (9.3)‡	21.5 (8.9)	20.7 (9.2)	0.02	19.4 (8.0)†‡	21.6 (9.2)§	23.0 (9.3)	<0.0001
Micronutrients																
Fibers (g)	17.4 (7.8)	17.4 (7.7)	17.5 (8.2)	0.99	17.5 (8.2)	17.4 (7.6)	17.3 (8.0)	0.90	18.1 (7.6)†	17.3 (7.4)	16.8 (8.8)	0.009	16.4 (7.5)‡	17.2 (7.7)§	18.5 (8.2)	0.001
Calcium (mg)	671 (373)†‡	854 (397)§	1096 (469)	<0.0001	752 (410)†‡	868 (419)§	1001 (459)	<0.0001	785 (436)†‡	859 (432)§	966 (420)	<0.0001	711 (328)†‡	847 (415)§	1012 (492)	<0.0001
≥1200 mg Calcium, n (%)	31 (7.9)	131 (16.0)	122 (33.0)	<0.0001	63 (13.8)	123 (16.1)	98 (27.1)	<0.0001	62 (14.5)	127 (16.5)	95 (24.6)	0.0003	25 (7.9)	136 (16.4)	123 (28.2)	<0.0001
Iron (mg)	11.0 (6.1)	11.0 (5.4)	11.3 (5.7)	0.63	11.0 (6.0)	11.1 (5.3)	10.9 (5.7)	0.38	11.7 (6.2)‡	11.0 (5.5)	10.5 (5.2)	0.001	10.1 (4.5)‡	10.9 (5.6)§	12.1 (6.2)	<0.0001
Phosphorus (mg)	998 (381)†‡	1093 (360)§	1272 (412)	<0.0001	1045 (401)†‡	1115 (369)§	1187 (407)	<0.0001	1089 (382)‡	1101 (391)	1157 (396)	0.03	980 (316)†‡	1094 (386)§	1241 (410)	<0.0001
Zinc (mg)	7.6 (7.0)	7.0 (6.1)§	8.1 (7.5)	0.02	8.4 (7.7)	6.8 (6.0)	7.5 (6.6)	0.07	7.2 (6.4)	7.3 (6.3)	7.8 (7.7)	0.99	6.8 (6.8)	7.4 (6.8)	7.9 (6.4)	<0.0001
Vit B1 (mg)	0.97 (0.44)‡	1.00 (0.42)§	1.10 (0.47)	0.0002	0.98 (0.44)	1.04 (0.45)	1.03 (0.43)	0.02	1.03 (0.45)	1.00 (0.42)	1.05 (0.47)	0.23	0.99 (0.43)	1.01 (0.44)	1.06 (0.44)	0.04
Vit B2 (mg)	1.36 (0.62)†‡	1.54 (0.67)§	1.81 (0.82)	<0.0001	1.45 (0.75)†‡	1.56 (0.66)§	1.71 (0.74)	<0.0001	1.44 (0.69)†‡	1.56 (0.70)§	1.69 (0.74)	<0.0001	1.48 (0.64)‡	1.52 (0.64)§	1.70 (0.86)	<0.0001

Table 2. Cont.

	Total Dairy Products (Time/Day)				Milk (Time/Day)				Fresh Dairy Products (Time/Day)				Cheese (Time/Day)			
	≤2 n = 394	2–4 n = 820	≥4 n = 370	p	0 n = 456	0–1 n = 766	>1 n = 362	p	<0.5 n = 428	0.5–1.5 n = 770	>1.5 n = 386	p	≤0.5 n = 317	0.5–1.5 n = 831	>1.5 n = 436	p
Vit PP (mg)	14.7 (6.9)	14.4 (6.6)	14.4 (7.3)	0.58	15.0 (7.5)	14.5 (6.5)	13.8 (6.7)	0.06	15.1 (6.8)	14.3 (6.5)	14.1 (7.5)	0.03	13.8 (6.3) ‡	14.3 (6.7) §	15.4 (7.4)	0.003
Vit B5 (mg)	3.6 (1.6) †‡	4.1 (1.6) §	4.7 (2.0)	<0.0001	3.7 (1.9) †‡	4.1 (1.6) §	4.5 (1.8)	<0.0001	3.9 (1.8) ‡	4.1 (1.6)	4.3 (1.9)	0.005	3.9 (1.6) ‡	4.0 (1.7) §	4.4 (2.0)	0.0002
Vit B6 (mg)	1.40 (0.58) ‡	1.43 (0.56) §	1.52 (0.62)	0.346	1.42 (0.61)	1.46 (0.58)	1.43 (0.56)	0.33	1.47 (0.59)	1.42 (0.53)	1.45 (0.65)	0.55	1.38 (0.54) ‡	1.41 (0.57) §	1.55 (0.61)	<0.0001
Vit B12 (μg)	5.5 (9.6)	5.3 (10.4)	6.2 (12.8)	0.03	6.1 (12.9)	5.2 (9.6)	5.6 (10.5)	0.49	5.8 (10.1)	5.7 (10.8)	5.1 (11.7)	0.002	4.9 (9.3) ‡	5.2 (9.3) §	6.8 (14.1)	<0.0001
Vit C (mg)	75.9 (57.2) ‡	83.6 (60.9)	87.6 (63.9)	0.02	78.2 (60.5)	83.2 (59.7)	86.9 (63.4)	0.03	77.6 (60.0)	82.9 (57.2)	87.5 (68.0)	0.04	86.3 (64.6)	81.5 (57.7)	82.0 (63.8)	0.44
Vit D (μg)	1.81 (3.31)	1.70 (2.60)	1.73 (2.49)	0.202	1.87 (3.34)	1.69 (2.54)	1.66 (2.40)	0.61	1.95 (3.44)	1.61 (2.32)	1.74 (2.75)	0.35	1.67 (2.58)	1.77 (3.07)	1.72 (2.26)	0.009
Vit E (mg)	6.4 (4.9)	6.5 (4.3)	6.5 (4.8)	0.15	6.5 (4.9)	6.7 (4.5)	6.0 (4.2)	0.02	6.5 (4.7)	6.4 (4.2)	6.6 (5.1)	0.99	6.2 (5.0)	6.5 (4.5)	6.5 (4.5)	0.01

Daily intakes are derived from the 24-h dietary recall and expressed as mean (Standard Deviation), except where mentioned Abbreviations: SFA Saturated Fatty Acids, MUFA Monounsaturated Fatty Acids, PUFA Polyunsaturated Fatty Acids, Vit Vitamin Missing values for (1–5)% regarding the consumption of proteins >1 g/d † mean value of low category was significantly different from middle category (pairwise comparisons Tukey–Kramer test) ‡ mean value of low category was significantly different from high category (pairwise comparisons Tukey–Kramer test) § mean value of middle category was significantly different from high category (pairwise comparisons Tukey–Kramer test).

Table 3. Mean weekly food groups' frequency consumption based on the daily frequency consumption of total dairy products and dairy product subtypes among elderly community dwellers from the 3C study, Bordeaux (France), 2001–2002, $n = 1584$.

Frequency (Time/Week)	Total Dairy Products (Time/Day)				Milk (Time/Day)				Fresh Dairy Products (Time/Day)				Cheese (Time/Day)			
	≤2 $n=394$	2–4 $n=820$	≥4 $n=370$	p	0 $n=456$	0–1 $n=766$	>1 $n=362$	p	<0.5 $n=428$	0.5–1.5 $n=770$	>1.5 $n=386$	p	≤0.5 $n=317$	0.5–1.5 $n=831$	>1.5 $n=436$	p
Milk	2.0 (3.2)†‡	6.3 (3.9)§	11.2 (6.5)	<0.0001	—	—	—	—	5.6 (5.3)†‡	6.6 (5.5)	6.6 (5.8)	0.003	6.9 (6.2)‡	6.5 (5.4)	5.7 (5.2)	0.02
Fresh DP	3.3 (3.6)†‡	7.2 (4.5)§	11.0 (5.0)	<0.0001	7.0 (4.9)§	7.0 (4.9)§	7.9 (5.0)	0.0005	—	—	—	—	8.6 (5.7)†‡	7.0 (4.6)§	6.1 (5.5)	<0.0001
Cheese	5.8 (4.2)†‡	7.4 (4.2)§	10.5 (4.7)	<0.0001	8.1 (4.9)	7.7 (4.5)	7.3 (4.5)	0.11	8.8 (4.7)†‡	7.4 (4.1)	7.1 (5.2)	<0.0001	—	—	—	—
Cereals, bread	18.2 (5.5)	18.6 (5.2)	18.9 (5.7)	0.18	18.2 (5.7)	18.7 (5.1)	18.8 (5.5)	0.23	18.8 (5.0)‡	18.9 (5.1)§	17.8 (6.2)	0.02	17.5 (6.0)†‡	18.4 (5.3)§	19.7 (4.8)	<0.0001
Pulses	0.6 (0.7)	0.6 (0.6)	0.6 (0.8)	0.92	0.6 (0.6)	0.6 (0.7)	0.6 (0.6)	0.91	0.7 (0.8)‡	0.6 (0.6)	0.5 (0.6)	0.001	0.5 (0.7)‡	0.6 (0.6)	0.7 (0.8)	0.002
Pasta	2.0 (1.5)	2.0 (1.4)	2.3 (1.8)	0.17	2.1 (1.5)	2.1 (1.4)	2.2 (1.7)	0.83	2.1 (1.5)	2.0 (1.4)	2.3 (1.8)	0.34	2.0 (1.4)‡	2.0 (1.4)§	2.4 (1.7)	0.004
Potatoes	2.4 (1.6)	2.7 (1.7)	2.7 (1.8)	0.07	2.6 (1.7)	2.6 (1.7)	2.7 (1.7)	0.62	2.7 (1.6)	2.6 (1.7)	2.5 (1.7)	0.40	2.4 (1.6)‡	2.6 (1.7)§	2.9 (1.8)	0.0002
Rice	1.3 (1.4)	1.3 (1.1)	1.4 (1.3)	0.09	1.4 (1.3)	1.3 (1.2)	1.3 (1.2)	0.83	1.2 (1.2)	1.3 (1.2)	1.4 (1.3)	0.04	1.2 (1.2)	1.3 (1.2)	1.4 (1.3)	0.02
Biscuits, cakes	1.7 (3.0)‡	2.2 (3.5)§	2.7 (4.1)	0.0006	1.8 (3.1)‡	2.2 (3.4)§	2.8 (4.2)	0.0003	2.0 (3.0)	2.2 (3.6)	2.4 (3.9)	0.18	2.0 (3.3)	2.3 (3.7)	2.2 (3.4)	0.39
Sweets, chocolate, soda	7.8 (6.2)‡	8.7 (6.7)§	10.2 (8.1)	0.0006	7.6 (6.2)†‡	8.8 (6.8)§	10.4 (8.0)	<0.0001	8.4 (6.2)	9.0 (6.9)	8.9 (7.8)	0.53	7.8 (6.6)‡	9.0 (7.0)	9.3 (7.1)	0.01
Pizza, sandwich	0.4 (0.7)	0.4 (0.8)	0.5 (0.9)	0.51	0.4 (0.8)	0.4 (0.7)	0.5 (0.9)	0.56	0.4 (0.6)	0.4 (0.8)	0.5 (1.0)	0.66	0.4 (0.7)	0.4 (0.8)	0.4 (0.7)	0.65
Raw vegetables, salad	8.7 (5.0)	9.3 (5.1)	8.9 (5.7)	0.07	9.1 (5.3)	9.1 (5.0)	9.0 (5.6)	0.62	8.9 (5.3)	9.0 (4.9)	9.3 (5.8)	0.70	8.7 (5.5)	9.2 (5.0)	9.1 (5.4)	0.12
Cooked vegetables	9.5 (4.3)†‡	10.3 (4.2)	10.6 (4.6)	0.01	9.8 (4.4)	10.4 (4.3)	10.1 (4.3)	0.09	9.8 (4.4)	10.2 (4.1)	10.5 (4.7)	0.15	10.2 (4.5)	10.0 (4.1)	10.4 (4.6)	0.56
Fruits	13.4 (7.1)	13.5 (6.5)	13.5 (7.3)	0.53	13.6 (7.3)	13.4 (6.5)	13.5 (7.0)	0.78	13.4 (6.5)	13.4 (6.9)	13.7 (7.2)	0.89	13.1 (6.6)	13.7 (7.0)	13.4 (6.8)	0.46
Charcuterie	1.9 (2.4)	1.6 (2.1)	1.5 (2.3)	0.03	1.9 (2.6)‡	1.7 (2.1)§	1.3 (1.8)	0.01	2.1 (2.5)‡§	1.6 (2.1)§	1.2 (2.1)	<0.0001	1.4 (2.0)‡	1.5 (2.0)§	2.0 (2.7)	0.0005
Fish, seafood	2.9 (1.8)	2.9 (1.7)	2.8 (1.8)	0.86	2.9 (1.9)	2.9 (1.7)	2.7 (1.7)	0.17	2.8 (1.7)	2.9 (1.7)	2.9 (1.9)	0.68	2.9 (1.9)	2.8 (1.7)	2.9 (1.8)	0.74
Eggs	1.4 (1.1)†	1.5 (1.1)	1.5 (1.2)	0.01	1.4 (1.0)	1.5 (1.2)	1.5 (1.1)	0.59	1.4 (1.2)	1.5 (1.1)	1.5 (1.1)	0.03	1.5 (1.1)	1.5 (1.0)	1.5 (1.3)	0.79
Meat	5.1 (2.7)†	4.6 (2.3)	4.9 (2.5)	0.04	5.0 (2.5)‡	4.8 (2.5)	4.5 (2.3)	0.04	5.4 (2.6)†‡	4.6 (2.3)	4.6 (2.4)	<0.0001	4.7 (2.6)‡	4.6 (2.3)§	5.3 (2.5)	<0.0001
Poultry	1.8 (1.2)	1.8 (1.2)	1.9 (1.4)	0.61	1.9 (1.3)	1.7 (1.2)	1.8 (1.3)	0.49	1.7 (1.2)	1.8 (1.3)	1.9 (1.4)	0.44	1.7 (1.3)‡	1.7 (1.2)§	2.0 (1.4)	0.02

Table 3. Cont.

Frequency (Time/Week)	Total Dairy Products (Time/Day)				Milk (Time/Day)				Fresh Dairy Products (Time/Day)				Cheese (Time/Day)			
	≤2 n = 394	2-4 n = 820	≥4 n = 370	p	0 n = 456	0-1 n = 766	>1 n = 362	p	<0.5 n = 428	0.5-1.5 n = 770	>1.5 n = 386	p	≤0.5 n = 317	0.5-1.5 n = 831	>1.5 n = 436	p
Coffee	6.9 (5.5) †‡	5.3 (4.7)	5.0 (5.3)	<0.0001	8.1 (5.6) †‡	4.8 (4.6)	4.5 (4.5)	<0.0001	5.4 (5.0)	5.7 (4.9)	6.0 (5.6)	0.46	5.3 (5.1)	5.7 (4.9)	5.9 (5.6)	0.33
Tea	2.8 (4.6)	2.6 (4.3)	2.9 (4.9)	0.92	3.3 (4.6) †	2.4 (4.4)	2.8 (4.7)	0.0006	2.2 (4.2) †‡	2.9 (4.6)	3.1 (4.7)	0.002	2.9 (4.7)	2.8 (4.6)	2.5 (4.2)	0.28
Alcohol	11.9 (13.6) †‡	9.6 (10.9)	8.9 (10.6)	0.009	11.4 (13.7) ‡	10.6 (11.2) §	6.9 (8.5)	<0.0001	13.4 (14.0) †‡	9.3 (10.5) §	7.6 (9.9)	<0.0001	8.0 (10.1) ‡	8.9 (9.9) §	13.4 (14.6)	<0.0001

† mean value of the low category was significantly different from the middle category (pairwise comparisons Tukey–Kramer test). ‡ mean value of the low category was significantly different from the high category (pairwise comparisons Tukey–Kramer test) § mean value of the middle category was significantly different from the high category (pairwise comparisons Tukey–Kramer test). Values are mean (Standard Deviation) Abbreviations: DP Dairy Products.

Table 4. Mean daily dairy product (and subtypes) intakes based on the weekly frequency consumption of total dairy products and dairy product subtypes among elderly community dwellers from the 3C study, Bordeaux (France), 2001–2002, $n = 1584$.

Daily Intake	Total Dairy Products (Time/Day) *				Milk (Time/Day)				Fresh Dairy Products (Time/Day)				Cheese (Time/Day)			
	≤2 n = 394	2-4 n = 820	≥4 n = 370	p	0 n = 456	0-1 n = 766	>1 n = 362	p	<0.5 n = 428	0.5-1.5 n = 770	>1.5 n = 386	p	≤0.5 n = 317	0.5-1.5 n = 831	>1.5 n = 436	p
Milk (mL)	43.1 (105.4) †‡	111.5 (145.1) §	186.7 (185.3)	<0.0001	78.8 (107.6) †‡	126.5 (147.2) §	217.3 (181.3)	<0.0001	97.7 (144.7)	119.3 (162.8)	113.5 (150.8)	0.02	122.4 (160.2)	111.3 (151.4)	105.9 (159.2)	0.10
Fresh Dairy Products (g)	43.2 (84.3) †‡	83.1 (99.0) §	122.6 (110.9)	<0.0001	42.9 (44.9)	80.1 (97.8)	91.9 (104.2)	0.04	15.1 (47.2) †‡	80.3 (88.4) §	161.3 (116.8)	<0.0001	103.0 (116.9) †‡	82.5 (96.7) §	67.1 (98.6)	<0.0001
Cheese (g)	33.1 (40.6) †‡	39.3 (39.4) §	52.6 (45.2)	<0.0001	11.4 (13.7)	39.6 (39.3)	41.1 (42.3)	0.83	48.1 (43.9) †‡	39.3 (40.9)	36.3 (39.6)	<0.0001	10.6 (22.4) †‡	38.1 (36.7) §	68.3 (44.1)	<0.0001

* French Dairy Products intake recommendations at the date of the dietary surveys (2001) were 3–4 servings of DP/day (whatever the subclass) Daily consumed amounts are derived from the 24-h dietary recall and expressed as mean (Standard Deviation) † mean value of the low category was significantly different from the middle category (pairwise comparisons Tukey–Kramer test) ‡ mean value of the low category was significantly different from the high category (pairwise comparisons Tukey–Kramer test) § mean value of the middle category was significantly different from the high category (pairwise comparisons Tukey–Kramer test).

3.2.2. Fresh DP

Second, regarding the frequency consumption of fresh DP, sex, marital status, and income were all significantly associated with fresh DP intake: participants with the highest fresh DP frequency consumption were more often women (72.0% for the highest fresh DP intake tertile vs. 47.0% for the lowest ones), widowed (40.5% for the highest fresh DP intake tertile vs. 27.3% for the lowest ones), and reported the lowest incomes. Among other characteristics, the frequency of consumption of fresh DP was significantly associated with social isolation and smoking status; the moderate consumers being less isolated (+1% of isolated participants with highest fresh DP intakes compared with the lowest ones), and the lowest fresh DP consumers more often being current or ex-smokers than the others (Table 1).

The frequency of fresh DP intake was significantly associated with the reported daily total energy intake of participants; a higher mean energy intake was reported among participants with the lowest frequency consumption of fresh DP. The consumption of total fatty acids, including MUFAs, total PUFAs, and omega-3 PUFAs, proteins from vegetable origins, and fiber were significantly lower among participants with the highest frequency consumption of fresh DP (Table 2). Again, the reported consumptions of calcium, phosphorus, and vitamin B2, in part provided by DP, were the highest when the frequency of fresh DP consumption was the highest. The higher frequency consumption of fresh DP was significantly associated with lower intakes of fiber, iron, and vitamins PP and B12 (Table 2). The proportion of participants in line with the RDA for calcium, but not for protein, significantly increased with the frequency of fresh DP intake.

The frequency consumption of fresh DP was significantly positively associated with that of milk while inversely associated with that of cheese (Table 3). The consumed amount of fresh DP was significantly higher among participants with the highest frequency consumption of milk and lower among participants with the highest frequency consumption of cheese, compared with the lowest frequency consumers (92 g/day vs. 43 g/day on average and 67 g/day vs. 103 g/day on average, respectively) (Table 4). In the study of all food groups recorded in the FFQ database, when the frequency of consumption of fresh DP was highest, the frequency consumption of rice, eggs, and tea was highest, while the frequency consumption of cereals, pulses, charcuterie, meat, and alcohol was lowest. The frequency consumption of all other food groups was not significantly associated with that of fresh DP (Table 3).

3.2.3. Cheese

Third, regarding the frequency of cheese intake, participants with the highest report were significantly more often men (48.2% men for the highest cheese intake tertile vs. 29.1% for the lowest ones) and married (59.9% married men for the highest cheese intake tertile vs. 50.5% for the lowest ones). The frequency consumption of cheese was significantly associated with social isolation and smoking status: participants with the highest frequency of cheese intake were less often isolated (5.5% for the highest cheese intake tertile vs. 10.2% for the lowest ones) and never smokers (58.0% for the highest cheese intake tertile vs. 67.2% for the lowest ones) (Table 1).

The reported daily total energy intake of participants was significantly associated with their cheese intake, as the highest consumers reported 370 kcal/day more than the lowest consumers on average. The consumption of carbohydrates, total and all sub-types fatty acids, proteins (from animal and vegetable sources), fiber, calcium, phosphorus, zinc, and all vitamins provided in part by DP were higher among participants with the highest frequency of cheese consumption (Table 2). The proportion of participants in line with the RDA for proteins and calcium significantly increased with the frequency of cheese intake. The frequency of consumption of cheese was inversely significantly associated with that of milk and fresh DP. This association was only observed regarding the consumed amount of fresh DP (67 g/day vs. 103 g/day on average for the highest vs. the lowest frequency of cheese consumption categories) but not the consumed amount of milk (Table 4).

When the frequency of consumption of cheese was highest, the frequency consumption of cereals, pulses, pasta, potatoes, rice, sweets, charcuterie, meat, poultry, and alcohol was highest. The frequency consumption of all other food groups was not significantly associated with that of cheese (Table 3).

4. Discussion

In this large sample of French elderly community dwellers, we observed that DP frequency consumption was associated with several socio-demographic, dietary characteristics, and lifestyle factors, with specificities according to each DP sub-type. Gender and smoking status appeared as key factors both associated with total DP and each DP sub-type intake, while marital status and social isolation were only associated with fresh DP and cheese frequency consumption, in the opposite direction. Overall, it appears from these results that cheese consumers differed from that of milk and fresh DP: a higher cheese frequency consumption was observed among men, married, less isolated, and more often smokers. Regarding dietary data, both food group and nutrient intakes differed according to the DP sub-type consumed. The fresh DP frequency consumers exhibited different dietary patterns than milk or cheese consumers as observed on the frequency consumptions of cereals, pulses, sweets and chocolate, eggs, and tea. As a consequence, these differences were also observed on a majority of nutrients, except for calcium, phosphorus, and vitamin B2, whose consumptions were always significantly higher regardless of the higher frequency of milk, fresh DP, or cheese consumption.

Few studies have characterized DP consumers, particularly among older adults in France [9,41]. In a previous study focusing on elderly people enrolled in a population-based cohort in south-east France and implemented in 2002 (i.e., at the same time as the present dietary survey) [29], DP consumption appeared as a major provider of both SFA and protein (mainly from animal sources) intakes. This was in accordance with results from the present study, while we added that among DP sub-types, the highest frequency consumption of fresh DP was not the main provider of these particular nutrients. Indeed, specific DP dietary patterns were observed here, since higher frequency consumers of fresh DP were also higher frequency consumers of milk, while higher frequency consumers of cheese were the lowest frequency consumers of milk and fresh DP.

Interestingly, from a recent national survey [42], it appeared that among participants aged 55 to 79 years in 2014, only 19% were aware of the French national guidelines, and 64% reported lower estimates than guidelines. The same results were reported earlier in another national sample of French elderly participants, suggesting that the advancement of knowledge, and possibly, as a consequence, of eating habits, may not yet have improved over time [43]. However, being high consumers of total DP or DP sub-types significantly increased the proportion of participants in line with the national total protein and calcium RDA. This would suggest (i) encouraging the consumption of total DP and particularly of milk and cheese, among this vulnerable population, to ensure adequate intake of protein and calcium [6], and (ii) modifying the guidelines about DP among older adults. However, this would also encourage a higher SFA intake, already above the recommendations among this sample as previously reported [36] and which may be not desired [25–27]. On the other hand, the various dietary patterns of DP consumers, whatever the sub-type, hence the multiple providers of SFA, complexed the picture further [23,28]. The best way to communicate about these recommendations on total DP, DP sub-types, and protein and calcium intake remains a public health challenge [5]. Indeed, when comparing the present results established on a sample of older people 67 years and over in 2001 with recent ones, we emphasize the secular trend for a decreased consumption of total DP over time in France. However, we also already described that the intake of major food groups appeared relatively stable during a follow-up in 3C-Bordeaux [44]. Despite the traditional French culinary cultural habits, two national surveys (i.e., the INCA2 and INCA3 studies) also reported that skipping breakfast (usually associated with a higher consumption of milk) becomes common, as well the simplification of main meals characterized by a single dish and therefore the absence of dessert, and possibly of yogurt [9,45].

Regarding the dietary patterns of the studied sample, we observed that the other recorded food groups' intake was distinctive features of each DP sub-type consumer. Briefly, the highest frequency consumers of milk faced a "biscuits and snacking" pattern, already identified among this cohort [35], of mainly women, who we could imagine dipping their biscuits in the milk. For the highest frequency consumers of fresh DP, we would be in the presence of a "low total energy intake" pattern, described as widowed and isolated women with low incomes. This can be compared with the "small eaters" pattern already characterized among this cohort [35]. For the highest frequency consumers of cheese, their overall dietary pattern referred to a "bon vivant" pattern, mainly characterized by men, who we could imagine consuming cheese in a friendly atmosphere, eating a piece of bread, a piece of sausage, drinking wine, and smoking. This last pattern could be compared with the "charcuterie-meat-alcohol" dietary pattern already identified by another statistical approach in this cohort but considering total DP intakes [35]. Here, it appears that "cheese" could be considered as the fourth component of such a dietary pattern, also known as the "traditional pattern" or "western diet" [46], and encourages a split of the total DP food group as separated components to build data-driven dietary patterns. It should be acknowledged that high cheese consumption is a hallmark of French dietary habits. Already, in 2009, Sofi et al. reported that Greece and France were countries from the Mediterranean basin with the highest consumption of cheese [47]. More recently, a report among the SHARE database reported considerable heterogeneity in DP consumption across Europe, with higher levels in central and northern countries and in Spain, and the lowest prevalence of dairy intake in eastern European countries [48]. Finally, the EFSA survey also reported that France and Italy were both countries with a large consumption of cheese, and that France is represented by low consumers of milk [49]. Altogether, the present results were in line with these previous observations.

As expected, several socio-demographic and lifestyle characteristics were associated with the consumption of total DP in the present studied sample, and our data added details on their associations with DP sub-types. Indeed, gender is a largely recognized factor associated with dietary habits, and our results confirmed that men were more likely high-frequency cheese consumers than women, who in turn were more often classified as higher milk and fresh DP frequency consumers in this sample [35,50]. An association between the frequency consumption of cheese and income would have been expected [51]: the maxim 'there is no good meal without cheese' appears as a key determinant of the dietary habits of these French participants, whatever the expensive costs of cheese [36]. Decreased perceived attractiveness of food with increased age in terms of taste, appetite, and palatability of food was also commonly admitted [52]. It may encourage elderly persons to choose more tasty cheese in addition to their traditional habits. Finally, smoking status was also differentially associated with the frequency of DP sub-types, as already observed in a previous study reporting that French and worldwide yogurt consumers, more often never smokers, had a better quality diet and lifestyle than non-consumers [53]. Across Europe, gender and age have also been associated with different total DP intakes, with women being greater consumers than men and older adults of 80 years and more being lesser consumers than their younger counterparts [48]. Among environmental factors, the influence of family relations on DP intakes has been reported, such as, for instance, the similarity between mothers and daughters in dairy-related dietary patterns [54]. In the present study, family relations were only assessed by marital status (including married, widowed, or separated and single people). The influence of family relations on DP intakes was illustrated by the fact that men were more often married and cheese consumers, and women more often widowed and fresh DP consumers.

We acknowledge that the accuracy of food intake assessment is crucial in dietary studies, and that performing a single 24-h dietary recall may have induced underestimations of nutrient intakes and intra-individual variations. This methodology also prevented us from assessing the possible loss of vitamins, minerals, and energy between the two surveys. However, a large sample size, even a single dietary survey, may be used to determine the average intake in defined subgroups of a population [55]. Moreover, results from the present study were in line with a previous national report (i.e., INCA 3 study, implemented in 2017 and using a quantitative dietary approach), where the consumed amounts of milk, fresh DP, and cheese were quite similar [42]. Finally, the 24-h dietary recall was administered at the same time as a comprehensive FFQ to collect weekly eating habits, and both surveys exhibited a high concordance between several food groups and nutrient intakes [37,44,56]. Since the present study is cross-sectional and observational, it prevented us from drawing definite conclusions on the associations between DP intakes and socio-demographics, lifestyle, or dietary data and some residual confounding could also explain our observations. The delay of 18 years between the 2 dietary surveys might have decreased the relevance of the present findings, while (i) the French RDA applied in 2001 for older adults is still in progress in 2020, (ii) the DP (and mainly cheese) intakes are part of the hallmark of French dietary habits [47], and it is unlikely that the characteristics of DP consumers have changed dramatically during this period, and (iii) understanding the correlates of DP consumption in year 2001–2002 can still inform today's DP consumption in the context of the life course approach of nutrition on health. Therefore, collecting this much data appears valuable and can still be informative. Finally, the representative nature of the sample needs to be established before our results can be extended to a larger sample of French elderly as a whole and conclusions drawn with regard to the prevention of inappropriate nutrient intake. Therefore, our results cannot be generalized to populations from different geographic areas with different socio-demographic backgrounds and/or cultural dietary habits. The strengths of the present study included the large sample size, the use of complementary dietary surveys, and the involvement of elderly community dwellers for whom DP recommendations appeared essential to prevent inadequate nutrient intake and possibly disease onset. Finally, this kind of study about non-dietary factors related to total DP and DP sub-type intakes remains strongly limited.

Thanks to the present cross-sectional study, it was possible to identify socio-demographic characteristics and lifestyle factors associated with quantitative and qualitative DP intakes in a French elderly group. It appears that each DP sub-type was also part of distinctive dietary patterns, which encourages individual consideration of these food groups in further analyses on nutrient adequacy among older adults.

Author Contributions: Conceptualization, C.F.; Formal analysis, H.P.; Funding acquisition, C.F.; Methodology, C.M., C.S. and C.F.; Supervision, C.F.; Validation, C.M. and C.S.; Writing – original draft, H.P. and C.F.; Writing–review and editing, C.M., C.S. and C.F. All authors have read and agreed to the published version of the manuscript.

Funding: This research was funded by the Centre National Interprofessionnel de l'Economie Laitière (CNIEL) grant number (2018-170).

Acknowledgments: The 3C Study is conducted under a partnership agreement between the Institut National de la Santé et de la Recherche Médicale (INSERM), the Victor Segalen–Bordeaux II University and the Sanofi-Synthélabo Company. The Fondation pour la Recherche Médicale funded the preparation and initiation of the study. The 3C-Study is also supported by the Caisse Nationale Maladie des Travailleurs Salariés, Direction Générale de la Santé, Conseils Régionaux of Aquitaine and Bourgogne, Fondation de France, Ministry of Research INSERM Programme 'Cohortes et collections de données biologiques', the Mutuelle Générale de l'Education Nationale, the Fondation Plan Alzheimer (FCS 2009–2012), and the Caisse Nationale pour la Solidarité et l'Autonomie (CNSA).

Conflicts of Interest: The authors declare no conflict of interest regarding the presented work.

References

1. DoEaS, U.N. World Population Prospects: The 2017 Revision, Key Findings and Advance Tables. Available online: https://population.un.org/wpp/Publications/Files/WPP2019_10KeyFindings.pdf (accessed on 17 May 2020).
2. World Health Organization. "Aging Well" Must Be a Global Priority. 2014. Available online: https://www.who.int/news-room/detail/06-11-2014--ageing-well-must-be-a-global-priority (accessed on 29 October 2020).
3. Di Ciaula, A.; Portincasa, P. The environment as a determinant of successful aging or frailty. *Mech. Ageing Dev.* **2020**, *188*, 111244. [CrossRef]
4. France, S.P. Manger Bouger. Available online: https://www.mangerbouger.fr/PNNS (accessed on 29 October 2020).
5. van Staveren, W.A.; de Groot, L.C. Evidence-based dietary guidance and the role of dairy products for appropriate nutrition in the elderly. *J. Am. Coll. Nutr.* **2011**, *30*, 429S–437S. [CrossRef] [PubMed]

6. ter Borg, S.; Verlaan, S.; Hemsworth, J.; Mijnarends, D.M.; Schols, J.M.; Luiking, Y.C.; de Groot, L.C. Micronutrient intakes and potential inadequacies of community-dwelling older adults: A systematic review. *Br. J. Nutr.* **2015**, *113*, 1195–1206. [CrossRef]
7. Quann, E.E.; Fulgoni, V.L., 3rd; Auestad, N. Consuming the daily recommended amounts of dairy products would reduce the prevalence of inadequate micronutrient intakes in the United States: Diet modeling study based on NHANES 2007–2010. *Nutr. J.* **2015**, *14*, 90. [CrossRef] [PubMed]
8. Pereira, P.C. Milk nutritional composition and its role in human health. *Nutrition* **2014**, *30*, 619–627. [CrossRef] [PubMed]
9. Coudray, B. The contribution of dairy products to micronutrient intakes in France. *J. Am. Coll. Nutr.* **2011**, *30*, 410S–414S. [CrossRef] [PubMed]
10. Estaquio, C.; Castetbon, K.; Kesse-Guyot, E.; Bertrais, S.; Deschamps, V.; Dauchet, L.; Peneau, S.; Galan, P.; Hercberg, S. The French National Nutrition and Health Program score is associated with nutritional status and risk of major chronic diseases. *J. Nutr.* **2008**, *138*, 946–953. [CrossRef] [PubMed]
11. Estaquio, C.; Kesse-Guyot, E.; Deschamps, V.; Bertrais, S.; Dauchet, L.; Galan, P.; Hercberg, S.; Castetbon, K. Adherence to the French Programme National Nutrition Sante Guideline Score is associated with better nutrient intake and nutritional status. *J. Am. Diet. Assoc.* **2009**, *109*, 1031–1041. [CrossRef]
12. Savaiano, D.A.; Hutkins, R.W. Yogurt, cultured fermented milk, and health: A systematic review. *Nutr. Rev.* **2020**. [CrossRef]
13. Gaucheron, F. Milk and dairy products: A unique micronutrient combination. *J. Am. Coll. Nutr.* **2011**, *30*, 400S–409S. [CrossRef]
14. Rizzoli, R. Dairy products, yogurts, and bone health. *Am. J. Clin. Nutr.* **2014**, *99*, 1256S–1262S. [CrossRef] [PubMed]
15. Dehghan, M.; Mente, A.; Rangarajan, S.; Sheridan, P.; Mohan, V.; Iqbal, R.; Gupta, R.; Lear, S.; Wentzel-Viljoen, E.; Avezum, A.; et al. Association of dairy intake with cardiovascular disease and mortality in 21 countries from five continents (PURE): A prospective cohort study. *Lancet* **2018**, *392*, 2288–2297. [CrossRef]
16. Trichia, E.; Luben, R.; Khaw, K.T.; Wareham, N.J.; Imamura, F.; Forouhi, N.G. The associations of longitudinal changes in consumption of total and types of dairy products and markers of metabolic risk and adiposity: Findings from the European Investigation into Cancer and Nutrition (EPIC)-Norfolk study, United Kingdom. *Am. J. Clin. Nutr.* **2020**, *111*, 1018–1026. [CrossRef] [PubMed]
17. Thorning, T.K.; Raben, A.; Tholstrup, T.; Soedamah-Muthu, S.S.; Givens, I.; Astrup, A. Milk and dairy products: Good or bad for human health? An assessment of the totality of scientific evidence. *Food Nutr. Res.* **2016**, *60*, 32527. [CrossRef] [PubMed]
18. Bhavadharini, B.; Dehghan, M.; Mente, A.; Rangarajan, S.; Sheridan, P.; Mohan, V.; Iqbal, R.; Gupta, R.; Lear, S.; Wentzel-Viljoen, E.; et al. Association of dairy consumption with metabolic syndrome, hypertension and diabetes in 147 812 individuals from 21 countries. *BMJ Open Diabetes Res. Care* **2020**, *8*. [CrossRef] [PubMed]
19. Guo, J.; Givens, D.I.; Astrup, A.; Bakker, S.J.L.; Goossens, G.H.; Kratz, M.; Marette, A.; Pijl, H.; Soedamah-Muthu, S.S. The Impact of Dairy Products in the Development of Type 2 Diabetes: Where Does the Evidence Stand in 2019? *Adv. Nutr.* **2019**, *10*, 1066–1075. [CrossRef]
20. Ding, M.; Li, J.; Qi, L.; Ellervik, C.; Zhang, X.; Manson, J.E.; Stampfer, M.; Chavarro, J.E.; Rexrode, K.M.; Kraft, P.; et al. Associations of dairy intake with risk of mortality in women and men: Three prospective cohort studies. *BMJ* **2019**, *367*, l6204. [CrossRef]
21. Drouin-Chartier, J.P.; Li, Y.; Ardisson Korat, A.V.; Ding, M.; Lamarche, B.; Manson, J.E.; Rimm, E.B.; Willett, W.C.; Hu, F.B. Changes in dairy product consumption and risk of type 2 diabetes: Results from 3 large prospective cohorts of US men and women. *Am. J. Clin. Nutr.* **2019**, *110*, 1201–1212. [CrossRef] [PubMed]
22. Guo, J.; Astrup, A.; Lovegrove, J.A.; Gijsbers, L.; Givens, D.I.; Soedamah-Muthu, S.S. Milk and dairy consumption and risk of cardiovascular diseases and all-cause mortality: Dose-response meta-analysis of prospective cohort studies. *Eur. J. Epidemiol.* **2017**, *32*, 269–287. [CrossRef] [PubMed]
23. Astrup, A.; Geiker, N.R.W.; Magkos, F. Effects of Full-Fat and Fermented Dairy Products on Cardiometabolic Disease: Food Is More Than the Sum of Its Parts. *Adv. Nutr.* **2019**, *10*, 924S–930S. [CrossRef]

24. Thorning, T.K.; Bertram, H.C.; Bonjour, J.P.; de Groot, L.; Dupont, D.; Feeney, E.; Ipsen, R.; Lecerf, J.M.; Mackie, A.; McKinley, M.C.; et al. Whole dairy matrix or single nutrients in assessment of health effects: Current evidence and knowledge gaps. *Am. J. Clin. Nutr.* **2017**, *105*, 1033–1045. [CrossRef]
25. Kris-Etherton, P.M.; Krauss, R.M. Public health guidelines should recommend reducing saturated fat consumption as much as possible: YES. *Am. J. Clin. Nutr.* **2020**. [CrossRef] [PubMed]
26. Krauss, R.M.; Kris-Etherton, P.M. Public health guidelines should recommend reducing saturated fat consumption as much as possible: NO. *Am. J. Clin. Nutr.* **2020**. [CrossRef]
27. Krauss, R.M.; Kris-Etherton, P.M. Public health guidelines should recommend reducing saturated fat consumption as much as possible: Debate Consensus. *Am. J. Clin. Nutr.* **2020**. [CrossRef]
28. Astrup, A.; Magkos, F.; Bier, D.M.; Brenna, J.T.; de Oliveira Otto, M.C.; Hill, J.O.; King, J.C.; Mente, A.; Ordovas, J.M.; Volek, J.S.; et al. Saturated Fats and Health: A Reassessment and Proposal for Food-Based Recommendations: JACC State-of-the-Art Review. *J. Am. Coll. Cardiol.* **2020**, *76*, 844–857. [CrossRef]
29. Carriere, I.; Delcourt, C.; Lacroux, A.; Gerber, M.; Group, P.S. Nutrient intake in an elderly population in southern France (POLANUT): Deficiency in some vitamins, minerals and omega-3 PUFA. *Int. J. Vitam. Nutr. Res.* **2007**, *77*, 57–65. [CrossRef]
30. Laird, E.; Casey, M.C.; Ward, M.; Hoey, L.; Hughes, C.F.; McCarroll, K.; Cunningham, C.; Strain, J.J.; McNulty, H.; Molloy, A.M. Dairy Intakes in Older Irish Adults and Effects on Vitamin Micronutrient Status: Data from the TUDA Study. *J. Nutr. Health Aging* **2017**, *21*, 954–961. [CrossRef]
31. Hiza, H.A.; Casavale, K.O.; Guenther, P.M.; Davis, C.A. Diet quality of Americans differs by age, sex, race/ethnicity, income, and education level. *J. Acad. Nutr. Diet.* **2013**, *113*, 297–306. [CrossRef]
32. Kang, M.; Park, S.Y.; Shvetsov, Y.B.; Wilkens, L.R.; Marchand, L.L.; Boushey, C.J.; Paik, H.Y. Sex differences in sociodemographic and lifestyle factors associated with diet quality in a multiethnic population. *Nutrition* **2019**, *66*, 147–152. [CrossRef]
33. Touvier, M.; Mejean, C.; Kesse-Guyot, E.; Vergnaud, A.C.; Hercberg, S.; Castetbon, K. Sociodemographic and economic characteristics associated with dairy intake vary across genders. *J. Hum. Nutr. Diet.* **2011**, *24*, 74–85. [CrossRef]
34. Group, C.S. Vascular factors and risk of dementia: Design of the Three-City Study and baseline characteristics of the study population. *Neuroepidemiology* **2003**, *22*, 316–325. [CrossRef]
35. Samieri, C.; Jutand, M.A.; Feart, C.; Capuron, L.; Letenneur, L.; Barberger-Gateau, P. Dietary patterns derived by hybrid clustering method in older people: Association with cognition, mood, and self-rated health. *J. Am. Diet. Assoc.* **2008**, *108*, 1461–1471. [CrossRef]
36. Feart, C.; Jutand, M.A.; Larrieu, S.; Letenneur, L.; Delcourt, C.; Combe, N.; Barberger-Gateau, P. Energy, macronutrient and fatty acid intake of French elderly community dwellers and association with socio-demographic characteristics: Data from the Bordeaux sample of the Three-City Study. *Br. J. Nutr.* **2007**, *98*, 1046–1057. [CrossRef]
37. Simermann, J.; Barberger-Gateau, P.; Berr, C. Validation d'un questionnaire de fréquence de consommation alimentaire dans une population âgée. In Proceedings of the 25e Congrès Annuel de la Société Française de Nutrition Clinique et Métabolisme SFNEP, Montpellier, France, 28–30 November 2007; pp. S69–S70.
38. Martin, A. The 'apports nutritionnels conseilles (ANC)' for the French population. *Reprod Nutr. Dev.* **2001**, *41*, 119–128. [CrossRef]
39. Rahi, B.; Colombet, Z.; Gonzalez-Colaco Harmand, M.; Dartigues, J.F.; Boirie, Y.; Letenneur, L.; Feart, C. Higher Protein but Not Energy Intake Is Associated With a Lower Prevalence of Frailty Among Community-Dwelling Older Adults in the French Three-City Cohort. *J. Am. Med. Dir. Assoc.* **2016**, *17*, 672 e677–672 e611. [CrossRef]
40. Cederholm, T.; Jensen, G.L.; Correia, M.; Gonzalez, M.C.; Fukushima, R.; Higashiguchi, T.; Baptista, G.; Barazzoni, R.; Blaauw, R.; Coats, A.; et al. GLIM criteria for the diagnosis of malnutrition—A consensus report from the global clinical nutrition community. *Clin. Nutr.* **2019**, *38*, 1–9. [CrossRef]
41. Dubuisson, C.; Lioret, S.; Touvier, M.; Dufour, A.; Calamassi-Tran, G.; Volatier, J.L.; Lafay, L. Trends in food and nutritional intakes of French adults from 1999 to 2007: Results from the INCA surveys. *Br. J. Nutr.* **2010**, *103*, 1035–1048. [CrossRef]
42. Agence Nationale de Sécurité Sanitaire de Lálimentation, INCA 3: Evolution des Habitudes et Modes de Consommation, de Nouveaux Enjeux en Matière de Sécurité Sanitaire et de Nutrition. Available online: https://www.anses.fr/fr/content/inca-3-evolution-des-habitudes-et-modes-de-consommation-de-nouveaux-enjeux-en-mati{[è]}re-de (accessed on 29 October 2020).

43. Escalon, H.; Bossard, C.; dir Beck, F. Baromètre santé nutrition 2008. In *Baromètres Santé, 2009*; Denis, S., Ed.; Institut national de préventionet d'éducation pour la santé(INPES): Saint-Denis, France, 2008; p. 424. ISBN 978-2-9161-9205.
44. Pelletier, A.; Barul, C.; Feart, C.; Helmer, C.; Bernard, C.; Periot, O.; Dilharreguy, B.; Dartigues, J.F.; Allard, M.; Barberger-Gateau, P.; et al. Mediterranean diet and preserved brain structural connectivity in older subjects. *Alzheimers Dement.* **2015**, *11*, 1023–1031. [CrossRef]
45. Charby, J.; Héble, P.; Vaudaine, S. Les produits laitiers en France: Évolution du marché et place dans la diète. *Cah. Nutr. Diététique* **2017**, *52*, S25–S34. [CrossRef]
46. Tucker, K.L.; Dallal, G.E.; Rush, D. Dietary patterns of elderly Boston-area residents defined by cluster analysis. *J. Am. Diet. Assoc.* **1992**, *92*, 1487–1491.
47. Sofi, F. The Mediterranean diet revisited: Evidence of its effectiveness grows. *Curr. Opin. Cardiol.* **2009**, *24*, 442–446. [CrossRef]
48. Ribeiro, I.; Gomes, M.; Figueiredo, D.; Lourenco, J.; Paul, C.; Costa, E. Dairy Product Intake in Older Adults across Europe Based On the SHARE Database. *J. Nutr. Gerontol. Geriatr.* **2019**, *38*, 297–306. [CrossRef]
49. European Food Safety Authority. EFSA Supporting Publication. *EN* **2017**, *1351*. [CrossRef]
50. Scali, J.; Richard, A.; Gerber, M. Diet profiles in a population sample from Mediterranean southern France. *Public Health Nutr.* **2001**, *4*, 173–182. [CrossRef]
51. Darmon, N.; Briend, A.; Drewnowski, A. Energy-dense diets are associated with lower diet costs: A community study of French adults. *Public Health Nutr.* **2004**, *7*, 21–27. [CrossRef]
52. Roberts, S.B.; Rosenberg, I. Nutrition and aging: Changes in the regulation of energy metabolism with aging. *Physiol. Rev.* **2006**, *86*, 651–667. [CrossRef]
53. Ginder-Coupez, V.; Hébel, P. Le yaourt, un marqueur "universel" de la qualité de la diète ? *Cah. Nutr. Diététique* **2017**, *52*, S35–S47. [CrossRef]
54. Wadolowska, L.; Ulewicz, N.; Sobas, K.; Wuenstel, J.W.; Slowinska, M.A.; Niedzwiedzka, E.; Czlapka-Matyasik, M. Dairy-Related Dietary Patterns, Dietary Calcium, Body Weight and Composition: A Study of Obesity in Polish Mothers and Daughters, the MODAF Project. *Nutrients* **2018**, *10*, 90. [CrossRef]
55. Willett, W. *Nutritional Epidemiology*, 2nd ed.; Oxford University Press: Oxford, NY, USA, 1998; p. 514.
56. Feart, C.; Peuchant, E.; Letenneur, L.; Samieri, C.; Montagnier, D.; Fourrier-Reglat, A.; Barberger-Gateau, P. Plasma eicosapentaenoic acid is inversely associated with severity of depressive symptomatology in the elderly: Data from the Bordeaux sample of the Three-City Study. *Am. J. Clin. Nutr.* **2008**, *87*, 1156–1162. [CrossRef]

Publisher's Note: MDPI stays neutral with regard to jurisdictional claims in published maps and institutional affiliations.

© 2020 by the authors. Licensee MDPI, Basel, Switzerland. This article is an open access article distributed under the terms and conditions of the Creative Commons Attribution (CC BY) license (http://creativecommons.org/licenses/by/4.0/).

Article

Kefir Peptides Prevent Estrogen Deficiency-Induced Bone Loss and Modulate the Structure of the Gut Microbiota in Ovariectomized Mice

Min-Yu Tu [1,2,3,4,†], Kuei-Yang Han [5,6,†], Gary Ro-Lin Chang [1,†], Guan-Da Lai [1,†], Ku-Yi Chang [5,6,7], Chien-Fu Chen [1,7], Jen-Chieh Lai [1,7], Chung-Yu Lai [8], Hsiao-Ling Chen [9,10] and Chuan-Mu Chen [1,11,12,*]

1. Department of Life Sciences, and Ph.D. Program in Translational Medicine, National Chung Hsing University, Taichung 402, Taiwan; du0807@yahoo.com.tw (M.-Y.T.); gary590422@yahoo.com.tw (G.R.-L.C.); gtrskyliner34@hotmail.com (G.-D.L.); balance1981@gmail.com (C.-F.C.); shoulia2001@yahoo.com.tw (J.-C.L.)
2. Aviation Physiology Research Laboratory, Kaohsiung Armed Forces General Hospital Gangshan Branch, Kaohsiung 820, Taiwan
3. Department of Health Business Administration, Meiho University, Pingtung 912, Taiwan
4. Department of Biomedical Engineering, Hungkuang University, Taichung 433, Taiwan
5. Department of Family Medicine, Jen-Ai Hospital, Dali Branch, Taichung 402, Taiwan; shinchen@livemail.tw (K.-Y.H.); smallhead1230@yahoo.com.tw (K.-Y.C.)
6. Department of Orthopedic Surgery, Jen-Ai Hospital, Dali Branch, Taichung 402, Taiwan
7. Department of Orthopedic Surgery, Taichung Armed Forces General Hospital, Taichung 411, Taiwan
8. Graduate Institute of Aerospace and Undersea Medicine, National Defense Medical Center, Taipei 114, Taiwan; multi0912@gmail.com
9. Department of Biomedical Sciences, Da-Yeh University, Changhua 515, Taiwan; bellchen@mail.dyu.edu.tw
10. Department of Bioresources, Da-Yeh University, Changhua 515, Taiwan
11. The iEGG and Animal Biotechnology Center, National Chung Hsing University, Taichung 402, Taiwan
12. Rong Hsing Research Center for Translational Medicine, National Chung Hsing University, Taichung 402, Taiwan
* Correspondence: chchen1@dragon.nchu.edu.tw; Tel.: +886-4-2285-6309
† These authors contributed equally to this study.

Received: 15 October 2020; Accepted: 3 November 2020; Published: 9 November 2020

Abstract: Osteoporosis is a major skeletal disease associated with estrogen deficiency in postmenopausal women. Kefir-fermented peptides (KPs) are bioactive peptides with health-promoting benefits that are produced from the degradation of dairy milk proteins by the probiotic microflora in kefir grains. This study aimed to evaluate the effects of KPs on osteoporosis prevention and the modulation of the composition of the gut microbiota in ovariectomized (OVX) mice. OVX mice receiving an 8-week oral gavage of 100 mg of KPs and 100 mg of KPs + 10 mg Ca exhibited lower trabecular separation (Tb. Sp), and higher bone mineral density (BMD), trabecular number (Tb. N) and bone volume (BV/TV), than OVX groups receiving Ca alone and untreated mice, and these effects were also reflected in bones with better mechanical properties of strength and fracture toughness. The gut microbiota of the cecal contents was examined by 16S rDNA amplicon sequencing. α-Diversity analysis indicated that the gut microbiota of OVX mice was enriched more than that of sham mice, but the diversity was not changed significantly. Treatment with KPs caused increased microbiota richness and diversity in OVX mice compared with those in sham mice. The microbiota composition changed markedly in OVX mice compared with that in sham mice. Following the oral administration of KPs for 8 weeks, the abundances of *Alloprevotella*, *Anaerostipes*, *Parasutterella*, *Romboutsia*, *Ruminococcus_1* and *Streptococcus* genera were restored to levels close to those in the sham group. However, the correlation of these bacterial populations with bone metabolism needs further investigation. Taken together, KPs prevent menopausal osteoporosis and mildly modulate the structure of the gut microbiota in OVX mice.

Keywords: kefir peptides; dairy milk protein; osteoporosis; ovariectomized (OVX) mice; 16S rDNA; gut microbiota

1. Introduction

Osteoporosis, which is characterized by low bone mass and the disruption of bone structure, is a major public health concern in postmenopausal women [1]. According to statistical data from the International Osteoporosis Foundation, osteoporotic fractures occur in one in three women worldwide older than 50 years during their lifetime. These fractures are usually accompanied by pain, disability and an increased mortality rate. In addition to the impact on health, osteoporosis also causes a huge economic burden. The cost is expected to increase to USD 25.3 billion in the US by 2025 [2]. Therefore, the prevention of osteoporosis becomes a critical issue in order to decrease the economic burden of managing osteoporosis and improving the life quality of patients.

Many studies have focused on finding safe, cost-effective and natural approaches to treat osteoporosis without side effects. Among these, the consumption of milk, yogurt and other fermented dairy products should be a good choice, because they are widely available and are considered a healthy lifestyle in many countries. These products are excellent sources of bioactive proteins, vitamins and minerals, as well as prebiotics or probiotics, with a range of health benefits, including bone health [3]. Kefir is a fermented milk similar to yogurt with a history of over one hundred years. It was first consumed in Russia and European countries, and became popular in Asia in recent years. Accumulating reports have indicated that the consumption of kefir is associated with many health benefits [4,5]. These benefits partly originate from the functions of some kefir peptides (KPs), which are produced during fermentation via the degradation of milk proteins by the microorganisms in kefir grains. The release and peptide profile of KPs are influenced by the different fermentation conditions and microbial communities of kefir grains [6]. To date, only a few peptides, mainly from the casein protein, were characterized with anti-hypertension, anti-microorganism, immune-modulation and opioid properties [7]. Most of the bioactive KPs remain to be identified for their functions.

The gut microbiota generally refers to the group of microorganisms that inhabit the intestines with symbiotic, commensal or pathogenic relationships with their hosts. Numerous studies have demonstrated that the gut microbiota can shape many aspects of host physiological processes, including metabolic functions, nutrient absorption, immune responses and hormone secretion. Because bone homeostasis is affected by metabolic pathways, immune systems and the hormone environment, the gut microbiota can also influence the bone metabolism balance via these pathways [8]. Most studies have reported that the modulation of the gut microbiota using probiotics, such as *Lactobacillus spp.* [9–13], or the products of the degradation of prebiotics, such as short-chain fatty acid [13,14], can increase bone BMD and promote bone formation, indicating that the consumption of dairy products may lead to a higher peak bone mass [8].

Few studies have explored the efficacy of KPs in the prevention of estrogen-associated bone loss [15,16]. The consumption of kefir or kefir-like dairy products may have a great impact on the structure of the gut microbiota [17,18], but the link between gut microbiota changes and bone health is limited. Thus, we used an ovariectomized (OVX) murine model to simulate estrogen-associated bone loss, and then fed the mice with KPs for 8 weeks. Afterward, the femoral bones were removed to examine the bone mass and bone structure, and the feces in the cecal segment of the large intestine was collected to analyze the gut microbiota.

2. Materials and Methods

2.1. Kefir Peptides (KPs) Preparation

KP powder (KEFPEP) was purchased from Phermpep Biotech Co. (Taichung, Taiwan). The kefir starter grains were firstly inoculated (5%, w/v) in sterilized goat milk at 20 °C for 20 h for activation. The grains were then retrieved through a sieve and then were re-inoculated (10%, w/v) in fresh sterilized goat milk at 20 °C for 20 h. After the grains were filtered, the fermented supernatant products were spray-dried into KP powder as described previously [19–21]. The peptide content was determined as triglycine equivalents in g per 100 g sample by the O-phthalaldehyde (OPA) method [19], which was 23.1 g/100 g within the sample. The commercial KEFPEP powder contained 23.1% peptides, 26.1% fat, ~50% carbohydrates, 0.28% sodium, and ~3% calcium [21].

2.2. Animal Experiments

Thirty 8-week-old female C57BL/6J mice were obtained from the National Laboratory Animal Center (Taipei, Taiwan). The mice were housed in an air-conditioned room (22 ± 2 °C/50 ± 10% humidity) with an automatic 12 h light-dark cycle, and were allowed free access to a regular rodent diet (Altromin® 1324 FORTI, Altromin GmbH, Lage, Germany) and water throughout the experiment. At 16 weeks of age, bilateral ovariectomy (24 mice) or sham surgery (6 mice) was performed. Two weeks later, the OVX mice were randomly divided into four groups (each $n = 6$) and treated as follows: (1) water/OVX, (2) Ca/OVX (10 mg of CaCO3 per kilogram of body weight), (3) KPs/OVX (100 mg of KPs per kilogram of body weight) and (4) KPs + Ca/OVX (100 mg of KPs + 10 mg of CaCO3 per kilogram of body weight). All the mice were subjected to daily oral gavages (0.1 mL) for 8 weeks. The body weight of the mice was recorded weekly. At the end of treatment, all the mice were sacrificed and their femoral bones were removed. The bones were immersed immediately in the fixation solution (4% formaldehyde in phosphate-buffered saline (PBS)) for one day and then were rinsed with PBS for subsequent characterization.

The cecal segments of the large intestines from the sham, OVX and KPs/OVX groups (each $n = 3$) were removed with their contents, placed immediately in liquid nitrogen, and then stored at −80 °C until ready for analysis. The present animal study was approved by the Institutional Animal Care and Use Committee (IACUC) of National Chung Hsing University (Taichung, Taiwan) with the IACUC No. 104-095.

2.3. Microcomputed Tomography (µ-CT)

The bone mineral density and trabecular microstructure of the right femur were examined using a high-resolution µ-CT scanner (Skyscan 1076 system; Bruker, Kontich, Belgium) at a resolution of 9 µm. The resulting image files were imported into CTAn software (Skyscan; Bruker, Kontich, Belgium) for three-dimensional (3D) image generation and the measurement of morphometric parameters, including the bone mineral density (BMD) and trabecular bone volume (BV/TV), thickness (Tb. Th), number (Tb. N) and separation (Tb. Sp). The structures of the trabecular bones were reconstituted using 100 µ-CT slices, which were approximately 0.9 mm in thickness from the growth plate of the distal femur.

2.4. Nanoindentation

The mechanical properties of the femur cortical bone were analyzed using a nanoindenter (TriboLab, Hysitron Inc., Minneapolis, MN, USA). A Berkovich diamond indenter with a tip radius of 50 nm was used to indent the polished surfaces of the cortical bone from the outer side to the inner side (near the bone marrow). For each sample, at least three series of indentation tests across the cortical bone shell (thickness of approximately 130–160 µm) were performed. The measurements obtained for the three parts of the residual indentation area of the cortical bone were averaged. The hardness and elastic modulus of the cortical bone at different locations were then calculated according to the indentation load–depth curves and Oliver–Pharr relationship, as described previously. To observe the fracture and estimate the fracture resistance, indentations were applied to the cortical bone surfaces under a high load of 500 mN. Scanning electron microscopy (SEM) was used to observe the fracture around the indented regions, and the residual indentation area was calculated according to the indentation load–depth curves.

2.5. Scanning Electron Microscopy

The distal femurs were trimmed in the sagittal plane and treated with a 5% sodium hypochlorite solution to expose the trabecular bone. All the skeletal samples were treated by dehydrating with acetone, air-drying, mounting on stubs, and coating with gold/palladium using an ion sputter (Hitachi E101, Tokyo, Japan), followed by examination using a scanning electron microscope (FEI Inspect S, Hillsboro, OR, USA).

2.6. Gut Microbiota

The collected cecal contents were submitted for gut microbiota analysis. From the DNA sampling to the final data, including nucleic acid extraction, PCR amplification, product purification, 16S rDNA amplicon sequencing and subsequent bioinformatics, the analyses were conducted by the Biotools Microbiome Research Center (Taipei, Taiwan). To guarantee the reliability of the data, quality control was performed at each step of the procedure. Briefly, the total genomic DNA from the samples was extracted using a QIAamp PowerFecal DNA Kit (Qiagen, Redwood, CA, USA). The DNA concentration was determined and adjusted to 5 ng/μL for PCR amplification. A specific primer set (319F: 5′-CCTACGGGNGGCWGCAG-3′ and 806R: 5′- GACTACHVGGGTATCTAATCC -3′) was used for PCR amplification of the V3-V4 region of the 16S rRNA gene. The PCR reaction was carried out with KAPA HiFi HotStart ReadyMix (Roche, Basel, Switzerland) under the following PCR conditions: 95 °C for 3 min; 25 cycles of 95 °C for 30 s, 55 °C for 30 s, 72 °C for 30 s; 72 °C for 5 min and hold at 4 °C. The PCR products were monitored on 1.5% agarose gel. Samples with a bright main strip around 500 bp were chosen and purified using the AMPure XP beads (Beckman Coulter, Indianapolis, IN, USA). A secondary PCR was performed by using the 16S rRNA V3-V4 region PCR amplicon and Nextera XT Index Kit with dual indices and Illumina sequencing adapters (Illumina, San Diego, CA, USA). The indexed PCR product quality was assessed on the Qubit 4.0 Fluorometer (Thermo Scientific, Alvarado, TX, USA) and Qsep100^{TM} system (Bioptic Inc., La Canada Flintridge, CA, USA). Equal amounts of the indexed PCR product were mixed to generate the sequencing library. At last, the library was sequenced on an Illumina MiSeq platform and paired 300 bp reads were generated.

2.7. Statistical Analysis

All the data were presented as means ± SEM (standard error of the mean) or means ± SD (standard deviations). Multiple group comparisons were performed using one-way ANOVA and Duncan's post hoc test, and the statistical significances ($p < 0.05$) were indicated by different letters. The relative abundances of the gut microbiota were compared by one-way ANOVA and Tukey's post hoc test, and the statistical significances were indicated by asterisks.

3. Results

3.1. Body Weight and Affected Organ Weight

The gain rates of the body weight were different between the OVX and the sham mice (Figure 1A). At the end of the study, the body weights of the OVX groups were on average higher than those of the sham group ($p < 0.01$). Differences were also observed in the deposits of kidney-surrounding fat. Due to ovariectomy, the untreated OVX mice accumulated more kidney-surrounding fat than the sham mice ($p < 0.05$); however, the treatments with KPs and KPs + Ca ($p < 0.05$) reduced the kidney-surrounding fat. By contrast, treatment with Ca alone showed little or no effect on the reduction of the fat deposits around the kidneys (Figure 1B).

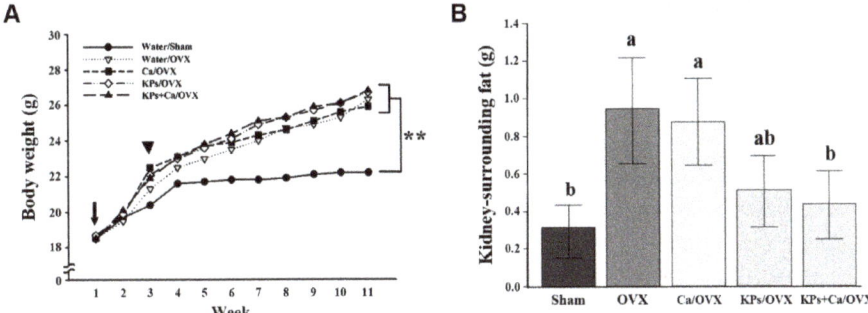

Figure 1. Body weight and kidney-surrounding fat. (**A**) The body weight was measured weekly from the onset of ovariectomy surgery to the end of the experiment. Water/Sham group: blank surgery mice fed with ddH$_2$O, Water/OVX group: ovariectomy surgery mice fed with ddH$_2$O; Ca/OVX group: ovariectomy surgery mice fed with 10 mg/kg CaCO3; KPs/OVX group: ovariectomy surgery mice fed with 100 mg/kg kefir peptides; KPs+Ca/OVX group: ovariectomy surgery mice fed with 100 mg/kg kefir peptides + 10 mg/kg CaCO3. In this panel, the arrow indicates the onset of surgery and the arrowhead indicates the beginning of oral administration. The mean body weights were higher in all the OVX groups than in the sham group (**, $p < 0.01$). (**B**) The kidney-surrounding fat was removed and measured immediately at the end of the experiment. The data are presented as means ± SEM ($n = 6$). Statistical analysis was performed by one-way ANOVA and Duncan's post hoc test, and statistical significances are indicated by different letters (a, b; $p < 0.05$).

3.2. Effects of KPs on BMD and Bone Structure

The effects of KPs on BMD and bone structure were examined by μ-CT. The OVX mice without treatment showed severe losses of trabecular bones, and mild attenuations in the thickness of the cortical bones, compared with the sham mice (Figure 2A). Although the administration of Ca alone conferred a slight protection from large-scale bone loss, larger spaces or a less-dense structure still appeared in the trabecular bones in the OVX mice receiving Ca alone. The OVX mice receiving the treatments of KPs and KPs + Ca exhibited trabecular bones with a healthy or dense appearance compared with the sham mice, suggesting that KPs or KPs combined with Ca can provide better protection than Ca alone.

The protection of KPs is reflected in the parametric changes of trabecular bones. The BMD of the untreated OVX group (0.34 g/cm^3) showed about a 19% reduction compared with the sham group (0.42 g/cm^3) ($p < 0.05$). The BMD values were 0.40, 0.48 and 0.47 g/cm^3 in the OVX groups following treatments with Ca, KPs and KPs + Ca, respectively (Figure 2B). Ovariectomy caused a relatively low trabecular bone volume (Tb. BV/TV) in the untreated OVX group (0.57%), showing a ~75% reduction compared with that in the sham group (2.32%) ($p < 0.001$). The bone volumes increased to 1.51%, 2.08% and 1.87% in the OVX groups receiving Ca, KPs and KPs + Ca, respectively (Figure 2C). The untreated OVX mice (0.14/mm^3) also showed a relatively lower trabecular bone number (Tb. N), with a ~67% reduction compared with the sham group (0.43/mm^3) ($p < 0.01$). However, following treatments with Ca, KPs and KPs + Ca, the Tb. N increased to 0.34, 0.47 and 0.47/mm^3, respectively, among which the groups treated with KPs and KPs + Ca were comparable to the sham group (Figure 2D). By contrast, the trabecular separation (Tb. Sp) increased in the untreated OVX mice (0.82 mm), showing a ~58% increase compared with the sham group (0.52 mm) ($p < 0.05$). However, following treatments with Ca, KPs and KPs + Ca, the Tb. Sp decreased to 0.57, 0.52 and 0.51 mm, respectively, in which the groups treated with KPs and KPs + Ca were comparable to the sham group ($p > 0.05$) (Figure 2E).

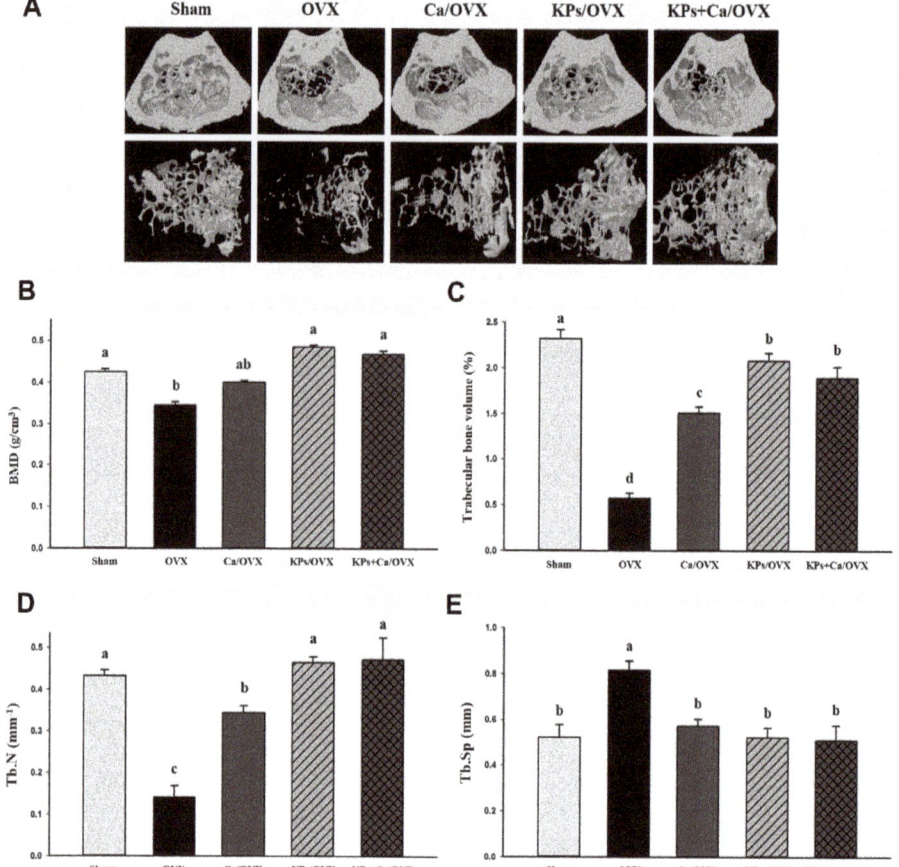

Figure 2. Micro-CT analysis of the femur. (**A**) Three-dimensional images of the distal femur. The images in the upper panel show the transverse section of the femur, and images in the lower panel show the structure of the trabecular bone. (**B**–**E**) show the morphological parameters of μ-CT analysis, including bone mineral density (**B**), trabecular bone volume (Tb. BV/TV) (**C**), trabecular number (Tb. N) (**D**) and trabecular separation (Tb. Sp) (**E**). The data are presented as means ± SEM (n = 6). Statistical analysis was performed by one-way ANOVA and Duncan's post hoc test, and statistical significances are indicated by different letters (a, b, c; $p < 0.05$).

3.3. Effect of KPs on the Mechanical Indices of the Cortical Bones

The mechanical indices (elastic modulus and hardness) of the cortical bones were examined by nanoindentation. The cortical elastic modulus was significantly decreased in the OVX group (18.6 GPa), showing a ~32% reduction compared with that in the sham group (27.2 GPa) (Figure 3A; $p < 0.01$). The cortical elastic modulus values substantially increased to 22.5, 26.4 and 27.0 GPa after treatments with Ca, KPs and KPs + Ca, respectively, in which the groups treated with KPs and KPs + Ca were comparable to the sham group. A similar trend was also observed in the change of the cortical hardness. The cortical hardness markedly decreased in the OVX group (0.61 GPa), showing a ~36% reduction compared with that in the sham group (0.95 GPa) (Figure 3B; $p < 0.01$). Treatments with Ca, KPs and KPs + Ca conferred the OVX mice with higher cortical hardness values up to 0.73, 0.83 and 0.91 GPa, respectively, in which the groups treated with KPs and KPs + Ca were comparable to the sham group.

The indented surfaces of the cortical bones were examined by SEM, and the residual indentation areas, marked with the triangles in Figure 3C, were measured. The residual indentation area of the untreated OVX group was 2.1-fold higher than that of the sham group (Figure 3D; $p < 0.05$). A higher residual indentation area indicates poor bone strength and fracture toughness. Treatment with Ca alone (1.8-fold higher than sham) conferred an insignificant change in the residual indentation areas. However, treatments with KPs (1.3-fold higher than sham) and KFPs + Ca (1.2-fold higher than sham) significantly decreased the residual indentation areas ($p < 0.05$), and both were comparable with the sham group.

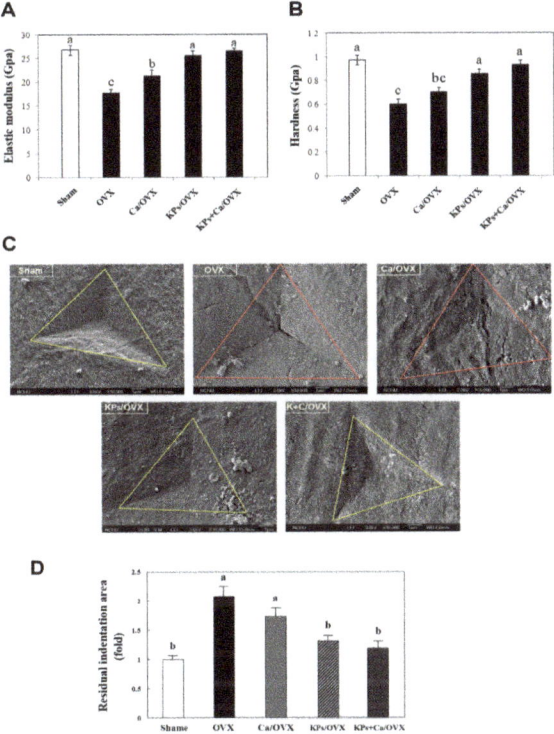

Figure 3. Nanoindentation analysis of the mechanical properties of cortical femoral bones at the end of oral administration. Each sample was analyzed by at least three series of outer-to-inner indentation tests (15 points) across the transverse section of the cortical bone shell (130–160 μm in thickness). The mechanical properties of the elastic modulus (**A**) and hardness (**B**) were compared among the groups. (**C**) The indented surfaces were further examined by scanning electron microscopy. The residual nanoindentation areas are marked by triangles and were quantitated using ImageJ software. The relative residual nanoindentation areas were calculated by comparison with the sham group (**D**). The data are presented as means ± SEM ($n = 6$). Statistical analysis was performed by one-way ANOVA and Duncan's post hoc test, and statistical significances are indicated by different letters ($p < 0.05$).

3.4. Effect of KPs on the Gut Microbiota

The gut microbiota of the cecal contents collected from the sham, OVX and KPs/OVX groups were evaluated by 16S amplicon sequencing. In total, 316 operational taxonomic units (OTUs), equal to 30,582 sequences/sample, with >97% identity were used for analysis. First, the α- and β-diversity were calculated to evaluate the effect of ovariectomy and KP treatment on the total abundance of the gut microbiota. Principal co-ordinates analysis (PCoA) showed distinct clustering of the gut microbiota from the sham, OVX and KPs/OVX groups, but comparisons via the β-diversity indices of unweighted uniFrac and weighted uniFrac indicated that the difference was not significant (Figure 4A and Table 1). Alpha-diversity indices were calculated to evaluate the change in the microbiota richness and diversity between various groups. The OVX and KPs/OVX groups showed significantly higher observed OTU, Ace and Chao1 indices than the sham group ($p < 0.05$), the KPs/OVX group showed significantly higher Shannon and Simpson indices than the sham group ($p < 0.05$), and no significant differences in the α-diversity indices were observed between the OVX and KPs/OVX groups ($p > 0.05$) (Table 1).

Next, we compared the microbiota structure at different hierarchical levels. At the phylum level, the gut microbiota of mice mainly comprised Bacteroidetes, Firmicutes and Deferribacteres, but the OVX groups had a lower relative abundance (RA) of Deferribacteres than the sham group (Figure 4B). The Firmicutes to Bacteroidetes (F/B) ratio, a known marker of obesity [20], was slightly higher in the OVX groups than in the sham group. The oral administration of KPs increased the F/B ratio, but the differences in the F/B ratio among the sham, OVX and KPs/OVX groups were not statistically significant (Figure 4C; $p > 0.05$). To identify specific microbial populations affected by oral KPs treatment, the RA of the gut microbiota was compared at the genus level. In this study, 148 of 316 OTUs were annotated to the genus level, accounting for 73 genera. A Venn diagram was constructed to represent the relationship among the three groups. The quantity of OTUs (genera) unique to each group was less than 0.6% in abundance, while the genera shared by the sham, OVX and KPs/OVX groups were more than 99% in abundance (Figure 5A). The identified genera with >1% abundance are shown in Figure 5B; the other genera accounted for a very small fraction of the total abundance. At the genus level, the gut microbiota of the sham group comprised Lachnospiraceae_NK4A136, Prevotellaceae_UCG_001, Mucispirillum, Oscillibacter and Alloprevotella predominantly (87% abundance in total); however, the abundance of the predominant genera decreased significantly in the OVX group (29% abundance). By contrast, the sham group contained relatively lower abundances of Bacteroides, Ruminiclostridium_9, Anaerostipes, Alistipes, Ruminiclostridium, Coprococcus_2, Parabacteroides, Acetatifactor, Parasutterella, Romboutsia, Ruminococcus_1 and Ruminococcaceae_UCG014 (<5% abundance); however, these genera were enriched importantly in the OVX group (63% abundance). Thus, the structure of the gut microbiota was markedly affected by ovariectomy. Moreover, we were interested in the bacterial genera affected by KPs treatment. The oral administration of KPs for 8 weeks in OVX mice revealed that the abundances of the genera Alloprevotella, Anaerostipes, Parasutterella, Romboutsia, Ruminococcus_1 and Streptococcus were restored to a level close to those in the sham group (Figure 5C). Among these bacterial populations, the abundances of the genera Anaerostipes, Ruminococcus_1 and Streptococcus in the KPs/OVX group were significantly different compared with those in the OVX group ($p < 0.05$).

Table 1. The mean α- and β-diversity indices and the P values for various group comparisons.

Group	α-Diversity Index					β-Diversity Index	
	Observed-OTU	Ace	Chao1	Shannon	Simpson	Unweighted uniFrac	Weighted uniFrac
Sham	160	177.0	180.6	4.938	0.932	0.199	0.038
OVX	217	233.1	232.9	5.657	0.965	0.080	0.034
KPs/OVX	230	244.1	248.3	5.790	0.968	0.327	0.042
Comparison	p value of α-diversity					p value of β-diversity	
Sham vs OVX	**0.004**	**0.002**	**0.006**	0.055	0.062	0.441	0.645
Sham vs KPs/OVX	**0.001**	**0.001**	**0.002**	**0.028**	**0.049**	0.395	0.761
OVX vs KPs/OVX	0.480	0.514	0.363	0.850	0.979	0.077	0.299

Statistical significances were analyzed by one-way ANOVA and Tukey's post-hoc test, and the bold values represent $p < 0.05$.

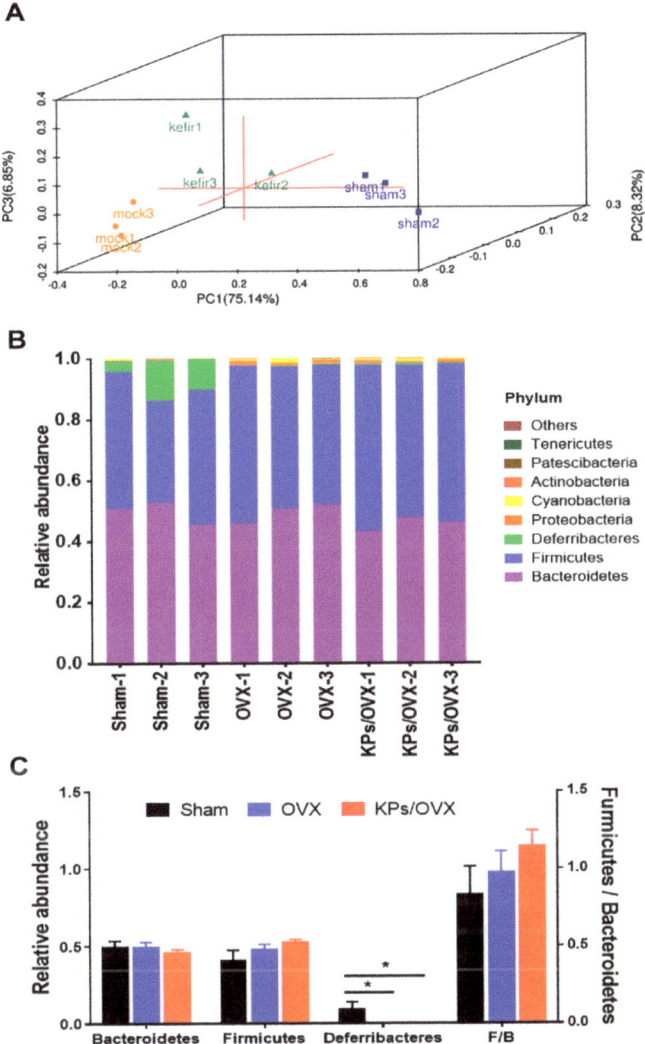

Figure 4. Analysis of the gut microbiota by 16S rDNA amplicon sequencing. (**A**) Principal co-ordinates analysis (PCoA), a common method of β-diversity index analysis, was used to evaluate the differential significance of the total abundance of the gut microbiota among the groups. PC1: Pricipal coordinate analysis axis 1; PC2: Pricipal coordinate analysis axis 2; PC3: Pricipal coordinate analysis axis 3. (**B**) Structural comparison of the gut microbiota at the phylum level. (**C**) Relative abundances of the phyla Bacteroidetes Firmicutes and Deferribacteres, and the ratio of Firmicutes to Bacteroidetes (F/B), a marker of obesity. Statistical analysis was performed by one-way ANOVA and Tukey's post hoc test. Statistical significances are indicated by asterisks (*, $p < 0.05$).

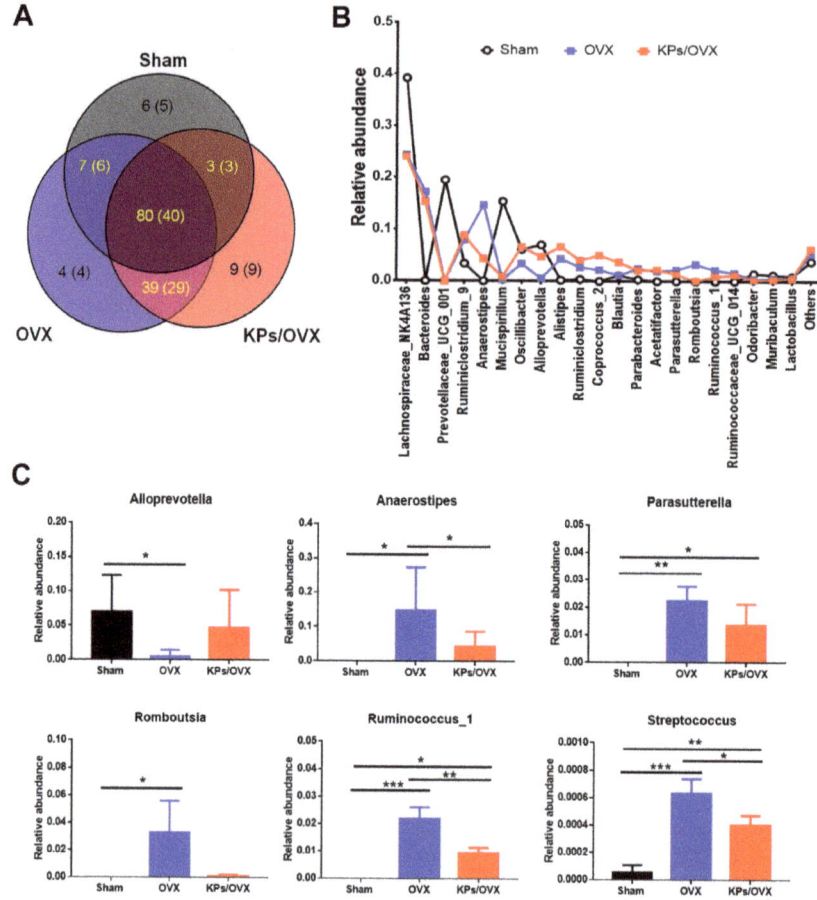

Figure 5. Structural comparison of the gut microbiota at the genus level. (**A**) Venn diagram comparing the observed operational taxonomic units in the gut microbiota of sham, OVX and KPs/OVX mice. (**B**) Major bacterial genera identified in this study. (**C**) Mean relative abundances of the genera *Alloprevotella, Anaerostipes, Parasutterella, Romboutsia, Ruminococcus_1* and *Streptococcus*. Compared with the sham group, these bacterial genera were inhibited or enriched in the OVX group but were reversed in the KPs/OVX group. The data are presented as means ± standard deviation (SD) ($n = 3$). Statistical analysis was performed by one-way ANOVA and Tukey's post hoc test. Statistical significances are indicated by asterisks (*, $p < 0.05$; **, $p < 0.01$ and ***, $p < 0.001$).

4. Discussion

Osteoporosis is a prevalent bone disease in the population of postmenopausal women. A safe and cost-effective natural product for osteoporosis prevention and treatment will help to decrease the incidence of osteoporosis in postmenopausal women. Kefir, fermented milk produced from kefir grains, has attracted widespread interest in the scientific community because it contains many bioactive peptides with health-promoting benefits. The association of kefir with osteoporosis was first published in our previous report [15], which demonstrated the benefits of kefir in preventing postmenopausal bone loss using an OVX rat model. Subsequently, we demonstrated the beneficial effects of kefir on bone mineral density and bone metabolism in a clinical trial of osteoporotic patients [16]. In this study, we further demonstrated the function of kefir peptides (KPs) on osteoporosis prophylaxis, and provided more evidence to gain insight into the association of KPs with the structural change of the gut microbiota in OVX mice.

OVX rats or mice are the widely used animal models of postmenopausal osteoporosis. The remarkable elevation of the body weight and kidney-surrounding fat in the untreated OVX mice indicated the success of the present animal model. At the end of this study, we found that the kidney-surrounding fat decreased in OVX mice with KPs treatment, suggesting the modulating activities of KPs on lipid metabolism. The functions of kefir or KPs in inhibiting lipogenesis were fully discussed in our previous reports [19,21–25].

Calcium carbonate and other calcium supplements are often used as an aid to prevent osteoporosis and to treat patients, along with the other medications. The present bone analysis indicated that supplementation with calcium carbonate alone provided little protection against OVX-induced bone loss, but greater protection followed KPs treatment and the combined use of KPs and calcium carbonate. The protection from OVX-induced bone loss also conferred the bones with higher mechanical strength to resist external pressure. Thus, the administration of KPs and KPs + Ca maintained the mechanical parameters of elastic moduli and hardness at a level similar to that in the sham group, with a marked difference compared with that in the untreated OVX mice or mice treated with Ca alone. The OVX mice treated with KPs or KPs + Ca also exhibited a lower residual indentation area than untreated OVX mice, suggesting the higher strength to resist bone fracture (Figure 3D). The microarchitectural and mechanical changes in the bones in OVX mice were consistent with those observed in our previous OVX rat model with different dosages of KPs [15]. In addition to structural and mechanical properties, the bone-protective effects of KPs can also be monitored by serum bone turnover markers, such as alkaline phosphatase (ALP) and C-terminal telopeptide fragment of type I collagen C-terminus (CTX-1). In this study, we did not evaluate the effects of KPs on serum bone turnover markers, but a reduced serum level of ALP and CTX-1 has been found in OVX rats treated with KPs. These results confirmed the outstanding effect of KPs in preventing postmenopausal osteoporosis. The recent studies using proteomic or peptidomic approaches to analyze kefir beverages have provided us with a comprehensive understanding of the released peptide composition during the fermentation process [6,7,26–28]. Hundreds to thousands of peptides have been identified in those studies, but only a few contained 100% sequence homology to known functional peptides [6,7,28]. Most KPs have not been investigated fully, and many are released as a precursor form of the functional peptide, in which the functional sequence may be contained within the released peptide [6,26–28]. Currently, the KPs with osteo-protective potential remain unclear. However, various studies have indicated the positive effect of bioactive peptides derived from casein and whey proteins (two major proteins of milk) on bone metabolism, a finding that may provide directions to understand the effects of active components of KPs on bone metabolism. Casein phosphopeptides (CPPs), a family of casein-derived peptides containing serine phosphate able to bind and solubilize calcium, may promote calcium absorption by the intestine and increase the bioavailability of calcium ions by other tissues in the body. There are two reports that have indicated the osteo-protective effect of CPPs in rats due to the increased calcium absorption caused by the administration of CPPs orally, or by supplementing CPPs in the rodent diet [29,30]. KPs were also shown to promote calcium absorption in Caco-2 cells in our previous study [15]. In the study of Ebner et al. [6], some identified peptides in bovine milk fermented with kefir grains showed similar sequence coverages of caseins with known CPPs, e.g., αs1-casein (59–79), αs1-casein (43–58), β-casein (1–25) and β-casein (33–48) [31]. However, the positive correlation of CPPs with the promotion of calcium absorption remains debatable, because some studies have reported the opposite results [32,33]. As such, further investigations are required to evaluate the functions of CPPs in calcium absorption and their correlation with bone metabolism. Casein-derived antioxidative peptide (β-casein (185–191): VLPVPQK), whey derived-antioxidative (β-lactoglobulin (60–64): YVEEL) and angiotensin-converting enzyme (ACE) inhibitory (β-lactoglobulin (120–123): YLLF) peptides have also been shown to exert osteo-protective properties under in vitro or in vivo conditions [34–37]. The osteo-protective properties of these antioxidative and ACE-inhibitory peptides originate from their ability to suppress the inflammatory status and RAS (renin-angiotensin system) activity, respectively. Peptides homologous to the sequence of known antioxidative or ACE-inhibitory peptides were also identified in the abovementioned proteomic studies. Again, most of their functions remain to be further investigated. We have isolated and identified several peptide candidates responsible for bone protection using our materials, and their effects on osteoclastogenesis and osteoblastogenesis are currently being studied.

Following 8 weeks of treatment, we used high-throughput sequencing of bacterial 16S rDNA to compare the compositions of the gut microbiota in different groups (Figures 4 and 5). According to α-diversity analysis, we found that the OVX group had significantly higher Chao1 and Ace indices than the sham group (Table 1). The Chao1 and Ace indices are usually used to indicate the richness of the gut microbiota; thus, our findings indicated that ovariectomy increased the richness of gut microbiota. Additionally, the Shannon and Simpson indices, which are used to represent the diversity of microbiota, were not significantly changed between the OVX and sham groups, suggesting the diversity of the gut microbiota was not affected significantly after ovariectomy (Table 1). Eight-week administration with KPs did not alter the α-diversity indices significantly compared with the OVX mice receiving mock treatment, indicating that the microbiota richness and diversity in OVX mice were not affected significantly by KPs treatment. However, significant differences were observed in the richness and diversity of the gut microbiota between the KPs/OVX and sham groups. The structure of the gut microbiota was changed markedly by ovariectomy, as indicated by the predominant bacterial populations being inhibited and many less-abundant bacterial populations being enriched at different hierarchical levels. These effects should be reasonable because estrogen plays important roles in the regulation of various metabolic pathways in women, particularly regarding its association with postmenopausal bone homeostasis [38]. At the cellular level, the central mechanism of estrogen deficiency-induced bone loss occurs via the promotion of osteoclast formation and the expansion of RANKL- and TNF-expressing cells [39,40]. The gut microbiota regulates bone homeostasis by influencing host metabolism, calcium absorption, the immune system and the endocrine system [8]. The interplay between estrogen and the gut microbiota is therefore apparent [41]. For example, an increased Firmicutes to Bacteroidetes (F/B) ratio in the gut microbiota has even been reported in OVX rats or mice [42–44]. In the present study, we observed a higher F/B ratio resulting from a mild increase in Firmicutes in both the OVX and KPs/OVX groups, although the differences did not reach a significant level. Consistent with the higher weight gain observed in all OVX mice (Figure 1A), our data supported using the F/B ratio as a microbiota marker of obesity [22]. The F/B ratio seemed to be higher in OVX mice with KPs treatment than sham mice, but the difference was not statistically important. At the phylum level, Deferribacteres exhibited decreased abundances in all OVX mice (Figure 4B,C). In healthy mice, Deferribacteres emerged as the third dominant phylum in the gut microbiota, and metagenomic and metatranscriptomic analysis revealed that the genes for cofactor, vitamin metabolism and amino acid metabolism were upregulated in the Deferribacteres family [45]. To further study the response of the gut microbiota to ovariectomy surgery and KPs treatment, the bacterial populations of all three groups were compared at the genus level. The predominant bacterial genera in the sham group included *Lachnospiraceae_NK4A136*, *Prevotellaceae_UCG_001*, *Mucispirillum*, *Oscillibacter* and *Alloprevotella*, which were decreased significantly in OVX mice (87% for sham vs. 29% for OVX). The minor populations *Bacteroides*, *Ruminiclostridium_9*, *Anaerostipes*, *Alistipes*, *Ruminiclostridium*, *Coprococcus_2*, *Parabacteroides*, *Acetatifactor*, *Parasutterella*, *Romboutsia*, *Ruminococcus_1* and *Ruminococcaceae_UCG_014* were significantly enriched in the OVX group (5% for sham vs. 63% for OVX), indicating that ovariectomy had a substantial effect on the alteration of the gut microbiota. At the genus level, the oral administration of KPs restored the abundances of six bacterial populations in OVX mice to a level close to that in sham mice. These genera were *Alloprevotella*, *Anaerostipes*, *Parasutterella*, *Romboutsia*, *Ruminococcus_1* and *Streptococcus* (Figure 5C). *Alloprevotella* was reported to be negatively correlated with nonalcoholic fatty acid liver and lipid accumulation [46]. The increased abundance of *Alloprevotella* may explain the reduction in kidney-surrounding fat accumulation in OVX mice receiving KP treatment (Figure 1B). Notably, the oral administration of KPs prevented nonalcoholic fatty acid liver and hyperlipidemia and obesity in our previous murine models [21,24]. *Parasutterella* and *Streptococcus* are potential harmful bacteria that were reported to be correlated with inflammatory bowel disease [47] and the progression of some tumors or cancers [48,49]. Additionally, *Streptococcus* has been reported to be positively correlated with body mass index in obese individuals [50]. *Romboutsia* were also reported to be obesity-related bacteria. Interestingly, KPs can reverse the alteration of the gut microbiota in OVX mice by enriching beneficial bacteria and decreasing potentially harmful pathogens. *Anaerostipes* and *Ruminococcus_1* are short-chain fatty acid (SCFA)-producing anaerobic bacteria. SCFAs, such as butyrate and propionate, are essential bacterial metabolites from carbohydrates in the gut because they evoke anti-inflammatory effects in the intestinal mucosa and promote bone metabolism [14,51,52]. We hypothesized that ovariectomy would reduce the abundance of *Anaerostipes* and *Ruminococcus* genera, and then KP treatment would reverse it. However, an opposite trend was observed in our study. High systemic concentrations of SCFAs were reported to be toxic and caused adverse effects in the host, mostly arising from the enhanced permeability of the gut barrier to increase the serum level of SCFAs [13,53]. These data imply that the increment in butyrate-producing bacteria after estrogen deficiency may produce excessive SCFAs in the intestine. Recently, Ma et al. [54] reported the correlation of a gut microbiota change with bone turnover parameters. In that report, *Ruminococcus*, *Clostridium*, *Coprococcus* and *Robinsoniella* were shown to be positively correlated with osteoclastic indicators (CTX, Tb.Sp) in OVX rats, but *Bacteroides* and *Butyrivibrio* showed the opposite patterns, and were negatively correlated with loss of bone mass. It is difficult to consider that one bacterial population will be absolutely correlated with specific physiological metabolic pathways or diseases. A healthy condition should be based on a more balanced structure of the microbiota in the host. The gut microbiota plays a critical role in the regulation of bone homeostasis by secreting various bacterial metabolites (SCFAs and bile acids) into the intestinal lumen, and influencing host intestinal barrier permeability, the immune system, hormone secretion and the gut–brain axis. Therefore, further mechanistic studies are needed to verify the causal role of gut microbiota in estrogen deficiency-induced osteoporosis.

5. Conclusions

In this study, we used the OVX mouse model to confirm the potential role of KPs in preventing menopausal osteoporosis, and indicated that the osteo-protective function of KPs is independent of calcium supplementation. Furthermore, our results suggested that the oral administration of KPs alters the structure of the gut microbiota in OVX mice by enriching the abundance of beneficial bacteria and reducing the abundance of potential pathogens.

Author Contributions: Conceptualization, H.-L.C. and C.-M.C.; methodology, M.-Y.T. and K.-Y.H.; validation, M.-Y.T., K.-Y.H., G.-D.L., K.-Y.C., and G.R.-L.C.; formal analysis, C.-Y.L., C.-F.C. and J.-C.L.; investigation, M.-Y.T., K.-Y.H., G.-D.L.; resources, C.-M.C. and H.-L.C.; data curation, G.R.-L.C.; writing—original draft preparation, M.-Y.T. and G.R.-L.C.; writing—review and editing, C.-M.C.; visualization, H.-L.C.; supervision, C.-M.C.; project administration, M.-Y.T.; funding acquisition, C.-M.C. All authors have read and agreed to the published version of the manuscript.

Funding: This research was funded by grant MOST-108-2313-B-005-039-MY3 from the Ministry of Science and Technology of Taiwan (C.-M.C.), and partially supported by the iEGG and Animal Biotechnology Center from the Feature Areas Research Center Program within the framework of the Higher Education Sprout Project by the Ministry of Education (MOE-109-S-0023) in Taiwan (C.-M.C.).

Acknowledgments: We thank our colleagues, Cheng-Wei Lai and Tung-Chou Tsai, in the Molecular Embryology & DNA Methylation Laboratory for their help with discussions and technical issues.

Conflicts of Interest: The authors declare no conflict of interest. The funders had no role in the design of the study; in the collection, analyses, or interpretation of data; in the writing of the manuscript, or in the decision to publish the results.

References

1. Sozen, T.; Ozisik, L.; Basaran, N.C. An overview and management of osteoporosis. *Eur. J. Rheumatol.* **2017**, *4*, 46–56. [CrossRef] [PubMed]
2. Burge, R.; Dawson-Hughes, B.; Solomon, D.H.; Wong, J.B.; King, A.; Tosteson, A. Incidence and Economic Burden of Osteoporosis-Related Fractures in the United States, 2005–2025. *J. Bone Miner. Res.* **2006**, *22*, 465–475. [CrossRef] [PubMed]
3. Fardellone, P. The effect of milk consumption on bone and fracture incidence, an update. *Aging Clin. Exp. Res.* **2019**, *31*, 759–764. [CrossRef] [PubMed]
4. Bourrie, B.C.T.; Willing, B.P.; Cotter, P.D. The Microbiota and Health Promoting Characteristics of the Fermented Beverage Kefir. *Front. Microbiol.* **2016**, *7*, 647. [CrossRef] [PubMed]
5. Guzel-Seydim, Z.B.; Kök-Taş, T.; Greene, A.K.; Seydim, A.C. Review: Functional Properties of Kefir. *Crit. Rev. Food Sci. Nutr.* **2011**, *51*, 261–268. [CrossRef]
6. Ebner, J.; Arslan, A.A.; Fedorova, M.; Hoffmann, R.; Küçükçetin, A.; Pischetsrieder, M. Peptide profiling of bovine kefir reveals 236 unique peptides released from caseins during its production by starter culture or kefir grains. *J. Proteom.* **2015**, *117*, 41–57. [CrossRef] [PubMed]
7. Dallas, D.C.; Citerne, F.; Tian, T.; Silva, V.L.M.; Kalanetra, K.M.; Frese, S.A.; Robinson, R.C.; Mills, D.A.; Barile, D. Peptidomic analysis reveals proteolytic activity of kefir microorganisms on bovine milk proteins. *Food Chem.* **2016**, *197*, 273–284. [CrossRef]
8. Behera, J.; Ison, J.; Tyagi, S.C.; Tyagi, N. The role of gut microbiota in bone homeostasis. *Bone* **2020**, *135*, 115317. [CrossRef]
9. Dar, H.Y.; Shukla, P.; Mishra, P.K.; Anupam, R.; Mondal, R.K.; Tomar, G.B.; Sharma, V.; Srivastava, R.K. Lactobacillus acidophilus inhibits bone loss and increases bone heterogeneity in osteoporotic mice via modulating Treg-Th17 cell balance. *Bone Rep.* **2018**, *8*, 46–56. [CrossRef]
10. Malik, T.A.; Chassaing, B.; Tyagi, A.M.; Vaccaro, C.; Luo, T.; Adams, J.; Darby, T.M.; Weitzmann, M.N.; Mulle, J.G.; Gewirtz, A.T.; et al. Sex steroid deficiency–associated bone loss is microbiota dependent and prevented by probiotics. *J. Clin. Investig.* **2016**, *126*, 2049–2063. [CrossRef]
11. Parvaneh, M.; Karimi, G.; Jamaluddin, R.; Ng, M.H.; Zuriati, I.; Muhammad, S.I. Lactobacillus helveticus (ATCC 27558) upregulates Runx2 and Bmp2 and modulates bone mineral density in ovariectomy-induced bone loss rats. *Clin. Interv. Aging* **2018**, *13*, 1555–1564. [CrossRef] [PubMed]
12. Zhang, J.; Motyl, K.J.; Irwin, R.; MacDougald, O.A.; Britton, R.A.; McCabe, L.R. Loss of Bone and Wnt10b Expression in Male Type 1 Diabetic Mice Is Blocked by the Probiotic Lactobacillus reuteri. *Endocrinology* **2015**, *156*, 3169–3182. [CrossRef]
13. Tyagi, A.M.; Yu, M.; Darby, T.M.; Vaccaro, C.; Li, J.-Y.; Owens, J.A.; Hsu, E.; Adams, J.; Weitzmann, M.N.; Jones, R.M.; et al. The Microbial Metabolite Butyrate Stimulates Bone Formation via T Regulatory Cell-Mediated Regulation of WNT10B Expression. *Immunity* **2018**, *49*, 1116–1131.e7. [CrossRef] [PubMed]
14. Lucas, S.; Omata, Y.; Hofmann, J.; Böttcher, M.; Iljazovic, A.; Sarter, K.; Albrecht, O.; Schulz, O.; Krishnacoumar, B.; Krönke, G.; et al. Short-chain fatty acids regulate systemic bone mass and protect from pathological bone loss. *Nat. Commun.* **2018**, *9*, 1–10. [CrossRef]
15. Chen, H.-L.; Tung, Y.-T.; Chuang, C.-H.; Tu, M.-Y.; Tsai, T.-C.; Chang, S.-Y.; Chen, C.-M. Kefir improves bone mass and microarchitecture in an ovariectomized rat model of postmenopausal osteoporosis. *Osteoporos. Int.* **2014**, *26*, 589–599. [CrossRef]
16. Tu, M.-Y.; Chen, H.-L.; Tung, Y.-T.; Kao, C.-C.; Hu, F.-C.; Chen, C.-M. Short-Term Effects of Kefir-Fermented Milk Consumption on Bone Mineral Density and Bone Metabolism in a Randomized Clinical Trial of Osteoporotic Patients. *PLoS ONE* **2015**, *10*, e0144231. [CrossRef] [PubMed]
17. Aslam, H.; Marx, W.; Rocks, T.; Loughman, A.; Chandrasekaran, V.; Ruusunen, A.; Dawson, S.L.; West, M.; Mullarkey, E.; Pasco, J.A.; et al. The effects of dairy and dairy derivatives on the gut microbiota: A systematic literature review. *Gut Microbes* **2020**, *12*, 1799533. [CrossRef]
18. Bellikci-Koyu, E.; Sarer-Yurekli, B.P.; Akyon, Y.; Kose, F.A.; Karagözlü, C.; Özgen, A.G.; Brinkmann, A.; Nitsche, A.; Ergünay, K.; Yilmaz, E.; et al. Effects of Regular Kefir Consumption on Gut Microbiota in Patients with Metabolic Syndrome: A Parallel-Group, Randomized, Controlled Study. *Nutrients* **2019**, *11*, 2089. [CrossRef]

19. Chen, H.-L.; Hung, K.-F.; Yen, C.-C.; Laio, C.-H.; Wang, J.-L.; Lan, Y.-W.; Chong, K.-Y.; Fan, H.-C.; Chen, C.-M. Kefir peptides alleviate particulate matter <4 mum (PM4.0)-induced pulmonary inflammation by inhibiting the NF-kappaB pathway using luciferase transgenic mice. *Sci. Rep.* **2019**, *9*, 11529. [CrossRef]
20. Chen, H.-L.; Tung, Y.-T.; Tsai, C.-L.; Lai, C.-W.; Lai, Z.-L.; Tsai, H.-C.; Lin, Y.-L.; Wang, C.-H.; Chen, C.-M. Kefir improves fatty liver syndrome by inhibiting the lipogenesis pathway in leptin-deficient ob/ob knockout mice. *Int. J. Obes.* **2013**, *38*, 1172–1179. [CrossRef]
21. Tung, Y.-T.; Chen, H.-L.; Wu, H.-S.; Ho, M.-H.; Chong, K.-Y.; Chen, C.-M. Kefir Peptides Prevent Hyperlipidemia and Obesity in High-Fat-Diet-Induced Obese Rats via Lipid Metabolism Modulation. *Mol. Nutr. Food Res.* **2018**, *62*, 1700505. [CrossRef]
22. Turnbaugh, P.J.; Ley, R.E.; Mahowald, M.A.; Magrini, V.; Mardis, E.R.; Gordon, J.I. An obesity-associated gut microbiome with increased capacity for energy harvest. *Nat. Cell Biol.* **2006**, *444*, 1027–1031. [CrossRef]
23. Choi, J.-W.; Kang, H.W.; Lim, W.-C.; Kim, M.-K.; Lee, I.-Y.; Cho, H.-Y. Kefir prevented excess fat accumulation in diet-induced obese mice. *Biosci. Biotechnol. Biochem.* **2017**, *81*, 958–965. [CrossRef] [PubMed]
24. Chen, H.L.; Tsai, T.C.; Tsai, Y.C.; Liao, J.W.; Yen, C.C.; Chen, C.M. Kefir peptides prevent high-fructose corn syrup-induced non-alcoholic fatty liver disease in a murine model by modulation of inflammation and the JAK2 signaling pathway. *Nutr. Diabetes* **2016**, *6*, e237. [CrossRef] [PubMed]
25. Ho, J.-N.; Choi, J.-W.; Lim, W.-C.; Kim, M.-K.; Lee, I.-Y.; Cho, H.-Y. Kefir inhibits 3T3-L1 adipocyte differentiation through down-regulation of adipogenic transcription factor expression. *J. Sci. Food Agric.* **2012**, *93*, 485–490. [CrossRef]
26. Santini, G.; Bonazza, F.; Pucciarelli, S.; Polidori, P.; Ricciutelli, M.; Klimanova, Y.; Silvi, S.; Polzonetti, V.; Vincenzetti, S. Proteomic characterization of kefir milk by two-dimensional electrophoresis followed by mass spectrometry. *J. Mass Spectrom.* **2020**, e4635. [CrossRef]
27. Amorim, F.G.; Coitinho, L.B.; Dias, A.T.; Friques, A.G.F.; Monteiro, B.L.; De Rezende, L.C.D.; Pereira, T.D.M.C.; Campagnaro, B.P.; De Pauw, E.; Vasquez, E.C.; et al. Identification of new bioactive peptides from Kefir milk through proteopeptidomics: Bioprospection of antihypertensive molecules. *Food Chem.* **2019**, *282*, 109–119. [CrossRef] [PubMed]
28. Izquierdo-González, J.J.; Amil-Ruiz, F.; Zazzu, S.; Sánchez-Lucas, R.; Fuentes-Almagro, C.A.; Rodríguez-Ortega, M.J. Proteomic analysis of goat milk kefir: Profiling the fermentation-time dependent protein digestion and identification of potential peptides with biological activity. *Food Chem.* **2019**, *295*, 456–465. [CrossRef]
29. Kim, J.-H.; Kim, M.S.; Oh, H.-G.; Lee, H.-Y.; Park, J.-W.; Lee, B.-G.; Park, S.-H.; Moon, D.-I.; Shin, E.-H.; Oh, E.-K.; et al. Treatment of eggshell with casein phosphopeptide reduces the severity of ovariectomy-induced bone loss. *Lab. Anim. Res.* **2013**, *29*, 70–76. [CrossRef]
30. Liu, G.; Sun, S.; Guo, B.; Miao, B.; Luo, Z.; Xia, Z.; Ying, D.; Liu, F.; Guo, B.; Tang, J.; et al. Bioactive peptide isolated from casein phosphopeptides promotes calcium uptake in vitro and in vivo. *Food Funct.* **2018**, *9*, 2251–2260. [CrossRef]
31. Meisel, H.; Olieman, C. Estimation of calcium-binding constants of casein phosphopeptides by capillary zone electrophoresis. *Anal. Chim. Acta* **1998**, *372*, 291–297. [CrossRef]
32. Teucher, B.; Majsak-Newman, G.; Dainty, J.R.; McDonagh, D.; Fitzgerald, R.J.; Fairweather-Tait, S.J. Calcium absorption is not increased by caseinophosphopeptides. *Am. J. Clin. Nutr.* **2006**, *84*, 162–166. [CrossRef]
33. Narva, M.; Kärkkäinen, M.; Poussa, T.; Lamberg-Allardt, C.; Korpela, R. Caseinphosphopeptides in milk and fermented milk do not affect calcium metabolism acutely in postmenopausal women. *J. Am. Coll. Nutr.* **2003**, *22*, 88–93. [CrossRef]
34. Pandey, M.; Kapila, S.; Kapila, R.; Trivedi, R.; Karvande, A. Evaluation of the osteoprotective potential of whey derived-antioxidative (YVEEL) and angiotensin-converting enzyme inhibitory (YLLF) bioactive peptides in ovariectomised rats. *Food Funct.* **2018**, *9*, 4791–4801. [CrossRef]
35. Pandey, M.; Kapila, R.; Kapila, S. Osteoanabolic activity of whey-derived anti-oxidative (MHIRL and YVEEL) and angiotensin-converting enzyme inhibitory (YLLF, ALPMHIR, IPA and WLAHK) bioactive peptides. *Peptides* **2018**, *99*, 1–7. [CrossRef] [PubMed]
36. Mada, S.B.; Reddi, S.; Kumar, N.; Kumar, R.; Kapila, S.; Kapila, R.; Trivedi, R.; Karvande, A.; Ahmad, N. Antioxidative peptide from milk exhibits antiosteopenic effects through inhibition of oxidative damage and bone-resorbing cytokines in ovariectomized rats. *Nutrients* **2017**, *43*, 21–31. [CrossRef]

37. Mada, S.B.; Reddi, S.; Kumar, N.; Kapila, S.; Kapila, R. Protective effects of casein-derived peptide VLPVPQK against hydrogen peroxide–induced dysfunction and cellular oxidative damage in rat osteoblastic cells. *Hum. Exp. Toxicol.* **2017**, *36*, 967–980. [CrossRef]
38. Weitzmann, M.N. Estrogen deficiency and bone loss: An inflammatory tale. *J. Clin. Investig.* **2006**, *116*, 1186–1194. [CrossRef]
39. D'Amelio, P.; Grimaldi, A.; Di Bella, S.; Brianza, S.Z.M.; Cristofaro, M.A.; Tamone, C.; Giribaldi, G.; Ulliers, D.; Pescarmona, G.P.; Isaia, G. Estrogen deficiency increases osteoclastogenesis up-regulating T cells activity: A key mechanism in osteoporosis. *Bone* **2008**, *43*, 92–100. [CrossRef]
40. Taxel, P.; Kaneko, H.; Lee, S.-K.; Aguila, H.L.; Raisz, L.G.; Lorenzo, J.A. Estradiol rapidly inhibits osteoclastogenesis and RANKL expression in bone marrow cultures in postmenopausal women: A pilot study. *Osteoporos. Int.* **2007**, *19*, 193–199. [CrossRef] [PubMed]
41. Chen, K.L.; Madak-Erdogan, Z. Estrogen and Microbiota Crosstalk: Should We Pay Attention? *Trends Endocrinol. Metab.* **2016**, *27*, 752–755. [CrossRef] [PubMed]
42. Kaliannan, K.; Robertson, R.C.; Murphy, K.; Stanton, C.; Kang, C.; Wang, B.; Hao, L.; Bhan, A.K.; Kang, J.X. Estrogen-mediated gut microbiome alterations influence sexual dimorphism in metabolic syndrome in mice. *Microbiome* **2018**, *6*, 1–22. [CrossRef]
43. Zhang, Z.; Chen, Y.; Xiang, L.; Wang, Z.; Xiao, G.G.; Hu, J. Effect of Curcumin on the Diversity of Gut Microbiota in Ovariectomized Rats. *Nutrients* **2017**, *9*, 1146. [CrossRef]
44. Xie, W.; Han, Y.; Li, F.; Gu, X.; Su, D.; Yu, W.; Li, Z.; Xiao, J. Neuropeptide Y1 Receptor Antagonist Alters Gut Microbiota and Alleviates the Ovariectomy-Induced Osteoporosis in Rats. *Calcif. Tissue Int.* **2019**, *106*, 444–454. [CrossRef]
45. Chung, Y.W.; Gwak, H.-J.; Moon, S.; Rho, M.; Ryu, J.-H. Functional dynamics of bacterial species in the mouse gut microbiome revealed by metagenomic and metatranscriptomic analyses. *PLoS ONE* **2020**, *15*, e0227886. [CrossRef]
46. Ning, K.; Lu, K.; Chen, Q.; Guo, Z.; Du, X.; Riaz, F.; Feng, L.; Fu, Y.; Yin, C.; Zhang, F.; et al. Epigallocatechin Gallate Protects Mice against Methionine–Choline-Deficient-Diet-Induced Nonalcoholic Steatohepatitis by Improving Gut Microbiota To Attenuate Hepatic Injury and Regulate Metabolism. *ACS Omega* **2020**, *5*, 20800–20809. [CrossRef]
47. Chen, Y.-J.; Wu, H.; Wu, S.; Lu, N.; Wang, Y.-T.; Liu, H.-N.; Dong, L.; Liu, T.-T.; Shen, X. Parasutterella, in association with irritable bowel syndrome and intestinal chronic inflammation. *J. Gastroenterol. Hepatol.* **2018**, *33*, 1844–1852. [CrossRef]
48. Jian, X.; Zhu, Y.; Ouyang, J.; Wang, Y.; Lei, Q.; Xia, J.; Guan, Y.; Zhang, J.; Guo, J.; He, Y.; et al. Alterations of gut microbiome accelerate multiple myeloma progression by increasing the relative abundances of nitrogen-recycling bacteria. *Microbiome* **2020**, *8*, 1–21. [CrossRef] [PubMed]
49. Alhinai, E.A.; Walton, G.E.; Commane, D.M. The Role of the Gut Microbiota in Colorectal Cancer Causation. *Int. J. Mol. Sci.* **2019**, *20*, 5295. [CrossRef]
50. Naderpoor, N.; Mousa, A.; Gomez-Arango, L.F.; Barrett, H.L.; Nitert, M.D.; De Courten, B. Faecal Microbiota Are Related to Insulin Sensitivity and Secretion in Overweight or Obese Adults. *J. Clin. Med.* **2019**, *8*, 452. [CrossRef]
51. D'Amelio, P.; Sassi, F. Gut Microbiota, Immune System, and Bone. *Calcif. Tissue Int.* **2018**, *102*, 415–425. [CrossRef]
52. Besten, G.D.; Van Eunen, K.; Groen, A.K.; Venema, K.; Reijngoud, D.-J.; Bakker, B.M. The role of short-chain fatty acids in the interplay between diet, gut microbiota, and host energy metabolism. *J. Lipid Res.* **2013**, *54*, 2325–2340. [CrossRef]
53. Bloemen, J.G.; Damink, S.W.O.; Venema, K.; Buurman, W.A.; Jalan, R.; DeJong, C.H. Short chain fatty acids exchange: Is the cirrhotic, dysfunctional liver still able to clear them? *Clin. Nutr.* **2010**, *29*, 365–369. [CrossRef]
54. Ma, S.; Qin, J.; Hao, Y.; Shi, Y.; Fu, L. Structural and functional changes of gut microbiota in ovariectomized rats and their correlations with altered bone mass. *Aging* **2020**, *12*, 10736–10753. [CrossRef]

Publisher's Note: MDPI stays neutral with regard to jurisdictional claims in published maps and institutional affiliations.

© 2020 by the authors. Licensee MDPI, Basel, Switzerland. This article is an open access article distributed under the terms and conditions of the Creative Commons Attribution (CC BY) license (http://creativecommons.org/licenses/by/4.0/).

Review

Milk and Dairy Products and Their Impact on Carbohydrate Metabolism and Fertility—A Potential Role in the Diet of Women with Polycystic Ovary Syndrome

Justyna Janiszewska, Joanna Ostrowska * and Dorota Szostak-Węgierek

Department of Clinical Dietetics, Faculty of Health Sciences, Medical University of Warsaw, E Ciołka Str. 27, 01-445 Warsaw, Poland; jjaniszewska@wum.edu.pl (J.J.); dorota.szostak-wegierek@wum.edu.pl (D.S.-W.)
* Correspondence: jostrowska@wum.edu.pl; Tel.: +48-22-572-09-31

Received: 22 October 2020; Accepted: 11 November 2020; Published: 13 November 2020

Abstract: Milk and dairy products are considered an important component of healthy and balanced diet and are deemed to exert a positive effect on human health. They appear to play a role in the prevention and treatment of carbohydrate balance disturbances. The products include numerous valuable components with a potential hypoglycemic activity, such as calcium, vitamin D, magnesium and probiotics. Multiple authors suggested that the consumption of dairy products was negatively associated with the risk of type 2 diabetes mellitus, insulin resistance and ovulation disorders. However, there are still numerous ambiguities concerning both the presumed protective role of dairy products in carbohydrate metabolism disorders, and the advantage of consuming low-fat dairy products over high-fat ones, especially in women with the risk of ovulation disorders. Therefore, this literature review aims at the presentation of the current state of knowledge concerning the relationship between dairy product consumption and the risk of insulin resistance, type 2 diabetes mellitus in women, and the potential effect on the course of polycystic ovary syndrome.

Keywords: milk; dairy products; type 2 diabetes mellitus; insulin resistance; polycystic ovary syndrome; fertility; ovulation

1. Introduction

Milk and dairy products have been considered as an important component of healthy and balanced diet for many years. According to Polish recommendations of the Food and Nutrition Institute [1], they should be included in everyday diet regardless of age. It is recommended that adults consume at least two glasses of milk daily. They may be replaced with yoghurt, kefir and, partially, cheese.

Cow milk contains 87% of water, 3–4% of lipids, 3.5% of protein, 5% of lactose and 1.2% of vitamins (B2, B12, A, D) and minerals (calcium, phosphorus, potassium, magnesium, zinc and selenium). Cow milk fat consists of 60% of saturated fatty acids, including mainly palmitic acid. The milk of ruminants also contains conjugated dienes of linoleic acids (CLAs) which present numerous health-promoting properties. However, the particularly nutritious value of milk is mostly due to high-quality protein which includes the whole set of exogenous amino acids necessary for the synthesis of body protein. Milk protein consists of 80% of casein, 20% of whey, which plays a role in short- and long-lasting regulation of food consumption via the induction of satiety signals, thereby promoting the maintenance of appropriate body weight. Bioactive milk peptides may exert a positive influence on human health through the regulation of physiological functions, a direct effect on metabolism and on some receptors. It was suggested that they presented antineoplastic, antihypertensive, antithrombotic and immunomodulatory properties. Milk and dairy products were also attributed favorable properties in the prevention and treatment of carbohydrate metabolism disorders [2–4]. Polycystic ovary syndrome (PCOS) is one of the most common endocrine disorders in women of reproductive age. It is accompanied by oligoovulation and/or the lack of ovulation, clinical and/or biochemical hyperandrogenism and the presence of polycystic ovaries in ultrasound examination [5]. It is estimated that even 90–95% of ovulatory infertility cases are caused by this medical condition. Due to the presence of endocrine and metabolic disorders, women with PCOS are a group that is particularly susceptible to the development of insulin resistance, secondary disorders of glucose tolerance and type 2 diabetes mellitus (T2DM), cardiovascular diseases and dyslipidemia [6]. Increasing attention has recently been paid to the significance of dairy products in the diet of women with PCOS, particularly comprising their influence on ovulation and fertility and the associated risk of carbohydrate metabolism disorders, such as insulin resistance or T2DM. The obtained results are frequently contradictory. Therefore, it is necessary to conduct a comprehensive overview of the most recent studies in this area.

2. Dairy Products and Insulin Resistance

The effect of milk and dairy products on carbohydrate metabolism is the subject of numerous studies. However, the results are still contradictory. It is known that protein consumption has the same capacity to stimulate insulin secretion as carbohydrate consumption. However, it was demonstrated that not all protein-containing products exerted the same effect on insulin secretion and modulated insulin sensitivity in tissues in various ways. Milk proteins exerted the strongest influence on the secretion of insulin and incretins compared to other animal proteins [7]. It is mostly attributed to the high content of branched-chain amino acids (leucines, isoleucines, valines) which activate various pathways associated with insulin resistance [8]. However, apart from protein components, such as insulinogenic amino acids and bioactive peptides, dairy products also contain calcium, magnesium, potassium and carbohydrates with low glycemic index, which all seem to have a favorable effect on the control of glycemia, insulin secretion, insulin sensitivity of tissues and the reduction in the risk of T2DM. Moreover, unsaturated trans fatty acids which naturally occur in milk fat modulate the expression of PPAR-γ (peroxisome proliferator-activated receptor γ) and PPAR-α (peroxisome proliferator-activated receptor α), which is also beneficial in glucose homeostasis. Furthermore, fermentation and enhancing dairy products with probiotics and vitamin D may improve their glucoregulatory activity [7,9].

According to some studies conducted in women and men, the consumption of milk and dairy products might be associated with higher tissue sensitivity to the activity of insulin and lower fasting insulin levels [10,11]. The observation was confirmed by a meta-analysis of 30 randomized clinical trials. It demonstrated that the consumption of dairy products, especially low-fat ones, was beneficial in terms of tissue insulin sensitivity [12]. However, the results of some studies suggested that only long-lasting consumption of dairy products might have a beneficial effect on insulin sensitivity in tissues. A systematic review of 10 interventional studies [13] was conducted to analyze the effect of dairy products consumption on insulin sensitivity in individuals without T2DM. It was demonstrated that improved sensitivity to insulin occurred only after 12 weeks of a diet higher in dairy content, while studies lasting below 8 weeks did not show any significant changes concerning insulin sensitivity in tissues. Similar results were obtained by Rideout et al. [14], who noticed that the values of HOMA-IR (Homeostatic Model Assessment-Insulin Resistance) markedly improved over 6 months in individuals who consumed higher amounts of low-fat dairy products (four servings of milk of yoghurt daily) compared to those who consumed less (less than two servings of milk or yoghurt daily). However, not all studies confirmed those observations. A systematic review and a meta-analysis of 44 randomized studies revealed that the increased supply of dairy products exerted no effect on fasting insulin concentrations and HOMA-IR index values in healthy diabetes-free individuals [15]. A randomized clinical trial conducted by O'Connor et al. [16] also showed no significant changes in insulin secretion and insulin sensitivity in adults with hyperinsulinemia who were characterized by high dairy consumption (>4 servings/day) compared to those who consumed small amounts of dairy ($\leq$2 servings/day). Interesting results were obtained by Eelderink et al. [17] who demonstrated that postprandial insulin concentrations in persons consuming a diet with low dairy content (≤1 serving of dairy per day) were not significantly different compared to participants who consumed high amounts of dairy products (five servings/day in women and six servings/day in men). However, significantly higher fasting insulin and HOMA-IR values were associated with a diet high in dairy products compared to low-dairy diet (2.21 ± 0.91 versus 1.99 ± 0.72; $p = 0.027$).

There is paucity of data regarding the correlation between the consumption of milk products and the risk of insulin resistance in women. However, it may be presumed that such products may increase the risk of insulin resistance in women at various ages. Lawlor et al. [18], who investigated the association between milk consumption, insulin resistance and metabolic syndrome in 4024 British women, observed that women who had never drunk milk had lower HOMA-IR values and developed metabolic syndrome less frequently than women who regularly drank milk. Similarly, a study conducted by Tucker et al. [19] showed that the values of HOMA-IR went up with increased milk consumption in the studied women. Factors such as age, body weight, adipose tissue amount or physical activity had no significant influence on the relationship between milk consumption and HOMA-IR values. The results underlay the conclusion that long-lasting hyperinsulinemia which occurred due to high dairy consumption may be a significant predictor of insulin resistance in women. Moreover, an 8-week interventional study by Phy et al. [20] demonstrated that diet low in starch and milk products resulted in an increased sensitivity to insulin (HOMA-IR reduction by 1.9 ± 1.2, $p < 0.001$), lowered fasting insulin level (−17.0 ± 13.6 µg/mL, $p < 0.001$) and a 75 g 2 h oral glucose tolerance test (−82.8 ± 177.7 µg/mL, $p = 0.03$) in women with PCOS. An unfavorable influence of milk products in women was also confirmed in a study by von Post-Skagegård et al. [21], who demonstrated that the 120 min ratios of insulin to glucose and insulin to peptide C were significantly higher after a meal containing milk proteins compared to a meal containing fish or soy protein. Furthermore, Turner et al. [22] noted that HOMA-IR was markedly lower in women who had consumed a diet including red meat compared to diet containing milk products. Moreover, women who consumed <1 portion of milk products daily were characterized by significantly lower fasting insulin levels and HOMA-IR compared to women whose diet included from four to six portions of low-fat milk products daily.

However, not all studies indicated a negative impact of dairy product consumption on the risk of insulin resistance in women. According to some authors, the influence of dairy products was neutral or even favorable in terms of sensitivity to insulin in women. A study by Drouin-Chartier et al. [23] revealed that a diet including milk had no effect on HOMA-IR and fasting insulin levels in postmenopausal women. The Coronary Artery Risk Development in Young Adults Study (The CARDIA Study) [24] showed that a daily increase in the consumption of milk products translated into the reduction in the risk of developing insulin resistance by 30% in Black women (Odds Ratio (OR) 0.70, 95% Confidence Interval (CI), 0.54–0.91, $p < 0.05$) and by 38% in White women (OR 0.62, 95% CI, 0.46–0.84, $p < 0.05$). Yoghurt appears to be particularly beneficial in the prophylaxis of insulin resistance in women. A study by Chen et al. [25] revealed that full-fat yoghurt significantly reduced HOMA-IR, fasting insulin levels and a 75 g 2 h oral glucose tolerance test compared to full-fat milk in women with metabolic syndrome and nonalcoholic fatty liver disease. However, the study showed significantly reduced HOMA-IR, fasting glucose and insulin levels in a group of women consuming milk, while the level of insulin in a 75 g 2 h oral glucose tolerance test significantly increased. Based on the results, the authors suggested that the unfavorable influence of milk consumption on carbohydrate metabolism was not associated with weakened insulin sensitivity, but only with the fact that milk might prolong postprandial insulin secretion.

Some authors pointed out particularly beneficial properties of probiotics. A randomized clinical trial conducted in a group of women with PCOS showed that supplementation with probiotics contributed to a considerable reduction in fasting glucose levels [26]. Other studies conducted in women showed that the consumption of yoghurts fortified both with vitamin D and probiotic bacteria was associated with a significantly higher reduction in HOMA-IR and fasting insulin compared to women consuming traditional low-fat yoghurt [27,28], so the favorable properties of yoghurt may be enhanced with the addition of probiotic bacteria and vitamin D. Therefore, the consumption of yoghurt (especially the fortified types) by women seems to have a beneficial effect on tissue insulin sensitivity. However, their positive properties may not be fully confirmed due to the paucity of studies in women with PCOS. Detailed results of studies on the effects of dairy consumption on insulin resistance in women are described in Table 1.

Basing on the observations described above, it cannot be clearly determined whether the consumption of milk and dairy products has a beneficial effect on insulin sensitivity in tissues in women, and due to the lack of studies conducted in women with PCOS, it is even more difficult to draw conclusions concerning their beneficial effect in this condition. It may even be assumed that the consumption of diet with a high dairy content may be a predictor of hyperinsulinemia and insulin resistance in women. However, it is worth emphasizing that the results of some studies suggested that only long-lasting consumption of dairy products might have a beneficial effect on reducing insulin resistance in tissues. Therefore, it is necessary to conduct more well-planned randomized clinical trials in women with PCOS to provide a clear answer concerning the significance of dairy product consumption in the prevention and treatment of insulin resistance in this condition.

Table 1. The influence of dairy product consumption on insulin resistance in women.

Author/Reference Number	Year	Study Design	Sample (n)	Outcome Measures	Result
Intake of Total Dairy Products					
Pereira et al. [24]	2002	Population-based prospective study Intake of total dairy products	3157 Black and White adults aged 18 to 30 years	Fasting plasma insulin and glucose	An increase in the daily intake of milk products reduced the risk of insulin resistance by 30% in black women (OR 0.70, 95% CI, 0.54–0.91, $p < 0.05$) and by 38% in white women (OR 0.62, 95% CI, 0.46–0.84, $p < 0.05$).
Tucker et al. [19]	2015	Cross-sectional study Intake of total dairy products; low intake—0 to 0.5 servings of dairy per day, moderate intake—0.6 to 1.5 servings of dairy per day, high intake—1.6 to 6 servings of dairy per day	272 middle-aged, nondiabetic and apparently healthy women	HOMA-IR score	Women who consumed diet high in dairy products had markedly higher HOMA-IR values (0.41 ± 0.53) compared to those who consumed moderate (0.22 ± 0.55) and low amounts of dairy (0.19 ± 0.58).
Intake of Low-Fat Dairy Products					
Turner et al. [22]	2015	Randomized crossover study Intake of low-fat dairy products or red meat	47 overweight and obese men and women > 20 years old	Fasting insulin, HOMA-IR score, Matsuda Index	Fasting insulin was significantly higher after a diet including milk products compared to diet including red meat (7.38 versus 5.62, $p = 0.02$). HOMA-IR was significantly higher after a diet including low-fat milk products compared to diet including red meat (1.71 versus 1.31 $p = 0.01$). Insulin sensitivity calculated with the Matsuda method was lower by 14.7% in women who had a diet including milk products compared to diet including red meat (6.81 versus 8.14, $p = 0.01$). Women who consumed <1 portion of milk products daily were characterized by significantly lower fasting insulin levels and HOMA-IR, and a significantly higher Matsuda index compared to women whose diet included from 4 to 6 portions of low-fat milk products daily (fasting insulin—6.16 versus 7.38, $p = 0.05$, HOMA—1.42 versus 1.71, $p = 0.05$ and Matsuda Index 8.61 versus 6.81, $p = 0.05$).
Intake of Milk and Milk Protein					
Lawlor et al. [18]	2005	Prospective cohort study Intake of milk versus no intake of milk	4024 British women aged 60–79 years	HOMA-IR score	Women who did not drink milk had their HOMA-IR lower by 13% compared to women who drank milk (1.49 versus 1.72).

Table 1. Cont.

Author/Reference Number	Year	Study Design	Sample (n)	Outcome Measures	Result
Intake of Total Dairy Products					
von Post-Skagegård et al. [21]	2006	A randomized study Intake of three meals with different types of protein (either cod protein, milk protein or soy protein)	17 healthy women, 30-65 years old	Blood glucose, serum insulin, C-peptide	The 120 min insulin to glucose ratio was higher after a meal including milk protein compared to meals including cod or soy protein (milk protein—4.36, cod protein—2.03, soy protein—2.78, $p = 0.0002$). The 120 min insulin to peptide C ratio was significantly higher in case of a meal including milk protein compared to meals including cod or soy protein (milk protein—0.008, cod protein—0.003, soy protein—0.005, $p = 0.001$).
Drouin-Chartier et al. [23]	2015	Randomized, crossover study, diet for 6 weeks, one with 3.2 servings/d of 2% fat milk per 2000 kcal and another without milk	27 postmenopausal women in good health with abdominal obesity, less than 70 years of age	Fasting glucose, fasting insulin, Matsuda Index	No effect of milk on fasting insulin levels and insulin sensitivity index. Both diets, with and without milk, significantly reduced fasting glucose levels (diet including milk—6.08 versus 5.77, $p < 0.001$, diet not including milk—5.98 versus 5.80, $p < 0.009$).
Intake of Yoghurt					
Madjd et al. [28]	2016	Randomized single-blind controlled trial Intake of low-fat yoghurt versus probiotic yoghurt	Overweight and obese women	Fasting plasma glucose, 2 h glucose, fasting plasma insulin, HOMA-IR score, HbA1c	A significantly higher reduction was observed as regards HOMA-IR, 2 h postprandial glucose and fasting insulin in a group of women consuming probiotic yoghurt. Fasting glucose levels, 2 h glucose level, HbA1c, fasting insulin and HOMA-IR significantly decreased in both groups.
Jafari et al. [27]	2016	Randomized, placebo-controlled, double-blind parallel-group clinical trial Intake of vitamin D fortified yoghurt versus plain low-fat yoghurt for 12 weeks	59 post-menopausal women with type 2 diabetes	HOMA-IR, QUICKI	Insulin sensitivity of tissues was increased in a group of women who consumed yoghurt fortified with vitamin D—HOMA-IR (3.32 versus 2.13, $p = 0.02$), QUICKI (0.331 versus 0.348, $p = 0.001$) and fasting insulin was reduced (7.71 versus 5.17, $p = 0.03$). The markers of carbohydrate metabolism deteriorated in a group of women consuming low-fat yoghurt.
No Dairy Products in the Diet					
Phy et al. [20]	2015	Intervention study 8-week diet without starch and dairy products	24 overweight and obese women (BMI ≥ 25 kg/m^2 and ≤ 45 kg/m^2) with PCOS	Fasting and 2 h glucose and insulin, HOMA-IR score	Diet without starch and milk products reduced fasting insulin by 52% (−17.0 ± 13.6 µg/mL, $p < 0.001$), 2 h insulin in the load test of 75 g glucose by 37% (−82.8 ± 177.7 µg/mL, $p = 0.03$) and HOMA-IR by 51% (−1.9 ± 1.2, $p < 0.001$).

HOMA-IR, Homeostatic Model Assessment—Insulin Resistance; QUICKI, Quantitative Insulin Sensitivity Check Index; BMI, body mass index; HbA1c, glycated hemoglobin; PCOS, polycystic ovary syndrome; OR, odds ratio; CI, confidence interval.

3. Dairy Products and Type 2 Diabetes Mellitus

As mentioned above, dairy products, due to their high content of whey proteins which are rich in branched-chain amino acids (leucine, isoleucine, valine) and lysine, may stimulate insulin secretion and reduce postprandial glycemia, which is particularly favorable in the prophylaxis of T2DM [8]. Conversely, the excessive amount of branched-chain amino acids in the diet is considered to lead to insulin resistance in tissues via the activation of mTOR (mammalian target of rapamycin) kinase, thereby increasing the risk of T2DM [29]. It was corroborated by a prospective study conducted in a cohort of Chinese women. The study showed that higher branched-chain amino acid content consumed with meat and dairy products in the second part of pregnancy was associated with the increase in the risk of gestational diabetes mellitus by approximately 95% [30]. According to some authors, whey proteins, by modifying gene expression, may affect glucose metabolism, also by its increased use in the liver. Moreover, the influence of dairy products on glucose metabolism and the risk of T2DM may depend on glucokinase genetic polymorphism which is specific for a particular person. Therefore, it is suggested that some individuals may find high dairy intake more beneficial than others [31]. Furthermore, it seems that hyperinsulinemia due to dairy intake may be favorable in glucose homeostasis regulation in patients with hyperglycemia and T2DM [32]. Systematic reviews and meta-analyses of observational and cohort studies in women and men [32–36] indicated that dairy product consumption was negatively associated with the risk of T2DM. Moreover, such a relationship was particularly intensified in cases of low-fat and fermented dairy products. A randomized study by Díaz-López et al. [37] also revealed a negative relationship between total dairy intake and the risk of T2DM. It was particularly visible in the case of low-fat dairy products. It is consistent with the results of a meta-analysis of 13 cohort studies. The meta-analysis revealed that increasing the consumption of low-fat dairy products by 200 g daily was linked to T2DM risk reduction by 4% (Relative Risk (RR) 0.96; 95% CI 0.92, 1.00; $p = 0.072$). In the case of full-fat dairy products no such correlation was observed (RR 0.98; 95% CI 0.93, 1.04; $p = 0.52$) [32]. Similar outcomes were obtained in the Lifelines Cohort Study [38], in which a 2% reduction in the risk of prediabetes was achieved with the intake of skimmed dairy products (RR 0.98; 95% CI 0.97, 1.00; $p = 0.02$) and fermented dairy products (RR 0.98; 95% CI 0.97, 0.99; $p = 0.004$) increased by 100 g daily. Conversely, the consumption of full-fat dairy products was associated with the increased risk (RR 1.03; 95% CI 1.01, 1.06; $p = 0.004$). A systematic review of meta-analyses by Drouin-Chartier et al. [39] revealed that current evidence obtained from scientific research indicated favorable or neutral interrelations between the consumption of dairy products and T2DM occurrence. However, recommendations concerning the advantage of low-fat product consumption over full-fat ones were confirmed by a low number of reliable scientific papers. Similar conclusions were reached by Yakoob et al. [40] and Guo et al. [41] indicating no convincing evidence to confirm the hypothesis stating that low-fat dairy intake was more effective in reducing the risk of type 2 diabetes compared to full-fat dairy products. Moreover, a systematic review of studies concerning the relationship between dairy product intake and the risk of cardiovascular disease showed that the consumption of full-fat, semi-skimmed and fermented dairy products was neutrally associated with the risk of T2DM, while the consumption of low-fat dairy was positively associated with the risk [36]. According to some authors, full-fat dairy products, despite the high content of saturated fatty acids, had a positive effect on human health. It was also stated that there was insufficient evidence to confirm that those fatty acids increased the risk of cardiometabolic pathologies, such as T2DM, and they might even present some protective properties [42–44]. It is consistent with the results of a cohort study by Korat et al. [45] who demonstrated no relationship between milk fat and the risk of T2DM both in the population of men and women.

It is considered that sex is one of the biological factors modulating the course and incidence of cardiometabolic diseases, including T2DM. An increasing amount of evidence confirmed the role of sex in the course and treatment of T2DM and its influence on the increased risk of the disease [46]. Research conducted in the populations of women indicated that a diet rich in milk products was associated with a lower risk of developing T2DM [47,48]. The observations were confirmed by systematic reviews and meta-analyses of observational and cohort studies [32,34,35], which indicated that milk product consumption was inversely correlated with the risk of T2DM in women. Moreover, Kirri et al. [49] observed that the beneficial correlation between milk product consumption and the risk of T2DM was statistically significant only in women. A prospective study by Liu et al. [50] showed that the risk of T2DM in women from the highest quintile of milk product consumption (>2.9 servings/day) was lower by 20% compared to women from the lowest quintile (<0.85 servings/day). Furthermore, each increment of the daily consumption by one serving was associated with T2DM risk reduction by 4% (RR 0.96, 95% CI, 0.93–1.0, $p < 0.05$). The beneficial effect of milk product consumption on the risk of T2DM in women is mainly attributed to low-fat milk products, while their high-fat equivalents may even increase the risk. It was confirmed by the results of a study by Margolis et al. [51] who demonstrated that the risk of T2DM in women from the highest quintile of low-fat milk product consumption was lower by 30% compared to women from the lowest quintile (RR 0.70, 95% CI, 0.64–0.77, $p < 0.0001$). However, no such correlation was demonstrated for high-fat milk products. A prospective cohort The Black Women's Health Study [52] also showed that the consumption of low-fat milk products was associated with the risk of T2DM lower by 13% in Black women. At the same time, no such correlation was observed for high-fat milk products. Another prospective The Nurses' Health Study [53] revealed a 25% lower risk of T2DM (RR 0.75; 95% CI 0.55, 1.02, $p = 0.03$) in women from the highest quintile of total milk product consumption compared to women from the lowest quintile. The beneficial influence of milk product consumption was observed both in the cases of low-fat and high-fat products. Moreover, constant high consumption of milk products continued in adulthood was also associated with a lower risk of T2DM, which might suggest that long-lasting milk product consumption might be beneficial in the context of the prophylaxis of T2DM in women.

Yoghurt appears to be a particularly important dairy product. It should be introduced into the diet of women with PCOS because of strong scientific evidence suggestive of the relationship between its consumption and lowering the risk of developing type 2 diabetes. Yoghurt consumption increases the concentrations of circulating anorexic peptides—glucagon-like peptide 1 (GLP-1) and peptide YY (PYY), whose activity is associated with the improvement of glucose homeostasis via the modulation of hepatic gluconeogenesis [54]. Fermented dairy product intake was associated with a lower risk of developing diabetes by the influence on intestinal microbiota and, thereby, on the insulin sensitivity of tissues and glucose tolerance [55]. Probiotics contained in such products may determine their favorable influence on T2DM risk [56]. A study by Liu et al. [50] revealed that the risk of T2DM was lower by 18% in women who consumed at least two servings of yoghurt weekly compared to women who consumed yoghurt less frequently than once a month (RR 0.82, 95% CI: 0.70–0.97, $p = 0.03$). Similar results were obtained in a study by Buziau et al. [57], in which the risk of developing T2DM was 19% lower in women from the highest tertile of yoghurt consumption than in women from the lowest tertile (OR 0.81; 95% CI: 0.67; 0.99; $p = 0.041$). It is consistent with the results obtained by Rosenberg et al. [58] and Margolis et al. [51] who confirmed a lower risk of developing T2DM in women consuming yoghurt. Detailed results of studies on the effects of dairy consumption on the risk of T2DM in women are described in Table 2.

Therefore, high dairy intake seems to reduce the risk of developing prediabetes and type 2 diabetes in women. It appears particularly beneficial to introduce yoghurt, fermented and low-fat dairy products into the diet. However, based on previous study results it may not be clearly confirmed whether the consumption of high-fat dairy products by women increased the risk of T2DM and had a negative impact on glycemia. Furthermore, due to the paucity of studies concerning the relationship between the consumption of milk products and the risk of T2DM in women with PCOS it seems justified to conduct a randomized clinical study in such a group of women in order to provide an explicit answer concerning the question of the influence of milk products on the course and treatment of PCOS.

Table 2. The influence of dairy product intake on the risk of type 2 diabetes mellitus (T2DM) in women.

Author/Reference Number	Year	Study Design	Sample (n)	Outcome Measures	Results
Pittas et al. [48]	2006	Prospective cohort study Intake of total dairy products	83,779 apparently healthy women, aged 30–55 years	T2DM	The risk of T2DM lower by 13% in women consuming higher amounts (>3 servings/day) of dairy products compared to women consuming small amounts (<1 serving/day).
Liu et al. [50]	2006	Prospective cohort study Intake of total dairy products, low-fat, full-fat and yoghurt	37,183 healthy, middle-aged and older women	T2DM	The risk of T2DM lower by 20% in women consuming higher amounts (>2.9 servings/day) of dairy products compared to women consuming small amounts (<0.85 serving/day). The risk of T2DM lower by 18% in women consuming higher amounts (>2 servings/day) of low-fat dairy products compared to women consuming small amounts (≤0.27 serving/day). The risk of T2DM lower by 18% in women consuming higher amounts (>2 servings/week) of yoghurt compared to women consuming small amounts (<1 serving/month).
van Dam et al. [52]	2006	Prospective cohort study Intake of total, low-fat and full-fat dairy products	41,186 women, aged 21–69	T2DM	The risk of T2DM lower by 25% in women consuming higher amounts (>2 servings/day) of total dairy products compared to women consuming small amounts (<1 serving/week). The risk of T2DM lower by 13% in women consuming higher amounts (>1 servings/day) of low-fat dairy products compared to women consuming small amounts (<1 serving/week). No significant correlation between the risk of T2DM and the consumption of full-fat dairy products in women.
Kirri et al. [49]	2009	Prospective cohort study Intake of total dairy products, milk, cheese and yoghurt	33,919 middle-aged and older women	T2DM	The risk of T2DM lower by 29% in women consuming higher amounts (≥300 g/day) of dairy products compared to women consuming small amounts (<50 g/day). No correlation between the consumption of milk, cheese and yoghurt and the risk of T2DM in women.

Table 2. *Cont.*

Author/Reference Number	Year	Study Design	Sample (n)	Outcome Measures	Results
Malik et al. [53]	2011	Prospective cohort study	116,671 female registered nurses aged 24–42	T2DM	The risk of T2DM lower by 25% in women from the highest quintile of total milk product consumption compared to women from the lowest quintile. The risk of T2DM lower by 26% in women from the highest quintile of low-fat milk product consumption compared to women from the lowest quintile. The risk of T2DM lower by 28% in women from the highest quintile of high-fat milk product consumption compared to women from the lowest quintile.
Margolis et al. [51]	2011	Prospective cohort study Intake of total, low-fat, full-fat dairy products and yoghurt	82,076 women, aged 50–79	T2DM	The risk of T2DM lower by 21% in women consuming higher amounts (>2.6 servings/day) of total dairy products compared to women consuming small amounts (<0.7 serving/day). The risk of T2DM lower by 30% in women consuming higher amounts (>1.9 servings/day) of low-fat dairy products compared to women consuming small amounts (<0.2 serving/day). The risk of T2DM lower by 54% in women consuming higher amounts (≥2 servings/week) of yoghurt compared to women consuming small amounts (<1 serving/month).
Aune et al. [34]	2013	Systematic review and dose-response meta-analysis of cohort studies Intake of total, low-fat, full-fat dairy products, milk and yoghurt	526,482 healthy men and women ≥ 20 years	T2DM	The risk of T2DM in women reduced by 34% with the increase in milk consumption by 200 g daily. The risk of T2DM in women reduced by 33% with the increase in yoghurt consumption by 200 g daily. No significant correlation with the total, full-fat, low-fat milk product consumption and cheese consumption in women.
Gijsbers et al. [32]	2016	A dose-response meta-analysis of observational studies Intake of total dairy products	579,832 healthy men and women, aged ≥ 20 years	T2DM	The risk of T2DM decreased by 3% with the increase in total dairy intake by 200 g daily. The risk of T2DM in women decreased by 8% with the increase in low-fat dairy intake by 200 g daily. The risk of T2DM in women increased by 2% with the increase in low-fat milk consumption by 200 g daily. The risk of T2DM in women reduced by 5% with the increase in high-fat milk consumption by 200 g daily. The risk of T2DM in women reduced by 11% with the increase in yoghurt consumption by 50 g daily. No correlation between the risk of T2DM and the total, high-fat milk product consumption and cheese consumption in women.

Table 2. Cont.

Author/Reference Number	Year	Study Design	Sample (n)	Outcome Measures	Results
Mishali et al. [35]	2019	Systematic review and meta-analysis of prospective cohort studies with subgroup analysis of men versus women Intake of total dairy products	545,677 men and women aged ≥ 18 years	T2DM	The risk of T2DM lower by 13% in women consuming higher amounts of dairy products compared to women consuming small amounts.
Buziau et al. [57]	2019	Prospective cohort study Intake of yoghurt	8748 Australian women, aged 45–50	T2DM	The risk of T2DM lower by 19% in women from the highest tertile of yoghurt consumption compared to women from the lowest tertile.
Rosenberg et al. [58]	2020	Prospective cohort study Total intake of yoghurt	59,000 U.S. Black women, aged 21–69	T2DM	The risk of T2DM lower by 18% in women consuming higher amounts (≥1 serving/day) of yoghurt compared to women consuming small amounts (<1 serving/month).

4. Dairy Products versus Ovulation and Fertility in Women

Research on the influence of dairy products on female fertility and ovulation has been conducted for many years. The results of animal studies suggested a potentially unfavorable influence of dairy products on reproductive functions due to high lactose content, which reduced ovulation in rats and led to premature ovarian insufficiency. Moreover, it was demonstrated that rats fed with high amounts of galactose were characterized by markedly lower concentrations of estradiol and elevated levels of FSH (follicle stimulating hormone) and LH (luteinizing hormone). Rats fed with lactose had considerably reduced progesterone concentrations, while no differences were confirmed in serum estradiol concentrations [59,60].

Changes in hormone levels resulting from dairy product intake were also observed in studies conducted in people. According to Kim et al. [61] each increase in the consumption of dairy products by one serving per day was associated with the reduction in serum estradiol concentrations by 4.6% and free estradiol by 4.0%. Conversely, the highest total dairy intake was linked to an increase in LH concentrations by 2.9% over the whole cycle compared to the lowest intake. However, a study by Greenlee et al. [62] revealed that dairy products supported female fertility, because the participants drinking over three glasses of milk daily were characterized by a 70% drop in the risk of infertility compared to women who did not drink milk at all. Wise et al. [63] compared the cohorts of women from Denmark and North America. In both groups they observed a positive association between milk consumption and fertility. Moreover, the authors found no significant differences between low- and full-fat milk consumption as regards the influence on fertility in either of the cohorts. Furthermore, higher lactose intake was associated with higher fertility in the study cohorts, which is inconsistent with the previous accepted view stating that lactose impaired fertility. Additionally, Afeiche et al. [64] conducted a study on women undergoing assisted reproductive technology procedures. It was demonstrated that the group of women aged ≥35 who were in the highest quartile of dairy product intake prior to the treatment was characterized by a considerably higher probability of delivering a live neonate than women in the lowest quartile. Notably, the fat content of dairy products consumed by the participants had no influence on the strength of such a relationship. Contradictory results were arrived at by Souter et al. [65], who assessed the relationship between milk protein intake and antral follicle count (AFC) in a prospective group of women of reproductive age. They concluded that higher total milk protein intake (≥5.24% of energy value or ≥2.3 glasses of milk daily) was associated with lower AFC. The authors deduced that the factors influencing the reduction in antral follicle count in women consuming dairy products might include: high amounts of steroid hormones and growth factors present in dairy products, contamination of dairy products with pesticides and chemical substances, which might markedly affect endocrine function and folliculogenesis. Furthermore, an increased dairy intake may be associated with higher concentrations of IGF-I (insulin-like growth factor I) in the blood, which also produces a negative effect on ovarian function and antral follicle count.

A prospective cohort Nurses' Health Study II (NHS II) did not reveal an association between total dairy intake and ovulatory infertility. However, increasing the consumption of low-fat milk products by one serving daily was linked to an increase in the risk of ovulatory infertility by 11%, while adding one serving of whole milk without increasing energy content was associated with reducing the risk by over 50%. According to the authors, it was due to the fact that high-fat dairy products included more estrogens and contributed to a lower-grade increase in IGF-I concentration in the serum compared to low-fat products. Moreover, based on the results, the authors hypothesized that the relationship between low-fat and full-fat dairy intake and infertility due to anovulation was stronger in women without certain clinical signs of PCOS than in women with those signs [66]. Notably, Adebamowo et al. [67] demonstrated that the consumption of skimmed milk was associated with a more common occurrence of acne, one of the clinical signs of PCOS, which may be explained by the presence of androgen precursors in milk. Rajaeieh et al. [68] studied the relationship between dairy products intake and the risk of developing polycystic ovary syndrome. They observed that each increase in milk consumption by one serving daily resulted in an increase in PCOS risk. Furthermore, women with this medical condition were characterized by a markedly higher consumption of low-fat or skimmed milk compared to healthy women. The authors noted a possible role in the pathogenesis of PCOS, because low-fat dairy products are characterized by a considerably higher strength of stimulating IGF-I secretion compared to full-fat products. Considering the low quality of evidence, it may not be explicitly concluded that the influence of dairy products on the risk of infertility and PCOS is unfavorable. Therefore, further research is necessary in this area [69].

5. Conclusions

It seems justified to include milk and dairy products into the diet of women with polycystic ovary syndrome because of the beneficial effect of those products on the risk of developing type 2 diabetes mellitus in women. Moreover, the products appear not to have a negative effect on ovulation and fertility in women. However, due to the lack of unambiguous evidence, the advantage of full-fat over low-fat dairy products may not be confirmed despite the fact that high-fat dairy intake seems to be more beneficial in polycystic ovary syndrome. Notably, studies concerning the influence of milk consumption in women with PCOS are scarce, so its beneficial effect may not be explicitly confirmed in this group of patients. Therefore, it is necessary to conduct well-designed extensive research in women with PCOS to lead to the final conclusion as to whether milk product consumption is beneficial in their case and which products should be selected: full-fat or skimmed ones.

Author Contributions: Conceptualization, J.J. and D.S.-W.; writing—original draft preparation, J.J.; writing—review and editing, J.J.; D.S.-W. and J.O. All authors have read and agreed to the published version of the manuscript.

Funding: This research received no external funding.

Conflicts of Interest: The authors declare no conflict of interest.

References

1. The Food and Nutrition Institute. Available online: http://www.izz.waw.pl/zasady-prawidowego-ywienia (accessed on 20 April 2020).
2. Marangoni, F.; Pellegrino, L.; Verduci, E.; Ghiselli, A.; Bernabei, R.; Calvani, R.; Cetin, I.; Giampietro, M.; Perticone, F.; Piretta, L.; et al. Cow's Milk Consumption and Health: A Health Professional's Guide. *J. Am. Coll. Nutr.* **2019**, *38*, 197–208. [CrossRef]
3. Davoodi, S.; Shahbazi, R.; Esmaeili, S.; Sohrabvandi, S.; Mortazavian, A.M.; Jazayeri, S.; Taslimi, A. Health-Related Aspects of Milk Proteins. *Iran. J. Pharm. Res.* **2016**, *15*, 573–591.
4. Mohanty, D.P.; Mohapatra, S.; Misra, S.; Sahu, P.S. Milk derived bioactive peptides and their impact on human health—A review. *Saudi J. Biol. Sci.* **2016**, *23*, 577–583. [CrossRef]
5. The Rotterdam ESHRE/ASRM–Sponsored PCOS consensus workshop group, Revised 2003 consensus on diagnostic criteria and long-term health risks related to polycystic ovary syndrome (PCOS). *Hum. Reprod.* **2004**, *19*, 41–47. [CrossRef]
6. Sahmay, S.; Mtahyk, B.A.; Sofiyeva, N.; Atakul, N.; Azemi, A.; Erel, T. Serum AMH levels and insulin resistance in women with PCOS. *Eur. J. Obstet Gynecol. Reprod. Biol.* **2018**, *224*, 159–164. [CrossRef]
7. Comerford, K.; Pasin, G. Emerging evidence for the importance of dietary protein source on glucoregulatory markers and type 2 diabetes: Different effects of dairy, meat, fish, egg, and plant protein foods. *Nutrients* **2016**, *8*, 446. [CrossRef]
8. Chartrand, D.; Da Silva, M.S.; Julien, P.; Rudkowska, I. Influence of amino acids in dairy products on glucose homeostasis: The clinical evidence. *Can. J. Diabetes* **2017**, *41*, 329–337. [CrossRef]
9. Tremblay, B.L.; Rudkowska, I. Nutrigenomic point of view on effects and mechanisms of action of ruminant trans fatty acids on insulin resistance and type 2 diabetes. *Nutr. Rev.* **2017**, *75*, 214–223. [CrossRef] [PubMed]
10. Drehmer, M.; Pereira, M.A.; Schmidt, M.I.; Del Carmen, B.; Molina, M.; Alvim, S.; Lotufo, P.A.; Duncan, B.B. Associations of dairy intake with glycemia and insulinemia, independent of obesity, in Brazilian adults: The Brazilian Longitudinal Study of Adult Health (ELSA-Brasil). *Am. J. Clin. Nutr.* **2015**, *101*, 775–782. [CrossRef] [PubMed]

11. Feeney, E.L.; O'Sullivan, A.; Nugen, A.P.; McNulty, B.; Walton, J.; Bibney, E.R. Patterns of dairy food intake, body composition and markers of metabolic health in Ireland: Results from the National Adult Nutrition Survey. *Nutr. Diabetes* **2017**, *7*, 1–8. [CrossRef]
12. Sochol, K.M.; Johns, T.J.; Buttar, R.S.; Randhawa, L.; Sanchez, E.; Gal, M.; Lestrade, K.; Merzkani, M.; Abramovitz, M.K.; Mossovar-Rahmani, Y.; et al. The Effects of Dairy Intake on Insulin Resistance: A Systematic Review and Meta-Analysis of Randomized Clinical Trials. *Nutrients* **2019**, *11*, 2237. [CrossRef] [PubMed]
13. Turner, K.M.; Keogh, J.B.; Clifton, P.M. Dairy consumption and insulin sensitivity: A systematic review of short- and long-term intervention studies. *Nutr. Metab. Cardiovasc. Dis.* **2015**, *25*, 3–8. [CrossRef] [PubMed]
14. Rideout, T.C.; Marinangeli, C.P.F.; Martin, H.; Browne, R.W.; Rempel, C.B. Consumption of low-fat dairy foods for 6 months improves insulin resistance without adversely affecting lipids or bodyweight in healthy adults: A randomized free-living cross-over study. *Nutr. J.* **2013**, *12*, 1–9. [CrossRef] [PubMed]
15. O'Connor, S.; Turcotte, A.F.; Gagnon, C.; Rudkowska, I. Increased dairy product intake modifies plasma glucose levels and glycated hemoglobin: A systematic review and meta-analysis of randomized controlled trials. *Adv. Nutr.* **2019**, *10*, 262–279. [CrossRef]
16. O'Connor, S.; Julien, P.; Weisnagel, S.J.; Gagnon, C.; Rutkowska, I. Impact of a High Intake of Dairy Product on Insulin Sensitivity in Hyperinsulinemic Adults: A Crossover Randomized Controlled Trial. *Curr. Dev. Nutr.* **2019**, *3*, 1–7. [CrossRef]
17. Eelderink, C.; Rietsema, S.; Van Vliet, I.M.; Loef, L.C.; Boer, T.; Koehorst, M.; Nolte, I.M.; Westerhuis, R.; Singh-Povel, C.M.; Geurts, J.M.W.; et al. The effect of high compared with low dairy consumption on glucose metabolism, insulin sensitivity, and metabolic flexibility in overweight adults: A randomized crossover trial. *Am. J. Clin. Nutr.* **2019**, *109*, 1555–1568. [CrossRef] [PubMed]
18. Lawlor, D.A.; Ebrahim, S.; Timpson, N.; Smith, G.D. Avoiding milk is associated with a reduced risk of insulin resistance and the metabolic syndrome: Findings from the British Women's Heart and Health Study. *Diabet. Med.* **2005**, *22*, 808–811. [CrossRef]
19. Tucker, L.A.; Erickson, A.; LeCheminant, J.D.; Bailey, B.W. Dairy Consumption and Insulin Resistance: The Role of Body Fat, Physical Activity, and Energy Intake. *J. Diabetes Res.* **2015**, *2015*, 1–11. [CrossRef]
20. Phy, J.; Pohlmeier, A.M.; Cooper, J.A.; Watkins, P.; Spallholz, J.; Harris, K.S.; Berenson, A.B.; Boylan, M. Low Starch/Low Dairy Diet Results in Successful Treatment of Obesity and Co-Morbidities Linked to Polycystic Ovary Syndrome (PCOS). *J. Obes. Weight Loss Ther.* **2015**, *5*, 1–12.
21. Von Post-Skagegård, M.; Vessby, B.; Karlström, B. Glucose and insulin responses in healthy women after intake of composite meals containing cod-, milk-, and soy protein. *Eur. J. Clin. Nutr.* **2006**, *60*, 949–954.
22. Turner, K.M.; Keogh, J.B.; Clifton, P.M. Red meat, dairy, and insulin sensitivity: A randomized crossover intervention study. *Am. J. Clin. Nutr.* **2015**, *101*, 1173–1179. [CrossRef] [PubMed]
23. Drouin-Chartier, J.P.; Gagnon, J.; Labonté, M.E.; Desroches, S.; Charest, A.; Grenier, G.; Dodin, S.; Lemieux, S.; Couture, P.; Lamarche, B. Impact of milk consumption on cardiometabolic risk in postmenopausal women with abdominal obesity. *Nutr. J.* **2015**, *14*, 12–22. [CrossRef] [PubMed]
24. Pereira, M.A.; Jacobs, D.R.; van Horn, L.; Slattery, M.L.; Kartashov, A.L.; Ludwig, D.S. Dairy Consumption, Obesity, and the Insulin Resistance Syndrome in Young Adults. The CARDIA Study. *JAMA* **2002**, *287*, 2081–2089. [CrossRef] [PubMed]
25. Chen, Y.; Feng, R.; Yang, X.; Dai, J.; Huang, M.; Ji, X.; Li, Y.; Okekunle, A.P.; Gao, G.; Onwuka, J.U.; et al. Yogurt improves insulin resistance and liver fat in obese women with nonalcoholic fatty liver disease and metabolic syndrome: A randomized controlled trial. *Am. J. Clin. Nutr.* **2019**, *109*, 1611–1619. [CrossRef]
26. Shoaei, T.; Heidari-Beni, M.; Tehrani, H.G.; Feizi, A.; Esmaillzadeh, A.; Askari, G. Effects of Probiotic Supplementation on Pancreatic β-cell Function and C-reactive Protein in Women with Polycystic Ovary Syndrome: A Randomized Double-blind Placebo-controlled Clinical Trial. *Int. J. Prev. Med.* **2015**, *6*, 27–33.
27. Jafari, T.; Faghihimani, E.; Feizi, A.; Iraj, B.; Javanmard, S.H.; Esmaillzadeh, A.; Fallah, A.A.; Askari, G. Effects of vitamin D-fortified low-fat yogurt on glycemic status, anthropometric indexes, inflammation, and bone turnover in diabetic postmenopausal women: A randomised controlled clinical trial. *Clin. Nutr.* **2016**, *35*, 67–76. [CrossRef]
28. Madjd, A.; Taylor, M.A.; Mousavi, N.; Delavari, A.; Malekzadeh, R.; Macdonald, I.A.; Farshchi, H.R. Comparison of the effect of daily consumption of probiotic compared with low-fat conventional yogurt on weight loss in healthy obese women following an energy-restricted diet: A randomized controlled trial. *Am. J. Clin. Nutr.* **2016**, *103*, 323–329. [CrossRef]

29. Lynch, C.J.; Adams, S.H. Branched-chain amino acids in metabolic signalling and insulin resistance. *Nat. Rev. Endocrinol.* **2014**, *10*, 723–736. [CrossRef]
30. Liang, Y.I.; Gong, Y.; Zhang, X.; Yang, D.; Zhao, D.; Quan, L.; Zhou, R.; Bao, W.; Cheng, G. Dietary Protein Intake, Meat Consumption, and Dairy Consumption in the Year Preceding Pregnancy and During Pregnancy and Their Associations With the Risk of Gestational Diabetes Mellitus: A Prospective Cohort Study in Southwest China 2018. *Front. Endocrinol.* **2018**, *9*, 1–9. [CrossRef]
31. Da Silva, M.S.; Chartrand, D.; Vohl, M.C.; Barbier, O.; Rudkowska, I. Dairy product consumption interacts with glucokinase (GCK) gene polymorphisms associated with insulin resistance. *J. Pers. Med.* **2017**, *7*, 8. [CrossRef]
32. Gijsbers, L.; Ding, E.; Malik, V.; Goede, J.; Geleijnse, J.M.; Soedamah-Muthu, S.S. Consumption of dairy foods and diabetes incidence: A dose-response meta-analysis of observational studies. *Am. J. Clin. Nutr.* **2016**, *103*, 1111–1124. [CrossRef] [PubMed]
33. Gao, D.; Ning, N.; Wang, C.; Wang, Y.; Li, Q.; Meng, Z.; Liu, Y.; Li, Q. Dairy products consumption and risk of type 2 diabetes: Systematic review and dose-response meta-analysis. *PLoS ONE* **2013**, *8*, 1–15. [CrossRef]
34. Aune, D.; Norat, T.; Romundstad, P.; Vatten, L.J. Dairy products and the risk of type 2 diabetes: A systematic review and dose-response meta-analysis of cohort studies. *Am. J. Clin. Nutr.* **2013**, *98*, 1066–1083. [CrossRef] [PubMed]
35. Mishali, M.; Prizant-Passal, S.; Avrech, T.; Shoenfeld, Y. Association between dairy intake and the risk of contracting type 2 diabetes and cardiovascular diseases: A systematic review and meta-analysis with subgroup analysis of men versus women. *Nutr. Rev.* **2019**, *77*, 417–429. [CrossRef] [PubMed]
36. Schwingshackl, L.; Hoffmann, G.; Lampousi, A.M.; Knuppel, S.; Iqbal, K.; Schwedhelm, C.; Bechthold, A.; Schlesinger, S.; Boeing, H. Food groups and risk of type 2 diabetes mellitus: A systematic review and meta-analysis of prospective studies. *Eur. J. Epidemiol.* **2017**, *32*, 363–375. [CrossRef]
37. Díaz-López, A.; Bulló, M.; Martínez-González, M.A.; Corella, D.; Estruch, R.; Fitó, M.; Gómez'Gracia, E.; Fiol, M.; Grácia de la Corte, F.J.; Ros, E.; et al. Dairy product consumption and risk of type 2 diabetes in an elderly Spanish Mediterranean population at high cardiovascular risk. *Eur. J. Nutr.* **2016**, *55*, 349–360. [CrossRef]
38. Brouwer-Brolsma, E.M.; Sluik, D.; Singh-Povel, C.M.; Feskens, E.J.M. Dairy product consumption is associated with pre-diabetes and newly diagnosed type 2 diabetes in the Lifelines Cohort Study. *Br. J. Nutr.* **2018**, *119*, 442–455. [CrossRef]
39. Drouin-Chartier, J.P.; Brassard, D.; Tessier-Grenier, M.; Cote, J.A.; Labonte, M.E.; Desroches, S.; Couture, P.; Lamarche, B. Systematic review of the association between dairy product consumption and risk of cardiovascular-related clinical outcomes. *Adv. Nutr.* **2016**, *7*, 1026–1040. [CrossRef]
40. Yakoob, M.Y.; Shi, P.; Willett, W.C.; Rexrode, K.M.; Campos, H.; Orav, E.J.; Hu, F.B.; Mozaffarian, D. Circulating biomarkers of dairy fat and risk of incident diabetes mellitus among US men and women in two large prospective cohorts. *Circulation* **2016**, *133*, 1645–1654. [CrossRef]
41. Guo, J.; Astrup, A.; Lovegrove, J.A.; Gijsbers, L.; Givens, D.I.; Soedamah-Muthu, S.S. Milk and dairy consumption and risk of cardiovascular diseases and all-cause mortality: Dose–response meta-analysis of prospective cohort studies. *Eur. J. Epidemiol.* **2017**, *32*, 269–287. [CrossRef]
42. Pfeuffer, M.; Watzl, B. Nutrition and health aspects of milk and dairy products and their ingredients. *Ernahr. Umsch.* **2018**, *65*, 22–33.
43. Thorning, T.K.; Bertram, H.C.; Bonjour, J.P.; de Groot, L.; Dupont, D.; Feeney, E.; Ipsen, R.; Lecerf, J.M.; Mackie, A.; McKinley, M.C.; et al. Whole dairy matrix or single nutrients in assessment of health effects: Current evidence and knowledge gaps. *Am. J. Clin. Nutr.* **2017**, *105*, 1033–1045. [CrossRef] [PubMed]
44. Astrup, A.; Geiker, N.R.W.; Magkos, M. Effects of Full-Fat and Fermented Dairy Products on Cardiometabolic Disease: Food Is More than the Sum of Its Parts. *Adv. Nutr.* **2019**, *10*, 924–930. [CrossRef] [PubMed]
45. Korat, A.V.; Li, Y.; Sacks, F.; Rosner, B.; Willett, W.C.; Hu, F.B.; Sun, Q. Dairy fat intake and risk of type 2 diabetes in 3 cohorts of US men and women. *Am. J. Clin. Nutr.* **2019**, *110*, 1192–1200. [CrossRef] [PubMed]
46. Huebschmann, A.G.; Huxley, R.R.; Kohrt, W.M.; Zeitler, P.; Regensteiner, J.G.; Reusch, J.E.B. Sex differences in the burden of type 2 diabetes and cardiovascular risk across the life course. *Diabetologia* **2019**, *62*, 1761–1772. [CrossRef]

47. Villegas, R.; Yang, G.; Gao, Y.-T.; Cai, H.; Li, H.; Zeng, W.; Shu, X.O. Dietary patterns are associated with lower incidence of type 2 diabetes in middle-aged women: The Shanghai Women's Health Study. *Int. J. Epidemiol.* **2010**, *39*, 889–899. [CrossRef]
48. Pittas, A.G.; Dawson-Hughes, B.; Li, T.; van Dam, R.; Willet, W.C.; Manson, J.E.; Hu, F.B. Vitamin D and Calcium Intake in Relation to Type 2 Diabetes in Women. *Diabetes Care* **2006**, *29*, 650–656. [CrossRef]
49. Kirri, K.; Mizoue, T.; Iso, H.; Takahashi, Y.; Kato, M.; Inoue, M.; Noda, M.; Tsugane, S. Calcium, vitamin D and dairy intake in relation to type 2 diabetes risk in a Japanese cohort. *Diabetologia* **2009**, *52*, 2542–2550. [CrossRef]
50. Liu, S.; Choi, H.K.; Ford, E.; Song, Y.; Klevak, A.; Buring, J.E.; Manson, J.E. A Prospective Study of Dairy Intake and the Risk of Type 2 Diabetes in Women. *Diabetes Care* **2006**, *29*, 1579–1584. [CrossRef]
51. Margolis, K.L.; Wei, F.; de Boer, I.H.; Howard, B.V.; Liu, S.; Manson, J.E.; Mossavar-Rahmani, Y.; Phillips, L.S.; Shikany, J.M.; Tinker, L.F. A Diet High in Low-Fat Dairy Products Lowers Diabetes Risk in Postmenopausal Women. *J. Nutr.* **2011**, *141*, 1969–1974.
52. van Dam, R.M.; Hu, F.B.; Rosenberg, L.; Krishnan, S.; Palmer, J.R. Dietary Calcium and Magnesium, Major Food Sources, and Risk of Type 2 Diabetes in U.S. Black Women. *Diabetes Care* **2006**, *29*, 2238–2243. [CrossRef] [PubMed]
53. Malik, V.S.; Rob, Q.S.; van Dam, R.B.; Rimm, E.B.; Willet, W.C.; Rosner, B.; Hu, F.B. Adolescent dairy product consumption and risk of type 2 diabetes in middle-aged women. *J. Am. Clin. Nutr.* **2011**, *94*, 854–861. [CrossRef] [PubMed]
54. Salas-Salvadó, J.; Guasch-Ferré, M.; Díaz-López, A.; Babio, N. Yogurt and Diabetes: Overview of Recent Observational Studies. *J. Nutr.* **2017**, *147*, 1452–1461. [CrossRef] [PubMed]
55. Wen, L.; Duffy, A. Factors influencing the gut microbiota, inflammation, and type 2 diabetes. *J. Nutr.* **2017**, *147*, 1468–1475. [CrossRef]
56. Mozaffarian, D.; Wu, J.H.Y. Flavonoids, dairy foods, and cardiovascular and metabolic health: A review of emerging biologic pathways. *Circ. Res.* **2018**, *122*, 369–384. [CrossRef]
57. Buziau, A.M.; Soedamah-Muthu, S.S.; Geleijnse, J.M.; Mishra, G.D. Total Fermented Dairy Food Intake Is Inversely Associated with Cardiovascular Disease Risk in Women. *J. Nutr.* **2019**, *149*, 1797–1804. [CrossRef]
58. Rosenberg, L.; Robles, Y.P.; Li, S.; Ruiz-Narvaez, E.A.; Palme, J.R. A prospective study of yogurt and other dairy consumption in relation to incidence of type 2 diabetes among black women in the USA. *Am. J. Clin. Nutr.* **2020**, *112*, 512–518. [CrossRef]
59. Liu, G.; Shi, F.; Blas-Machado, U.; Duong, Q.; Davis, V.L.; Foster, W.G.; Hughes, C.L. Ovarian effects of a high lactose diet in the female rat. *Reprod. Nutr. Dev.* **2005**, *45*, 185–192. [CrossRef]
60. Bandyopadhyay, S.; Chakrabarti, J.; Banerjee, S.; Pal, A.K.; Goswami, S.K.; Chakravarty, B.N.; Kabir, S.N. Galactose toxicity in the rat as a model for premature ovarian failure: An experimental approach readdressed. *Hum. Reprod.* **2003**, *18*, 2031–2038. [CrossRef]
61. Kim, K.; Wactawski-Wende, J.; Michels, K.A.; Plowden, T.C.; Chaljub, E.N.; Sjaarda, L.A.; Mumford, S.L. Dairy Food Intake Is Associated with Reproductive Hormones and Sporadic Anovulation among Healthy Premenopausal Women. *J. Nutr.* **2017**, *147*, 218–226. [CrossRef]
62. Greenlee, A.R.; Arbuckle, T.E.; Chyou, P.H. Risk factors for female infertility in an agricultural region. *Epidemiology* **2003**, *14*, 429–436. [CrossRef]
63. Wise, L.A.; Wesselink, A.K.; Mikkelsen, E.M.; Cueto, H.; Hahn, K.A.; Rothman, K.J.; Tucker, K.L.; Sorensen, H.T.; Hatch, E.E. Dietary fat intake and fecundability in 2 preconception cohort studies. *Am. J. Epidemiol.* **2017**, *105*, 100–110. [CrossRef] [PubMed]
64. Afeiche, M.C.; Chiu, Y.H.; Gaskins, A.J.; Williams, P.L.; Souter, I.; Wright, D.L.; Hauser, R.; Chavarro, J.E. Dairy intake in relation to in vitro fertilization out- comes among women from a fertility clinic. *Hum. Reprod.* **2016**, *31*, 563–571. [CrossRef] [PubMed]
65. Souter, I.; Chiu, Y.H.; Batsis, M.; Afeiche, M.C.; Williams, P.L.; Hauser, R.; Chavarro, J.E. The association of protein intake (amount and type) with ovarian antral follicle counts among infertile women: Results from the EARTH prospective study cohort. *BJOG Int. J. Obstet. Gynaecol.* **2017**, *124*, 1547–1555. [CrossRef] [PubMed]
66. Chavarro, J.E.; Rich-Edwards, J.W.; Rosner, B.; Willett, W.C. A prospective study of dairy foods intake and anovulatory infertility. *Hum. Reprod.* **2007**, *22*, 1340–1347. [CrossRef]
67. Adebamowo, C.A.; Spiegelman, D.; Danby, F.W.; Frazier, A.L.; Willett, W.C.; Holmes, M.D. High school dietary dairy intake and teenage acne. *J. Am. Acad. Dermatol.* **2005**, *52*, 207–214. [CrossRef]

68. Rajaeieh, G.; Marasi, M.; Shahshahan, Z.; Hassenbeigi, F.; Safavi, S.M. The Relationship between intake of dairy products and polycystic ovary syndrome in women who referred to Isfahan University of Medical Science Clinics in 2013. *Int. J. Prev. Med.* **2014**, *5*, 687–693. [PubMed]
69. Gaskins, A.J.; Chavarro, J.E. Diet and Fertility: A Review. *Am. J. Obstet. Gynecol.* **2018**, *218*, 379–389. [CrossRef] [PubMed]

Publisher's Note: MDPI stays neutral with regard to jurisdictional claims in published maps and institutional affiliations.

© 2020 by the authors. Licensee MDPI, Basel, Switzerland. This article is an open access article distributed under the terms and conditions of the Creative Commons Attribution (CC BY) license (http://creativecommons.org/licenses/by/4.0/).

Article

Association of Milk Consumption and Vitamin D Status in the US Population by Ethnicity: NHANES 2001–2010 Analysis

Moises Torres-Gonzalez [1,*], Christopher J. Cifelli [1], Sanjiv Agarwal [2] and Victor L Fulgoni III [3]

1. National Dairy Council, Rosemont, IL 60018, USA; Chris.Cifelli@dairy.org
2. NutriScience LLC, East Norriton, PA 19403, USA; agarwal47@yahoo.com
3. Nutrition Impact LLC, Battle Creek, MI 49014, USA; VIC3RD@aol.com
* Correspondence: Moises.torres-gonzalez@dairy.org

Received: 17 November 2020; Accepted: 1 December 2020; Published: 2 December 2020

Abstract: Vitamin D has been identified as a nutrient of public health concern, and higher intake of natural or fortified food sources of vitamin D, such as milk, are encouraged by the 2015–2020 Dietary Guidelines for Americans. We, therefore, examined the association of milk consumption and vitamin D status in the United States (US) population. Twenty-four-hour dietary recall data and serum 25(OH)D concentrations were obtained from the National Health and Nutrition Examination Survey 2001–2010 and were analyzed by linear and logistic regression after adjusting for anthropometric and demographic variables. Significance was set at $p < 0.05$. Approximately 57–80% children and 42–60% adults were milk consumers. Milk intake (especially reduced-fat, low fat and no-fat milk) was positively associated ($p_{\text{linear trend}} < 0.05$) with serum vitamin D status and with a 31–42% higher probability of meeting recommended serum vitamin D (>50 nmol/L) levels among all age groups. Serum vitamin D status was also associated with both type and amount of milk intake depending upon the age and ethnicity. In conclusion, the results indicate that milk consumers consistently have higher serum vitamin D levels and higher probability of meeting recommended levels. Therefore, increasing milk intake may be an effective strategy to improve the vitamin D status of the US population.

Keywords: 25-hydroxyvitamin D (25(OH)D); Mexican–American; Other Hispanic; non-Hispanic White; non-Hispanic Black

1. Introduction

Vitamin D (calciferol) is a fat-soluble vitamin, photosynthesized in the skin by the action of solar ultraviolet (UV) B radiation. It is naturally found in only a few foods, such as fish-liver oils, fatty fishes, mushrooms, egg yolks, and liver [1]. Vitamin D is known to regulate calcium and phosphorus absorption and, therefore, it has been traditionally associated with skeletal health, and its deficiency increases the risk of rickets in children, and osteoporosis, fractures and falls in adults [1–5]. Emerging evidence suggests that vitamin D deficiency may also be linked to the development of cardiovascular disease, hypertension, diabetes, metabolic syndrome, cancer, multiple sclerosis, rheumatoid arthritis, and sarcopenia [6–13].

The United States (US) Centers for Disease Control and Prevention (CDC) Second National Report on Biochemical Indicators of Diet and Nutrition in the US population reported that, in 2003–2006, approximately 8% of the population aged 1 year and older were at risk for vitamin D deficiency (VDD), which varied by age, gender, or race/ethnicity, and was as high as 31% in non-Hispanic blacks [14]. More recent analysis of NHANES 2011–2014 data, which oversampled Asian, non-Hispanic black, and Hispanic individuals to obtain reliable estimates for these population subgroups, indicated that 18.3% Americans aged 1+ years were at risk of vitamin D inadequacy (VDI) based on serum levels [15].

While there were no significant gender differences, there was a quadratic trend for risk of VDI by age and was higher for adults 20–39 years than for children 1–5 years and for seniors ≥ 60 years [15]. VDI also varied by ethnicity and was lowest among non-Hispanic White, followed by Hispanics and Asians, and was highest among non-Hispanic Blacks [15]. Vitamin D levels in humans are assessed by the determining serum 25-hydroxyvitamin D (25(OH)D) concentrations. Serum 25(OH)D levels of less than 30 nmol/L (12 ng/mL) are considered at risk for deficiency, serum levels between 30 and less than 50 nmol/L (12 to less than 20 ng/mL) are considered at risk for inadequacy; serum levels between 50–75 nmol/L (20–30 ng/mL) are considered sufficient; serum levels greater than 125 nmol/L (50 ng/mL) may be of potential concern [1].

Although, vitamin D is produced endogenously by exposure to sunlight, seasonal variations, cultural practices, and physiologic factors can impair sunlight-induced synthesis of vitamin D. The current usual dietary intakes of vitamin D among US adults aged 19+ years are 5.3 µg/d for males and 4.1 µg/d for females, and 92% males and over 97% females are below the Estimated Average Requirement (EAR) [16]. Similarly, vitamin D intakes of 90 to 93% in male and 95 to over 97% in female children aged 4–18 years are also below the EAR [16]. Accordingly, the 2015–2020 Dietary Guidelines for Americans (DGA) identified vitamin D as a "nutrient of public health concern" as it is under-consumed to an extent that may lead to adverse health outcomes [17]. The Scientific Report of the 2020 Dietary Guidelines Advisory Committee reaffirmed vitamin D is under-consumed and is of public health concern [18]. The EAR of vitamin D is 400 IU (10 µg) for ages 1+ years and the Recommended Dietary Allowance (RDA) is 600 IU (15 µg) for ages 1–70 years and 800 IU (20 µg) for ages 70+ years [1]. Vitamin D can be acquired from fortified foods and dietary supplements [1,17]. In the American diet, fortified foods are a main source of the vitamin D [1,19,20]. Varieties of foods fortified with vitamin D in the US include dairy products (mostly milk), cereals and fruit juices. Milk is voluntarily fortified with 400 IU per quart (or 385 IU/L) of vitamin D [19] and almost all fluid milks are fortified with vitamin D in the US market [21]. Indeed, DGA encourages a higher intake of food sources of vitamin D, such as milk, to meet the requirements [17].

The objective of the present investigation was to determine the association of milk consumption and vitamin D status in the US population, and to examine if milk consumers have better vitamin D status as compared to non-consumers. We hypothesized that higher milk consumption is associated with better vitamin D status and that milk consumers, regardless of the type of milk consumed, have better vitamin D status compared to non-milk consumers across all age groups.

2. Materials and Methods

2.1. Database & Subjects

Data from five separate cycles of What We Eat In America (WWEIA), the dietary intake component of the National Health and Nutrition Examination Survey (NHANES), a continuous survey conducted by the National Center for Health Statistics (NCHS), were used (2001–2010). NHANES data are collected using a complex stratified multistage cluster sampling probability design. A detailed description of the subject recruitment, survey design, and data collection procedures are available online [22] and all data obtained for this study are publicly available at: http://www.cdc.gov/nchs/nhanes/. Dietary intake data with reliable 24-h recall dietary interviews (day 1 data only) from 33,672 participants (17,132 male and 16,540 female; 4061 aged 2–8 years, 8700 aged 9–18 years, 17,457 aged 19–70 years and 3454 aged 71–99 years; 7827 Mexican American, 2094 Other Hispanic, 14,525 non-Hispanic White, 7739 non-Hispanic Black, 1487 of other ethnicity) were used after with exclusions for unreliable data ($n = 5690$), aged < 2 years ($n = 3079$), pregnant or lactating females ($n = 1272$), missing serum vitamin D data ($n = 5733$), missing Poverty Index Ratio (PIR; $n = 2808$), missing Body Mass Index (BMI; $n = 836$) or zero calorie intake ($n = 2$). All participants or proxies (i.e., parents or guardians) provided written informed consent and the Research Ethics Review Board at the NCHS approved the survey protocol.

This study was a secondary data analysis which lacked personal identifiers and, therefore, did not require Institutional Review Board review.

2.2. Estimation of Dietary Intake

Intake of milk was assessed using the sum of Food Patterns Equivalents Database (FPED) variable "D_Milk" as cup equivalents/day from associated WWEIA categories:

- Whole Milk—1002 (Milk, Whole), 1202 (Flavored milk, whole)
- Reduced-fat Milk—1004 (Milk, reduced-fat), 1204 (Flavored milk, reduced-fat)
- Low-fat Milk—1006 (Milk, low-fat), 1206 (Flavored milk, low-fat)
- Non-fat Milk—1008 (Milk, nonfat), 1208 (Flavored milk, non-fat)
- All Milk—sum of whole, reduced-fat, low-fat and non-fat milks

Non-consumers were defined as subjects not consuming any specific type of milk during the 24-h recall. Consumers of a specific type of milk were defined as subjects consuming that type of milk and no other milk during the 24-h recall. Subjects consuming more than one type of milk during the 24-h recall were designated as "Mixed Milk-Consumers". Consumer intake tertiles were calculated within each age/gender group (Table 1).

Table 1. Milk intake tertiles (cup eq/day) by milk type in different age groups, NHANES 2001–2010.

Milk Type	Age 2–8 Years	Age 9–18 Years	Age 19–70 Years	Age 71+ Years
Total Milk				
Tertile 1	<1.01	<1.02	<0.75	<0.50
Tertile 2	1.01 to <2.00	1.02 to <2.00	0.75 to <1.50	0.50 to <1.20
Tertile 3	≥2.00	≥2.00	≥1.50	≥1.20
Whole Milk				
Tertile 1	<0.92	<0.99	<0.50	<0.47
Tertile 2	0.92 to <1.62	0.99 to <1.62	0.50 to <1.41	0.47 to <1.06
Tertile 3	≥1.62	≥1.62	≥1.41	≥1.06
Reduced-Fat Milk				
Tertile 1	<0.96	<1.00	<0.75	<0.50
Tertile 2	0.96 to <1.82	1.00 to <1.78	0.75 to <1.44	0.50 to <1.04
Tertile 3	≥1.82	≥1.78	≥1.44	≥1.04
Low-Fat Milk				
Tertile 1	<0.99	<0.98	<0.74	<0.61
Tertile 2	0.99 to <1.39	0.98 to <1.87	0.74 to <1.50	0.61 to <1.29
Tertile 3	≥1.39	≥1.87	≥1.50	≥1.29
Non-Fat Milk				
Tertile 1	<0.76	<1.00	<0.63	<0.50
Tertile 2	0.76 to <1.45	1.00 to <1.43	0.63 to <1.43	0.50 to <1.13
Tertile 3	≥1.45	≥1.43	≥1.43	≥1.13

Mean usual intake ± standard error (SE). Intake of milk was assessed using the sum of Food Patterns Equivalents Database (FPED) variable "D_Milk" as cup equivalents/day from associated What We Eat In America (WWEIA) categories: Whole Milk—1002 (Milk, Whole), 1202 (Flavored milk, whole); Reduced-fat Milk—1004 (Milk, reduced-fat), 1204 (Flavored milk, reduced-fat); Low-fat Milk—1006 (Milk, low-fat), 1206 (Flavored milk, low-fat); Non-fat Milk—1008 (Milk, nonfat), 1208 (Flavored milk, non-fat); All Milk—sum of whole, reduced-fat, low-fat and non-fat milks.

2.3. Serum Vitamin D

Serum 25(OH)D concentrations were obtained from NHANES laboratory files [22]. Briefly, NHANES measured serum 25(OH)D, using a standardized liquid chromatography–tandem mass (LC-MS/MS) method for 2007–2010 cycles and using a DiaSorin RIA kit for 2001–2006 cycles. RIA measurements of 25(OH)D concentration were later converted to LC-MS/MS method equivalent measurements adjusting for assay drifts, due to concerns about imprecision and bias in the method [23].

2.4. Statistics

All analyses were performed using SAS 9.2, 9.4 and SUDAAN 11. Day 1 weights were used all analyses and the data were adjusted for the complex sampling design of NHANES, using appropriate survey weights, strata, and primary sampling units. Separate analyses were conducted for the ages 2–8, 9–18, 19–70, and 71+ years. Least Square Means (LSM) were generated from models for each age/gender/ethnic group using linear regression across tertiles of milk intake and different types of milk intake after adjusting for age, gender, ethnicity, poverty income ratio and BMI or BMI Z-score when the population being analyzed was <19 years. Significance was set at $p < 0.05$. Logistic regression analysis was used to assess odds ratios (OR) and 95% confidence limits (Lower confidence limit (LCL); Upper confidence limit (UCL)) of meeting recommended levels of serum vitamin D (>50 nmol/L) associated with milk intake with non-consumers as reference group after adjusting for age, gender, ethnicity, poverty income ratio and BMI or BMI Z-score when the population being analyzed was <19 years old. Additionally, vitamin D from dietary supplements, seafood, and other non-milk vitamin D sources were also subsequently added to models to assess whether the intake of these variables impacted the association of milk with serum vitamin D (serum 25(OH)D) levels.

3. Results

Approximately 80% children aged 2–8 years, 57% children aged 9–18 years, 42% adults aged 19–70 years and 60% adults aged 71+ years were milk consumers. All milk (sum of all milk types) consumption was higher in child consumers (2–18 years), compared to adult consumers (19+ years). Adult consumers aged 71+ years consumed about 37% less milk than children consumers aged 9–18 years (1.12 cup eq/d vs. 1.77 cup eq/d, respectively) (Table 2). Consumption of whole milk compared to non-fat milk was about 21% greater among children consumers aged 2–8 years but was 19% less among adult consumers aged 71+ years. Consumption of whole milk was similar to that of non-fat milk for children consumers aged 9–18 years and for adult consumers aged 19–70 years (Table 2). Depending upon ethnicity, milk contributed about 61–71% among those aged 2–8 years, 49–62% among those aged 9–18 years, 24–42% among those aged 19–70 years, and 28–48% among those aged 71+ years of total vitamin D intake (Table 3).

Serum vitamin D (serum 25(OH)D) was positively associated with milk intake in children aged 9–18 years and adults aged 19+ years (p linear trend < 0.05). The association was also significant for both males and females of age 9+ years (p linear trend < 0.05), except for 9–18-year-old males and for 71+ year-old females (p linear trend > 0.05) (Table 4). In children aged 2–8 years, consumers of whole milk, reduced fat milk, and low-fat milk had significantly higher ($p < 0.05$) serum vitamin D (serum 25(OH)D) levels than non-consumers. Similarly, in children aged 9–18 years and adults aged 19+ years, consumers of reduced-fat milk, low-fat milk, and non-fat milk had higher ($p < 0.05$) serum vitamin D (serum 25(OH)D) levels than non-consumers of same age group. Children (aged 9–18 years) and adult (aged 19–70 years) consumers of mixed milk also had higher ($p < 0.05$) serum vitamin D (serum 25(OH)D) than respective non-consumers (Table 5). Adjusting for data in Tables 2 and 3 for dietary supplements, seafood, and other non-milk vitamin D sources did not change these results (Tables 4 and 5).

Table 2. Mean intake of milk (cup eq/day) by milk type in different age and gender groups, NHANES 2001–2010.

Milk Type	Age 2–8 Years		Age 9–18 Years		Age 19–70 Years		Age 71+ Years	
	Total Population	Consumer	Total Population	Consumer	Total Population	Consumer	Total Population	Consumer
All Milk								
All	1.40 ± 0.03	1.71 ± 0.03	1.11 ± 0.03	1.77 ± 0.04	0.62 ± 0.02	1.41 ± 0.03	0.69 ± 0.02	1.12 ± 0.03
Male	1.51 ± 0.04	1.82 ± 0.04	1.33 ± 0.05	2.00 ± 0.06	0.72 ± 0.02	1.63 ± 0.04	0.78 ± 0.08	1.24 ± 0.05
Female	1.28 ± 0.04	1.58 ± 0.04	0.89 ± 0.03	1.50 ± 0.04	0.52 ± 0.02	1.19 ± 0.03	0.63 ± 0.02	1.03 ± 0.03
Whole Milk								
All	0.46 ± 0.03	1.50 ± 0.04	0.27 ± 0.02	1.58 ± 0.06	0.15 ± 0.01	1.38 ± 0.05	0.12 ± 0.01	0.94 ± 0.05
Male	0.49 ± 0.03	1.59 ± 0.06	0.34 ± 0.02	1.75 ± 0.08	0.19 ± 0.01	1.60 ± 0.08	0.16 ± 0.02	1.18 ± 0.07
Female	0.43 ± 0.03	1.40 ± 0.05	0.20 ± 0.01	1.34 ± 0.06	0.12 ± 0.01	1.12 ± 0.04	0.09 ± 0.01	0.73 ± 0.05
Reduced-Fat Milk								
All	0.61 ± 0.04	1.55 ± 0.05	0.47 ± 0.02	1.62 ± 0.05	0.23 ± 0.01	1.36 ± 0.04	0.27 ± 0.01	1.04 ± 0.04
Male	0.67 ± 0.05	1.65 ± 0.06	0.58 ± 0.04	1.87 ± 0.08	0.28 ± 0.01	1.55 ± 0.06	0.31 ± 0.02	1.15 ± 0.06
Female	0.53 ± 0.03	1.43 ± 0.05	0.36 ± 0.02	1.32 ± 0.05	0.17 ± 0.01	1.13 ± 0.04	0.24 ± 0.02	0.94 ± 0.04
Low-Fat Milk								
All	0.17 ± 0.02	1.36 ± 0.06	0.16 ± 0.01	1.65 ± 0.07	0.09 ± 0.01	1.37 ± 0.06	0.12 ± 0.01	1.14 ± 0.05
Male	0.17 ± 0.02	1.42 ± 0.09	0.17 ± 0.02	1.69 ± 0.09	0.09 ± 0.01	1.60 ± 0.11	0.12 ± 0.01	1.15 ± 0.06
Female	0.17 ± 0.02	1.29 ± 0.06	0.15 ± 0.02	1.60 ± 0.10	0.08 ± 0.01	1.18 ± 0.05	0.12 ± 0.01	1.09 ± 0.06
Non-Fat Milk								
All	0.16 ± 0.02	1.24 ± 0.08	0.20 ± 0.02	1.55 ± 0.09	0.15 ± 0.01	1.37 ± 0.05	0.19 ± 0.01	1.16 ± 0.05
Male	0.18 ± 0.03	1.26 ± 0.10	0.23 ± 0.03	1.78 ± 0.14	0.16 ± 0.01	1.60 ± 0.08	0.19 ± 0.02	1.26 ± 0.07
Female	0.15 ± 0.03	1.21 ± 0.04	0.17 ± 0.02	1.31 ± 0.08	0.14 ± 0.01	1.18 ± 0.05	0.19 ± 0.01	1.09 ± 0.06

Mean usual intake ± standard error (SE). Intake of milk was assessed using the sum of Food Patterns Equivalents Database (FPED) variable "D_Milk" as cup equivalents/day from associated What We Eat In America (WWEIA) categories: Whole Milk—1002 (Milk, Whole), 1202 (Flavored milk, whole); Reduced-fat Milk—1004 (Milk, reduced-fat), 1204 (Flavored milk, reduced-fat); Low-fat Milk—1006 (Milk, low-fat), 1206 (Flavored milk, low-fat); Non-fat Milk—1008 (Milk, nonfat), 1208 (Flavored milk, non-fat); All Milk—sum of whole, reduced-fat, low-fat and non-fat milks. N = 4061 aged 2–8 years; 8700 aged 9–18 years; 17,457 aged 19–70 years; 3454 aged 71–99 years.

Table 3. Estimated mean dietary intake of vitamin D (μg/day) by milk type and by age and ethnicity. NHANES 2001–2010 Day 1 dietary data.

	Mexican American	Other Hispanic	Non-Hispanic White	Non-Hispanic Black	Other
2–8 years					
All Milk	4.81 ± 0.13	5.47 ± 0.31	4.47 ± 0.12	3.29 ± 0.12	4.22 ± 0.17
Whole Milk	2.43 ± 0.15	2.83 ± 0.27	1.28 ± 0.10	1.72 ± 0.11	1.82 ± 0.21
Reduced-Fat	1.71 ± 0.10	1.75 ± 0.25	1.96 ± 0.11	1.17 ± 0.09	1.90 ± 0.22
Low-Fat Milk	0.36 ± 0.05	0.70 ± 0.18	0.62 ± 0.06	0.23 ± 0.05	0.33 ± 0.09
Non-Fat Milk	0.31 ± 0.05	0.20 ± 0.05	0.61 ± 0.08	0.17 ± 0.03	0.17 ± 0.06
Not Milk	2.16 ± 0.06	2.43 ± 0.17	1.87 ± 0.06	2.14 ± 0.07	2.08 ± 0.12
9–18 years					
All Milk	3.03 ± 0.13	2.90 ± 0.21	3.71 ± 0.14	2.13 ± 0.07	2.52 ± 0.23
Whole Milk	1.00 ± 0.07	1.38 ± 0.20	0.75 ± 0.07	0.98 ± 0.07	0.71 ± 0.12
Reduced-Fat	1.27 ± 0.07	0.87 ± 0.10	1.61 ± 0.10	0.75 ± 0.05	1.21 ± 0.18
Low-Fat Milk	0.36 ± 0.04	0.27 ± 0.05	0.62 ± 0.06	0.18 ± 0.03	0.21 ± 0.05
Non-Fat Milk	0.39 ± 0.07	0.38 ± 0.11	0.73 ± 0.07	0.22 ± 0.03	0.39 ± 0.08
Not Milk	2.25 ± 0.06	2.36 ± 0.16	2.32 ± 0.06	2.26 ± 0.07	2.40 ± 0.20
19–70 years					
All Milk	1.83 ± 0.08	1.49 ± 0.08	2.05 ± 0.06	0.92 ± 0.05	1.46 ± 0.17
Whole Milk	0.82 ± 0.06	0.59 ± 0.07	0.43 ± 0.03	0.46 ± 0.04	0.49 ± 0.08
Reduced-Fat	0.75 ± 0.06	0.58 ± 0.06	0.74 ± 0.03	0.36 ± 0.03	0.70 ± 0.15
Low-Fat Milk	0.12 ± 0.02	0.15 ± 0.03	0.32 ± 0.03	0.06 ± 0.01	0.15 ± 0.06
Non-Fat Milk	0.14 ± 0.02	0.17 ± 0.05	0.55 ± 0.04	0.04 ± 0.01	0.12 ± 0.02
Not Milk	2.63 ± 0.09	2.60 ± 0.16	2.85 ± 0.06	2.84 ± 0.11	2.79 ± 0.19
71+ years					
All Milk	1.99 ± 0.14	2.04 ± 0.27	2.17 ± 0.07	1.16 ± 0.12	1.46 ± 0.31
Whole Milk	0.72 ± 0.09	1.01 ± 0.23	0.34 ± 0.03	0.45 ± 0.09	0.27 ± 0.11
Reduced-Fat	0.90 ± 0.12	0.53 ± 0.15	0.83 ± 0.05	0.50 ± 0.06	0.65 ± 0.15
Low-Fat Milk	0.12 ± 0.04	0.06 ± 0.04	0.41 ± 0.04	0.12 ± 0.03	0.09 ± 0.05
Non-Fat Milk	0.26 ± 0.05	0.44 ± 0.15	0.59 ± 0.04	0.10 ± 0.02	0.46 ± 0.27

Mean usual intake ± standard error (SE). Milk type was assessed using the sum of Food Patterns Equivalents Database (FPED) variable "D_Milk" as cup equivalents/day for What We Eat In America (WWEIA) categories: Whole Milk—1002 (Milk, Whole), 1202 (Flavored milk, whole); Reduced-fat Milk—1004 (Milk, reduced-fat), 1204 (Flavored milk, reduced-fat); Low-fat Milk—1006 (Milk, low-fat), 1206 (Flavored milk, low-fat); Non-fat Milk—1008 (Milk, nonfat), 1208 (Flavored milk, non-fat); All Milk—sum of whole, reduced-fat, low-fat and non-fat milks. $N = 42,154$.

Table 4. Serum Vitamin D (serum 25(OH)D) status (nmol/L) by all milk intake in different age and gender groups, NHANES 2001–2010.

	All	Non-Consumers	Consumer Tertile 1	Consumer Tertile 2	Consumer Tertile 3	Plinear trend	Plinear trend [a]
2–8 years							
All	74.5 ± 0.7	71.7 ± 1.6	74.5 ± 1.1	75.4 ± 1.0 *#	75.4 ± 0.8 *#	0.1721	0.1541
Male	75.0 ± 0.8	71.8 ± 2.0	74.9 ± 1.1	76.9 ± 1.3 *#	75.7 ± 1.0	0.6085	0.5164
Female	73.9 ± 0.9	71.7 ± 2.0	73.2 ± 1.6	74.9 ± 1.2	74.8 ± 1.0	0.0609	0.0624
9–18 years							
All	65.7 ± 0.8	63.5 ± 1.1	65.7 ± 1.0 *	66.1 ± 1.0 *#	69.3 ± 0.9 *#	0.0034	0.0049
Male	67.3 ± 0.8	64.1 ± 1.3	67.7 ± 1.3 *#	68.2 ± 1.2 *#	71.1 ± 1.1 *#	0.0505	0.0478
Female	63.9 ± 0.8	62.6 ± 1.2	64.2 ± 1.1	64.6 ± 1.4	66.3 ± 1.2 *	0.0123	0.0267
19–70 years							
All	64.4 ± 0.5	62.9 ± 0.6	64.3 ± 0.8	66.8 ± 0.7 *#	67.8 ± 0.6 *#	<0.0001	<0.0001
Male	63.9 ± 0.5	62.5 ± 0.6	64.0 ± 0.9	65.8 ± 0.8 *#	67.2 ± 0.6 *#	0.0004	0.0002
Female	64.8 ± 0.6	63.1 ± 0.7	65.4 ± 1.0 *#	67.5 ± 1.1 *#	68.5 ± 0.9 *#	0.0109	0.0128
71+ years							
All	65.8 ± 0.9	63.0 ± 1.2	67.2 ± 1.6 *#	65.4 ± 1.1	70.2 ± 1.1 *#	0.0039	0.0001
Male	65.7 ± 0.9	64.3 ± 1.3	65.2 ± 1.7	64.5 ± 1.3	70.1 ± 1.4 *#	0.0049	0.0010
Female	65.9 ± 1.0	62.0 ± 1.4	68.4 ± 1.8 *#	66.3 ± 1.5 *	70.1 ± 1.5 *#	0.1658	0.0229

Mean serum vitamin D concentration ± standard error (SE). Least Square Means (LSM) were modeled for each age/gender group using linear regression across tertiles after adjusting the data for age, gender, ethnicity, poverty income ratio and BMI or BMI Z-score when the population being analyzed was < 19 years. * Significant difference from non-consumer at p < 0.05. # Significant difference from non-consumer at p < 0.05 after additionally adjusting data for vitamin D intake from dietary supplements, sea food, and other non-milk dietary sources.
[a] p linear trend after additionally adjusting data for vitamin D intake from non-milk sources, supplements and sea food. N = 4061 aged 2–8 years; 8700 aged 9–18 years; 17,457 aged 19–70 years; 3454 aged 71–99 years.

Table 5. Serum vitamin D (serum 25(OH)D) status (nmol/L) by different types of milk intake in different age groups, NHANES 2001–2010.

Type of Milk	Age 2–8 Years	Age 9–18 Years	Age 19–70 Years	Age 71+ Years
Total Population	74.5 ± 0.7	65.7 ± 0.8	64.4 ± 0.5	65.8 ± 0.9
Non-Consumer	71.7 ± 1.6	63.6 ± 1.1	62.9 ± 0.6	63.0 ± 1.2
Whole Milk Consumer	75.4 ± 1.2 *#	65.3 ± 1.2	63.7 ± 0.8	65.7 ± 2.0
Reduced-Fat Milk Consumer	74.9 ± 0.9 *#	66.5 ± 0.9 *#	66.4 ± 0.6 *#	66.8 ± 1.3 *#
Low-Fat Milk Consumer	78.2 ± 2.2 *#	68.0 ± 1.4 *#	67.8 ± 1.1 *#	70.2 ± 1.5 *#
Non-Fat Milk Consumer	73.4 ± 1.9	68.1 ± 1.6 *#	68.3 ± 0.8 *#	68.7 ± 1.2 *#
Mixed Milk Consumer	73.2 ± 1.3	69.3 ± 1.0 *#	67.6 ± 1.9 *#	65.5 ± 2.3

Mean serum vitamin D concentration ± standard error (SE). Least Square Means (LSM) were modeled for each age group using linear regression across different types of milk intake after adjusting gender combined data for age, gender, ethnicity, poverty income ratio and BMI or BMI Z-score when the population being analyzed was < 19 years. * Significant difference from non-consumer at $p < 0.05$. # Significant difference from non-consumer at $p < 0.05$ after additionally adjusting data for vitamin D intake from dietary supplements, sea food, and other non-milk dietary sources. N = 4061 aged 2–8 years; 8700 aged 9–18 years; 17,457 aged 19–70 years; 3454 aged 71–99 years.

Consumption of all milk (for all ages—i.e., 2+ years), whole milk (for ages 19–70 years), reduced-fat milk (for ages 9–70 years), low-fat milk (for those aged 19+ years), and non-fat milk (for those aged 9+ years) was associated with significantly higher probability of meeting serum vitamin D recommendations when the analysis was conducted by milk amount (Table 6). Consumers of whole milk (of aged 2–8 years), reduced-fat milk (of all ages), low-fat milk (of all ages), non-fat milk (of aged 19+ years) and mixed milk (of aged 2–70 years) had a higher probability of meeting serum vitamin D recommendations in the analysis by milk type (Table 6).

Table 7 shows the serum vitamin D (serum 25(OH)D) status by tertiles of all milk intake among different ethnic populations. Serum vitamin D (serum 25(OH)D) was positively associated with milk intake among children aged 9–18 years and adults aged 19+ years of Mexican American and of "other" ethnicity (p linear trend < 0.05). The increase in serum vitamin D (serum 25(OH)D) status with increasing milk intake was also significant (p linear trend < 0.05) among non-Hispanic Whites aged 19+ years, and non-Hispanic Blacks aged 2–18 years. Additional adjustment for dietary supplements, seafood, and other non-milk vitamin D sources did not change the results (Table 7).

Table 8 presents the data on vitamin D status by milk type across different ethnic groups. Significantly ($p < 0.05$) higher levels of serum vitamin D (serum 25(OH)D) levels were observed for consumers of whole milk, reduced fat milk, low-fat milk, non-fat milk and mixed milk among the different age and ethnic groups examined (Table 6). Adjusting the data for dietary supplements, seafood, and other non-milk vitamin D sources did not modify the results (Table 8).

Table 6. Odds ratios (OR), 95% confidence limits (Lower confidence limit (LCL)/Upper confidence limit (UCL)) for meeting the recommended serum vitamin D (>50 nmol/L), NHANES 2001–2010.

	OR (LCL, UCL) by Milk Amount	OR (LCL, UCL) by Milk Type
2–8 years		
All Milk	1.42 (1.19, 1.69)	–
Whole Milk Consumer	1.21 (0.98, 1.50)	2.06 (1.46, 2.89)
Reduced-Fat Milk Consumer	1.24 (0.97, 1.57)	1.88 (1.21, 2.91)
Low-Fat Milk Consumer	1.40 (0.80, 2.45)	3.15 (1.34, 7.39)
Non-Fat Milk Consumer	1.29 (0.81, 2.05)	1.67 (0.88, 3.17)
Mixed Milk	–	3.77 (2.06, 6.90)
9–18 years		
All Milk	1.31 (1.20, 1.44)	—
Whole Milk Consumer	1.07 (0.95, 1.21)	1.22 (0.96, 1.54)
Reduced-Fat Milk Consumer	1.34 (1.19, 1.50)	1.77 (1.38, 2.27)
Low-Fat Milk Consumer	1.20 (0.90, 1.59)	1.48 (1.02, 2.14)
Non-Fat Milk Consumer	1.36 (1.08, 1.70)	1.49 (0.90, 2.47)
Mixed Milk	—	2.66 (1.91, 3.71)
19–70 years		
All Milk	1.31 (1.23, 1.41)	—
Whole Milk Consumer	1.12 (1.04, 1.20)	1.20 (1.00, 1.43)
Reduced-Fat Milk Consumer	1.27 (1.16, 1.39)	1.59 (1.39, 1.82)
Low-Fat Milk Consumer	1.58 (1.24, 2.03)	2.22 (1.61, 3.06)
Non-Fat Milk Consumer	1.31 (1.06, 1.63)	1.90 (1.53, 2.36)
Mixed Milk	—	2.10 (1.43, 3.09)
71+ years		
All Milk	1.35 (1.17, 1.56)	-
Whole Milk Consumer	1.08 (0.92, 1.27)	1.13 (0.87, 1.47)
Reduced-Fat Milk Consumer	1.18 (0.96, 1.44)	1.46 (1.10, 1.93)
Low-Fat Milk Consumer	1.67 (1.24, 2.24)	2.33 (1.59, 3.43)
Non-Fat Milk Consumer	1.29 (1.05, 1.59	1.77 (1.31, 2.39)
Mixed Milk	-	1.38 (0.73, 2.59)

Gender combined data. OR were estimated using logistic regressions to model meeting recommended serum vitamin d (>50 nmol/L) on milk intake (OR Amount) or on 6 types of milk consumers (OR Type). Non-consumers were the reference group in both estimations. $N = 4061$ aged 2–8 years; 8700 aged 9–18 years; 17,457 aged 19–70 years; 3454 aged 71–99 years.

Table 7. Serum vitamin D (serum 25(OH)D) status (nmol/L) by all milk intake in ethnic population groups, NHANES 2001–2010.

	All	Non-Consumers	Consumer Tertile 1	Consumer Tertile 2	Consumer Tertile 3	$P_{\text{linear trend}}$	$P_{\text{linear trend}}$ [a]
Mexican American							
2–8 years	67.6 ± 0.7	63.5 ± 1.3	67.8 ± 1.3 *#	68.0 ± 1.0 *#	69.6 ± 1.1 *#	0.0929	0.1458
9–18 years	57.4 ± 0.8	54.1 ± 1.0	57.2 ± 1.1 *#	59.7 ± 1.1 *#	61.0 ± 1.1 *#	0.0004	0.0010
19–70 years	54.2 ± 0.8	53.4 ± 0.9	53.8 ± 1.3	53.9 ± 1.2	57.8 ± 1.3 *#	0.0024	0.0064
71+ years	56.0 ± 1.5	55.9 ± 2.3	50.4 ± 2.4 #	56.6 ± 2.5	60.6 ± 2.2 #	0.0018	<0.0001
Other Hispanics							
2–8 years	69.9 ± 1.3	64.8 ± 2.0	67.1 ± 2.1	73.9 ± 2.8 *#	72.2 ± 1.9 *#	0.0199	0.0348
9–18 years	60.5 ± 1.2	58.2 ± 1.5	59.5 ± 1.9	61.9 ± 2.2	64.2 ± 3.0	0.0755	0.0517
19–70 years	57.1 ± 1.2	55.9 ± 1.6	57.1 ± 1.2	60.2 ± 2.1	58.2 ± 1.6	0.5415	0.3929
71+ years	63.7 ± 2.2	57.8 ± 2.7	61.4 ± 3.9	66.0 ± 4.9	74.7 ± 6.9 *#	<0.0001	<0.0001
Non-Hispanic White							
2–8 years	80.4 ± 1.2	78.1 ± 2.6	81.8 ± 1.7	80.7 ± 1.7	80.3 ± 1.2	0.8450	0.9377
9–18 years	72.7 ± 1.1	71.2 ± 1.5	73.1 ± 1.5	72.6 ± 1.4	75.3 ± 1.2 *#	0.0645	0.0843
19–70 years	70.0 ± 0.6	68.4 ± 0.7	69.6 ± 1.0	72.6 ± 0.9 *#	73.4 ± 0.7 *#	0.0006	0.0006
71+ years	67.6 ± 0.9	64.7 ± 1.3	69.5 ± 1.7 *#	67.4 ± 1.2	71.5 ± 1.2 *#	0.0217	0.0006
Non-Hispanic Black							
2–8 years	61.4 ± 0.8	59.7 ± 1.6	57.4 ± 1.3	63.1 ± 1.4	64.6 ± 1.3 *#	<0.0001	<0.0001
9–18 years	48.3 ± 0.8	43.9 ± 1.2	49.0 ± 1.0 *#	48.7 ± 1.0 *#	55.0 ± 1.2 *#	0.0288	0.0332
19–70 years	44.2 ± 0.8	41.8 ± 0.9	46.8 ± 1.3 *#	47.7 ± 1.3 *#	47.2 ± 1.3 *#	0.6355	0.4175
71+ years	51.9 ± 1.8	48.7 ± 2.0	54.5 ± 3.1	50.6 ± 2.8	57.1 ± 4.9	0.6749	0.5408
Other							
2–8 years	70.1 ± 1.6	65.5 ± 3.1	68.7 ± 2.8	73.4 ± 1.8 *#	71.4 ± 2.1	0.1000	0.0419
9–18 years	56.5 ± 1.2	54.8 ± 2.1	52.2 ± 2.7	57.0 ± 2.5	63.0 ± 2.2 *#	0.0019	0.0011
19–70 years	52.8 ± 1.1	51.7 ± 1.4	50.6 ± 2.2	53.4 ± 2.8	58.5 ± 2.4 *#	0.0004	<0.0001
71+ years	56.9 ± 2.6	56.7 ± 3.6	51.8 ± 5.1	51.5 ± 6.4	67.6 ± 4.9	0.0067	0.0218

Mean serum vitamin D concentration ± standard error (SE). Least Square Means (LSM) were modeled for each age/ethnic group using linear regression across tertiles of milk intake after adjusting gender combined data for age, gender, poverty income ratio and BMI or BMI Z-score when the population being analyzed was < 19 years. * Significant difference from non-consumer at $p < 0.05$. # Significant difference from non-consumer at $p < 0.05$ after additionally adjusting data for vitamin D intake from non-milk sources, supplements and sea food. [a] $p_{\text{linear trend}}$ after additionally adjusting data for vitamin D intake from non-milk sources, supplements and sea food. N = 7827 Mexican American; 2094 Other Hispanic; 14,525 non-Hispanic White; 7739 non-Hispanic Black; 1487 Other.

Table 8. Serum vitamin D (serum 25(OH)D) status (nmol/L) by milk type in ethnic population groups, NHANES 2001–2010.

	Mexican American	Other Hispanic	Non-Hispanic White	Non-Hispanic Black	Other
2–8 years					
Non-Consumer	63.4 ± 1.3	64.8 ± 2.0	78.2 ± 2.6	59.7 ± 1.6	65.5 ± 3.1
Whole Milk Consumer	69.2 ± 1.0 *#	71.1 ± 2.6 *#	81.6 ± 2.0	60.8 ± 1.4	69.7 ± 1.9
Reduced-Fat Milk Consumer	68.5 ± 1.1 *#	74.6 ± 2.9 *#	80.3 ± 1.4	60.7 ± 1.3	72.8 ± 3.3 *#
Low-Fat Milk Consumer	71.4 ± 1.8 *#	68.5 ± 3.5	84.7 ± 3.8	65.1 ± 3.4	72.2 ± 5.7
Non-Fat Milk Consumer	61.4 ± 2.9	70.3 ± 6.0	82.5 ± 2.9	53.0 ± 2.9	67.0 ± 1.9
Mixed Milk Consumer	67.3 ± 1.1 *#	66.3 ± 1.8	78.0 ± 2.1	64.2 ± 1.7 *#	69.2 ± 2.7
9–18 years					
Non-Consumer	54.0 ± 1.0	58.2 ± 1.5	71.2 ± 1.5	43.9 ± 1.2	54.8 ± 2.1
Whole Milk Consumer	57.5 ± 1.2 *#	64.1 ± 2.3 *#	71.6 ± 1.8	49.0 ± 1.0 *#	55.3 ± 2.4
Reduced-Fat Milk Consumer	59.3 ± 1.0 *#	58.9 ± 1.5	72.5 ± 1.2	52.0 ± 1.0 *#	60.3 ± 2.4
Low-Fat Milk Consumer	59.5 ± 1.6 *#	58.9 ± 2.0	76.3 ± 2.0 *#	49.0 ± 2.4	52.4 ± 6.3
Non-Fat Milk Consumer	62.9 ± 1.7 *#	60.9 ± 4.0	75.8 ± 2.3	47.7 ± 1.7 *#	53.3 ± 6.6
Mixed Milk Consumer	59.7 ± 1.7 *#	64.3 ± 3.7	75.9 ± 1.6 *#	56.1 ± 1.8 *#	56.7 ± 4.0
19–70 years					
Non-Consumer	53.3 ± 0.9	55.8 ± 1.6	68.4 ± 0.7	41.8 ± 0.9	51.7 ± 1.4
Whole Milk Consumer	53.5 ± 1.3	58.0 ± 1.7	68.8 ± 1.0	45.0 ± 0.9 *#	52.8 ± 2.6
Reduced-Fat Milk Consumer	56.6 ± 1.2 *#	57.1 ± 1.6	71.9 ± 0.8 *#	47.8 ± 1.5 *#	53.9 ± 2.3
Low-Fat Milk Consumer	56.1 ± 1.8	54.8 ± 2.9	73.0 ± 1.1 *#	52.9 ± 2.8 *#	56.4 ± 5.8
Non-Fat Milk Consumer	55.3 ± 2.4	65.0 ± 3.4 *#	73.2 ± 0.9 *#	51.2 ± 3.0 *#	60.2 ± 4.2
Mixed Milk Consumer	56.5 ± 2.6	61.0 ± 3.2	74.3 ± 2.5 *#	49.7 ± 3.5 *#	35.7 ± 4.4 *#
71+ years					
Non-Consumer	55.8 ± 2.3	57.6 ± 2.8	64.7 ± 1.3	48.6 ± 2.0	56.7 ± 3.6
Whole Milk Consumer	53.7 ± 1.8	65.8 ± 5.9	67.7 ± 2.3	49.3 ± 3.0	55.9 ± 6.4
Reduced-Fat Milk Consumer	57.9 ± 2.0	63.8 ± 3.7	68.3 ± 1.4 *#	58.3 ± 3.0 *#	53.0 ± 5.5
Low-Fat Milk Consumer	53.2 ± 9.0	70.1 ± 8.2	72.3 ± 1.5 *#	53.8 ± 4.4	67.1 ± 8.4
Non-Fat Milk Consumer	57.6 ± 4.9	77.9 ± 7.3 *#	71.1 ± 1.3 *#	43.6 ± 3.5	63.5 ± 6.8
Mixed Milk Consumer	55.1 ± 10.8	70.2 ± 6.8	65.6 ± 2.6	71.3 ± 6.2 *#	59.4 ± 4.4

Mean serum vitamin D concentration ± standard error (SE). Least Square Means (LSM) were modeled for each age/ethnic group using linear regression across different types of milk intake after adjusting gender combined data for age, gender, poverty income ratio and BMI or BMI Z-score when the population being analyzed was < 19 years. * Significant difference from non-consumer at $p < 0.05$. # Significant difference from non-consumer at $p < 0.05$ after additionally adjusting data for vitamin D intake from non-milk sources. $N = 7827$ Mexican American; 2094 Other Hispanic; 14,525 non-Hispanic White; 7739 non-Hispanic Black; 1487 Other.

4. Discussion

The current cross-sectional analysis of data from the NHANES 2001–2010 demonstrated a significant association between milk consumption and serum vitamin D (serum 25(OH)D) status. Additionally, the results showed that the probability of meeting the vitamin D recommendations was greater in milk consumers vs. non-consumers. To the best of our knowledge, this is first analysis of nationally representative, non-institutionalized population of US children and adults examining the association of milk intake with vitamin D levels.

Poor vitamin D status (low serum 25(OH)D levels) is a global public health concern as over 50% of population has less than adequate serum vitamin D status [24]. In the US, about 18% of the US population aged 1+ years had insufficient serum vitamin D levels and were at risk of inadequacy (less than 50 nmol/L) according a recent analysis of NHANES 2011–2014 [15]. Liu et al. [25] estimated the prevalence of inadequate serum vitamin D levels in US adults to be 28.9% for VDD and 41.4% for VDI, from analysis of NHANES 2001–2010 and using the criteria recommended by the Endocrinology Society to define VDD as 25(OH)D <50 nmol/L and VDI as 25(OH)D <75 nmol/L [26]. The present analysis showed that the average serum vitamin D levels ranged from 64 to 75 nmol/L depending on age and gender in representative population of US children and adults aged 2+ years. These average serum vitamin D levels are well with in the 50–75 nmol/L range and are considered sufficient by IOM definition [1]. Vitamin D is a "nutrient of public health concern" as it is under-consumed to an extent that may lead to adverse health outcomes and higher intake of food sources are encouraged by DGA [17].

In the present analysis, serum Vitamin D levels were significantly associated with the intake of milk, depending on the type. Milk consumers, especially those of low fat and reduced fat milk, had higher probability of meeting >50 nmol/L serum vitamin D level benchmark set by IOM [1] than non-consumers; however, the mean serum vitamin D levels were always higher than 50 nmol/L. Although milk contains a low amount of naturally occurring vitamin D, almost all milk in the US is fortified with 100 IU/cup vitamin D irrespective of the type of milk [1,19,21]. Effectiveness of milk and other fortified foods in improving serum vitamin D status has been demonstrated in both clinical and observational studies (see [27,28] for reviews). A recent systematic review and meta-analysis of randomized controlled trials showed that vitamin D fortified foods (mostly milk and dairy products) increased serum vitamin D levels by 1.2 nmol/L for each 1 µg/d increased intake of vitamin D [27]. A cup of milk/d provides ~2.5 µg of vitamin D. A review of observational studies also concluded that the intake of vitamin D fortified milk products was positively associated with vitamin D intake and serum vitamin D status and the association was stronger in countries with a national vitamin D fortification policy [28]. However, this review included only five studies from US which had small sample sizes and included only certain population groups. In our present analysis, the intake of certain milk types (especially of whole milk) was not associated with an increase in serum vitamin D levels in all population sub-groups, which is not immediately understood.

Age and ethnicity have been shown to affect serum vitamin D status [14,15,29,30]. Vitamin D serum levels generally decrease with age, and non-Hispanic Blacks have the lowest vitamin D levels or highest prevalence of VDD, followed by Hispanics and Asians [14,15,29,30]. In contrast, non-Hispanic Whites have the highest vitamin D levels or lowest prevalence of VDD [14,15,29,30]. Lower intake of milk with age, which was also observed in our study, could potentially explain the inverse association of age with vitamin D. In regard to ethnicity and vitamin D status, differences in milk and overall vitamin D intake as well as skin pigmentation and other factors are potentially responsible for the ethnic differences in vitamin D status [31]. For instance, studies have shown lower milk intake among non-Hispanic Blacks compared to non-Hispanic Whites [32]. In the present analysis, serum vitamin D status was associated with both type and amount of milk intake depending upon the age and ethnicity. However, the effect of gender on serum vitamin D status has been reported to be insignificant or inconsistent [14,15,30], but the associations between amount of milk intake and serum vitamin D status were mostly significant for both males and females in the present analysis. Therefore, continuing to encourage an increase in milk intake, especially among populations with VDD or VDI, could be an effective strategy to improve vitamin D status.

In addition to providing vitamin D, milk and dairy products, make significant nutrient contributions including nutrients under-consumed by most Americans—calcium and potassium—as well as magnesium, phosphorus, zinc, vitamin A, vitamin B12, riboflavin (B2), choline, high-quality protein and saturated fat; as such, the inclusion of dairy foods into healthy dietary patterns is associated with improving diet quality and reducing risk of obesity and chronic diseases [17,18,33–36].

A major limitation of our study is the inability to determine a cause–effect relationship due to the cross-sectional design of NHANES. Additionally, as with any study based on self-reported data, under- or over-reporting cannot be ruled out. Additionally, the results from this study may not specifically reflect the effect of milk consumption on vitamin D status, although we used vitamin D from dietary supplements, seafood, and other non-milk sources as covariates to adjust some of our results. Strengths of this study included the use of a large nationally representative sample achieved through combining several sets of NHANES data releases and adjusting for numerous covariates, but even with these covariates, some residual confounding may still exist.

5. Conclusions

In conclusion, the results of this study indicate that milk consumers consistently have higher serum vitamin D levels and higher probability of meeting the recommended levels. Vitamin D has been identified as a "nutrient of public health concern in the US" and, therefore, increasing the intake of milk (especially low-fat and reduced-fat) should be encouraged. Other sources of vitamin D may also help in improving vitamin D status.

Author Contributions: The authors' responsibilities were as follows: M.T.-G. and C.J.C.: conceived the project, designed research, developed the overall research plan, and participated in revising the manuscript; S.A.: participated in interpretation of the data, prepared the first draft of the manuscript, and participated in revising the manuscript; V.L.F.III: designed research, developed overall research plan, analyzed data and performed statistical analysis, and participated in the interpretation of the data and revising the manuscript. All authors have read and agreed to the published version of the manuscript.

Funding: The study and the writing of the manuscript were supported by National Dairy Council, Rosemont, IL.

Conflicts of Interest: M.T.-G. and C.J.C. are employees of National Dairy Council, Rosemont, IL; S.A. as Principal of NutriScience, L.L.C. performs consulting and database analyses for various food and beverage companies and related entities; V.L.F.III is Senior Vice President of Nutrition Impact and received a research grant from National Dairy Council to conduct these analyses.

References

1. Institute of Medicine, Food and Nutrition Board. *Dietary Reference Intakes for Calcium and Vitamin D*; National Academy Press: Washington, DC, USA, 2011.
2. Holick, M.F. Vitamin D deficiency. *N. Engl. J. Med.* **2007**, *357*, 266–281. [CrossRef]
3. Holick, M.F. Resurrection of vitamin D deficiency and rickets. *J. Clin. Investig.* **2006**, *116*, 2062–2072. [CrossRef]
4. Tang, B.M.; Eslick, G.D.; Nowson, C.; Smith, C.; Bensoussan, A. Use of calcium or calcium in combination with vitamin D supplementation to prevent fractures and bone loss in people aged 50 years and older: A meta-analysis. *Lancet* **2007**, *370*, 657–666. [CrossRef]
5. Bischoff-Ferrari, H.A.; Dawson-Hughes, B.; Staehelin, H.B.; Orav, J.E.; Stuck, A.E.; Theiler, R.; Wong, J.B.; Egli, A.; Kiel, D.P.; Henschkowski, J. Fall prevention with supplemental and active forms of vitamin D: A meta-analysis of randomized controlled trials. *BMJ* **2009**, *339*, b3692. [CrossRef]
6. Giovannucci, E.; Liu, Y.; Hollis, B.W.; Rimm, E.B. 25-hydroxyvitamin D and risk of myocardial infarction in men: A prospective study. *Arch. Intern. Med.* **2008**, *168*, 1174–1180. [CrossRef]
7. Vaidya, A.; Forman, J.P. Vitamin D and Hypertension: Current Evidence and Future Directions. *Hypertension* **2010**, *56*, 774–779. [CrossRef]
8. Ford, E.S.; Ajani, U.A.; McGuire, L.C.; Liu, S. Concentrations of serum vitamin D and the metabolic syndrome among U.S. adults. *Diabetes Care* **2005**, *28*, 1228–1230. [CrossRef]

9. Munger, K.L.; Zhang, S.M.; O'Reilly, E.; Hernan, M.A.; Olek, M.J.; Willett, W.C.; Ascherio, A. Vitamin D intake and incidence of multiple sclerosis. *Neurology* **2004**, *62*, 60–65. [CrossRef]
10. Yin, L.; Ordonez-Mena, J.M.; Chen, T.; Schottker, B.; Arndt, V.; Brenner, H. Circulating 25-hydroxyvitamin D serum concentration and total cancer incidence and mortality: A systematic review and meta-analysis. *Prev. Med.* **2013**, *57*, 753–764. [CrossRef]
11. Merlino, L.A.; Curtis, J.; Mikuls, T.R.; Cerhan, J.R.; Criswell, L.A.; Saag, K.G. Iowa Women's Health Study. Vitamin D intake is inversely associated with rheumatoid arthritis: Results from the Iowa Women's Health Study. *Arthritis Rheum* **2004**, *50*, 72–77. [CrossRef]
12. Hirani, V.; Cumming, R.G.; Naganathan, V.; Blyth, F.; Le Couteur, D.G.; Hsu, B.; Handelsman, D.J.; Waite, L.M.; Seibel, M.J. Longitudinal Associations Between Vitamin D Metabolites and Sarcopenia in Older Australian men: The Concord Health and Aging in Men Project. *J. Gerontol. A Biol. Sci. Med. Sci.* **2017**, *73*, 131–138. [CrossRef]
13. Wei, M.Y.; Giovannucci, E.L. Vitamin D and multiple health outcomes in the Harvard cohorts. *Mol. Nutr. Food Res.* **2010**, *54*, 1114–1126. [CrossRef]
14. U.S. Centers for Disease Control and Prevention. *Second National Report on Biochemical Indicators of Diet and Nutrition in the U.S. Population 2012*; National Center for Environmental Health: Atlanta, GA, USA. Available online: http://www.cdc.gov/nutritionreport/pdf/nutrition_book_complete508_final.pdf (accessed on 16 October 2019).
15. Herrick, K.A.; Storandt, R.J.; Afful, J.; Pfeiffer, C.M.; Schleicher, R.L.; Gahche, J.J.; Potischman, N. Vitamin D status in the United States, 2011–2014. *Am. J. Clin. Nutr.* **2019**, *110*, 150–157. [CrossRef]
16. USDA, Agricultural Research Service. Usual Nutrient Intake from Food and Beverages, by Gender and Age, What We Eat in America, NHANES 2013–2016. Available online: https://www.ars.usda.gov/ARSUserFiles/80400530/pdf/usual/Usual_Intake_gender_WWEIA_2013_2016.pdf (accessed on 22 September 2020).
17. U.S. Department of Health and Human Services; U.S. Department of Agriculture. *2015–2020 Dietary Guidelines for Americans*, 8th ed.; USDA: Washington, DC, USA, 2015. Available online: http://health.gov/dietaryguidelines/2015/guidelines/ (accessed on 16 October 2019).
18. Dietary Guidelines Advisory Committee. *Scientific Report of the 2020 Dietary Guidelines Advisory Committee: Advisory Report to the Secretary of Agriculture and the Secretary of Health and Human Services*; Department of Agriculture, Agricultural Research Service: Washington, DC, USA. Available online: https://www.dietaryguidelines.gov/2020-advisory-committee-report (accessed on 22 September 2020).
19. Calvo, M.S.; Whiting, S.J. Vitamin D Fortification in North America: Current Status and Future Considerations. In *The Handbook of Food Fortification from Concepts to Public Health Applications*; Preedy, R.V., Srirajaskanthan, R., Patel, V., Eds.; Springer Science Business Media: New York, NY, USA, 2013; Volume 2, pp. 259–271.
20. Calvo, M.S.; Whiting, S.J.; Barton, C.N. Vitamin D fortification in the United States and Canada: Current status and data needs. *Am. J. Clin. Nutr.* **2004**, *80*, 1710S–1716S. [CrossRef]
21. Yetley, E.A. Assessing the vitamin D status of the US population. *Am. J. Clin. Nutr.* **2008**, *88*, 558S–564S. [CrossRef]
22. Centers for Disease Control and Prevention (CDC), National Center for Health Statistics. *National Health and Nutrition Examination Survey*; National Center for Health Statistics: Hyattsville, MD, USA. Available online: https://www.cdc.gov/nchs/nhanes/index.htm (accessed on 28 March 2018).
23. Schleicher, R.L.; Sternberg, M.R.; Lacher, D.A.; Sempos, C.T.; Looker, A.C.; Durazo-Arvizu, R.A.; Yetley, E.A.; Chaudhary-Webb, M.; Maw, K.L.; Pfeiffer, C.M.; et al. A Method-bridging Study for Serum 25-hydroxyvitamin D to Standardize Historical Radioimmunoassay Data to Liquid Chromatography-Tandem Mass Spectrometry. *Natl. Health Stat. Rep.* **2016**, *93*, 1–16.
24. Van Schoor, N.; Lips, P. Global Overview of Vitamin D Status. *Endocrinol. Metab. Clin. N. Am.* **2017**, *46*, 845–870. [CrossRef]
25. Liu, X.; Baylin, A.; Levy, P.D. Vitamin D deficiency and insufficiency among US adults: Prevalence, predictors and clinical implications. *Br. J. Nutr.* **2018**, *119*, 928–936. [CrossRef]
26. Holick, M.F.; Binkley, N.C.; Bischoff-Ferrari, H.A.; Gordon, C.M.; Hanley, D.A.; Heaney, R.P.; Murad, M.H.; Weaver, C.M. Evaluation, treatment, and prevention of vitamin D deficiency: An Endocrine Society Clinical Practice Guideline. *J. Clin. Endocrin. Metab.* **2011**, *96*, 1191–1930. [CrossRef]
27. Black, L.J.; Seamans, K.M.; Cashman, K.D.; Kiely, M. An updated systematic review and meta-analysis of the efficacy of vitamin D food fortification. *J. Nutr.* **2012**, *142*, 1102–1108. [CrossRef]

28. Itkonen, S.T.; Erkkola, M.; Lamberg-Allardt, C.J.E. Vitamin D Fortification of Fluid Milk Products and Their Contribution to Vitamin D Intake and Vitamin D Status in Observational Studies-A Review. *Nutrients* **2018**, *10*, 1054. [CrossRef]
29. Forrest, K.Y.; Stuhldreher, W.L. Prevalence and correlates of vitamin D deficiency in US adults. *Nutr. Res.* **2011**, *31*, 48–54. [CrossRef]
30. Parva, N.R.; Tadepalli, S.; Singh, P.; Qian, A.; Joshi, R.; Kandala, H.; Nookala, V.K.; Cheriyath, P. Prevalence of Vitamin D Deficiency and Associated Risk Factors in the US Population (2011–2012). *Cureus* **2018**, *10*, e2741. [CrossRef]
31. O'Neill, C.M.; Kazantzidis, A.; Kiely, M.; Cox, L.; Meadows, S.; Goldberg, G.; Prentice, A.; Kift, R.; Webb, A.R.; Cashman, K.D. A predictive model of serum 25-hydroxyvitamin D in UK white as well as black and Asian minority ethnic population groups for application in food fortification strategy development towards vitamin D deficiency prevention. *J. Steroid. Biochem. Mol. Biol.* **2017**, *173*, 245–252. [CrossRef]
32. Sebastian, R.S.; Goldman, J.D.; Wilkinson Enns, C.; LaComb, R.P. *Fluid Milk Consumption in the United States: What We Eat in America, NHANES 2005–2006*; Food Surveys Research Group: Beltsville, MD, USA, 2010. Available online: http://ars.usda.gov/Services/docs.htm?docid=19476 (accessed on 16 October 2019).
33. O'Neil, C.E.; Nicklas, T.A.; Fulgoni, V.L. Food sources of energy and nutrients of public health concern and nutrients to limit with a focus on milk and other dairy foods in children 2 to 18 years of age: National Health and Nutrition Examination Survey, 2011–2014. *Nutrients* **2018**, *10*, 1050. [CrossRef]
34. O'Neil, C.E.; Keast, D.R.; Fulgoni, V.L.; Nicklas, T.A. Food sources of energy and nutrients among adults in the US: NHANES 2003–2006. *Nutrients* **2012**, *4*, 2097–2120. [CrossRef]
35. Hess, J.M.; Cifelli, C.J.; Fulgoni, V.L., III. Energy and nutrient intake of Americans according to meeting current dairy recommendations. *Nutrients* **2020**, *12*, 3006. [CrossRef]
36. Thorning, T.K.; Raben, A.; Tholstrup, T.; Soedamah-Muthu, S.S.; Givens, I.; Astrup, A. Milk and dairy products: Good or bad for human health? An assessment of the totality of scientific evidence. *Food Nutr. Res.* **2016**, *60*, 32527. [CrossRef]

Publisher's Note: MDPI stays neutral with regard to jurisdictional claims in published maps and institutional affiliations.

© 2020 by the authors. Licensee MDPI, Basel, Switzerland. This article is an open access article distributed under the terms and conditions of the Creative Commons Attribution (CC BY) license (http://creativecommons.org/licenses/by/4.0/).

Article

Milk Containing A2 β-Casein ONLY, as a Single Meal, Causes Fewer Symptoms of Lactose Intolerance than Milk Containing A1 and A2 β-Caseins in Subjects with Lactose Maldigestion and Intolerance: A Randomized, Double-Blind, Crossover Trial

Monica Ramakrishnan, Tracy K. Eaton, Omer M. Sermet and Dennis A. Savaiano *

Department of Nutrition Science, College of Health and Human Sciences, Purdue University, West Lafayette, IN 47907, USA; ramakrm@purdue.edu (M.R.); tkeaton@purdue.edu (T.K.E.); omer_sermet@hotmail.com (O.M.S.)
* Correspondence: savaiano@purdue.edu

Received: 12 November 2020; Accepted: 15 December 2020; Published: 17 December 2020

Abstract: Acute-feeding and multiple-day studies have demonstrated that milk containing A2 β-casein only causes fewer symptoms of lactose intolerance (LI) than milk containing both A1 and A2 β-caseins. We conducted a single-meal study to evaluate the gastrointestinal (GI) tolerance of milk containing different concentrations of A1 and A2 β-casein proteins. This was a randomized, double-blind, crossover trial in 25 LI subjects with maldigestion and an additional eight lactose maldigesters who did not meet the QLCSS criteria. Subjects received each of four types of milk (milk containing A2 β-casein protein only, Jersey milk, conventional milk, and lactose-free milk) after overnight fasting. Symptoms of GI intolerance and breath hydrogen concentrations were analyzed for 6 h after ingestion of each type of milk. In an analysis of the 25 LI subjects, total symptom score for abdominal pain was lower following consumption of milk containing A2 β-casein only, compared with conventional milk ($p = 0.004$). Post hoc analysis with lactose maldigesters revealed statistically significantly improved symptom scores ($p = 0.04$) and lower hydrogen production ($p = 0.04$) following consumption of milk containing A2 β-casein only compared with conventional milk. Consumption of milk containing A2 β-casein only is associated with fewer GI symptoms than consumption of conventional milk in lactose maldigesters.

Keywords: A1 beta-casein; A2 beta-casein; beta-casomorphin; gastrointestinal intolerance; hydrogen breath test; lactose challenge; lactose intolerance symptoms; milk intolerance; Qualifying Lactose Challenge Symptom Score

1. Introduction

Approximately 30% of cows' milk protein is β-casein [1], of which two genetic variants exist: A1 and A2 [2]. A1 β-casein includes histidine at the 67th position in the peptide chain, whereas A2 β-casein includes proline at this position [3]. Although some cattle breeds maintain the A2 β-casein variant, a single nucleotide polymorphism in modern western cattle breeds means that they exhibit mixed A1 and A2 β-casein variants [3–5]. Digestive enzymes act on A1 β-casein and hydrolyze it, releasing beta-casomorphin-7 (BCM-7) [6–10]. The histidine residue in A1 β-casein allows cleavage to form BCM-7, whereas the proline residue in A2 β-casein limits such cleavage and BCM-7 formation [11].

In animal studies, BCM-7 is both pro-inflammatory and associated with slower gastrointestinal (GI) transit [12]. In intestinal and neuronal cells, BCM-7 downregulates the glutathione (GSH) levels [13], which is an important antioxidant in the body for combating oxidative stress, which otherwise can result in inflammation [14]. Oxidative stress has been shown to induce epigenetic changes, especially on genes that are important mediators of inflammation, leading to increased GI symptoms.

Multiple-day and acute-feeding studies in Chinese and Australian populations have shown that milk containing only the A2 β-casein protein caused fewer symptoms of lactose intolerance (LI) than milk containing both A1 and A2 β-casein protein [15–18]. However, these studies were conducted in subjects with self-reported LI, and no relevant blinded studies have been reported in verified LI individuals; this is a notable omission, given that self-reported LI can be unreliable [19]. Moreover, only one study to date has examined the effects of the administration of variable ratios of A1 and A2 β-casein [20]. Manifestations of LI can be both acute and chronic: they can be long-term due to epigenetic changes or genetic mutation [21]; they also appear to be a single-meal event, with symptoms occurring between 30 min and 6 h after exposure [22,23]. These symptoms and their cause are distinct from milk allergy, which results from an immune reaction to milk proteins.

We conducted a randomized, double-blind, single-meal feeding trial with four types of milk, which varied in A1/A2 β-casein protein ratio: milk containing A2 β-casein only, Jersey milk (containing 25%/75% A1/A2 β-casein), conventional milk (containing 75%/25% A1/A2 β-casein), and lactose-free milk. The principal study objective was to determine, via a hydrogen breath test (HBT) in lactose maldigester individuals living in the Midwest United States and who had verified LI following a blinded milk challenge, if lactose digestion and GI tolerance were affected by the four different milk types within 6 h after ingestion. We hypothesized that a single meal of milk containing A2 β-casein only would be better tolerated, producing fewer GI symptoms and less maldigestion during the 6 h study, than conventional milk containing both A1 and A2 β-casein. We also hypothesized that Jersey milk would produce an intermediate response regarding the HBT and occurrence of GI symptoms as a result of its higher level of A2 β-casein and lower level of A1 β-casein.

2. Materials and Methods

2.1. Subject Selection and Inclusion Criteria

This was a randomized, double-blind, crossover trial in subjects aged 18–65 years. Subjects were recruited through flyers and advertisements in local (West Lafayette, IN, USA) and campus (Purdue University, West Lafayette, IN, USA) newspapers, and Purdue Today email. Study recruitment started in February 2018 and was suspended in February 2020 due to COVID-19 restrictions by our institutional review board (IRB). A total of 853 people indicated interest in the study and contacted the investigators via email or phone. Of these, 258 subjects successfully completed phone screening and were then categorized as eligible or ineligible to participate in the trial (Appendix A). After the phone screens, all eligible subjects signed a consent form and agreed to participate in the study. We queried all interested subjects regarding demographic information, current medication use, and height and weight for calculating body mass index (BMI). Participants were assigned an identification number upon signing the informed consent form. Identification numbers were block pre-randomized using randomization.com. Staff performed the randomization, enrolled participants, and provided milk type information to our Clinical Research Center (CRC) kitchen via sealed security envelopes. Participants and study staff assessing outcomes over the 6 h were blinded to the milk types consumed.

Eligible subjects had to have avoided dairy for at least 1 month prior to screening and were included in the study only if they agreed to refrain from dairy and all treatments or products used for dairy intolerance (e.g., Lactaid® dietary supplements; McNeil Nutritionals, LLC, Ft. Washington, PA, USA) throughout the trial. Subjects had to have a history of perceived dairy intolerance. Perceived LI was then confirmed by the Qualifying Lactose Challenge Symptom Score (QLCSS) during a 6 h HBT after consumption of a commercial milk containing a high amount of A1 β-casein [24]. Lactose maldigesters, defined as producing more than 20 parts per million (ppm) hydrogen at any time point following the baseline commercial milk challenge [25], were eligible for the intervention portion of the study.

Abdominal pain, bloating, flatulence, and diarrhea are typical symptoms of LI [26]. Although other studies of A1 and A2 β-caseins reported scores for stool frequency and stool consistency as indicators of LI [15–17], in the current study, qualifying scores for abdominal pain, bloating, flatulence, diarrhea, and fecal urgency were each recorded by subjects using a six-point Likert scale. Symptoms were ranked from 0 to 5, where 0 was for no symptoms, 1 for slight, 2 for mild, 3 for moderate, 4 for moderately severe, and 5 for severe symptoms. If subjects met one of the following three criteria regarding the QLCSS, they were classified as lactose intolerant: a score of 4 or 5 for an individual symptom; a score of 3 for at least two symptoms; or a score of 3 for one symptom at two time points in the study.

2.2. Exclusion Criteria

Subjects were excluded from the study for the following reasons: allergy to milk; pregnancy or lactation; cigarette smoking, or use of tobacco or nicotine-containing products within 3 months of screening; diagnosis of abnormal GI motility; a history of GI tract surgery; the presence of any medical condition with symptoms that could confound collection of data about adverse events; ulcer; diabetes mellitus; congestive heart failure; HIV, hepatitis B, or hepatitis C virus infection; BMI > 35 kg/m^2; use of products to treat dairy intolerance within 7 days of screening; use of antacids and/or proton pump inhibitors; use of antibiotics or colonic enemas within 30 days prior to screening; any concurrent disease or symptoms that may interfere with assessment of the cardinal symptoms of dairy intolerance; use of ethanol (alcohol) and/or drug abuse in the past month; chemotherapy; or use of any investigational drug, or participation in any investigational study, within 30 days prior to screening.

2.3. Interventions

Four types of milk were evaluated in the study: milk containing A2 β-casein only (The a2 Milk Company, Boulder, CO, USA); Jersey milk (containing 25%/75% A1/A2 β-casein; American Jersey Cattle Association, Crockett, VA, USA); conventional milk (containing 75%/25% A1/A2 β-casein; Kroger® 2% reduced fat; The Kroger Co., Indianapolis, IN, USA); and lactose-free milk (containing 60%/40% A1/A2 β-casein; Lactaid®; McNeil Nutritionals, LLC). Jersey milk was shipped to our study laboratory (Purdue University, West Lafayette, IN, USA) from the American Jersey Cattle Association (Crockett, VA, USA). All other milk products were purchased from Payless (West Lafayette, IN, USA); if they were unavailable at Payless, then they were purchased from Fresh Thyme (West Lafayette, IN, USA). All four milk types were procured every two weeks and administered prior to their expiry dates.

2.4. A1/A2 Analysis

The ratio of A1/A2 β-casein in the four types of milk was analyzed using mass spectrometry (MS) at Purdue Proteomics Facility. Protein extraction from 100 μL of each milk was performed by denaturation with 400 μL of 8 M urea and 10 mM dithiothreitol (DTT) and vortexing for 15 min at room temperature to remove fat. This was followed by protein alkylation with 400 μL of 8 M urea and 10 mM DTT, and then digestion with pepsin using a 1:20 enzyme to substrate ratio for 1 h at room temperature. The peptides, which were cleaned/desalted via a C18 Silica MicroSpin column (The Nest Group Inc., Southborough, MA, USA) after digestion, were analyzed with the Dionex UltiMate 3000 RSLC nano System combined with the Q-Exactive High-Field Hybrid Quadrupole Orbitrap MS (Thermo Fisher Scientific, Waltham, MA, USA). Peptides were then re-suspended in 3% acetonitrile/0.1% formic acid/96.9% Milli-Q water, and 5 μL (1 μg) were used for liquid chromatography (LC)-MS/MS analysis. A trap (300 μm internal diameter (ID) × 5 mm packed with 5 μm 100 Å PepMap C18 medium; Thermo Fisher Scientific) was used to separate peptides and a 120 min gradient method with a flow rate of 300 nL/min was used for the analytical columns (75 μm ID × 15 cm long packed with 3 μm of 100 Å PepMap C18 medium). Mobile phase A contained 0.1% formic acid in water and mobile phase B contained 0.1% formic acid in 80% acetonitrile. The linear gradient started at 5% B and reached 30% B in 80 min, 45% B in 91 min, and 100% B in 93 min. The column was held at 100% B for 5 min and then brought back to 5% B. The column was held at 5% B for 20 min to equilibrate at 37 °C. The top 20 data-dependent MS/MS scan method was used to acquire MS data with a maximum injection time of 100 ms and a resolution of 120,000 at 200 m/z. High-energy C-trap dissociation with the normalized collision energy of 27 eV was used to fragment precursor ions. MS/MS scans were acquired at a resolution of 15,000 at 200 m/z. Repeated scanning of identical peptides was avoided by setting the dynamic exclusion at 20 s.

LC-MS/MS data were analyzed using MaxQuant software (version 1.6.0.1; Max Planck Institute of Biochemistry, Martinsried, Germany). The combined non-redundant *Bos taurus* protein sequence database downloaded from UniProt (www.uniprot.org) in January 2017 was used for protein identification and label-free relative quantitation. The following parameters were used for database searches: precursor mass tolerance of 10 ppm; enzyme pepsin allowing up to two missed cleavages; oxidation of methionine as a variable modification and iodoethanol as a fixed modification. The false discovery rate of peptide spectral match and protein identification was set to 0.01. Only proteins with a label-free quantitation value of 0 and MS/MS spectral counts of ≥2 were considered as a true identification before being used for further analysis.

2.5. Sugar, Protein, and Fat Analyses

Total sugars, fat, and protein were analyzed by Eurofins Food Integrity and Innovation (Eurofins Food Chemistry Testing US, Inc., Madison, WI, USA) [27–29]. The sugar profile was determined using 10 g of each milk type (with the exception of lactose-free milk), and sugars were extracted with a 50:50 methanol:water solution. Inert gas was used to dry each sample, which was derivatized prior to analysis, and the analysis was conducted via gas chromatography with flame ionization detection.

Because of its very low lactose content, lactose-free milk was analyzed using a different procedure. A 10 g milk sample was extracted with dilute HCl and centrifuged. The supernatant was filtered using a strong cation exchange cartridge, and an OnGuard II syringe filter (Thermo Fisher Scientific) was used for neutralization. Applicable amounts of dilutions were injected into a high-performance anion exchange chromatography system equipped with pulsed amperometric detection (Thermo Fisher Scientific).

Fat in the samples was analyzed by base hydrolysis, and protein was analyzed using the Dumas method [27].

2.6. Study Procedures

The subjects reported to our clinical research facility (Purdue University) for four visits, with at least six days between any two consecutive visits. Subjects consumed a low-fiber dinner and fasted for 12 h prior to visits. Subjects consumed a different randomized milk product at 8 a.m. on the day of each visit. Each milk meal, except the lactose-free milk, contained ~4.5 g of lactose/per 100 mL. The amount of milk consumed was calculated as approximately 0.5 g of lactose times bodyweight in kg, divided by 11 g (the normal amount of lactose in a cup of regular milk), then multiplied by 245 mL (one cup).

$$\text{Amount of milk consumed} = \frac{0.5 \text{ g of lactose} \times \text{bodyweight (kg)}}{11 \text{ g of lactose}} \times 245 \text{ mL of milk}$$

Subjects always consumed the same quantity of fluid (mL) and lactose (g), despite small variations in the lactose content.

2.7. Study Endpoints

The primary study endpoints were the occurrence of GI symptoms (abdominal pain, bloating, flatulence, diarrhea, and fecal urgency), and measurement of hydrogen in breath samples (a standard measure of maldigestion), for up to 6 h after consumption of each milk type. Breath samples were collected, and GI symptoms were recorded by the subjects, at 0, 0.5, 1, 1.5, 2, 3, 4, 5, and 6 h after commercial milk ingestion. To measure hydrogen in the breath samples, a hydrogen microanalyzer (QuinTron BreathTracker Digital Microlyzer, model SC; QuinTron Instrument Company, Inc., Milwaukee, WI, USA) was used. An increase of 20 ppm hydrogen between any two timepoints in the study indicated lactose maldigestion [26]. Symptoms were scored using a six-point Likert scale (as described in the inclusion criteria).

2.8. Study Ethics

The study (ClinicalTrials.gov #NCT03713346) and its protocol were approved by the Purdue IRB (IRB #1710019781). The trial was conducted in accordance with the Helsinki Declaration of 1975 as revised in 1983. The study was also conducted in accordance with International Conference on Harmonization Good Clinical Practice guidelines.

2.9. Statistical Analyses

The initial power calculation indicated a sample size requirement of 26, which was determined based on the selection of a 20% decrease in area under the curve for change in breath hydrogen as the minimal difference that would be clinically significant. Power calculations indicated that completion of the protocol, with a crossover study design, by 26 subjects would be adequate to demonstrate 80% statistical power, consistent with biological relevance using $\alpha = 0.05$ to detect a 20% change in breath hydrogen. The sample size for symptoms was derived from a previous study in Chinese preschoolers aged 5 to 6 years [18]. The COVID-19 pandemic caused us to suspend the study in February 2020 with 25 verified LI subjects and an additional eight maldigesters who did not meet the LI criteria. However, we observed that not all LI subjects met the criteria for LI after a second commercial milk dose, suggesting that the symptom criteria were arbitrary and inconsistent. A post hoc analysis indicated only 15 of 25 LI subjects met the LI criteria after receiving the second commercial milk challenge as part of the randomized intervention; therefore, we included maldigesters who did not meet the criteria for LI in the study in June 2019, to better understand the potential effect of milk containing A2 β-casein only and Jersey milk on all maldigesters. As a result, an analysis was conducted with 25 LI subjects, and a post hoc analysis was conducted with the addition of eight maldigesters (a total of 33 subjects).

GI symptoms, hydrogen at each timepoint, and total hydrogen during each 6 h study period were analyzed using the paired t-test. Two-tailed p-values were compared with a significance level of 0.05. Descriptive statistics were used to calculate mean and standard error values for GI symptoms and breath hydrogen. All statistical analyses were conducted using Microsoft® Excel and the Statistical Package for the Social Sciences (IBM SPSS Statistics for Windows, Version 26.0; IBM Corp., Armonk, NY, USA).

Lactose-free milk was used as a negative control. Conventional milk was compared with Jersey milk and milk containing A2 β-casein only, by measuring the occurrence of GI symptoms and breath hydrogen. Subject symptom scores for abdominal pain, bloating, flatulence, and diarrhea were summed over 6 h after each milk consumption, and total symptoms for each subject were calculated as the sum of all the total symptom scores for abdominal pain, bloating, flatulence, and diarrhea; symptoms of fecal urgency were analyzed separately. For each subject, baseline breath hydrogen concentration was subtracted from the breath hydrogen concentration produced at each timepoint (0, 0.5, 1, 1.5, 2, 3, 4, 5, and 6 h) to correct for residual hydrogen.

3. Results

3.1. Baseline and Demographic Characteristics

Of the 258 subjects who were phone-screened, 111 were ineligible due to medical conditions or lack of milk avoidance, and five subjects were ineligible because of Lactaid® use. A total of 142 subjects were eligible for the HBT screening, but only 94 chose to participate; the other 48 subjects did not respond to attempts to schedule the baseline screening. A total of 35 subjects met the maldigestion criteria and were randomized to one of four sequences for receiving the four milk products. Thirty-three completed the four study visits and had GI symptoms and hydrogen production in HBTs recorded for 6 h after each milk treatment. Of the 33 subjects, 25 met the symptom criteria and were classified as LI and an additional eight subjects were maldigesters without LI; two subjects were unable to complete the protocol owing to COVID-19 restrictions implemented by our IRB (Figure 1).

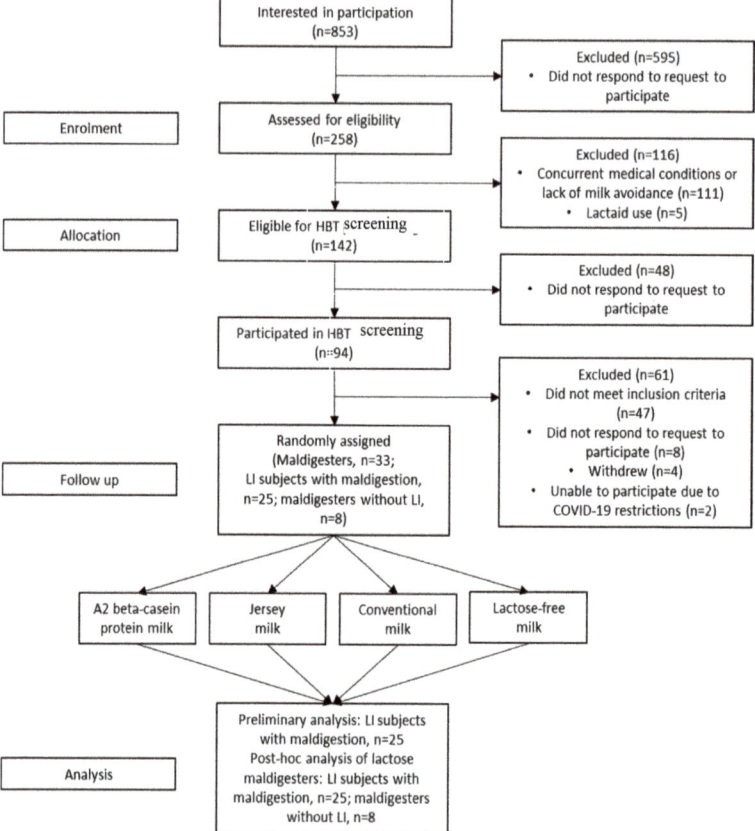

Figure 1. Study enrollment, randomization and analyses. HBT, hydrogen breath test; LI, lactose intolerance/intolerant.

Overall, 15 male and 18 female subjects with a mean age of 25 (range 19–50) years, and a mean BMI of 24 (range 18–33) kg/m^2, completed the study. The study population comprised 14 individuals who identified as Asian, four African Americans, 14 Caucasians, and one American Indian. All participants resided in the United States: five were Hispanic, 26 were non-Hispanic, and two participants did not disclose ethnicity (Hispanic/non-Hispanic) (Table 1). Nutrient composition for each of the four milk products evaluated is shown in Table 2.

Table 1. Baseline and demographic characteristics.

Age, mean (range); years	25 (19–50)
Bodyweight, mean (range); kg	71
Height, mean (range); cm	170
BMI, mean (range); kg/m^2	24
Male/female, n/n	15/18
Lactose intolerant maldigesters (meeting QLCSS; hydrogen > 20 ppm), n	25
Lactose tolerant maldigesters (hydrogen ≤ 20 ppm), n	8
Race, n	
Asian	14
African American	4
Caucasian	14
American Indian	1
Ethnicity, n	
Hispanic	5
Non-Hispanic	26
Unknown	2

Baseline and demographic characteristics of maldigesters and lactose intolerant subjects (n = 33) enrolled in this randomized, double-blinded trial comparing conventional milk with Jersey milk and milk containing A2 β-casein only. BMI, body mass index; ppm, parts per million; QLCSS, Qualifying Lactose Challenge Symptom Score.

Table 2. Nutrient composition of the four milk treatments.

Nutrient	Milk Containing A2 β-Casein Only	Jersey Milk	Conventional Milk	Lactose-Free Milk
Protein (g/serving)	3.14	3.95	3.30	3.21
Fat (g/serving)	2.10	2.00	1.90	2.00
Lactose (g/serving)	4.70	4.40	4.60	0.13
Carbohydrate (g/serving)	4.70	4.40	4.60	N/A
Calories (kcal/serving)	0.0541	0.0500	0.0500	0.0500
A1 β-casein protein (%)	0.00	25.00	75.00	60.00
A2 β-casein protein (%)	100.00	75.00	25.00	40.00

Subjects were fed approximately 4.5 g lactose/100 mL of each milk after an overnight fast, in random order, with six days between treatments.

3.2. GI Symptoms

3.2.1. LI Subjects

The total symptom score for abdominal pain during the 6 h after the consumption of milk containing A2 β-casein only, in LI subjects (n = 25), was significantly lower than that following consumption of conventional milk (112 vs. 146; $p = 0.004$); however, in contrast, the total score for abdominal pain after consumption of Jersey milk was not significantly different from that for conventional milk (135 vs. 146; $p = 0.63$). Regarding total symptom scores for bloating, flatulence, diarrhea, and fecal urgency, no significant differences were evident between those who had consumed conventional milk and those who had consumed either Jersey milk or milk containing A2 β-casein only (Figure 2). With respect to the combined total symptom scores for abdominal pain, bloating, flatulence, and diarrhea reported by subjects, there were no significant differences between conventional milk and Jersey milk or milk containing A2 β-casein only (Table 3 and Figure 2).

3.2.2. All Maldigesters

The total symptom score for abdominal pain after the consumption of milk containing A2 β-casein only, in all maldigesters (n = 33), was significantly lower than that following consumption of conventional milk (126 vs. 175; $p = 0.001$); however, the total score for abdominal pain after the consumption of Jersey milk was not significantly different from that for conventional milk (170 vs. 175; $p = 0.83$). Total symptom score for bloating was higher when consuming Jersey milk compared with conventional milk (293 vs. 240; $p = 0.05$). Total symptom scores for flatulence, diarrhea, and fecal urgency were similar in subjects consuming milk containing A2 β-casein only, Jersey milk, and conventional milk (Figure 3). The combined total symptom scores for abdominal pain, bloating, flatulence, and diarrhea showed there were fewer symptoms with milk containing A2 β-casein only (601 vs. 737; $p = 0.04$) compared with conventional milk, whereas consumption of Jersey milk or conventional milk produced similar symptom scores (790 vs. 737; $p = 0.44$) (Table 3 and Figure 3).

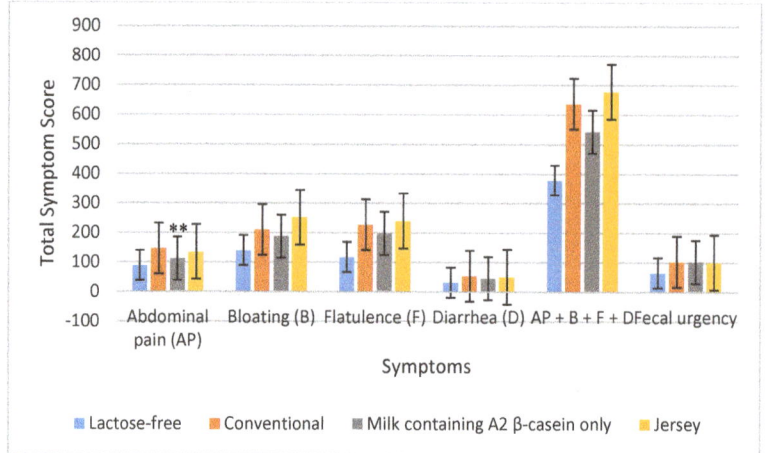

Figure 2. Total symptoms reported during the 6 h after consuming the four milk products in 25 lactose intolerant subjects. ** $p = 0.004$ for abdominal pain due to milk containing A2 β-casein only vs. conventional milk.

Table 3. Comparison of total hydrogen produced, and symptoms reported.

Criteria	Pairs	LI Subjects ($n = 25$)		Lactose Maldigesters ($n = 33$; 25 LI Subjects + 8 Maldigesters)	
		Total	p-Values	Total	p-Values
Total hydrogen produced per subject (ppm)	Conventional milk Milk containing A2 β-casein only	11,935 10,892	0.31	16,460 13,771	0.04
	Conventional milk Jersey milk	11,935 10,533	0.09	16,460 15,079	0.44
Total symptom scores [a]	Conventional milk Milk containing A2 β-casein only	637 543	0.13	737 601	0.04
	Conventional milk Jersey milk	637 678	0.55	737 790	0.17

Comparison of total hydrogen produced, and symptoms reported for six hours following consumption of conventional milk versus milk containing A2 β-casein only and conventional milk versus Jersey milk using paired t-tests. LI, lactose intolerant; ppm, parts per million; [a] abdominal pain + bloating + flatulence + diarrhea.

3.3. HBT Results

3.3.1. LI Subjects

Hydrogen breath concentration was analyzed in 25 LI subjects. The total quantity of hydrogen produced was not significantly different during the 6 h after the consumption of Jersey milk or milk containing A2 β-casein only when compared with consumption of conventional milk (Table 3 and Figure 4).

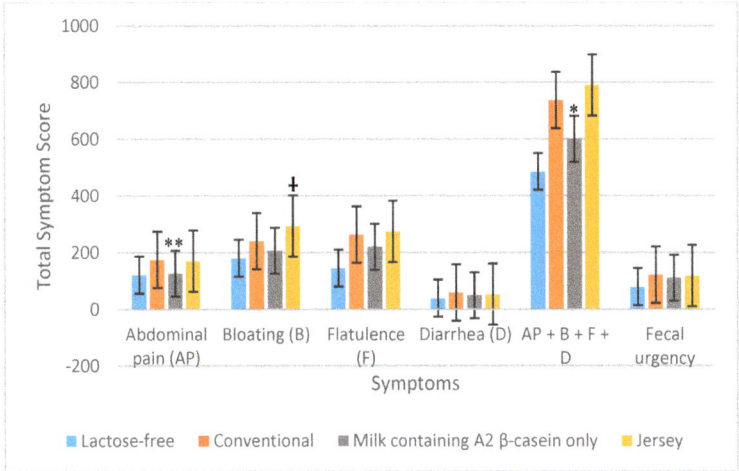

Figure 3. Total symptoms reported during the 6 h after consuming the four milk products in 33 lactose maldigesters. ** $p = 0.001$ for abdominal pain and * $p = 0.04$ for total symptoms (abdominal pain + bloating + flatulence + diarrhea) due to milk containing A2 β-casein only vs. conventional milk; † $p = 0.05$ for bloating due to Jersey milk versus conventional milk.

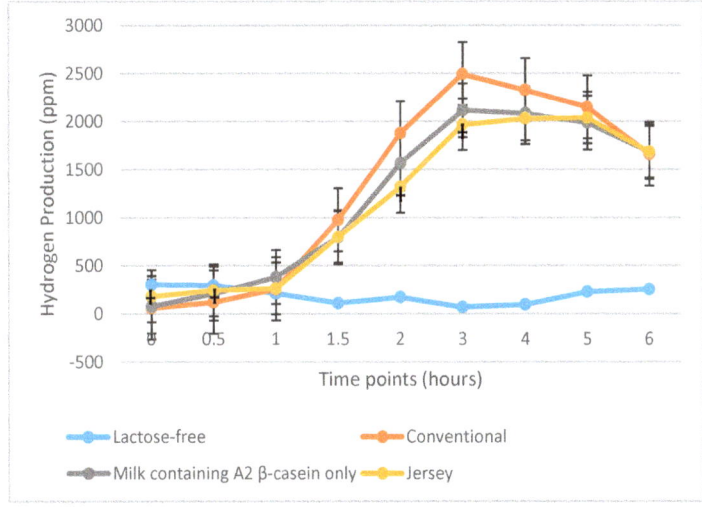

Figure 4. Total hydrogen produced during the 6 h after consuming the four milk products in 25 lactose intolerant subjects. ppm, parts per million. † $p = 0.05$, † $p = 0.03$, † $p = 0.01$, and † $p = 0.05$ for Jersey milk vs. commercial milk at 0, 0.5, 2, and 3 h, respectively.

3.3.2. All Maldigesters

Total hydrogen produced by 33 maldigesters following consumption of milk containing A2 β-casein only was significantly lower compared with hydrogen produced by subjects following consumption of conventional milk (13,771 vs. 16,460 ppm; $p = 0.04$). However, hydrogen production following consumption of Jersey milk was not significantly different from that following consumption of conventional milk (15,079 vs. 16,460 ppm; $p = 0.17$) (Table 3 and Figure 5).

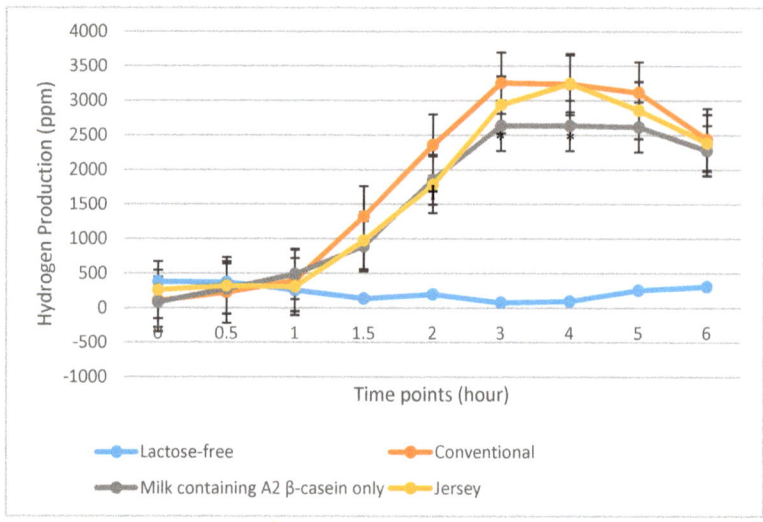

Figure 5. Total hydrogen produced during the 6 h after consuming the four milk products in 33 lactose maldigesters. ppm, parts per million. * $p = 0.05$, * $p = 0.03$ for milk containing A2 β-casein only vs. conventional milk at 3 and 4 h, respectively; † $p = 0.03$ for Jersey milk versus conventional milk at 2 h.

3.4. Adverse Events

There were no adverse events or unintended harmful effects reported by subjects due to the consumption of the four different types of milk.

4. Discussion

The results of our study indicate that the consumption of milk containing A2 β-casein only produced fewer GI symptoms in lactose maldigesters compared with consumption of conventional milk. On the other hand, Jersey milk did not reduce GI symptoms, compared with conventional milk. In LI subjects and lactose maldigesters, milk containing only A2 β-casein significantly decreased abdominal pain compared with the consumption of conventional milk. Conversely, the consumption of Jersey milk was not associated with reduced abdominal pain.

The effects of the milk treatments on GI symptoms may be related to GI effects due to longer transit time in the colon by milk containing A1 β-casein [17]. A study in Wistar rats and some human clinical trials showed that A1 β-casein increased GI transit time and colonic activity of the inflammatory marker myeloperoxidase [17,18,30]. These effects, which were counteracted by the opioid blocker naloxone, might be initiated and mediated by the opioid peptide BCM-7, which is formed after the ingestion of A1 β-casein [30]. Further, bovine casein-derived opioid peptides can inhibit cysteine uptake in both GI epithelial and neuronal cells, resulting in elevated oxidative stress and altered DNA methylation, including on genes that are important for mediating inflammation [31].

Consistent with the results reported herein, in Chinese subjects with self-reported LI, consumption of conventional milk (equivalent to the conventional milk tested in the present study) produced more GI symptoms than did milk containing A2 β-casein only [17]. Moreover, increased GI transit times and concentrations of serum inflammatory markers IgG and IL-4 were noted after the consumption of milk containing A1/A2 β-casein rather than milk containing A2 β-casein only. Therefore, the increase in GI symptoms may be due to inflammation and GI transit time, suggesting a need for further investigation.

The presence of LI should be confirmed by recording symptom scores for abdominal pain, bloating, flatulence, and diarrhea [26], and previous studies of A1 and A2 β-caseins in LI individuals did not specifically use QLCSS to screen subjects [15–18]; as a result, these earlier studies selected subjects with perceived LI for evaluation, many of whom may not have been truly intolerant. In contrast, we verified LI via symptom scores during screening. Thus, our study is the first to demonstrate that verified LI individuals are able to better tolerate a single meal of milk containing only A2 β-casein compared with conventional milk containing both A1 and A2 β-casein.

Because our study included only 25 LI subjects, the results may not be generalizable to larger populations, although this limitation is offset by the racial and ethnic diversity of the study population. Notably, the strict QLCSS inclusion criteria contributed to the small sample size. However, given the greater statistical significance when including the eight maldigesters without symptoms of intolerance, this rigorous inclusion criterion might not be important in the population we studied. Furthermore, the fact that only 35 of 94 eligible milk avoiders/perceived intolerant individuals met the maldigestion criteria suggests that the number of people in our population with perceived or self-reported LI markedly exceeds the number with actual LI verified by symptom scores and maldigestion.

We did not categorize the subjects into age groups. Among the 25 LI study participants, 22 were in the 19- to 35-year-old age group, and three were in the 36- to 50-year-old age group. However, in a study of 600 participants aged 20–50 years who were stratified into two groups (20–35 years and 36–50 years), age had no effect on GI symptoms after milk consumption [15]; nonetheless, adults aged >50 years might respond differently to milk ingestion and this requires further evaluation.

The BMI of subjects in our study ranged from 18–33 kg/m^2. Normal-weight, overweight, and obese subjects were included in the study, but severely obese individuals were excluded. The impact of BMI differences on GI symptoms was reduced by providing subjects with calculated quantities of milk with respect to bodyweight, something that was not done in previous studies [15–18].

The effects of A1 β-casein, or its digestive by-product BCM-7, appear to be acute in our study. That is, the effects of a single milk challenge in our study were monitored over a short period (30 min to 6 h), and results may have been different if a multi-meal or multi-day feeding trial had been conducted in the same population. Long-term feeding effects might worsen GI symptoms and prior studies have shown sustained inflammatory effects with A1 β-casein, which could worsen symptoms [17,18,32]. Furthermore, there are also changes in microbial metabolites such as butanoic acid, acetic acid, and propanoic acid in adults and children following consumption of A1 β-casein [17,18], showing that A1 β-casein affects the microbiota in the gut.

5. Conclusions

In summary, results of the analysis of 25 LI subjects revealed significantly lower abdominal pain after the consumption of milk containing only A2 β-casein compared with the consumption of conventional milk. Total breath hydrogen produced by LI subjects was not significantly different from that after the consumption of conventional milk, possibly because our sample size was too small to detect differences in breath hydrogen production. The reduction in abdominal pain after the consumption of milk containing A2 β-casein only, compared with the consumption of conventional milk, was consistent with the results of another clinical trial [15]. Among the eight maldigesters tested since June 2019, none met the LI criteria during screening or intervention. In the post hoc analysis, symptoms of intolerance were not reduced after the consumption of Jersey milk compared with conventional milk, potentially because of the presence of some A1 β-casein in Jersey milk. However, there was a significant reduction in symptoms among these 25 individuals and eight additional lactose maldigesters following the consumption of milk containing only A2 β-casein. These findings warrant confirmation in larger study populations.

Author Contributions: Conceptualization, D.A.S. and O.M.S.; formal analysis, M.R., D.A.S., and O.M.S.; data curation, M.R., O.M.S., and T.K.E.; writing—original draft preparation, M.R.; writing—review and editing, D.A.S. and T.K.E. All authors have read and agreed to the published version of the manuscript.

Funding: This study was co-funded by the a2 Milk Company Ltd. and National All-Jersey Inc.

Acknowledgments: Eurofins conducted the analysis of major nutrients in the four milk treatments; Purdue Proteomics Facility analyzed the A1/A2 β-casein ratio using mass spectrometry. We would also like to thank the Statistics Department at Purdue University, West Lafayette, IN, USA for helping with the power calculation, and staff members in the metabolic kitchen and Clinical Space at Purdue University. We thank David Murdoch, BSc (Hons) of Edanz Evidence Generation for providing editorial support.

Conflicts of Interest: The authors declare no potential conflicts of interest during the conduct of this study.

Appendix A

Inclusion and Exclusion Criteria for Phone Screening

Inclusion criteria:

1. Ability/desire to provide informed consent
2. Aged 18–65 years at screening
3. Current or recent history of intolerance to and avoidance of dairy for at least 1 mo (by self-report and self-reported symptoms).
4. Agreement to refrain from all other treatments and products used for dairy intolerance (e.g., Lactaid® dietary supplements; McNeil Nutritionals, LLC, Ft. Washington, PA, USA) during study involvement
5. Willing to return for all study visits and complete all study related procedures, including fasting before and during the hydrogen breath tests (HBTs)
6. Qualifying Lactose Challenge Symptom Score. Four symptom categories with severity measured from 0–5, as defined by one of the following:

    a. At least one score of "moderately severe" or "severe" on a single symptom during the 6 h HBT
    b. A score of "moderate" or greater for a single symptom at least two timepoints during the 6 h HBT
    c. At least one "moderate" score or greater for each of two symptoms during the 6 h HBT

7. Able to understand and provide written informed consent in English.

Exclusion criteria:

1. Allergic to milk
2. Currently pregnant
3. Currently lactating
4. Cigarette smoking, or other use of tobacco or nicotine-containing products within 3 mo of screening
5. Diagnosed with any of the following disorders known to be associated with abnormal gastrointestinal (GI) motility: gastroparesis, amyloidosis, neuromuscular diseases (including Parkinson's disease), collagen vascular diseases, alcoholism, uremia, malnutrition, or untreated hypothyroidism
6. History of surgery that alters normal GI tract function, including but not limited to: GI bypass surgery, bariatric surgery, gastric banding, vagotomy, fundoplication, pyloroplasty (N.B. history of uncomplicated abdominal surgeries such as removal of an appendix >12 months prior to screening will not be excluded)
7. Past or present: organ transplant, chronic pancreatitis, pancreatic insufficiency, symptomatic biliary disease, celiac disease, chronic constipation, diverticulosis, inflammatory bowel disease, ulcerative colitis, Crohn's disease, small intestine bacterial overgrowth syndrome, gastroparesis, gastro-esophageal reflux disease, irritable bowel syndrome, or any other medical condition with symptoms that could confound collection of adverse events
8. Active ulcers, or history of severe ulcers

9. Diabetes mellitus (type 1 and type 2)
10. Congestive heart failure
11. Human immunodeficiency virus, hepatitis B, or hepatitis C
12. Body mass index > 35 kg/m^2
13. Recent bowel preparation for endoscopic or radiologic investigation within 4 weeks of screening (e.g., colonoscopy preparation)
14. Use of concurrent therapy(ies) or other products (e.g., laxatives, stool softeners, Pepto Bismol®, Lactaid® dietary supplements) used for symptoms of dairy intolerance within 7 days of screening
15. Chronic antacid and/or proton pump inhibitor use
16. Recent use of systemic antibiotics, defined as use within 30 days prior to screening
17. Recent high colonic enema, defined as use within 30 days prior to screening
18. Any concurrent disease or symptoms that may interfere with assessment of the cardinal symptoms of dairy intolerance (i.e., gas, diarrhea, bloating, cramps, stomach pain)
19. History of ethanol (alcohol) and/or drug abuse in the past 12 months
20. Currently undergoing chemotherapy
21. Use of any investigational drug or participation in any investigational study within 30 days prior to screening
22. Prior enrollment in this study
23. Any other conditions/issues noted by the study staff and/or Principal Investigator that would impact participation and/or protocol compliance.

References

1. Phelan, M.; Aherne, A.; Fitzgerald, R.J.; O'Brien, N.M. Casein-derived bioactive peptides: Biological effects, industrial uses, safety aspects and regulatory status. *Int. Dairy J.* **2009**, *19*, 643–654. [CrossRef]
2. Formaggioni, P.; Summer, A.; Malacarne, M.; Mariani, P. Milk protein polymorphism: Detection and diffusion of the genetic variants in Bos genus. *Ann. Fac. Med. Vet. Univ. Parma* **1999**, *19*, 127–165. [CrossRef]
3. Ng-Kwai-Hang, K.F.; Grosclaude, F. Genetic polymorphism of milk proteins. In *Advanced Dairy Chemistry—1 Proteins*; Fox, P.F., McSweeney, P.L.H., Eds.; Springer: Boston, MA, USA, 2003; pp. 739–816. [CrossRef]
4. Bradley, D.G.; MacHugh, D.E.; Cunningham, P.; Loftus, R.T. Mitochondrial diversity and the origins of African and European cattle. *Proc. Natl. Acad. Sci. USA* **1996**, *93*, 5131–5135. [CrossRef] [PubMed]
5. MacHugh, D.E.; Shriver, M.D.; Loftus, R.T.; Cunningham, P.; Bradley, D.G. Microsatellite DNA variation and the evolution, domestication and phylogeography of taurine and zebu cattle (Bos taurus and Bos indicus). *Genetics* **1997**, *146*, 1071–1086. [PubMed]
6. De Noni, I. Release of β-casomorphins 5 and 7 during simulated gastrointestinal digestion of bovine β-casein variants and milk-based infant formulas. *Food Chem.* **2008**, *110*, 897–903. [CrossRef]
7. De Noni, I.; Cattaneo, S. Occurrence of β-casomorphins 5 and 7 in commercial dairy products and in their digests following in vitro simulated gastrointestinal digestion. *Food Chem.* **2010**, *119*, 560–566. [CrossRef]
8. Jinsmaa, Y.; Yoshikawa, M. Enzymatic release of neocasomorphin and beta-casomorphin from bovine beta-casein. *Peptides* **1999**, *20*, 957–962. [CrossRef]
9. Ul Haq, M.R.; Kapila, R.; Kapila, S. Release of beta-casomorphin-7/5 during simulated gastrointestinal digestion of milk beta-casein variants from Indian crossbred cattle (Karan Fries). *Food Chem.* **2015**, *168*, 70–79. [CrossRef] [PubMed]
10. Boutrou, R.; Gaudichon, C.; Dupont, D.; Jardin, J.; Airinei, G.; Marsset-Baglieri, A.; Benamouzig, R.; Tome, D.; Leonil, J. Sequential release of milk protein-derived bioactive peptides in the jejunum in healthy humans. *Am. J. Clin. Nutr.* **2013**, *97*, 1314–1323. [CrossRef]
11. Ul Haq, M.R.; Kapila, R.; Shandilya, U.K.; Kapila, S. Impact of milk derived β-casomorphins on physiological functions and trends in research: A review. *Int. J. Food Prop.* **2014**, *17*, 1726–1741. [CrossRef]
12. Brooke-Taylor, S.; Dwyer, K.; Woodford, K.; Kost, N. Systematic review of the gastrointestinal effects of A1 compared with A2 beta-casein. *Adv. Nutr.* **2017**, *8*, 739–748. [CrossRef] [PubMed]

13. Deth, R.; Clarke, A.; Ni, J.; Trivedi, M. Clinical evaluation of glutathione concentrations after consumption of milk containing different subtypes of beta-casein: Results from a randomized, cross-over clinical trial. *Nutr. J.* **2016**, *15*, 82. [CrossRef] [PubMed]
14. Pizzorno, J. Glutathione! *Integr. Med. (Encinitas)* **2014**, *13*, 8–12. [PubMed]
15. He, M.; Sun, J.; Jiang, Z.Q.; Yang, Y.X. Effects of cow's milk beta-casein variants on symptoms of milk intolerance in Chinese adults: A multicentre, randomised controlled study. *Nutr. J.* **2017**, *16*, 72. [CrossRef]
16. Ho, S.; Woodford, K.; Kukuljan, S.; Pal, S. Comparative effects of A1 versus A2 beta-casein on gastrointestinal measures: A blinded randomised cross-over pilot study. *Eur. J. Clin. Nutr.* **2014**, *68*, 994–1000. [CrossRef]
17. Jianqin, S.; Leiming, X.; Lu, X.; Yelland, G.W.; Ni, J.; Clarke, A.J. Effects of milk containing only A2 beta casein versus milk containing both A1 and A2 beta casein proteins on gastrointestinal physiology, symptoms of discomfort, and cognitive behavior of people with self-reported intolerance to traditional cows' milk. *Nutr. J.* **2015**, *15*, 35. [CrossRef]
18. Sheng, X.; Li, Z.; Ni, J.; Yelland, G. Effects of conventional milk versus milk containing only A2 beta-casein on digestion in Chinese children: A randomized study. *J. Pediatr. Gastroenterol. Nutr.* **2019**, *69*, 375–382. [CrossRef]
19. Suarez, F.L.; Savaiano, D.A.; Levitt, M.D. A comparison of symptoms after the consumption of milk or lactose-hydrolyzed milk by people with self-reported severe lactose intolerance. *N. Engl. J. Med.* **1995**, *333*, 1–4. [CrossRef]
20. Milan, A.M.; Shrestha, A.; Karlström, H.J.; Martinsson, J.A.; Nilsson, N.J.; Perry, J.K.; Day, L.; Barnett, M.P.G.; Cameron-Smith, D. Comparison of the impact of bovine milk β-casein variants on digestive comfort in females self-reporting dairy intolerance: A randomized controlled trial. *Am. J. Clin. Nutr.* **2020**, *111*, 149–160. [CrossRef]
21. Misselwitz, B.; Butter, M.; Verbeke, K.; Fox, M.R. Update on lactose malabsorption and intolerance: Pathogenesis, diagnosis and clinical management. *Gut* **2019**, *68*, 2080–2091. [CrossRef]
22. Di Costanzo, M.; Berni Canani, R. Lactose Intolerance: Common misunderstandings. *Ann. Nutr. Metab.* **2018**, *73* (Suppl. 4), 30–37. [CrossRef]
23. Hertzler, S.R.; Savaiano, D.A. Colonic adaptation to daily lactose feeding in lactose maldigesters reduces lactose intolerance. *Am. J. Clin. Nutr.* **1996**, *64*, 232–236. [CrossRef] [PubMed]
24. Savaiano, D.A.; Ritter, A.J.; Klaenhammer, T.R.; James, G.M.; Longcore, A.T.; Chandler, J.R.; Walker, W.A.; Foyt, H.L. Improving lactose digestion and symptoms of lactose intolerance with a novel galacto-oligosaccharide (RP-G28): A randomized, double-blind clinical trial. *Nutr. J.* **2013**, *12*, 160. [CrossRef] [PubMed]
25. Rezaie, A.; Buresi, M.; Lembo, A.; Lin, H.; McCallum, R.; Rao, S.; Schmulson, M.; Valdovinos, M.; Zakko, S.; Pimentel, M. Hydrogen and methane-based breath testing in gastrointestinal disorders: The North American consensus. *Am. J. Gastroenterol.* **2017**, *112*, 775–784. [CrossRef]
26. Wilt, T.J.; Shaukat, A.; Shamliyan, T.; Taylor, B.C.; MacDonald, R.; Tacklind, J.; Rutks, I.; Schwarzenberg, S.J.; Kane, R.L.; Levitt, M. Lactose intolerance and health. *Evid. Rep. Technol. Assess (Full Rep.)* **2010**, *192*, 410.
27. AOAC International. *Official Methods of Analysis of AOAC International*, 18th ed.; Methods 989.05, 932.05, 986.25, 945.48B, 968.06 and 992.15; AOAC International: Gaithersburg, MD, USA, 2005.
28. Mason, B.S.; Slover, H.T. A Gas chromatographic method for the determination of sugars in foods. *J. Agric. Food Chem.* **1971**, *19*, 551–554. [CrossRef]
29. Brobst, K.M. *Gas-Liquid Chromatography of Trimethylsilyl Derivatives, Methods in Carbohydrate Chemistry*; Academic Press: New York, NY, USA, 1972; Volume 6, pp. 3–8. [CrossRef]
30. Barnett, M.P.; McNabb, W.C.; Roy, N.C.; Woodford, K.B.; Clarke, A.J. Dietary A1 beta-casein affects gastrointestinal transit time, dipeptidyl peptidase-4 activity, and inflammatory status relative to A2 beta-casein in Wistar rats. *Int. J. Food Sci. Nutr.* **2014**, *65*, 720–727. [CrossRef]
31. Trivedi, M.S.; Shah, J.S.; Al-Mughairy, S.; Hodgson, N.W.; Simms, B.; Trooskens, G.A.; Van Criekinge, W.; Deth, R.C. Food-derived opioid peptides inhibit cysteine uptake with redox and epigenetic consequences. *J. Nutr. Biochem.* **2014**, *25*, 1011–1018. [CrossRef]

32. Yadav, S.; Yadav, N.D.S.; Gheware, A.; Kulshreshtha, A.; Sharma, P.; Singh, V.P. Oral feeding of cow milk containing A1 variant of β casein induces pulmonary inflammation in male Balb/c mice. *Sci. Rep.* **2020**, *10*, 8053. [CrossRef]

Publisher's Note: MDPI stays neutral with regard to jurisdictional claims in published maps and institutional affiliations.

© 2020 by the authors. Licensee MDPI, Basel, Switzerland. This article is an open access article distributed under the terms and conditions of the Creative Commons Attribution (CC BY) license (http://creativecommons.org/licenses/by/4.0/).

Article

β-Lactoglobulin Elevates Insulin and Glucagon Concentrations Compared with Whey Protein—A Randomized Double-Blinded Crossover Trial in Patients with Type Two Diabetes Mellitus

Stine B. Smedegaard [1,*], Maike Mose [1], Adam Hulman [1], Ulla R. Mikkelsen [2], Niels Møller [3], Gregers Wegener [4,5], Niels Jessen [1] and Nikolaj Rittig [1,3]

1. Steno Diabetes Center Aarhus, Aarhus University Hospital, Hedeager 3, 8200 Aarhus, Denmark; maikemose@clin.au.dk (M.M.); adahul@rm.dk (A.H.); niels.jessen@biomed.au.dk (N.J.); nikolaj.rittig@clin.au.dk (N.R.)
2. Arla Foods Ingredients Group P/S, Soenderhoej 10, 8260 Viby, Denmark; ulrmk@arlafoods.com
3. Department of Diabetes and Hormone Diseases, Aarhus University Hospital, Palle Juul-Jensens Blv. 99, 8200 Aarhus, Denmark; niels.moeller@clin.au.dk
4. Translational Neuropsychiatry Unit, Department of Clinical Medicine, Aarhus University, Noerrebrogade 44, Entrance 2B, 8000 Aarhus, Denmark; wegener@clin.au.dk
5. Centre of Excellence for Pharmaceutical Sciences, North-West University, 11 Hoffman Street, Potchefstroom 2531, South Africa
* Correspondence: stinsmed@rm.dk; Tel.: +45-7845-0000

Abstract: Whey protein is an insulinotropic fraction of dairy that reduces postprandial glucose levels in patients with type 2 diabetes mellitus (T2DM). We have recently shown that β-lactoglobulin (BLG), the largest protein fraction of whey, elevates insulin concentrations compared with iso-nitrogenous whey protein isolate (WPI) in healthy individuals. We therefore hypothesized that BLG pre-meals would lower glucose levels compared with WPI in patients with T2DM. We investigated 16 participants with T2DM using a randomized double-blinded cross-over design with two pre-meal interventions, (i) 25 g BLG and (ii) 25 g WPI prior to an oral glucose tolerance test (OGTT), followed by four days of continuous glucose monitoring (CGM) at home. BLG increased concentrations of insulin with 10%, glucagon with 20%, and glucose with 10% compared with WPI after the OGTT (all $p < 0.05$). Both BLG and WPI reduced the interstitial fluid (ISF) glucose concentrations (using CGM) with 2 mM and lowered glycemic variability with 10–15%, compared with tap-water ($p < 0.05$), and WPI lowered the ISF glucose with 0.5 mM compared with BLG from 120 min and onwards ($p < 0.05$). In conclusion, BLG pre-meals resulted in higher insulin, glucagon, and glucose concentrations compared with WPI in participants with T2DM. Pre-meal servings of WPI remains the most potent protein in terms of lowering postprandial glucose excursions.

Keywords: type 2 diabetes mellitus; whey; glucose; glycemic variability; beta-lactoglobulin; pre-meal; CGM

Citation: Smedegaard, S.B.; Mose, M.; Hulman, A.; Mikkelsen, U.R.; Møller, N.; Wegener, G.; Jessen, N.; Rittig, N. β-Lactoglobulin Elevates Insulin and Glucagon Concentrations Compared with Whey Protein—A Randomized Double-Blinded Crossover Trial in Patients with Type Two Diabetes Mellitus. *Nutrients* **2021**, *13*, 308. https://doi.org/10.3390/nu13020308

Received: 22 December 2020
Accepted: 20 January 2021
Published: 22 January 2021

Publisher's Note: MDPI stays neutral with regard to jurisdictional claims in published maps and institutional affiliations.

Copyright: © 2021 by the authors. Licensee MDPI, Basel, Switzerland. This article is an open access article distributed under the terms and conditions of the Creative Commons Attribution (CC BY) license (https://creativecommons.org/licenses/by/4.0/).

1. Introduction

Pre-meals of whey protein have shown promising effects on the subsequent glucose trajectories in both healthy participants and patients with type 2 diabetes mellitus (T2DM) [1,2]. Whey given 15–30 min before a meal mediates a rise in insulin concentration and results in lower postprandial blood glucose concentrations [1,3]. The underlying mechanisms behind the insulinotropic properties observed following whey protein consumption are complex and not fully understood. Whey is especially rich in the branched chained amino acid (BCCA), leucine, which has direct insulin stimulating effect on the beta cell of the pancreas [4]. Whey protein also increases the concentration of the incretin hormones glucose-dependent insulinotropic polypeptide (GIP) [1,5,6] and glucagon-like peptide-1 (GLP-1) [1,5,7], which are also known to stimulate insulin secretion. Data from mouse pancreatic islets suggests that the exposure to an amino acid mixture and GIP [4], rather than one specific amino acid, has the greatest insulinotropic effects on the beta-cell.

Milk protein consists of around 80% casein and 20% whey [8]. Whey protein consists of 50–60% β-lactoglobulin (BLG), 17% α-lactalbumin, 10% immunoglobulins, 5% albumin and other polypeptides [9]. Recent data from our group show that BLG increases the serum(s)-concentration of insulin 23% more than a regular iso-nitrogenous whey protein isolate (WPI) in individuals without prior health issues. This observation led us to the hypothesis that pre-meal servings of BLG would stimulate insulin secretion and lower glucose trajectories compared with WPI in patients with T2DM. A more potent protein may lower protein and excessive calorie intake and improve compliance in prolonged protein pre-meal treatment regimes. Therefore, we performed a randomized double-blinded cross-over trial to investigate the effects of BLG and WPI pre-meals in patients with T2DM.

2. Materials and Methods

2.1. Study Approval

The study complied with the Declaration of Helsinki and was approved by the regional research ethics committee (1-10-72-226-19), registered at ClinicalTrials.gov (NCT04166760), and applied to the regulations of the Danish Data Protection Agency. All participants gave their written informed consent before inclusion in the study.

2.2. Participants

Participants were eligible for inclusion if they had T2DM, were between 18 and 80 years old, had a BMI between 20 and 35 kg/m^2, had hemoglobin (Hb)-A1c between 40 and 69 mmol/L, and C-peptide between 370 and 1200 pmol/L. Recruitment was performed through social media (Facebook) and local newspapers. Exclusion criteria were milk allergies, daily intake of protein supplements, anti-glycemic medication other than metformin, or inability to speak or understand Danish. All participants were screened with a blood test panel of HbA1c, creatinine, thyrotropin, C-reactive-protein, sodium, potassium, albumin, alanine aminotransferase, alkaline phosphatase, bilirubin, hemoglobin, and C-peptide before inclusion.

2.3. Design and Protocol

The study was a randomized, double-blinded, cross-over trial with two interventions. Study days were identical except for interventions and consisted of an oral glucose tolerance test (OGTT) performed in our laboratory and four days of monitoring at home. The study was performed at the Steno/Medical laboratory, Aarhus University Hospital, Denmark. The two interventions consisted of: (i) BLG and (ii) whey protein isolate (WPI). There was a minimum washout period of one week and a maximum of six weeks between the two OGTTs. Participants and investigators were blinded in regard to the interventions. For an overview of the design and randomization, see Figure 1.

Before attending the laboratory, participants were asked to eat according to Danish nutritional guidelines (15% fat, 30% protein, and 55% carbohydrates) for 48 h and to avoid strenuous physical activity before and during each of the investigations (laboratory and home monitoring). If participants received metformin, this treatment was discontinued for five days before and during the investigations. All participants arrived following a 10-h overnight fast. During each study day, an intravenous catheter was placed in an antecubital vein for blood sampling. The participants consumed either 25 g of WPI or BLG 30 min before a 75 g OGTT was performed. Blood samples were collected consecutively in the three following hours.

Following the OGTT investigation, participants were equipped with a continuous glucose monitor (CGM), an activity monitor, four standardized breakfasts, a protein drink shaker, and four small plastic bags with 25 grams of the protein intervention. The first 24 h of CGM and activity recordings were used to calibrate equipment and excluded from analyses. Each participant was randomized to consume the protein pre-meals 30 min before the standardized breakfast and dinner during days two and three or during days four and five (Figure 1). Participants consumed an iso-voluminous amount of tap-water (CTR) 30 min before breakfast and dinner during the days without protein pre-meals. They were asked to avoid strenuous physical activity and eat similarly during the days of home-monitoring. The participants filled out a food-diary with timestamps for pre-meals and meals to ensure compliance and perform CGM analyses.

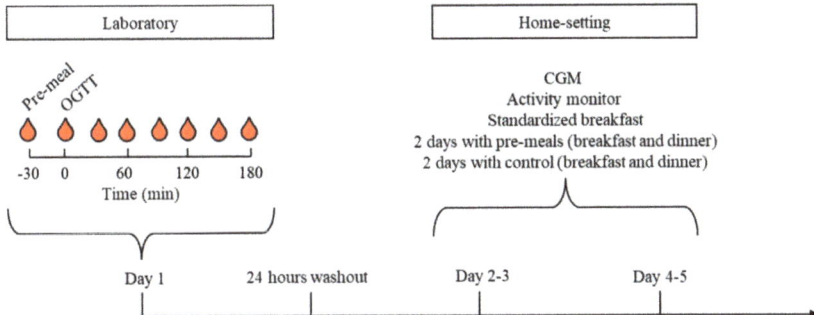

Figure 1. Flowchart of the study investigations. Participants were randomized to consume one of two pre-meals, (i) β-lactoglobulin (BLG) or (ii) whey protein isolate (WPI), 30 min before a 75 g oral glucose tolerance test (OGTT) in our laboratory or before breakfast and dinner at home. Participants were equipped with a continuous glucose monitor (CGM), an activity monitor, and standardized breakfast meals. Participants were also randomized to consume pre-meals before breakfast and dinner on days 2–3 or days 4–5 and control (tap-water) on the other two days. The experiment was repeated after 1 to 6 weeks from the OGTT.

2.4. Interventions and Meals

The primary investigator enrolled and assigned participants to the sequence of interventions using www.randomizer.org [10]. The WPI (Lacprodan DI-9213) and BLG were provided by Arla Foods Ingredients Group P/S, Viby J, Denmark. The interventions were similar in appearance and taste. Two persons without relation to the investigations dosed and blinded 25 g of protein in small, labeled plastic bags. The proteins were dissolved in 200 mL of tap-water and served as a shake. The standardized breakfast consisted of 50 g cornflakes (Vores Cornflakes 500 g), 31 g raisins (Svansoe Rosiner 1500 g), and 250 mL skimmed milk (Arla® Skummetmaelk 0.1% 250 mL) equivalent to 77.6 g carbohydrates, 13.9 g protein, 1 g fat/375 kCal. The characteristics of WPI and BLG are shown in Table 1. Participants and all persons involved in the trials, including the outcome assessors, remained blinded until statistical analyses had been performed. There were no adverse events reported.

Table 1. Composition of the interventions; β-lactoglobulin (BLG) and whey protein isolate (WPI).

Nutritional Content	BLG /100 g Product	WPI /100 g Product
Total energy, kCal	375	355
Fat, g	0.1	0.1
Carbohydrate, g	0.1	0.1
Protein, g	93.5	88.3
No-calorie flavor, %	0.56	0.56
Amino Acids	BLG g/100 g Protein	WPI g/100 g Protein
Alanine	7.0	6.0
Arginine	2.8	2.3
Aspartic acid	12.1	11.8

Table 1. Cont.

Cysteine	3.1	2.6
Glutamic acid	20.1	19.7
Glycine	1.3	1.6
Histidine	1.7	1.7
Hydroxyproline	<0.1	<0.1
Isoleucine	6.3	7.0
Leucine	16.1	11.7
Lysine	12.3	10.6
Methionine	2.8	2.4
Ornithine	<0.1	<0.1
Phenylalanine	3.6	3.1
Proline	5.4	6.8
Serine	3.9	5.2
Threonine	5.3	8.0
Tryptophan	2.2	1.9
Tyrosine	3.7	2.9
Valine	6.1	6.4
Sum	115.9	111.8

The composition of amino acid analysis was done by Eurofins (GLP (Good Laboratory Practice) certified), and the nutritional content analysis was done by Danmark Protein (Arla Foods Ingredients).

2.5. Blood Analysis

Blood samples were drawn at −30, 0, 10, 20, 30, 40, 50, 60, 90, 120, 150, and 180 min following the OGTT. Plasma(p)-glucose was measured immediately using YSI 2300 model Stat Plus glucose analyzer (YSI Incorporated, Yellow Springs, OH). Blood for the remaining analyses was centrifuged at 4 °C, frozen at −20 °C, stored at −80 °C, and analyzed on the same assay after both arms of the study were completed for all participants. S-insulin, s-C-peptide, and p-glucagon concentrations were measured with an enzyme-linked immunosorbent assay (ELISA) technique using a commercial kit (Mercodia Insulin ELISA, Mercodia Glucagon ELISA, Mercodia C-peptide ELISA, Sweden). S-free fatty acids (FFA) were measured using the in vitro enzymatic colorimetric method assay NEFA-HR(2), which quantifies the concentration of non-esterified fatty acids (FUJIFILM Wako Chemicals Europe GmbH, Germany). P-amino acids (AA) concentrations were measured by high-pressure liquid chromatography (HPLC) method using a Thermo Scientific Ultimate 3000 system, as earlier described [11]. Briefly, the samples were diluted 1.11x by adding 2 M Perchloric acid ($HClO_4$) and then centrifuged at $14,000\times g$ at 4 °C for 10 min. The supernatant was removed and filtered through a spin filter (0.22 µm) at $14,000\times g$ for 1 min; then diluted 50× with 0.2 M $HClO_4$ to a final dilution factor of 55.5. Hereafter, the samples were injected into the HPLC. For separation, a Kinetex EVO C18 2.6 µm 4.6 × 150 mm column from Phenomenex, U.S., was used. Detection was done by fluorometric detection with excitation on 337 nm and emission on 442 nm. Samples for p-GIP and p-GLP-1 were extracted in final concentrations of 70% ethanol before analyses. Samples were analyzed on radioimmunoassays using antiserum #89390 for GLP-1 and antiserum #80867 for GIP targeting the C-terminal end of the hormones reacting equally with the intact hormone and the primary metabolites (N-terminally truncated) [12,13].

2.6. Continuous Glucose Monitoring

Continuous measurement of glucose concentrations in the interstitial fluid (ISF) was performed using a CGM device (NordicInfu Care Denmark, Dexcom G6, Dexcom Inc., San Diego, CA, USA). The device measures glucose every five minutes via a subcutaneous sensor. The participants wore the device on the abdomen and were unaware of their glucose level as the receiver was blinded. Data were uploaded to and analyzed in the software CLARITY (Dexcom CLARITY, v3.32.0, Dexcom Inc., San Diego, CA, USA). The mean glucose ± standard deviation (SD), daily maximum glucose level, and the coefficient of variation (CV) was used as outcome measures.

2.7. Activity Monitoring

A combined accelerometer and heart rate (HR) monitor (Actiheart 5 (AH), CamNtech Limited, Cambridge, UK) was used to evaluate the activity and estimate energy expenditure during the home investigation period. The AH unit was worn on and connected to the chest with two self-adhesive electrodes—one below the sternum and one under the left pectoral muscle. Data on accelerometry was collected at 32 Hz, and HR was collected as inter-beat-intervals. Data from the unit was uploaded to and analyzed using AH software (Actiheart software, version 5.1.10, camNtech ltd., Cambridge, UK). The software, processing of data, and validation of the system have been described in detail elsewhere [14]. Briefly, the AH software has a built-in function to correct missing beats and clean noise from the HR data. The software uses the cleaned HR data and data on activity in the integrated branched chained model "Group Cal JAP 2007" to estimate activity energy expenditure (AEE). In the case of missing HR data >5 min, the AEE is solely based on activity. The software provides the total energy expenditure (TEE) from a model using weight, height, age, sleeping heart rate (resting heart rate—10), AEE, and diet-induced thermogenesis. The software provides variables on HR, maximum HR, activity counts, AEE, and TEE.

2.8. Statistical Analysis

Statistical analyses and figures were conducted using the nlme (version 3.1-142), Epi (version 2.37) packages in R (R Foundation for Statistical Computing, Vienna, Austria, version 3.6.2) and SigmaPlot (San Jose, CA, USA, version 14.0). Trajectories on substrates and hormones in relation to the OGTT were fitted using random-effects models with a natural cubic spline specification for time. The number and position of the knots are different for those outcomes measured at a different set of time points. Interaction terms were included for each time term and a binary variable coding the two interventions. This, in combination with appropriate contrast matrices, allowed us to estimate trajectories for both interventions and their difference at any time point during the investigation. The differences between trajectories were expressed as percentages, as the outcomes were log-transformed (natural logarithm) before running the models due to their skewed distributions. Individual specific random intercepts and slopes were included in the models to account for the dependence within the data due to its repeated measurement nature. The same method was used to assess glucose trajectories during the three hours following breakfast and dinner for BLG, WPI, and CTR. For this analysis, measurements were included if their time points were after, but within three hours of, the recorded time of breakfast and dinner. The incremental area under the curve (iAUC) was calculated using the trapezoidal approach [15], and a paired t-test or one-way RM ANOVA was used for comparison of each outcome.

CGM-based summary measures and activity characteristics were compared between interventions and controls using random effects models with individual specific random intercepts to account for the cross-over design of the experiment. Differences between groups and their 95% confidence intervals (CIs) were estimated using the appropriate contrast matrices.

A pre-study power calculation with a significance level of 0.05 and a power of 80% was performed. We expected to detect a 25% difference in iAUC in insulin concentration (which was the primary outcome) between BLG and WPI with a 23% SD following the OGTT. This resulted in a sample size of 14. We expected a dropout rate of 10% and therefore included 16 participants.

3. Results

3.1. Participants

Sixty-five individuals were initially screened by the primary investigator over the phone (Figure 2). Sixteen participants were included and completed the studies between January 2020 and June 2020. One participant was unable to complete the home monitoring program. Patient characteristics are shown in (Table 2). There was a median washout-period of 9 days (range 7–23 days) between laboratory investigations.

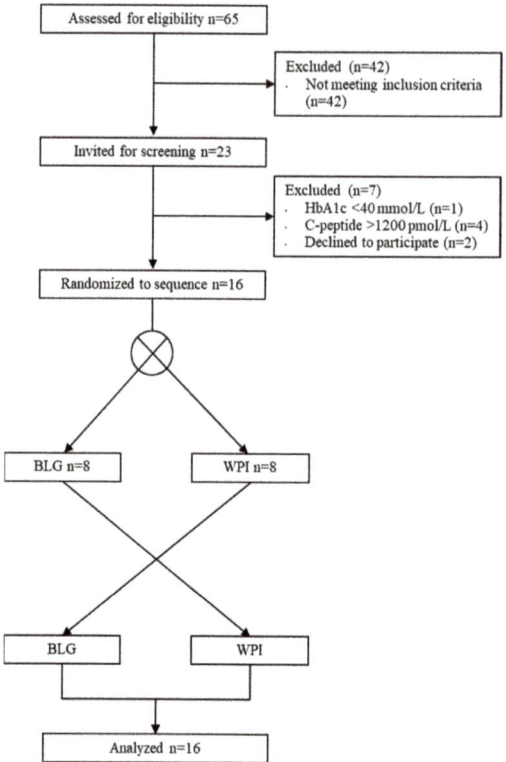

Figure 2. Flow diagram of inclusion in the randomized cross-over trial. WPI, whey protein isolate; BLG, β-lactoglobulin.

Table 2. Demographic characteristics.

Characteristics	$n = 16$
Age, years	67.5 (40–78)
BMI, kg/m^2	27.0 (21.2–32.9)
Women, %	56.2
Metformin treatment, n	14
HbA1c, mmol/mol	50 (43–55)
Fasting c-peptide, nmol/L	932 (499–1155)
Fasting insulin, pmol/L	40 (20–94)

Data are presented as absolute numbers or medians (ranges).

3.2. Oral Glucose Tolerance Test (OGTT)

3.2.1. Substrate and Hormone Concentrations

The p-glucose concentration was higher 120 min following BLG compared with WPI ingestion and reached a maximum difference of 10% 180 min following the OGTT (Figure 3A). The s-insulin concentration was elevated with 10% at 30–60 min and p-glucagon with 20% at 60–90 min after the OGTT following BLG compared with WPI (Figure 3B,C). Both BLG and WPI elevated s-C-peptide concentrations with no difference between BLG and WPI (Figure 3D). The WPI led to a higher insulin/glucagon ratio at 60 min (Figure 4A). Both proteins elevated p-GIP and p-GLP-1 concentrations and suppressed s-FFA concentrations with no difference between BLG and WPI (Figure 4B–D).

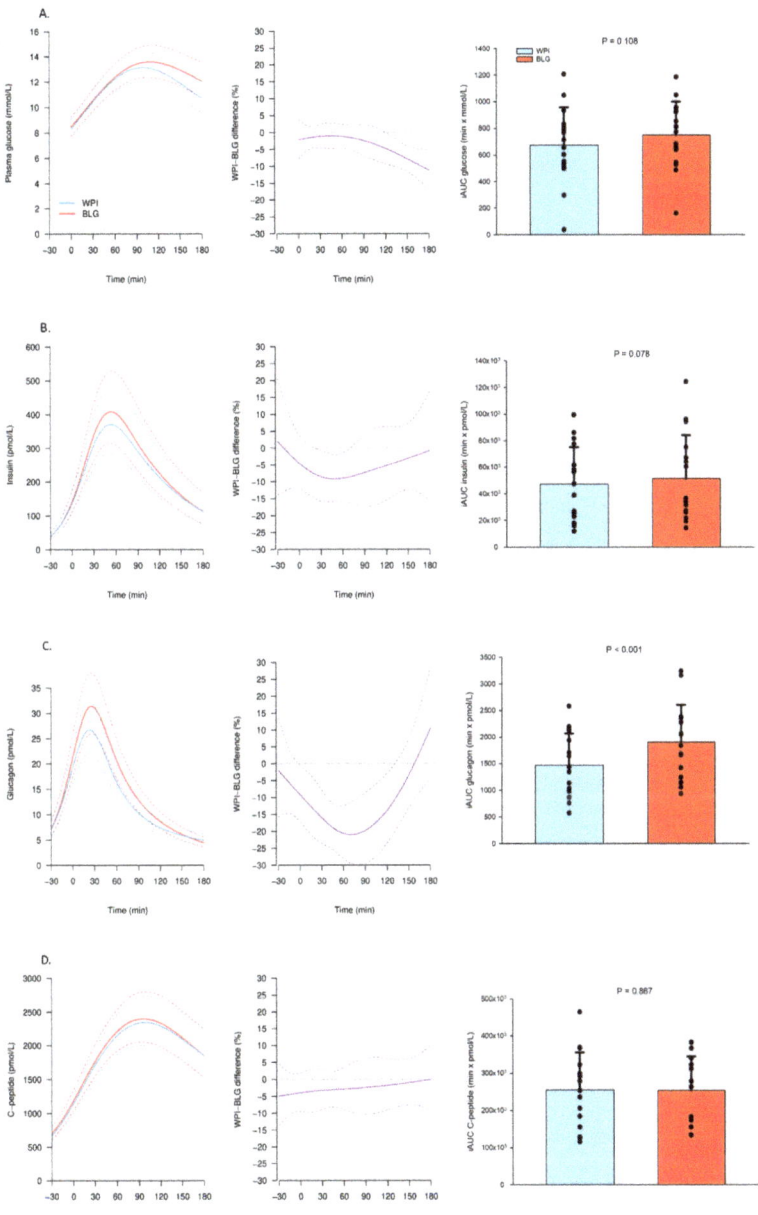

Figure 3. Plasma (p) and serum (s) concentrations of hormones and substrates after β-lactoglobulin (BLG) and whey protein isolate (WPI) pre-meals 30 min before an OGTT (0 min). Panels to the left show trajectories of the mean concentration (solid lines) with 95% confidence intervals (95% CIs) (dashed lines) of (**A**) p-glucose, (**B**) s-insulin, (**C**) p-glucagon, (**D**) s-C-peptide after WPI (blue) and BLG (red) consumption. The mean relative difference (solid line, purple) with 95% CIs (dashed lines) between the two interventions is shown in the middle panels. Panels to the right show the individual incremental area under the curve (iAUC) with a bar plot showing the mean ± standard deviation after WPI (blue) and BLG (red) consumption. $n = 16$.

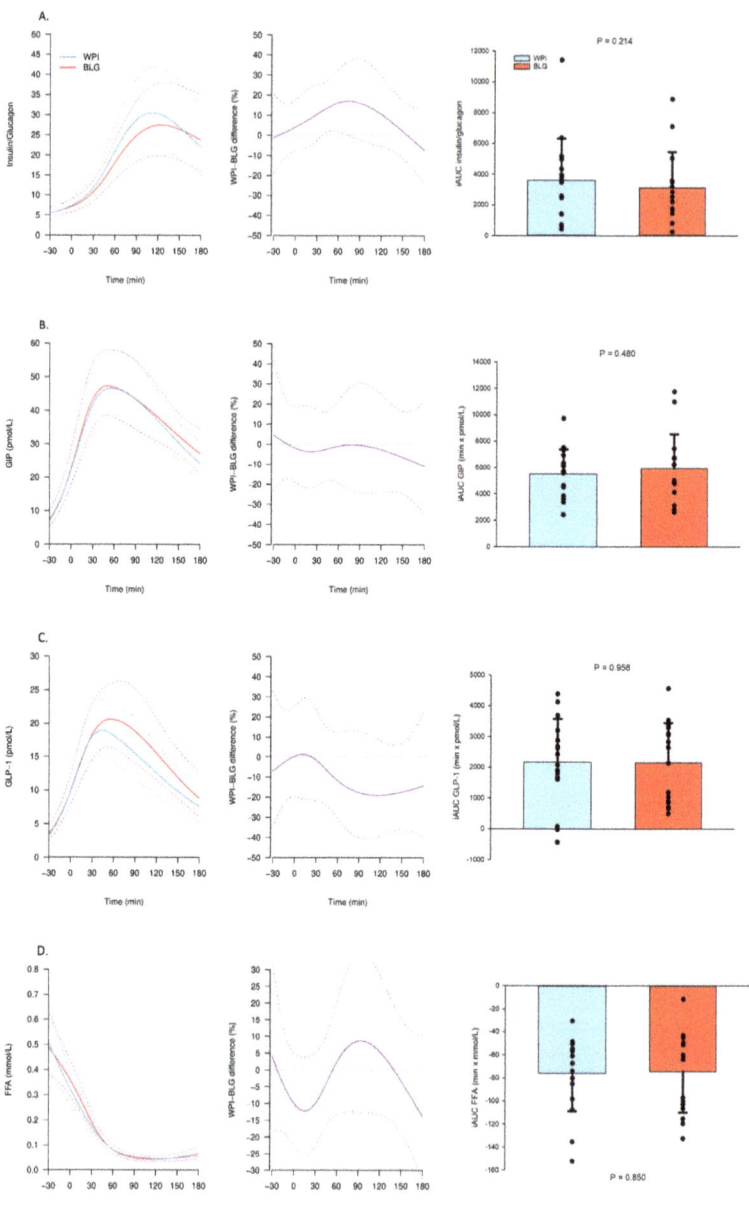

Figure 4. Plasma (p) and serum (s) concentrations of hormones and substrates after β-lactoglobulin (BLG) and whey protein isolate (WPI) pre-meals 30 min before an OGTT (0 min). Panels to the left show trajectories of the mean concentration (solid lines) with 95% confidence intervals (95% CIs) (dashed lines) of (**A**) s-insulin/p-glucagon ratio, (**B**) p-glucose-dependent insulinotropic polypeptide (GIP), (**C**) p-glucagon-like peptide-1 (GLP-1), (**D**) s-free fatty acids (FFA) after WPI (blue) and BLG (red) consumption. The mean relative difference (solid line, purple) with 95% CIs (dashed lines) between the two interventions is shown in the middle panels. Panels to the right show the individual incremental area under the curve (iAUC) with a bar plot showing the mean ± standard deviation after WPI (blue) and BLG (red) consumption. $n = 16$.

3.2.2. Amino Acids

BLG elevated the p-concentration of aspartate, glutamate, leucine, lysine, methionine, phenylalanine, proline, and tyrosine compared with WPI (Figure S1). WPI elevated the p-concentration of glycine, isoleucine, serine, and threonine compared with BLG (Figure S1).

3.3. Home-Monitoring with Continuous Glucose Monitoring (CGM)

3.3.1. CGM Glucose Trajectories Following Breakfast

Both protein pre-meals lowered postprandial ISF-glucose concentration following breakfast with the largest difference of 15% (WPI) and 17% (BLG) (2 mM) around 60 min compared with CTR (Figure 5). In alignment with our results from the OGTT, the ISF-glucose was 7% (0.5 mM) lower after 150 min following WPI compared with BLG, but 4% higher around breakfast consumption (Figure 5).

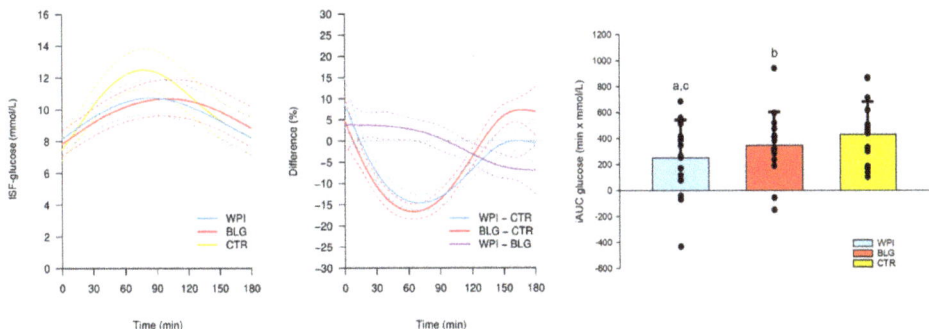

Figure 5. Interstitial fluid concentration of glucose (ISF-glucose) in the 180 min following intake of the pre-meals β-lactoglobulin (BLG), whey protein isolate (WPI), and control: tap-water (CTR) (−30 min) and standardized breakfast at home (0 min). The panel to the left shows trajectories of the mean ISF-glucose (solid lines) with 95% confidence intervals (95% CIs) (dashed lines) after WPI (blue), BLG (red) and CTR (yellow) consumption. The mean relative differences (solid lines) with 95% CIs (dashed lines) between the WPI and CTR (blue), BLG and CTR (red) and WPI and BLG (purple) are shown in the middle panel. The panel to the right shows the individual incremental area under the curve (iAUC) with a bar plot showing the mean ± standard deviation after WPI (blue), BLG (red) and CTR (yellow) consumption. One-way repeated measure ANOVA, $p = 0.002$, and post hoc (Student–Newman–Keuls) paired t-tests: a, WPI vs. CTR: $p = 0.002$; b, BLG vs. CTR: $p = 0.077$; c, WPI vs. BLG: $p = 0.052$. $n = 15$.

3.3.2. CGM and Summary Statistics

There was no difference in mean ISF-glucose between BLG, WPI, and CTR. The glycemic variability expressed as the CV was lower by 10% during WPI and by 15% during BLG, and the SD was lower by 9% during WPI and by 13% during BLG, compared with CTR. Additionally, after breakfast, the maximum glucose concentration was lower by 13% during WPI and 12% during BLG compared with CTR. The daily maximum glucose level was lower by 7% during WPI and 5% during BLG compared with CTR. No statistically significant differences were detected between BLG and WPI in any of the CGM summary variables (Table 3).

3.4. Energy Expenditure

TEE and AEE were higher on days with BLG compared with WPI. There was no significant difference between days with protein compared with CTR. Participants had similar activity counts, HR, and maximum HR on days with protein and days with CTR (Table 4).

Table 3. Summary variables on continuous glucose monitoring.

n = 15	Mean Glucose, mmol/L	SD, mmol/L	CV, %
CTR	8.7 (7.9; 9.5)	2.0 (1.8; 2.2)	22.7 (21.3; 24.1)
WPI	8.8 (8.0; 9.7)	1.8 (1.6; 2.0)	20.6 (18.9; 22.2)
BLG	8.6 (8.0; 9.7)	1.7 (1.5; 1.9)	19.2 (17.5; 20.9)
WPI–CTR	0.1 (−0.2; 0.4)	−0.2 (−0.3; −0.0) **	−2.2 (−3.9; −0.5) *
BLG–CTR	0.2 (−0.1; 0.5)	−0.3 (−0.4; −0.1) **	−3.5 (−5.2; −1.8) **
WPI–BLG	−0.1 (−0.4; 0.3)	0.1 (−0.1; 0.3)	1.3 (−0.6; 3.3)
n = 15	Max after Breakfast, mmol/L	Daily max, mmol/L	Max after Dinner, mmol/L
CTR	14.4 (13.1; 15.6)	14.5 (13.2; 15.9)	10.8 (9.8; 11.8)
WPI	12.5 (11.2; 13.9)	13.5 (12.2; 14.9)	10.9 (9.7; 12.0)
BLG	12.7 (11.4; 14.0)	13.8 (12.5; 15.2)	11.0 (9.8; 12.1)
WPI–CTR	−1.8 (−2.4; −1.3) *	−1.0 (−1.6; −0.4) **	0.1 (−0.8; 1.0)
BLG–CTR	−1.7 (−2.2; −1.1) **	−0.7 (−1.3; −0.1) *	0.2 (−0.7; 1.1)
WPI–BLG	−0.2 (−0.8; 0.5)	−0.3 (−1.0; 0.4)	−0.1 (−1.1; 0.9)

Coefficient of variation (CV) and standard deviation (SD) as parameters on glycemic variability. Maximum (max) after breakfast and dinner is the peak in the postprandial glucose concentration during three hours following the meals. Values are presented as means with 95% confidence intervals. * $p < 0.05$, ** $p < 0.01$, statistically significant differences are highlighted (bold). CTR, control; WPI, whey protein isolate; BLG, beta-lactoglobulin.

Table 4. Activity measurements and energy expenditure.

n = 15	TEE, kCal	AEE, kCal	Activity, counts/min	Mean HR, BPM	Maximum HR, BPM
CTR	2446 (2234; 2658)	659 (514; 804)	34 (26; 41)	74 (70; 79)	104 (97; 110)
WPI	2338 (2110; 2565)	559 (396; 721)	28 (19; 36)	75 (70; 79)	103 (96; 111)
BLG	2499 (2277; 2722)	708 (551; 865)	34 (25; 42)	76 (71; 80)	105 (98; 112)
WPI–CTR	−109 (−243; 26)	−100 (−221; 20)	−7 (−13; 0) *	0 (−2; 3)	0 (−5; 5)
BLG–CTR	53 (−74; 180)	49 (−66; 163)	0 (−6; 6)	2 (−1; 4)	2 (−3; 7)
WPI–BLG	−161 (−313; −10) *	−149 (−286; −13) *	−6 (−14; 1)	−1 (−4; 1)	−2 (−8; 4)

Data are expressed as means with 95% confidence intervals. * $p < 0.05$, statistically significant differences are highlighted (bold). TEE, total energy expenditure; kCal, kilocalories; AEE, activity energy expenditure; HR, heart rate; BPM, beats per minute; CTR, control; WPI, whey protein isolate; BLG, beta-lactoglobulin.

4. Discussion

In this study, we showed how a BLG pre-meal served 30 min before an OGTT resulted in higher concentrations of insulin, glucagon, and glucose compared with WPI in patients with T2DM. The study was originally designed to investigate the insulinotropic properties of BLG with the hypothesis that elevated insulin concentrations would lower postprandial glucose excursions compared with WPI. We confirmed that BLG elevates insulin concentrations compared with WPI, but the simultaneous glucagonotropic effect also associated with BLG most likely explains why ISF-glucose concentrations were slightly higher (0.5 mM) following BLG compared with WPI, opposing our original hypothesis. Despite the similarity between the two dairy products, BLG contained more leucine and phenylalanine than WPI, which was also present in the p-concentrations of these specific AA following interventions. Both leucine and phenylalanine have been shown to stimulate insulin secretion [4,16] which, to some extent, may explain the insulinotropic properties. Also, p-concentrations of methionine and tyrosine have been shown to correlate with glucagon concentrations in humans [17], and perfusion studies in dogs and rodents have shown that aspartate, glutamate, lysine, and proline stimulate glucagon secretion [18,19]. These AA were all significantly higher after BLG consumption compared with WPI. Glucagon release is potently stimulated by GIP [20], but plasma concentrations of GIP following interventions were comparable between interventions. Hence, we suggest that the insulinotropic and glucagonotropic effects associated with BLG may relate to its specific AA composition.

Our study is the first to show glucose-lowering effect of pre-meal whey protein in a home-setting using CGM. Both interventions lowered glucose excursions with 2 mM following a standardized breakfast compared with tap-water (CTR). The effect is in line with other studies investigating similar doses of whey protein pre-meal servings [6,21]. To our knowledge, only one other study has investigated whey pre-meals in individuals with T2DM using CGM in a home-setting [22]. This study compared whey protein with a mixture of indigestible potato starch (carbohydrate rich) and could not show statistically significant effects on mean glucose levels, glucose trajectories, or glycemic variability following meals. It is well known that ingestion of small amounts of carbohydrates preceding a glucose load lowers the following glucose excursion, an effect referred to as the Staub-Traugott effect [23,24] that may explain why no significant differences were found in this study.

Only one study on long-term pre-meal whey exposure in T2DM has been performed (12 weeks) [21]. This study showed a small significant reduction in HbA1c (−1 mmol/mol). It should be noted that the participants were already well regulated with an HbA1c of 49 mmol/mol, which may have affected the size of the outcome. However, HbA1c does not necessarily reflect postprandial glucose excursions [25], and results might have been more substantial on glycemic variability. Large glucose excursions and high glycemic variability have been associated with risk of cardiovascular disease [26] as well as impaired cognitive function [27]. This emphasizes that minimizing postprandial glucose excursions may be important in the management of T2DM. We showed a reduction in glycemic variability, maximum glucose levels, and lower glucose excursions after consuming the pre-meal proteins compared with tap-water. Future long-term studies on pre-meal whey protein in participants with T2DM should preferably include investigations on glycemic variability.

Our study was limited, as the CTR intervention (tap-water) was unblinded. However, morning glucose concentrations were comparable between conditions, the participants were given the same standardized breakfast meal, and activity levels were similar between days with protein interventions and CTR. Activity and energy expenditure were comparable between groups, but showed a minor statistically significant elevation in TEE and AEE during BLG compared with WPI. These differences were small and only strengthen our findings showing lower glucose levels during WPI compared with BLG. We instructed participants to eat according to the Danish national recommendations (55% carbohydrates, 30% fat, and 15% protein) and to eat similar portion sizes during the home monitoring period. Still, the food diaries were generally of poor quality and lacked information. We did not include a control condition (e.g., tap-water) in the OGTT experiment because the primary aim of the study was to compare BLG and WPI. Pre-meals of whey have, in many previous studies, already proven effective in lowering glucose concentrations [2,3,21,28], but direct comparisons to other proteins are sparse.

A major strength of our study is the combination of investigations in a controlled laboratory and home setting. We included both men and women in our trial and both CGM and activity monitors in our investigations. The cross-over design eliminated any inter-individual differences.

In conclusion, a pre-meal of BLG elevates insulin, glucagon, and glucose concentration compared with WPI following an OGTT in patients with T2DM. Both WPI and BLG lowered glycemic variability and glucose trajectories compared with tap-water. WPI remains the most potent pre-meal in the management of postprandial glucose excursions.

Supplementary Materials: The following are available online at https://www.mdpi.com/2072-6643/13/2/308/s1, Figure S1: Plasma concentrations amino acids (AA) after β-lactoglobulin (BLG) and whey protein isolate (WPI) pre-meals 30 min before an OGTT (0 min). Panels to the left show trajectories of the mean concentration (solid lines) with 95% confidence intervals (95%CIs) (dashed lines) of (A) alanine, (B) arginine, (C) aspartate, (D) glutamate, (E) glycine, (F) histidine, (G) isoleucine, (H) leucine, (I) lysine, (J) methionine, (K) phenylalanine, (L) proline, (M) serine, (N) threonine, (O) tyrosine, (P) valine, (Q) total AA after WPI (blue) and BLG (red) consumption. The mean relative difference (solid line, purple) with 95%CIs (dashed lines) between the two interventions is shown in the middle panel. Panels to the right show the individual incremental area under the curve (iAUC) with a bar plot showing the mean ± standard deviation after WPI (blue) and BLG (red) consumption. N = 16.

Author Contributions: Conceptualization, M.M., U.R.M., N.J., and N.R.; data curation, S.B.S.; formal analysis, S.B.S. and A.H.; funding acquisition, M.M. and N.R.; investigation, S.B.S.; project administration, S.B.S.; resources, N.M., G.W., and N.J.; software, A.H.; supervision, N.R.; visualization, S.B.S.

and A.H.; writing—original draft, S.B.S.; writing—review and editing, N.R. All authors have read and agreed to the published version of the manuscript.

Funding: This research was funded by Arla Food Ingredients Group P/S, Soenderhoej 10-12, 8260 DK-Viby, CVR 33372116.

Institutional Review Board Statement: The study was conducted according to the guidelines of the Declaration of Helsinki, and approved by the Institutional Review Board (or Ethics Committee) of The Central Denmark Region Committees on Health Research Ethics (1-10-72-226-19, November 5, 2019).

Informed Consent Statement: Informed consent was obtained from all participants involved in the study.

Data Availability Statement: The data presented in this study are available on request from the corresponding author. The data are not publicly available due to privacy.

Acknowledgments: We are grateful for the outstanding technical support performed by Elsebeth Hornemann, Hanne Fjeldsted Petersen, Lene Albaek, Lisa Buus and Per Fuglsang Mikkelsen. In addition, we would like to thank Arla Foods Ingredients P/S for providing the protein interventions and funding the project. Last, we thank all participants for their participation in the study.

Conflicts of Interest: The study was supported by Arla Foods Ingredients Group P/S. U.R.M. is employed as a nutrition research scientist at Arla Foods Ingredients Group P/S. Arla Foods amba, Soenderhoej 14, 8260 Viby J, Denmark, has a pending patent application regarding BLG, with N.R., M.M., N.M., A.H., U.R.M., and S.B.S. as co-inventors. N.R., M.M., N.J. and U.R.M. took initiative to and designed the study together. U.R.M. and Arla Food Ingredients P/S had no role in the collection, analyses, or interpretation of data, in the writing of the manuscript, or in the decision to publish the results.

References

1. Ma, J.; Stevens, J.E.; Cukier, K.; Maddox, A.F.; Wishart, J.M.; Jones, K.L.; Clifton, P.M.; Horowitz, M.; Rayner, C.K. Effects of a protein preload on gastric emptying, glycemia, and gut hormones after a carbohydrate meal in diet-controlled type 2 diabetes. *Diabetes Care* **2009**, *32*, 1600–1602. [CrossRef] [PubMed]
2. Akhavan, T.; Luhovyy, B.L.; Brown, P.H.; Cho, C.E.; Anderson, G.H. Effect of premeal consumption of whey protein and its hydrolysate on food intake and postmeal glycemia and insulin responses in young adults. *Am. J. Clin. Nutr.* **2010**, *91*, 966–975. [CrossRef] [PubMed]
3. Bjørnshave, A.; Holst, J.J.; Hermansen, K. Pre-Meal Effect of Whey Proteins on Metabolic Parameters in Subjects with and without Type 2 Diabetes: A Randomized, Crossover Trial. *Nutrients* **2018**, *10*, 122. [CrossRef] [PubMed]
4. Salehi, A.; Gunnerud, U.; Muhammed, S.J.; Ostman, E.; Holst, J.J.; Björck, I.; Rorsman, P. The insulinogenic effect of whey protein is partially mediated by a direct effect of amino acids and GIP on β-cells. *Nutr. Metab.* **2012**, *9*, 48. [CrossRef] [PubMed]
5. Nilsson, M.; Holst, J.J.; Björck, I.M. Metabolic effects of amino acid mixtures and whey protein in healthy subjects: Studies using glucose-equivalent drinks. *Am. J. Clin. Nutr.* **2007**, *85*, 996–1004. [CrossRef]
6. Frid, A.H.; Nilsson, M.; Holst, J.J.; Björck, I.M. Effect of whey on blood glucose and insulin responses to composite breakfast and lunch meals in type 2 diabetic subjects. *Am. J. Clin. Nutr.* **2005**, *82*, 69–75. [CrossRef]
7. Jakubowicz, D.; Froy, O.; Ahrén, B.; Boaz, M.; Landau, Z.; Bar-Dayan, Y.; Ganz, T.; Barnea, M.; Wainstein, J. Incretin, insulinotropic and glucose-lowering effects of whey protein pre-load in type 2 diabetes: A randomised clinical trial. *Diabetologia* **2014**, *57*, 1807–1811. [CrossRef]
8. Zheng, H.; Clausen, M.R.; Dalsgaard, T.K.; Bertram, H.C. Metabolomics to Explore Impact of Dairy Intake. *Nutrients* **2015**, *7*, 4875–4896. [CrossRef]
9. Layman, D.K.; Lönnerdal, B.; Fernstrom, J.D. Applications for α-lactalbumin in human nutrition. *Nutr. Rev.* **2018**, *76*, 444–460. [CrossRef]
10. Urbaniak, G.C.; Plous, S. Research Randomizer (Version 4.0) [Computer Software]. Available online: http://www.randomizer.org/ (accessed on 9 January 2020).
11. Liebenberg, N.; Jensen, E.; Larsen, E.R.; Kousholt, B.S.; Pereira, V.S.; Fischer, C.W.; Wegener, G. A Preclinical Study of Casein Glycomacropeptide as a Dietary Intervention for Acute Mania. *Int. J. Neuropsychopharmacol.* **2018**, *21*, 473–484. [CrossRef]
12. Orskov, C.; Rabenhøj, L.; Wettergren, A.; Kofod, H.; Holst, J.J. Tissue and plasma concentrations of amidated and glycine-extended glucagon-like peptide I in humans. *Diabetes* **1994**, *43*, 535–539. [CrossRef] [PubMed]
13. Lindgren, O.; Carr, R.D.; Deacon, C.F.; Holst, J.J.; Pacini, G.; Mari, A.; Ahrén, B. Incretin hormone and insulin responses to oral versus intravenous lipid administration in humans. *J. Clin. Endocrinol. Metab.* **2011**, *96*, 2519–2524. [CrossRef] [PubMed]
14. CamNtech Ltd. *Actiheart User Manual 5.1.14*; CamNtech Ltd.: Fenstanton, UK, 2020.
15. Brouns, F.; Bjorck, I.; Frayn, K.N.; Gibbs, A.L.; Lang, V.; Slama, G.; Wolever, T.M. Glycaemic index methodology. *Nutr. Res. Rev.* **2005**, *18*, 145–171. [CrossRef] [PubMed]

16. Van Loon, L.J.; Saris, W.H.; Verhagen, H.; Wagenmakers, A.J. Plasma insulin responses after ingestion of different amino acid or protein mixtures with carbohydrate. *Am. J. Clin. Nutr.* **2000**, *72*, 96–105. [CrossRef]
17. Calbet, J.A.L.; MacLean, D.A. Plasma Glucagon and Insulin Responses Depend on the Rate of Appearance of Amino Acids after Ingestion of Different Protein Solutions in Humans. *J. Nutr.* **2002**, *132*, 2174–2182. [CrossRef]
18. Rocha, D.M.; Faloona, G.R.; Unger, R.H. Glucagon-stimulating activity of 20 amino acids in dogs. *J. Clin. Investig.* **1972**, *51*, 2346–2351. [CrossRef]
19. Galsgaard, K.D.; Jepsen, S.L.; Kjeldsen, S.A.S.; Pedersen, J.; Wewer Albrechtsen, N.J.; Holst, J.J. Alanine, arginine, cysteine, and proline, but not glutamine, are substrates for, and acute mediators of, the liver-α-cell axis in female mice. *Am. J. Physiol. Endocrinol. Metab.* **2020**, *318*, E920–E929. [CrossRef]
20. Lund, A.; Vilsbøll, T.; Bagger, J.I.; Holst, J.J.; Knop, F.K. The separate and combined impact of the intestinal hormones, GIP, GLP-1, and GLP-2, on glucagon secretion in type 2 diabetes. *Am. J. Physiol. Endocrinol. Metab.* **2011**, *300*, E1038–E1046. [CrossRef]
21. Watson, L.E.; Phillips, L.K.; Wu, T.; Bound, M.J.; Checklin, H.L.; Grivell, J.; Jones, K.L.; Clifton, P.M.; Horowitz, M.; Rayner, C.K. A whey/guar "preload" improves postprandial glycaemia and glycated haemoglobin levels in type 2 diabetes: A 12-week, single-blind, randomized, placebo-controlled trial. *Diabetes Obes. Metab.* **2019**, *21*, 930–938. [CrossRef]
22. Almario, R.U.; Buchan, W.M.; Rocke, D.M.; Karakas, S.E. Glucose-lowering effect of whey protein depends upon clinical characteristics of patients with type 2 diabetes. *BMJ Open Diabetes Res. Care* **2017**, *5*, e000420. [CrossRef]
23. Staub, H. Untersuchungen über den Zuckerstoffwechsel des Menschen (Studies on sugar metabolism of humans). *Z. Klin. Med.* **1921**, *91*, 44–60.
24. Traugott, K. Über das Verhalten des Blutzuckerspiegels bei Wiederholter und Verschiedener Art Enteraler Zuckerzufuhr und Dessen Bedeutung für die Leberfunktion (About the behavior of blood sugar level in repeated and different types of enteral sugar intake and their impact on liverfunction). *Klin. Wochenschr.* **2005**, *1*, 892–894.
25. Battelino, T.; Danne, T.; Bergenstal, R.M.; Amiel, S.A.; Beck, R.; Biester, T.; Bosi, E.; Buckingham, B.A.; Cefalu, W.T.; Close, K.L.; et al. Clinical Targets for Continuous Glucose Monitoring Data Interpretation: Recommendations from the International Consensus on Time in Range. *Diabetes Care* **2019**, *42*, 1593–1603. [CrossRef] [PubMed]
26. Cavalot, F.; Pagliarino, A.; Valle, M.; Di Martino, L.; Bonomo, K.; Massucco, P.; Anfossi, G.; Trovati, M. Postprandial blood glucose predicts cardiovascular events and all-cause mortality in type 2 diabetes in a 14-year follow-up: Lessons from the San Luigi Gonzaga Diabetes Study. *Diabetes Care* **2011**, *34*, 2237–2243. [CrossRef]
27. Xia, W.; Luo, Y.; Chen, Y.C.; Chen, H.; Ma, J.; Yin, X. Glucose Fluctuations Are Linked to Disrupted Brain Functional Architecture and Cognitive Impairment. *J. Alzheimer's Dis.* **2020**, *74*, 603–613. [CrossRef] [PubMed]
28. King, D.G.; Walker, M.; Campbell, M.D.; Breen, L.; Stevenson, E.J.; West, D.J. A small dose of whey protein co-ingested with mixed-macronutrient breakfast and lunch meals improves postprandial glycemia and suppresses appetite in men with type 2 diabetes: A randomized controlled trial. *Am. J. Clin. Nutr.* **2018**, *107*, 550–557. [CrossRef]

Article

Bovine Milk-Derived Emulsifiers Increase Triglyceride Absorption in Newborn Formula-Fed Pigs

Kristine Bach Korsholm Knudsen [1,2], Christine Heerup [3], Tine Røngaard Stange Jensen [1], Xiaolu Geng [4], Nikolaj Drachmann [5], Pernille Nordby [6], Palle Bekker Jeppesen [7], Inge Ifaoui [2], Anette Müllertz [3], Per Torp Sangild [1], Marie Stampe Ostenfeld [5] and Thomas Thymann [1,*]

[1] Department of Veterinary and Animal Science, University of Copenhagen, 68 Dyrlægevej, DK-1870 Frederiksberg, Denmark; kristine.bach.korsholm.knudsen@regionh.dk (K.B.K.K.); tine.stange@hotmail.com (T.R.S.J.); pts@sund.ku.dk (P.T.S.)
[2] Rigshospitalet, Department of Pediatric Surgery, 9 Blegdamsvej, DK-2100 Copenhagen, Denmark; inge.ifaoui@regionh.dk
[3] Department of Pharmacy, University of Copenhagen, 2 Universitetsparken, DK-2100 Copenhagen, Denmark; christine.heerup@sund.ku.dk (C.H.); anette.mullertz@sund.ku.dk (A.M.)
[4] Department of Food Science, University of Copenhagen, 26 Rolighedsvej, DK-1958 Frederiksberg, Denmark; xiage@arlafoods.com
[5] Arla Foods Ingredients, 10-12 Soenderhoej, DK-8260 Viby, Denmark; nidra@arlafoods.com (N.D.); mstos@arlafoods.com (M.S.O.)
[6] DTU Bioengineering, Technical University of DK, Søltofts Plads, DK-2800 Kongens Lyngby, Denmark; nordby.p@gmail.com
[7] Rigshospitalet, Department of Gastroenterology, 9 Blegdamsvej, DK-2100 Copenhagen, Denmark; Palle.Bekker.Jeppesen@regionh.dk
* Correspondence: thomas.thymann@sund.ku.dk; Tel.: +45-3533-2622

Abstract: Efficient lipid digestion in formula-fed infants is required to ensure the availability of fatty acids for normal organ development. Previous studies suggest that the efficiency of lipid digestion may depend on whether lipids are emulsified with soy lecithin or fractions derived from bovine milk. This study, therefore, aimed to determine whether emulsification with bovine milk-derived emulsifiers or soy lecithin (SL) influenced lipid digestion in vitro and in vivo. Lipid digestibility was determined in vitro in oil-in-water emulsions using four different milk-derived emulsifiers or SL, and the ultrastructural appearance of the emulsions was assessed using electron microscopy. Subsequently, selected emulsions were added to a base diet and fed to preterm neonatal piglets. Initially, preterm pigs equipped with an ileostomy were fed experimental formulas for seven days and stoma output was collected quantitatively. Next, lipid absorption kinetics was studied in preterm pigs given pure emulsions. Finally, complete formulas with different emulsions were fed for four days, and the post-bolus plasma triglyceride level was determined. Milk-derived emulsifiers (containing protein and phospholipids from milk fat globule membranes and extracellular vesicles) showed increased effects on fat digestion compared to SL in an in vitro digestion model. Further, milk-derived emulsifiers significantly increased the digestion of triglyceride in the preterm piglet model compared with SL. Ultra-structural images indicated a more regular and smooth surface of fat droplets emulsified with milk-derived emulsifiers relative to SL. We conclude that, relative to SL, milk-derived emulsifiers lead to a different surface ultrastructure on the lipid droplets, and increase lipid digestion.

Keywords: preterm neonates; fat; gastric lipase; absorption; intestine; milk; emulsions; vegetable oil; soy lecithin

1. Introduction

Exclusive breastfeeding is, according to WHO, the preferred nutrition from birth to the age of six months [1], yet worldwide this is only accomplished for approximately 41% of all infants [2]. Reasons for this include mothers who are unable or chose not to breastfeed or have complications such as preterm birth [3]. Whereas donor milk may be an alternative, particularly for preterm infants [4], high-quality formulas are required when donor milk is not available to ensure survival and normal development. Relative to breastfed infants, formula-fed infants have an increased risk of developing atopic diseases [5], respiratory infections [6], necrotizing enterocolitis [7], and other gastrointestinal complications [8]. There is also evidence of a higher risk of reduced neurodevelopment in formula-fed infants than in infants fed mothers' milk, even after adjusting for important confounders [9,10]. These effects on cognitive outcomes have been observed in both preterm and term infants [9], where formula-fed infants have lower IQ and a lower score for cognitive functions [10,11], which may persist into later life [12,13].

Lipid supplementation to the brain is essential for normal neurological development. Accumulation of lipids in the brain begins in the third trimester and continues the first two years of postnatal life [14]. Especially long-chain polyunsaturated fatty acids (LC-PUFAs) are important, as they represent essential cell membrane components in the brain. A lower concentration of LC-PUFAs in the brain has been observed in formula-fed infants and they also have poorer neurological outcomes [15–17]. There are differences between breast milk and infant formula with regard to the bioavailability of LC-PUFAs, which may partly explain the differences observed in neurodevelopment [18]. However, compensating for lower bioavailability by supplementing infant formula with more LC-PUFAs has not shown any cognitive improvement [19,20].

One reason for the lack of improved neurological outcomes from formulas high in LC-PUFA may be low intestinal absorption leading to low delivery of LC-PUFA to the brain. Fat absorption in infants is generally less efficient relative to adults, and this is even more pronounced in preterm infants who have an absorption rate of 70–80% relative to 95% in adults. This is mainly due to the involvement of different lipases in infants than in adults and thereby different digestive capacity [21]. Moreover, formula-fed infants have reduced fat digestion and absorption relative to infants fed mother's milk [22,23]. Accordingly, lipid digestion in formula-fed infants should be improved to approximate breastfed infants' absorption levels. In mother´s milk, lipids in the form of triglycerides, are mainly transported as milk fat globules surrounded by a milk fat globule membrane (MFGM) [24]. Milk fat globules are secreted into the milk from the mammary gland by a unique mechanism giving MFGM a triple-layered membrane. This is a highly complex membrane containing several classes of phospholipids (including sphingomyelin), and glycosphingolipids, cholesterol, and unique membrane proteins, many of which are highly glycosylated [25]. MFGM forms a hydrophilic layer around the triglyceride core, making the fat globules water-soluble in the milk's water matrix [26,27]. Another phospholipid-rich source in milk is extracellular vesicles (EVs) [28]. EVs are also secreted from the mammary gland and consist of a lipid bilayer membrane comprised of phospholipids, glycerosphingolipids, cholesterol, and membrane proteins, but are devoid of a central triglyceride core. The EV membrane is rich in sphingomyelin and cholesterol as well as tetraspanins CD9, CD63, and CD81.

In infant formula today, soy lecithin (SL) is commonly used as an emulsifier and stabilizer. SL mainly consists of phosphatidylcholine and does not contain sphingomyelin or cholesterol, both of which play a major role in forming the lipid rafts found in MFGM [27]. Thus SL provides a different surface structure on the lipid droplets in infant formula [29]. Moreover, human milk lipids are more readily digested in preterm infants compared with infant formula [23]. Lipid digestion in infants has been examined using an in vitro model, simulating infant gastrointestinal conditions. This model used human gastric aspirate as a source of gastric lipase and porcine pancreatin as a source of pancreatic lipase. A higher in vitro gastric lipolysis rate was found when emulsifying lipids with milk phospholipids relative to SL, and this was also the case for intestinal lipolysis rate [30].

On this background, we hypothesized that emulsification with bovine phospholipid sources such as whey protein concentrate enriched in phospholipids (WPC-PL) or WPC from acid whey-enriched in extracellular vesicles (WPC-A-EV) with their unique composition of glycolipids, phospholipids, sphingomyelin, and glycosylated proteins joined in a complex membrane structure, would enhance digestion of dietary lipids compared to SL. Accordingly, the objective was to emulsify vegetable-based oils with either SL or WPC-PL or WPC-A-EV and determine their effect on the in vitro rate of lipolysis, and in vivo digestion and absorption of triglycerides. To determine any influence of emulsifiers on lipid digestion, we chose to use a preterm neonatal piglet model. This was from the assumption that prematurity per se would associate with lower digestive capacity, thereby making any potential improvements of lipid digestion more clear.

2. Materials and Methods

2.1. In Vitro Lipolysis

For the in vitro lipolysis studies, five oil-in-water emulsions were made using either SL (AAK, Karlshamn, Sweden), or bovine whey protein concentrate enriched in phospholipids (WPC-PL), whey protein concentrate from acid whey enriched in triglycerides (WPC-A-TAG), whey protein concentrate from acid whey enriched in EVs (WPC-A-EV), or whey protein concentrate from acid whey enriched in soluble whey protein (WPC-A-WP). All bovine products were kindly donated by Arla Foods Ingredients Group P/S, and the composition of the emulsifiers is provided in Table 1.

Table 1. Emulsifier composition of protein, neutral fat, and phospholipids.

	WPC-PL	WPC-A-TAG	WPC-A-EV	WPC-A-WP	SL
Percent of total:					
Proteins	72.7	53	76	89.3	N/A
Neutral fat	17.8	41	18	1.87	N/A
Phospholipids (PL)	7.1	12.4	8.9	0.75	43.3
Percent of PL:					
PC	27.4	26	27.3	28	31.2
PE	29	29	28	28	16.6
PI	7.2	5.5	5.6	5.3	26.2
PS-Na	6	10.6	9.8	9.3	0.6
SM	30.2	26.9	27.5	24	0
Other	0.2	2	1.8	5.4	25.4

PL: phospholipids; PC: phosphatidylcholine; PE: phosphatidylethanolamine; PI: phosphatidylinositol; PS-Na: phosphatidylserine-sodium; SM: sphingomyelin, N/A: not assessed; WPC-A-EV: whey protein concentrate from acid whey-enriched in extracellular vesicles (WPC-A-EV), WPC enriched in phospholipids; WPC-A-TAG: whey protein concentrate from acid whey enriched in triglycerides; WPC-A-WP: whey protein concentrate from acid whey enriched in soluble whey protein; SL: soy lecithin.

The emulsions were prepared using the emulsification method, composition, and oil-blend described in Heerup et al. (submitted) [31]. In brief, the emulsions were made with 0.35% emulsifier and 3.5% oil-blend (98.92% Akonino NS (AAK, Karlshamn, Sweden), 0.97% MEG-3 (DSM, Mulgrave, NS, Canada), and 0.51% Arasco (DSM)) in an aqueous 11.5 mM $CaCl_2$ and 8.5 mM NaCl solution.

The emulsions were digested using the in vitro pediatric gastro-intestinal digestion model described in Heerup et al. In brief, the model consisted of a 50 min gastric step at pH 6.4 with 3.75 TBU/mL recombinant human gastric lipase (rHGL) kindly donated by Bioneer A/S (Hørsholm, Denmark) and 126 U/mL pepsin purchased from Sigma Aldrich (St. Louis, MO, USA), a 90 min intestinal step at pH 6.5 with 26.5 TBU/mL pancreatin (Sigma Aldrich), and a back titration to pH 9. Since the SL emulsion was not stable in the 11.5 mM $CaCl_2$ and 8.5 mM NaCl solution, $CaCl_2$ and NaCl were instead added as part of the gastric medium. Table 2 shows the final gastric and intestinal assay composition, including the contribution of $CaCl_2$ and NaCl from the emulsions. The degree of lipolysis over time was measured indirectly by continuous titration of ionized free fatty acids with

0.2 mM NaOH using a Metrohm Titrando pH Stat (Metrohm, Glostrup, Denmark). The particle size distribution of the undigested emulsions was measured on the day of lipolysis.

Table 2. Final in vitro gastric- and intestinal digestion assay concentrations (mM). The contribution of $CaCl_2$ and NaCl from the emulsions is included in the shown concentrations.

	Final Assay Composition	
Compound [1]	Simulated Gastric Digestion, mM	Simulated Intestinal Digestion, mM
NaCl	10.4	51.8
Tris	2.0	2.2
Maleic acid	2.0	2.2
$CaCl_2$	10.1	5.9
Sodium taurocholate	0.0	0.5
Phospholipid	0.0	0.1

[1] NaCl was purchased from VWR (Darmstadt, Germany), Tris from ICN Biomedicals (Santa Ana, CA, USA), phospholipids (phosphatidylcholine) from Lipoid (Köln, Germany), maleic acid, and sodium taurocholate from Sigma Aldrich.

2.2. Microstructure of Emulsions

To determine structural differences we selected a subfraction of emulsions, i.e., SL, WPC-PL, and WPC-A-EV. These emulsions were visualized with transmission electron microscopy (TEM) and cryo-scanning electron microscopy (SEM). The emulsions used for TEM and SEM were prepared according to the method described by Heerup et al., although with no $CaCl_2$ or NaCl added to any of the emulsions. The emulsions were mixed 1:1 with 2% agarose (Carl Roth, Germany) at 37 °C and left at room temperature for solidification. Several small pieces of the solidified sample were cut and fixed in the 2% glutaraldehyde phosphate buffer (pH 7.2) for 30 min at room temperature, followed by washing and postfix in 1% w/v OsO_4 with 0.05 M $K_3Fe(CN)_6$ in 0.12 M phosphate buffer (pH 7.2) for 2 h. After that, a standard procedure for dehydration, embedding, and sectioning was applied. Finally, the ultra-thin sectioned sample (~60 nm) were collected on copper grids with Formvar supporting membranes, stained with uranyl acetate and lead citrate, and examined by a Philips CM-100 electron microscope (Philips, Eindhoven, The Netherlands) operated at 100 kV. For SEM, specimens were sandwiched in 2 × 100 μm planchettes and cryopreserved by high-pressure freezing (HPM100, Leica, Vienna, Austria). The sandwiched planchettes were mounted in a planchette holder (Leica) under liquid N_2 and transferred to a vitreous cryo transfer shuttle (VCT100, Leica). The samples were cracked and sputter-coated (approximately 6 nm) (MED020, Leica) with carbon/platinum. Specimens were examined with an FEI Quanta 3D scanning electron microscope operated at an accelerating 2 kV voltage.

2.3. Preparation of Emulsions for In Vivo Studies

From the initial five emulsifiers tested in the in vitro system, we selected WPC-PL and SL for in vivo study 1 and 2. While WPC-PL was chosen because it had shown a beneficial effect in infant formula [32], we chose SL as it is a common emulsifier and stabilizer often used in infant formulas. In the in vivo study, three were used as the most promising experimental emulsifier (WPC-A-EV) along with WPC-PL and SL to validate the improved digestibility observed in vitro. Ideally, all three emulsifiers could have been studied in each in vivo experiment but it was not feasible to include so many groups as the studies were very labor-intensive. To prepare the emulsions, we used a Rannie homogenizer (APV, Copenhagen, Denmark) at pressure 25 bar/250 bar, instead of a microfluidizer. The oil-in-water emulsions for the in vivo studies were made with 10% oil (w/w%) using an oil-blend containing 91.66% Akonino NS, 5.46% MEG-3, and 2.88% Arasco, and 1% emulsifier. Different experimental diets were used for each of three in vivo studies: In in vivo study 1, the experimental diet consisted of complete formulas based on 10% oil-in-water emulsions with either SL (VWR, Darmstadt, Germany) or WPC-PL mixed with a base-diet to achieve a final fat concentration of 5.1%. The base-diet was made from whey protein (Lacprodan® DI-9224), casein (Miprodan® 40), lactose, and minerals (Variolac® 855, all Arla Foods Ingredients) and was designed to meet the nutritional needs of pigs. In in vivo study 2, the experimental diet consisted of pure 10% oil-in-water emulsions with either SL (VWR) or WPC-PL. Finally, in in vivo study 3, the experimental diet consisted of complete formulas based on 10% oil-in-water emulsions with either SL (AAK, Aarhus, Denmark), WPC-PL, or WPC-A-EV mixed with the base diet to a final fat concentration of 10%.

2.4. In Vivo Lipid Digestibility of Complete Formulas (Study 1)

All procedures were approved by the Danish Animal Experiments Inspectorate (license number 2014-15-0201-00418), which follows the guidelines from Directive 2010/63/EU of the European Parliament and the ARRIVE guidelines [33]. We used cesarean-derived preterm neonatal piglets as they were assumed to have lower fat digestive capacity relative to term pigs, thereby making any effect of emulsifiers more detectable. In brief, one litter of preterm piglets, $n = 22$, (Landrace × Large white × Duroc, Gadstrup, Denmark) was born by cesarean section at day 113 of gestation and reared in preheated and oxygenated incubators as described previously [34]. Immediately after birth, the pigs were equipped with oral and vascular catheters to allow enteral and parenteral feeding. See supplemental for further information. On the second day, the pigs were surgically equipped with a jejunostomy to allow the quantitative collection of stoma output. Details for housing, feeding, surgery, post-surgical care, and sample collection are provided in the supplemental material. The piglets were block-randomized according to bodyweight into two groups receiving complete formulas emulsified with either SL ($n = 5$–9) or WPC-PL ($n = 6$–9). Enteral feeding was initiated as quickly as possible postoperatively at a rate of 6 mL/kg every three hours, gradually increasing to 15 mL/kg every three hours on day seven and eight. The personnel were blinded to the treatment groups. Stoma output was collected quantitatively on days 3, 4, and 7, and following measurement of fat concentration in the stoma output, intestinal fat absorption was calculated as described earlier [35].

Fat accumulation in the tissues of the small intestine was measured. A piece of the proximal part of the small intestine was fixed in a cryo cassette (Tissue-Tek Cryomold, Sakura Finetek, Zoeterwouder, Holland, The Netherlands) with tissue O.C.T. (Tissue-Tek, Sakura, Finetek Zoeterwoude, Holland) and frozen in liquid nitrogen. The tissue was sliced with a cryostat (Leica CM 1950, Leica Biosystems, Wetzlar Germany), and stained with Oil-Red-O (Sigma-Aldrich, Darmstadt, Germany). Six digital images were taken at 20× magnification from six different regions using a microscope (Olympus BX45, Tokyo, Japan), and the degree of fat infiltration was scored.

2.5. In Vivo Fat Absorption Kinetics of Pure Emulsions (Study 2)

Subsequently, we examined the triglyceride absorption kinetics after dosing of pure 10% oil-in-water emulsions. One litter of preterm piglets, $n = 19$, was born by cesarean section at day 106 of gestation, making the piglets 7 days more premature relative to the first study which further sensitizes the gut toward low digestive capacity. We have used this degree of prematurity in many previous experiments and based on sensitivity to develop prematurity-related complications like necrotizing enterocolitis, which may correspond to week 25–28 in human pregnancy. Procedures for cesarean section, postnatal rearing, and provision of parenteral nutrition were identical to the previous study. On day two, the piglets were blocked randomized according to body weight to receive an enteral bolus (6 mL/kg body weight) of pure emulsion with either SL or WPC-PL. Using a cross-over design, the piglets received a second pure emulsion bolus on day three, such that each piglet had been exposed to both emulsions. Blood samples from the arterial umbilical catheter were collected on both days at t = 0, 90, 180, 270, 360, and 540 min after the bolus was given and stored in heparinized tubes. Plasma was isolated after centrifugation (1270× g, 4 °C, 10 min), and the concentration of triglycerides was analyzed using an automated ADVIA 1800 Chemistry System (Siemens Healthcare A/S, Ballerup, Denmark). Following blood sampling, all pigs were euthanized.

2.6. In Vivo Lipid Absorption Kinetics from Complete Formulas (Study 3)

In a final in vivo experiment, the triglyceride absorption kinetics following ingestion of complete formulas was determined. A total of 49 preterm piglets from three pregnant sows were born by cesarean section at day 106 of gestation. They were stabilized and received parenteral nutrition as described above and were block-randomized according to birthweight into three groups, all receiving the complete formula, with a fat content of 10%. Collectively for all three in vivo studies, we ensured to have pigs from each litter equally represented in all treatment groups, allowing us to correct for any variance between litters and their specific genotype, in the analysis of variance. The complete formulas were made with the emulsions based on SL, WPC-PL, or WPC-A-EV. To stimulate the intestinal absorptive function, we initially fed the pure base-diet without emulsions during the first 24 h, at a rate of 3 mL/kg every three hours. On day two, the complete formulas including the emulsions were given at a bolus dose of 9 mL/kg, following a 6 hr fasting period. Blood samples were collected via the umbilical catheter at t = 0, 30, 60, 90 min after the bolus. Following this, the piglets were again fed with increasing amounts of enteral diet (6–9 mL/kg every three hours). On day four, the piglets were again fed a test meal of 9 mL/kg after a fasting period of 3.5 h, and blood samples were collected at t = 0, 30, 60, 90, and 120 min. Blood was drawn via the jugular vein in cases where the umbilical catheter was dysfunctional. Plasma was isolated, and triglyceride concentration was measured.

The pigs were euthanized on day four following a standardized feeding regimen to ensure an equal amount of gastric content at the time of euthanasia. Specifications for recordings and sample collection are provided in the supplementary section. Gastric content was weighed, and gastric lipase activity in gastric content was measured using the method described earlier with slight modifications [36]. The assay was carried out using the same pH Stat equipment as for the in vitro lipolysis. It was initiated by mixing 14.5 mL assay medium containing 1.5 µM Bovine Serum Albumin (AppliChem, Darmstadt, Germany), 150 mM NaCl (VWR), and 2 mM sodium taurodeoxycholate (Sigma Aldrich) with 0.5 mL tributyrin (Sigma Aldrich) and 0.5 mL stomach content. The enzymatic activity, calculated as U/mL was based on the measured titration rate of NaOH by a Metrohm Titrando pH Stat over five minutes of digestion at pH 5.5. As the butyric acid was not being fully titratable at pH 5.5, a correction factor of 1.12 was multiplied to the calculated activity. The total amount of collected stomach content, as well as the pH, was also measured.

3. Statistics

Statistical analyses for the in vitro digestions were performed using GraphPad Prism version 7.0 for Windows (GraphPad Software, San Diego, CA, USA, www.graphpad.com). Group comparisons were made using unpaired t-tests based on the area under the curve (AUC).

Statistical analyses for the in vivo studies were performed using R (version 3.5.0, R Foundation for Statistical Computing, Vienna, Austria). Continuous data were analyzed with a linear mixed model using the lm function (lme4 package). Group comparisons were made with an ANOVA (lme4 package) and Post Hoc Tukey test with the glht function (multcomp package). Normal distribution and homoscedasticity of residuals were visualized for model validation. Birthweight was used as a covariate, sex as a fixed variable, and litter as a random variable in all the models. Repeated blood samples were analyzed using SAS (SAS Software 9.4, SAS Institute, Cary, NC, USA). In all repeated measures models, birthweight and baseline triglyceride levels were included as covariates, whereas sex and litter were included as fixed and random effects respectively. Survival curves of basic motor skills were evaluated in GraphPad (Prism version 7.0) using the Logrank test for the trend. Fat accumulation scores were analyzed as ordinal data with the clmm function (ordinal package), and group comparison was made with the post hoc Tukey test. Group comparisons of the gastric lipase activity were made with unpaired t-tests in the GraphPad (Prism version 7.0).

p-values < 0.05 were regarded significant and p-values < 0.1 as a tendency to effect. Data are presented as means ± standard deviation unless otherwise stated.

4. Results

4.1. In Vitro Lipolysis

The in vitro lipolysis assessed as the amount of released free fatty acids (FFAs) titrated over time during digestion of each of the five 3.5% fat emulsions, using SL, WPC-PL, WPC-A-EV, WPC-A-TAG, and WPC-A-WP, is presented in Figure 1. After 50 min of in vitro digestion 28–44 µmol of FFAs were titrated depending on the emulsifier (SL: 28.1 µmol ± 0.5, WPC-A-WP: 29.3 µmol ± 1.6, WPC-A-TAG: 38.0 µmol ± 1.7, WPC-PL: 42.2 µmol ± 0.6, and WPC-A-EV: 44.3 µmol ± 6.6). After 140 min 145–190 µmol of FFAs were titrated (SL: 145.2 µmol ± 6.4, WPC-A-TAG: 148.6 µmol ± 13.7, WPC-PL: 176.8 µmol ± 1.1, WPC-A-WP: 180.6 µmol ± 8.0, and WPC-A-EV: 190.2 µmol ± 2.0). The in vitro digestion of emulsions based on WPC-A-EV, WPC-PL, and WPC-A-WP showed a higher lipolysis rate relative to SL. This was found, both when considering the AUC of the gastric and intestinal step separately, as well as the two combined (all $p < 0.01$), whereas WPC-A-TAG showed values similar to SL.

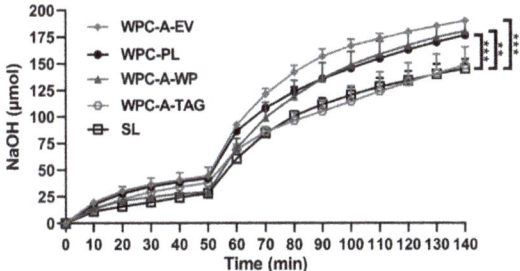

Figure 1. Titration of free fatty acids during simulated human gastric digestion with recombinant human gastric lipase (rHGL) and pepsin at pH 6.4 (0–50 min) and pH 6.5 (50–140 min) of 3.5% oil-in-water emulsions using either whey protein concentrate from acid whey-enriched in extracellular vesicles (WPC-A-EV) ($n = 3$), WPC enriched in phospholipids (WPC-PL) ($n = 3$), whey protein concentrate from acid whey enriched in triglycerides (WPC-A-TAG) ($n = 3$), whey protein concentrate from acid whey enriched in soluble whey protein (WPC-A-WP) ($n = 3$), and soy lecithin (SL) ($n = 2$ due to removal of outliers). Values are presented as mean ± SD. ** $p < 0.001$. *** $p < 0.0001$.

4.2. In Vitro Study 2—Microstructure of Emulsions

The droplet size distribution for each of the five 3.5% fat emulsions, SL, WPC-PL, WPC-A-EV, WPC-A-TAG, and WPC-A-WP, are presented in Figure 2, Panel 1 as the volume mean diameter. All emulsions had droplet sizes within the range of 0.04–100 µm, with the smallest droplet size being observed for the SL emulsion, which had a unimodal distribution centered at 0.16 µm. WPC-A-EV and WPC-PL emulsions had similar droplet size distributions, and both had a primary population with modes centered at 0.5–0.7 µm and a smaller population around 7 µm.

The droplet sizes in the WPC-A-TAG and WPC-A-WP emulsions were similar and showed two distributions of nearly uniform height, with the first at 0.6 µm for both WPC-A-TAG and WPC-A-WP and the second mode around 4 µm for WPC-A-TAG and 17 µm for WPC-A-WP. The microstructure of WPC-PL, WPC-A-EV, and SL emulsion was further characterized by TEM (Figure 2, Panel 2). In general, the lipids droplets in the WPC-PL (subpanels A and D) and WPC-A-EV (subpanels B and E) emulsions had a similar structure with clear edges and relatively round and smooth appearances compared with droplets in the SL emulsion (subpanels C and F). Some of the droplets in the WPC-PL emulsions were partially covered by a thick dark layer (subpanel A), which might be composed of milk protein (mainly from aggregated whey proteins) from the WPC-PL product. These thick layers were less frequently observed for the droplet made of WPC-A-EV (subpanel B). Due to the higher content of native whey proteins present in WPC-A-EV relative to WPC-PL (communication with Arla Foods Ingredients), less aggregated protein complex can be associated with the interface of the lipid droplet presumably due to lower protein content in WPC-A-EV relative to WPC-PL. The SL lipid droplets had a different morphology than that of the other two emulsions, with irregular and uneven appearances (subpanel C). Besides the fluffy and thick layer of phospholipids on the droplet's surface, we also observed a white layer of particles that either lies on the surface of the lipids droplets or are incorporated within the droplets. These droplets may be nano-sized liposomes generated during homogenization. Further, the cryo-SEM images provided a 3D cross-section view of the emulsions, which indicated that the surface of lipid droplets in the WPC-PL and WPC-A-EV emulsions were thinner and smoother relative to the SL emulsion droplets.

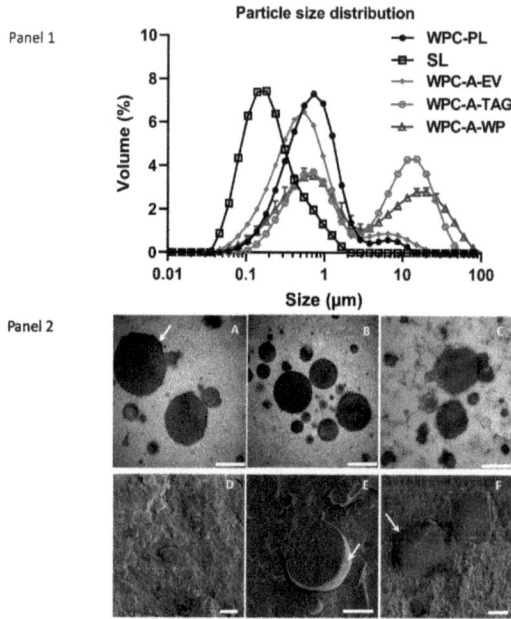

Figure 2. Particle size distribution and transmission electron microscopy (**Panel 1**): Volume mean diameter for each of the 3.5% oil-in-water emulsions, WPC-A-EV, WPC-PL, WPC-A-TAG, WPC-A-WP, and soy lecithin (SL). $n = 3$. Values are presented as means $\pm$ SD. (**Panel 2**): Micrographs from transmission electron microscopy (subpanels A, B, C) and cryo-scanning electron microscopy (subpanels D, E, F) of three emulsions. A and D: WPC-PL emulsion, B and E: WPC-A-EV emulsion, C and F: SL emulsion. The scale bar for micrographs A–C is 500 nm, and D–F is 2 µm. The arrow in A points out the dark layer, in E the arrow points out the lipid droplet's smooth surface, and in F the arrow points out the thick and uneven surface of the lipid droplet.

4.3. In Vivo Study 1—In Vivo Fat Digestibility of Complete Emulsions

Of the initial 22 piglets, three were euthanized due to post-surgical complications, and later two died from poor clinical conditions. The final number of piglets included in each group was $n = 9$ (SL) and $n = 8$ (WPC-PL). Birthweight was similar between WPC-PL and SL piglets, with a mean weight of 1530 g $\pm$ 235 and 1465 g $\pm$ 306, respectively. Changes in body weight in the immediate postnatal period and following period with an ileostomy were generally characterized by weight loss in both groups, albeit with lower weight loss in WPC-PL versus SL (-12.4 ± 12.3 versus -17.5 ± 14.9 g/(kg·day), $p > 0.05$). We were able to quantify stoma output reliably in a subfraction of the piglets ($n = 5$–10 piglets), see Figure 3. Based on this, we found that the fat absorption coefficient was similar for WPC-PL and SL (95.4% ± 2.5 in WPC-PL versus 95.8% ± 2.3 in SL) on day three, declining to 85.7% ± 9.4 in WPC-PL versus 83.2% ± 14.26 in SL on day seven.

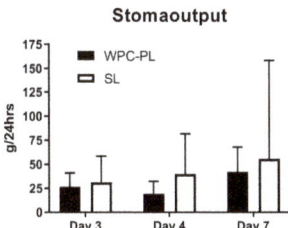

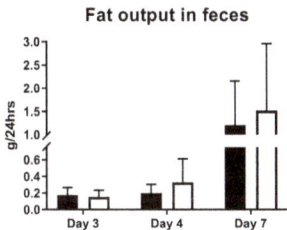

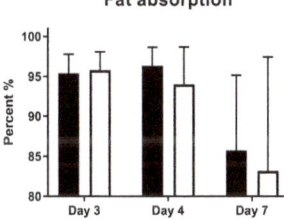

Figure 3. Total fecal stoma output (**upper panel**), total fat output in the feces (**middle panel**) and fat absorption percentages (**lower panel**) on day 3, 4, and 7 (WPC-PL n = 6–9, SL n = 5–9). Values presented as mean ± SD.

Due to poor clinical conditions, three piglets did not receive their last feeding, and they were excluded. Plasma triglyceride levels one hour postprandial did not differ between groups, reaching a level in the WPC-PL piglets (n = 7) of 0.25 mmol/L ± 0.12 and in the SL piglets (n = 6) of 0.15 mmol/L ± 0.03. Organ weights and histological fat accumulation score in the proximal intestine were similar between the groups. Details are provided in Supplementary Table S1 and Figure S1.

4.4. In Vivo Study 2—In Vivo Fat Absorption Kinetics of Pure Emulsions

Using the cross-over design, the final group sizes were n = 16 for WPC-PL and n = 18 for SL. Mean body weight at birth was 933 g ± 277 (WPC-PL) and 969 g ± 226 (SL). Triglyceride levels in plasma were similar for WPC-PL and SL on both days two and three, and for the pooled values across both days (Figure 4).

Peak plasma triglyceride level was reached between 60–120 min after bolus, with no noticeable difference between the groups. Other plasma measurements, including cholesterol, glucose, ALAT, ASAT, and creatinine, were also similar for WPC-PL and SL. The only exception from this was the WPC-PL piglets' baseline level of total bilirubin, which was significantly higher relative to SL, p = 0.03. All biochemistry data are available in Supplementary Table S2.

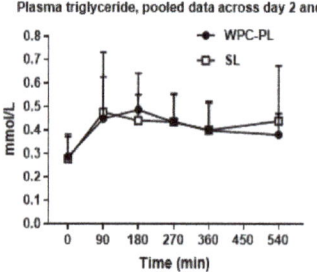
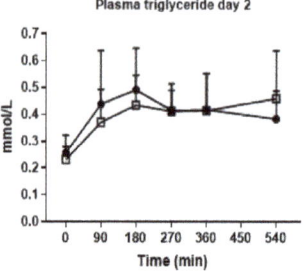
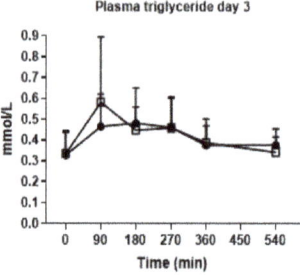

Figure 4. Triglyceride levels in plasma, mmol/L, as pooled data across day two and three in upper panel (WPC-PL n = 16, SL n = 18), and data from day two in middle panel (WPC-PL n = 9, SL = 9) and three in lower panel (WPC-PL n = 9, SL n = 7) separately. Values presented as means ± SD.

4.5. In Vivo Study 3—In Vivo Fat Absorption Kinetics from Complete Formulas

Final number of piglets were n = 17 (SL), n = 14 (WPC-PL) and n = 15 (WPC-A-EV). Birthweight was similar between SL (1124 g ± 264), WPC-PL (1077 g ± 225) and WPC-A-EV (1108 g ± 260) fed piglets, weight gain was similar for SL (−5.39 g/(kg·day) ± 11.6), WPC-PL (−2.19 g/(kg·day) ± 10.1) and WPC-A-EV (−1.24 g/(kg·day) ± 7.4). Three piglets were euthanized due to clinical complications unrelated to the diets.

Following the bolus administration on day two, plasma triglyceride levels increased similarly for all three groups over the entire 90 min sampling time relative to the common baseline level. (Figure 5, upper panel). When the bolus test was repeated on day four, plasma triglyceride level again peaked between 60–90 min, and importantly, the WPC-A-EV and WPC-PL groups had significantly higher plasma triglyceride levels compared to the SL group, $p < 0.01$ and $p < 0.001$ respectively. The values for WPC-A-EV and WPC-PL were similar (Figure 5, lower panel).

4.6. Gastric Fat Concentration and Lipase Activity—Related to In Vivo Study 3

As a result of the standardized feeding regimen before euthanasia, the gastric residuals were very similar across the groups, i.e., 15–16 g per pig (Supplementary Table S3). However, the fat concentration in the residuals was significantly higher in the SL piglets:

157 ± 27 mg/g, compared with WPC-PL: 108 ± 17 mg/g and WPC-A-EV: 103 ± 13 mg/g, $p < 0.001$ in both cases. Due to higher fat content in the SL group gastric residuals, the subsequent analysis of gastric lipase showed the highest activity in the SL group relative to WPC-PL and SL, which may partly be due to the higher availability of substrate (i.e., fat) for the lipase assay. Specific numbers were SL: 0.92 U/mL ± 0.63; WPC-PL: 0.27 U/mL ± 0.15 and WPC-A-EV: 0.30 U/mL ± 0.34 ($p < 0.01$ for WPC-PL and WPC-A-EV relative to SL). The gastric pH was similar between groups (Figure 6).

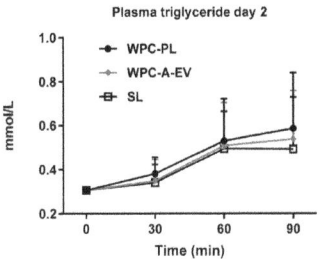

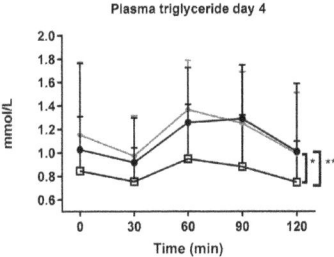

Figure 5. Triglyceride levels in plasma at day two (WPC-A-EV $n = 16$, WPC-PL $n = 16$, SL $n = 17$) and four (WPC-A-EV $n = 15$, WPC-PL $n = 14$, SL $n = 17$), baseline sample and 30, 60, 90 and 120 min after a test bolus was given. Presented as mean ± SD. * $p < 0.01$. ** $p < 0.001$.

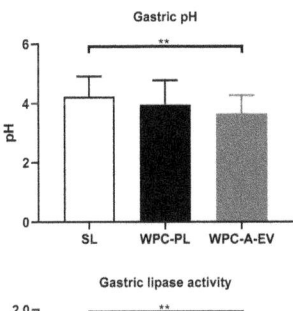

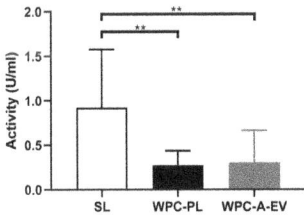

Figure 6. (**Top panel**): Gastric pH for WPC-A-EV ($n = 14$), WPC-PL ($n = 11$), and SL ($n = 17$) in gastric content collected post euthanasia. (**Bottom panel**): Gastric lipase activity for WPC-A-EV ($n = 13$), WPC-PL ($n = 11$), and SL ($n = 16$) in gastric content collected post euthanasia. ** $p < 0.01$.

Further, circulatory levels of albumin, liver enzymes, creatinine, creatine kinase, glucose, phosphate, and urea, were largely similar across the groups (Supplementary Table S4).

5. Discussion

The most important finding was that piglets fed complete formula with WPC-PL and WPC-A-EV emulsifiers showed higher plasma triglyceride levels relative to the SL group when studying fat absorption kinetics in vivo. This notion was further substantiated in the in vitro lipolysis assays. MFGM surrounds the milk fat globule in human milk, and this membrane structure has its unique content of phospholipids, glycerolipids, cholesterol, and glycosylated membrane proteins. This provides physical properties allowing lipases to have high catalytic efficiency during gut luminal fat digestion [26]. In contrast, the fat fraction in infant formula is emulsified primarily by SL and milk proteins like β-lactoglobulin and caseins, which creates a different lipid-water interface compared with MFGM in human milk [29]. Therefore, we hypothesized that the utility of naturally occurring dairy emulsifiers isolated industrially from bovine whey, here referred to as WPC-PL, WPC-A-TAG, WPC-A-WP, and WPC-A-EV, has superior absorptive effects relative to SL in vitro and in vivo. These absorptive effects could potentially result in higher availability of lipids for brain development. Gastric lipase activity is particularly important in early life, and surprisingly our ex vivo analysis of gastric lipase activity in the stomach content showed the highest activity in the SL group relative to WPC-PL and WPC-A-EV. This may, however, partly be explained by a higher residue of undigested lipids in the SL group gastric residuals, leading to more substrate available for the ex vivo lipolysis assay. However, the gastric fat content was approximately 50% higher in the SL group relative to WPC-PL and WPC-A-EV. We normalized the gastric lipase activity to gastric fat content and found that it does not fully account for the higher lipase activity seen in the SL group, and we speculate that other regulatory mechanisms play a role.

An adequate supply of lipids, including LC-PUFA to the brain during early life, helps secure synaptogenesis and brain maturation and gives lasting effects on cognition [21]. Therefore, infant formulas should ensure LC-PUFA is available to the brain. Although efforts have been made to customize infant formulas to mimic human milk, including fortifying formula with LC-PUFA, the results have been inconclusive with regard to cognitive effects [20]. Fat absorption is lower in formula-fed newborn infants than babies fed mother's milk, and assuming that the level of endogenous digestive enzyme activity is similar between breastfed and formula-fed infants, this indicates that factors that relate to the fat fraction per se, determine the level of lipolysis and absorption.

The interfacial layer of the lipid droplets affects the fat digestion in milk, and our results indicate that when the interfacial layer is composed mainly of phosphatidylcholine (i.e., the SL emulsions), the lipolysis is slower compared to lipid droplets emulsified with bovine dairy emulsifiers of more complex polar lipid and protein composition. This agrees with the findings from earlier studies suggesting that the pancreatic lipase has a higher affinity for a lipid surface covered with casein or whey proteins [37]. This notion is supported by the TEM and cryo-SEM microstructure micrographs, which indicate that the surface of WPC-PL and WPC-A-EV emulsified droplets is thin and smooth, which may favor the lipase to penetrate and reach the lipids. In contrast, in SL emulsions, the formation of liposomes may additionally reduce the lipase efficacy.

We used a neonatal preterm piglet model to determine the effects of WPC-PL, WPC-A-EV, and SL emulsifiers on the absorption of triglycerides. Appropriate animal models offer the potential of eliminating important confounders that exist in the comparison of breastfeeding versus formula feeding, and we suggest that preterm pigs with their resemblance to preterm infants in terms of gut and brain development [34] and post-surgical responses [38], is a valuable tool to determine these effects. Our initial study of triglyceride digestibility using an ileostomy model showed similar triglyceride absorption levels between the WPC-PL and SL groups. Plasma triglyceride levels measured one hour postprandial were numerically higher in the WPC-PL group, but this did not reach significance with the limited sample size. The preterm pigs were born on gestational day 113, i.e., 4–5 days before the expected termination date. It is possible that the degree of prematurity was too small to sensitize the digestive system toward reduced-fat uptake. In the two subsequent piglet studies, we chose to further sensitize the model by delivering the piglets on day 106 of gestation, i.e., 11–12 days before the expected term date. Yet, histological lipid stainings showed only numerically higher scores for WPC-PL and WPC-A-EV relative to SL. From the notion that fat uptake is indeed higher in WPC-PL and WPC-A-EV relative to SL, the apparently similar fat level in the mucosa may indicate that the transport of fat as chylomicrons into lymph vessels is an equally efficient process relative to absorption from the gut lumen, resulting in no net fat accumulation in the mucosa.

Several randomized controlled trials have examined the safety and efficacy of dairy phospholipid-enriched infant formulas. In one multicenter randomized controlled trial, they used two commercial phospholipid-rich ingredients and investigated its safety in infant formula [39]. It was proven safe to use in healthy term-born infants, with eczema being the only adverse effect reported. Likewise, Xionan and colleagues [40], found that a commercial dairy phospholipid-rich ingredient was safe and well-tolerated, and found no difference in skin effects relative to neither standard formula nor breastfeeding. In another randomized controlled trial, a standard formula was compared with an experimental formula enriched with the same dairy phospholipid-rich ingredient and showed improved cognition score in the experimental group using Bayley Scales of Infant and Toddler Development III at 12 months of age [32] and lower incidence of acute otitis media [41]. Whereas the dairy phospholipid-enriched group generally performed similar to a breastfed reference group, the breastfed infants performed better than both formula-fed groups on a verbal subscale. In the same infants, an analysis of lipidomics profile in serum/plasma and erythrocyte membranes at four, six, and twelve months of age showed significant differences between the dairy phospholipid-enriched formula and the standard formula groups. The difference in serum did, however, disappear six months after the intervention [42]. The discrepancy in the lipidomics profile was mainly accounted for by sphingomyelin, likely to be explained by the contribution from the dairy phospholipid ingredient. The provision of sphingomyelin into the blood circulation may provide the developing brain with this essential nutrient, yet the direct causality remains to be determined. In support of the notion that dairy phospholipids are important for brain development, Gurnida and colleagues showed positive effects of glycosphingolipids on hand-eye coordination, performance score, and total score measured with the Griffiths Mental Developmental Scale (GMDS) in term-born infants [43].

We conclude that the milk-derived emulsifiers WPC-PL and WPC-A-EV increase fat digestion and absorption of triglycerides relative to SL. The effects are seen in vivo when the emulsions are an integrated part of a complete diet, but not when the pigs were fed pure emulsions. Higher levels of lipid hydrolysis, as indicated in vitro, suggest that the higher absorptive rates in vivo result from increased lipase activity when WPC-PL and WPC-A-EV are used as emulsifiers relative to SL. It remains to be determined whether emulsification with these milk-derived polar lipids can affect brain development.

Supplementary Materials: The following are available online at https://www.mdpi.com/2072-6643/13/2/410/s1, Table S1. Organ weight of pigs equipped with an ileostomy and fed milk diets with SL or WPC-PL for seven days; Table S2. Circulatory markers in pigs with boluses of pure emulsion; Table S3. Organ weight in pigs fed complete formulas with either SL, WPC-PL, or WPC-A-EV; Table S4. Circulatory markers in pigs fed complete formulas with either SL, WPC-PL, or WPC-A-EV as an emulsifier; Figure S1. Higher score indicates higher fat infiltration, as assessed subjectively by several independent observers who were blinded to the treatment groups [44].

Author Contributions: Conceptualization: K.B.K.K., C.H., P.N., A.M., P.T.S., M.S.O. and T.T.; Methodology: K.B.K.K., C.H., T.R.S.J., X.G., P.B.J., I.I. and T.T. Resources: N.D., P.N. and M.S.O. Data analysis: K.B.K.K., C.H., T.R.S.J., X.G. and T.T.; Writing original draft: K.B.K.K., C.H., T.R.S.J. and T.T.; Writing, review and editing: all. Project administration: University of Copenhagen; Funding aqusition: Denmark; T.T. had primary responsibility for final content. All authors have read and agreed to the published version of the manuscript.

Funding: This research was funded by The Innovation Foundation Denmark (Grant No.: 5158-00014B), the Danish Dairy Research Foundation and Arla Foods Ingredients.

Institutional Review Board Statement: Danish Animal Experiments Inspectorate (license number 2014-15-0201-00418).

Informed Consent Statement: Not applicable.

Data Availability Statement: All generated data are stored on pass word protected servers at University of Copenhagen, and can be made available upon request.

Acknowledgments: We thank Elin Skytte, Jane Povlsen, Kristina Møller, Michelle Christiansen, Rikke Stolberg and Anne Heckmann for their valuable contributions. The late Lars Ingvar Hellgren, Technical University of Denmark, is acknowledged for his pas-sionate initiation and steering of the Infant Brain project until his health condition prevented further involvement.

Conflicts of Interest: Authors M.S.O., X.G., N.D. are employees of Arla Foods Ingredients. Remaining authors have no conflicts of interest.

References

1. WHO. *The Optimal Duration of Exclusive Breastfeeding, Report of an Expert Consultation*; World Health Organization: Geneva, Switzerland, 2001.
2. UNICEF. Infant and Young Child Feeding Data. Available online: https://data.unicef.org/topic/nutrition/infant-and-young-child-feeding/ (accessed on 27 January 2021).
3. Simmons, L.E.; Rubens, C.E.; Darmstadt, G.L.; Gravett, M.G. Preventing preterm birth and neonatal mortality: Exploring the epidemiology, causes, and interventions. *Semin. Perinatol.* **2010**, *34*, 408–415. [CrossRef] [PubMed]
4. Arslanoglu, S.; Corpeleijn, W.; Moro, G.; Braegger, C.; Campoy, C.; Colomb, V.; Decsi, T.; Domellof, M.; Fewtrell, M.; Hojsak, I.; et al. Donor human milk for preterm infants: Current evidence and research directions. *J. Pediatr. Gastroenterol. Nutr.* **2013**, *57*, 535–542. [CrossRef] [PubMed]
5. Gdalevich, M.; Mimouni, D.; David, M.; Mimouni, M. Breast-feeding and the onset of atopic dermatitis in childhood: A systematic review and meta-analysis of prospective studies. *J. Am. Acad. Dermatol.* **2001**, *45*, 520–527. [CrossRef] [PubMed]
6. Ip, S.; Chung, M.; Raman, G.; Trikalinos, T.A.; Lau, J. A summary of the Agency for Healthcare Research and Quality's evidence report on breastfeeding in developed countries. *Breastfeed. Med.* **2009**, *4* (Suppl. 1), S17–S30. [CrossRef]
7. Chowning, R.; Radmacher, P.; Lewis, S.; Serke, L.; Pettit, N.; Adamkin, D.H. A retrospective analysis of the effect of human milk on prevention of necrotizing enterocolitis and postnatal growth. *J. Perinatol.* **2016**, *36*, 221–224. [CrossRef]
8. Petit, V.; Sandoz, L.; Garcia-Rodenas, C.L. Importance of the regiospecific distribution of long-chain saturated fatty acids on gut comfort, fat and calcium absorption in infants. *Prostaglandins Leukot. Essential Fatty Acids* **2017**, *121*, 40–51. [CrossRef]
9. Anderson, J.W.; Johnstone, B.M.; Remley, D.T. Breast-feeding and cognitive development: A meta-analysis. *Am. J. Clin. Nutr.* **1999**, *70*, 525–535. [CrossRef]
10. Horta, B.L.; Loret de Mola, C.; Victora, C.G. Breastfeeding and intelligence: A systematic review and meta-analysis. *Acta Paediatr.* **2015**, *104*, 14–19. [CrossRef]
11. Kramer, M.S.; Aboud, F.; Mironova, E.; Vanilovich, I.; Platt, R.W.; Matush, L.; Igumnov, S.; Fombonne, E.; Bogdanovich, N.; Ducruet, T.; et al. Breastfeeding and child cognitive development: New evidence from a large randomized trial. *Arch. Gen. Psychiatry* **2008**, *65*, 578–584. [CrossRef]
12. Victora, C.G.; Horta, B.L.; Loret de Mola, C.; Quevedo, L.; Pinheiro, R.T.; Gigante, D.P.; Goncalves, H.; Barros, F.C. Association between breastfeeding and intelligence, educational attainment, and income at 30 years of age: A prospective birth cohort study from Brazil. *Lancet Glob. Health* **2015**, *3*, e199–e205. [CrossRef]
13. Kafouri, S.; Kramer, M.; Leonard, G.; Perron, M.; Pike, B.; Richer, L.; Toro, R.; Veillette, S.; Pausova, Z.; Paus, T. Breastfeeding and brain structure in adolescence. *Int. J. Epidemiol.* **2013**, *42*, 150–159. [CrossRef] [PubMed]
14. Janssen, C.I.; Kiliaan, A.J. Long-chain polyunsaturated fatty acids (LCPUFA) from genesis to senescence: The influence of LCPUFA on neural development, aging, and neurodegeneration. *Prog. Lipid Res.* **2014**, *53*, 1–17. [CrossRef] [PubMed]
15. Jamieson, E.C.; Farquharson, J.; Logan, R.W.; Howatson, A.G.; Patrick, W.J.; Weaver, L.T.; Cockburn, F. Infant cerebellar gray and white matter fatty acids in relation to age and diet. *Lipids* **1999**, *34*, 1065–1071. [CrossRef]
16. Makrides, M.; Neumann, M.A.; Byard, R.W.; Simmer, K.; Gibson, R.A. Fatty acid composition of brain, retina, and erythrocytes in breast- and formula-fed infants. *Am. J. Clin. Nutr.* **1994**, *60*, 189–194. [CrossRef] [PubMed]
17. Neuringer, M.; Connor, W.E.; Lin, D.S.; Barstad, L.; Luck, S. Biochemical and functional effects of prenatal and postnatal omega 3 fatty acid deficiency on retina and brain in rhesus monkeys. *Proc. Natl. Acad. Sci. USA* **1986**, *83*, 4021–4025. [CrossRef] [PubMed]
18. Delplanque, B.; Gibson, R.; Koletzko, B.; Lapillonne, A.; Strandvik, B. Lipid Quality in Infant Nutrition: Current Knowledge and Future Opportunities. *J. Pediatr. Gastroenterol. Nutr.* **2015**, *61*, 8–17. [CrossRef] [PubMed]
19. Qawasmi, A.; Landeros-Weisenberger, A.; Leckman, J.F.; Bloch, M.H. Meta-analysis of long-chain polyunsaturated fatty acid supplementation of formula and infant cognition. *Pediatrics* **2012**, *129*, 1141–1149. [CrossRef]

20. Smithers, L.G.; Collins, C.T.; Simmonds, L.A.; Gibson, R.A.; McPhee, A.; Makrides, M. Feeding preterm infants milk with a higher dose of docosahexaenoic acid than that used in current practice does not influence language or behavior in early childhood: A follow-up study of a randomized controlled trial. *Am. J. Clin. Nutr.* **2010**, *91*, 628–634. [CrossRef]
21. Lindquist, S.; Hernell, O. Lipid digestion and absorption in early life: An update. *Curr. Opin. Clin. Nutr. Metab. Care* **2010**, *13*, 314–320. [CrossRef]
22. Chappell, J.E.; Clandinin, M.T.; Kearney-Volpe, C.; Reichman, B.; Swyer, P.W. Fatty acid balance studies in premature infants fed human milk or formula: Effect of calcium supplementation. *J. Pediatr.* **1986**, *108*, 439–447. [CrossRef]
23. Armand, M.; Hamosh, M.; Mehta, N.R.; Angelus, P.A.; Philpott, J.R.; Henderson, T.R.; Dwyer, N.K.; Lairon, D.; Hamosh, P. Effect of human milk or formula on gastric function and fat digestion in the premature infant. *Pediatr. Res.* **1996**, *40*, 429–437. [CrossRef] [PubMed]
24. Koletzko, B. Human Milk Lipids. *Ann. Nutr. Metab.* **2016**, *69* (Suppl. 2), 28–40. [CrossRef]
25. Lee, H.; Padhi, E.; Hasegawa, Y.; Larke, J.; Parenti, M.; Wang, A.; Hernell, O.; Lonnerdal, B.; Slupsky, C. Compositional Dynamics of the Milk Fat Globule and Its Role in Infant Development. *Front. Pediatr.* **2018**, *6*, 313. [CrossRef] [PubMed]
26. Hernell, O.; Timby, N.; Domellof, M.; Lonnerdal, B. Clinical Benefits of Milk Fat Globule Membranes for Infants and Children. *J. Pediatr.* **2016**, *173*, S60–S65. [CrossRef] [PubMed]
27. Thompson, A.K.; Hindmarsh, J.P.; Haisman, D.; Rades, T.; Singh, H. Comparison of the structure and properties of liposomes prepared from milk fat globule membrane and soy phospholipids. *J. Agric. Food Chem.* **2006**, *54*, 3704–3711. [CrossRef]
28. Zempleni, J.; Aguilar-Lozano, A.; Sadri, M.; Sukreet, S.; Manca, S.; Wu, D.; Zhou, F.; Mutai, E. Biological Activities of Extracellular Vesicles and Their Cargos from Bovine and Human Milk in Humans and Implications for Infants. *J. Nutr.* **2017**, *147*, 3–10. [CrossRef] [PubMed]
29. Gallier, S.; Vocking, K.; Post, J.A.; Van De Heijning, B.; Acton, D.; Van Der Beek, E.M.; Van Baalen, T. A novel infant milk formula concept: Mimicking the human milk fat globule structure. *Colloids Surf. B Biointerfaces* **2015**, *136*, 329–339. [CrossRef]
30. Mathiassen, J.H.; Nejrup, R.G.; Frøkiær, H.; Nilsson, Å.; Ohlsson, L.; Hellgren, L.I. Emulsifying triglycerides with dairy phospholipids instead of soy lecithin modulates gut lipase activity. *Eur. J. Lipid Sci. Technol.* **2015**, *117*, 1522–1539. [CrossRef]
31. Heerup, C.; Ebbesen, M.F.; Geng, X.; Madsen, S.F.; Berthelsen, R.; Müllertz, A. Effects of recombinant human gastric lipase and pancreatin during in vitro pediatric gastro-intestinal digestion. *Food Funct.* **2021**. submitted for publication.
32. Timby, N.; Domellof, E.; Hernell, O.; Lonnerdal, B.; Domellof, M. Neurodevelopment, nutrition, and growth until 12 mo of age in infants fed a low-energy, low-protein formula supplemented with bovine milk fat globule membranes: A randomized controlled trial. *Am. J. Clin. Nutr.* **2014**, *99*, 860–868. [CrossRef] [PubMed]
33. McGrath, J.C.; Drummond, G.B.; McLachlan, E.M.; Kilkenny, C.; Wainwright, C.L. Guidelines for reporting experiments involving animals: The ARRIVE guidelines. *Br. J. Pharm.* **2010**, *160*, 1573–1576. [CrossRef] [PubMed]
34. Sangild, P.T.; Thymann, T.; Schmidt, M.; Stoll, B.; Burrin, D.G.; Buddington, R.K. Invited review: The preterm pig as a model in pediatric gastroenterology. *J. Anim. Sci.* **2013**, *91*, 4713–4729. [CrossRef] [PubMed]
35. Jeppesen, P.B.; Hoy, C.E.; Mortensen, P.B. Differences in essential fatty acid requirements by enteral and parenteral routes of administration in patients with fat malabsorption. *Am. J. Clin. Nutr.* **1999**, *70*, 78–84. [CrossRef]
36. Moreau, H.; Gargouri, Y.; Lecat, D.; Junien, J.L.; Verger, R. Purification, characterization and kinetic properties of the rabbit gastric lipase. *Biochim. Biophys. Acta* **1988**, *960*, 286–293. [CrossRef]
37. Berton, A.; Sebban-Kreuzer, C.; Rouvellac, S.; Lopez, C.; Crenon, I. Individual and combined action of pancreatic lipase and pancreatic lipase-related proteins 1 and 2 on native versus homogenized milk fat globules. *Mol. Nutr. Food Res.* **2009**, *53*, 1592–1602. [CrossRef]
38. Aunsholt, L.; Thymann, T.; Qvist, N.; Sigalet, D.; Husby, S.; Sangild, P.T. Prematurity Reduces Functional Adaptation to Intestinal Resection in Piglets. *JPEN J. Parenter. Enter. Nutr.* **2015**, *39*, 668–676. [CrossRef]
39. Billeaud, C.; Puccio, G.; Saliba, E.; Guillois, B.; Vaysse, C.; Pecquet, S.; Steenhout, P. Safety and tolerance evaluation of milk fat globule membrane-enriched infant formulas: A randomized controlled multicenter non-inferiority trial in healthy term infants. *Clin. Med. Insights Pediatr.* **2014**, *8*, 51–60. [CrossRef]
40. Li, X.; Peng, Y.; Li, Z.; Christensen, B.; Heckmann, A.B.; Stenlund, H.; Lonnerdal, B.; Hernell, O. Feeding Infants Formula With Probiotics or Milk Fat Globule Membrane: A Double-Blind, Randomized Controlled Trial. *Front. Pediatr.* **2019**, *7*, 347. [CrossRef]
41. Timby, N.; Hernell, O.; Vaarala, O.; Melin, M.; Lonnerdal, B.; Domellof, M. Infections in infants fed formula supplemented with bovine milk fat globule membranes. *J. Pediatr. Gastroenterol. Nutr.* **2015**, *60*, 384–389. [CrossRef]
42. Grip, T.; Dyrlund, T.S.; Ahonen, L.; Domellof, M.; Hernell, O.; Hyotylainen, T.; Knip, M.; Lonnerdal, B.; Oresic, M.; Timby, N. Serum, plasma and erythrocyte membrane lipidomes in infants fed formula supplemented with bovine milk fat globule membranes. *Pediatr. Res.* **2018**, *84*, 726–732. [CrossRef]
43. Gurnida, D.A.; Rowan, A.M.; Idjradinata, P.; Muchtadi, D.; Sekarwana, N. Association of complex lipids containing gangliosides with cognitive development of 6-month-old infants. *Early Hum. Dev.* **2012**, *88*, 595–601. [CrossRef] [PubMed]
44. Andersen, A.D.; Sangild, P.T.; Munch, S.L.; van der Beek, E.M.; Renes, I.B.; Ginneken, C.; Greisen, G.O.; Thymann, T. Delayed growth, motor function and learning in preterm pigs during early postnatal life. *Am. J. Physiol.-Regul. Integr. Comp. Physiol.* **2016**, *310*, R481–R492. [CrossRef] [PubMed]

Article

Acute Effects of Cheddar Cheese Consumption on Circulating Amino Acids and Human Skeletal Muscle

Naomi M.M.P. de Hart [1,†], Ziad S. Mahmassani [2,†], Paul T. Reidy [3], Joshua J. Kelley [2], Alec I. McKenzie [4], Jonathan J. Petrocelli [2], Michael J. Bridge [5], Lisa M. Baird [6], Eric D. Bastian [7], Loren S. Ward [8], Michael T. Howard [6] and Micah J. Drummond [1,2,*]

1. Department of Nutrition and Integrative Physiology, University of Utah, 250 S 1850 E, Salt Lake City, UT 84112, USA; Naomi.DeHart@utah.edu
2. Department of Physical Therapy and Athletic Training, University of Utah, 520 Wakara Way, Salt Lake City, UT 84108, USA; Ziad.Mahmassani@health.utah.edu (Z.S.M.); joshua.kelley@utah.edu (J.J.K.); jonathan.petrocelli@utah.edu (J.J.P.)
3. Department of Kinesiology, Nutrition and Health, Miami University, 420 S Oak St., Oxford, OH 45056, USA; reidypt@miamioh.edu
4. Geoge E. Wahlen Department of Veterans Affairs Medical Center, Geriatric Research, Education, and Clinical Center, 500 Foothill Dr., Salt Lake City, UT 84148, USA; alec.mckenzie@utah.edu
5. Cell Imaging Facility, University of Utah, 30 N 2030 E, Salt Lake City, UT 84112, USA; Mike.Bridge@m.cc.utah.edu
6. Department of Human Genetics, 15 N 2030 E, Salt Lake City, UT 84112, USA; lbaird@genetics.utah.edu (L.M.B.); mhoward@genetics.utah.edu (M.T.H.)
7. Dairy West Innovation Partnerships, 195 River Vista Place #306, Twin Falls, ID 83301, USA; ebastian@dairywest.com
8. Glanbia Nutritionals Research, 450 Falls Avenue #255, Twin Falls, ID 83301, USA; LWARD@glanbia.com
* Correspondence: micah.drummond@hsc.utah.edu; Tel.: +1-(801)-585-1310
† Co-first authors.

Abstract: Cheddar cheese is a protein-dense whole food and high in leucine content. However, no information is known about the acute blood amino acid kinetics and protein anabolic effects in skeletal muscle in healthy adults. Therefore, we conducted a crossover study in which men and women ($n = 24$; ~27 years, ~23 kg/m^2) consumed cheese (20 g protein) or an isonitrogenous amount of milk. Blood and skeletal muscle biopsies were taken before and during the post absorptive period following ingestion. We evaluated circulating essential and non-essential amino acids, insulin, and free fatty acids and examined skeletal muscle anabolism by mTORC1 cellular localization, intracellular signaling, and ribosomal profiling. We found that cheese ingestion had a slower yet more sustained branched-chain amino acid circulation appearance over the postprandial period peaking at ~120 min. Cheese also modestly stimulated mTORC1 signaling and increased membrane localization. Using ribosomal profiling we found that, though both milk and cheese stimulated a muscle anabolic program associated with mTORC1 signaling that was more evident with milk, mTORC1 signaling persisted with cheese while also inducing a lower insulinogenic response. We conclude that Cheddar cheese induced a sustained blood amino acid and moderate muscle mTORC1 response yet had a lower glycemic profile compared to milk.

Keywords: dairy; ribo-seq; muscle protein synthesis; anabolism; insulin

1. Introduction

Aminoacidemia from the digestion of protein sources is a major stimulator of skeletal muscle protein anabolism and important for maintenance of muscle mass and overall muscle health. Circulating amino acid kinetics and acute skeletal muscle protein anabolic responses have been extensively evaluated following ingestion of dairy proteins such as casein and whey protein isolate [1–5]. Though these data have provided fundamental information in understanding how muscle responds to protein, it is less generalizable to the community since most dietary protein sources contain a mixed-macronutrient profile, contain many micronutrients within their matrix, and are more complex during digestion.

Recent protein metabolism studies have evaluated blood amino acid kinetics and muscle anabolic responses to protein-enriched, nutrient-complex foods such as beef, egg, and pork [6–14] and as a result, have demonstrated unique amino acid and protein anabolic responses. For example, consumption of 18 g of protein from whole egg after a bout of exercise increased protein synthesis more so than egg whites in spite of similar post absorptive plasma leucine levels [14]. This suggests protein-dense whole foods have utility to promote protein anabolism not simply predicted by the amount of protein or level of aminoacidemia, which is in contrast to what has been observed with isolated protein products [5]. Therefore, there is a continued need to characterize whole food products to identify high quality protein sources that encourage human health.

To our knowledge, no studies have evaluated the amino acid pattern in plasma or muscle anabolic response to cheese ingestion. Cheddar cheese, is a low carbohydrate, high-fat, protein-rich food that is a regular dietary component of the U.S. diet [15]. Cheddar cheese has a well characterized amino acid profile with a high content of leucine (~10%) and is considered low glycemic. Moreover, the protein in Cheddar cheese is partially hydrolyzed due to aging/ripening [16], and therefore is likely to speed up digestion and promote the appearance of amino acids in the circulation [4,17]. Cheddar cheese is also composed of many other underappreciated nutrients within its food matrix [18] that are beneficial for human health and could further enhance protein anabolism.

Therefore, the primary purpose of this study was to characterize the amino acid response following 65 g (20 g protein) of Cheddar cheese, an amount of protein capable of increasing blood amino acid levels from a whole dairy product [7,9,19]. In addition, to gain insight on the protein anabolic processes in skeletal muscle, we evaluated mTORC1 localization and cellular signaling following cheese ingestion, given that mTORC1 intracellular signaling is highly responsive to acute protein intake particularly to sources that are rich in leucine [20,21]. We also complimented mTORC1 signaling with a unique 'omics approach of ribosome profiling [22] to capture key information regarding which mRNAs are translated after cheese ingestion. Finally, to provide context in comparison to a well-described whole food, we conducted a within subject crossover study comparing to an isonitrogenous amount of milk [19]. We hypothesized that a single dose of Cheddar cheese in young male and female adults, equivalent to 20 g of protein, would acutely increase the blood branched-chain amino acids (particularly leucine) and induce a translational program characterized by mTORC1 signaling.

2. Methods

2.1. Subjects

Twenty-four young male ($n = 12$) and female ($n = 12$) subjects participated in this study (Table 1; 27 ± 4 years; BMI 23.1 ± 3.5 kg/m^2). Interested subjects were notified of the study through posted flyers on campus and in areas around the university and were also contacted through the University of Utah PEAK Health and Fitness registry. Subjects were screened (self-report) based on the following exclusion criteria: history of cardiovascular disease, endocrine or metabolic disease (e.g., hypo/hyperthyroidism, diabetes), kidney disease or failure, liver disease, respiratory disease (acute upper respiratory infection, chronic lung disease), stroke with motor disability, use of anticoagulant therapy (e.g., Coumadin, heparin) including aspirin and fish oils within 7 days (d) of the first metabolic experiment, elevated systolic blood pressure > 150 or a diastolic blood pressure > 100, smoking, recent anabolic or corticosteroids use (within 12 weeks of first biopsy), pregnancy, lactose intolerance, and irregular menstruation. Enrolled participants read and signed the informed consent document, which was approved by the University of Utah Institutional Review Board (IRB #110963) and in agreement with the Declaration of Helsinki. This study is registered at clinicaltrials.gov (NCT04660877).

Table 1. Subject Characteristics.

	Pooled	Male	Female
Sample Size (N)	24	12	12
Age (year)	27 ± 4	27 ± 4	26 ± 4
Height (cm)	175 ± 8	181 ± 5 *	169 ± 7
Body Mass (kg)	71 ± 14	80 ± 12 *	63 ± 10
BMI (kg/m^2)	23.1 ± 3.5	24.6 ± 3.8	21.9 ± 2.8
Lean Mass (kg)	57.5 ± 11.9	67.9 ± 8.7 *	48.1 ± 4.9
Fat Mass (kg)	13.5 ± 7	12.5 ± 8	14.9 ± 6
Body Fat (%)	18.7 ± 8	14.8 ± 8	22.8 ± 6
Daily Protein Intake (g/kg/day)	1.32 ± 0.49	1.40 ± 0.61	1.25 ± 0.35
Steps/Day	8798 ± 3444	7364 ± 2845 *	10,122 ± 3514

Mean ± SD, * Different from Female ($p < 0.05$).

2.2. Experimental Design

After enrollment, participants completed baseline testing which included a dietary assessment, body composition and habitual activity levels. Body composition (lean and fat mass) was assessed using a Bod Pod instrument (conducted prior to Metabolic Study #1). Physical activity was tracked for a 7 days period between the Metabolic Study visits. Additionally, a 3 d daily dietary record (ASA24) was recorded before each Metabolic Study visit. The daily dietary record was averaged between all recorded days and reported in Table 1.

Each subject took part in two metabolic studies (Figure 1) with each designed to test the acute blood and muscle response to an ingested amount of either Cheddar cheese or milk matched for protein (Table 2). Approximately, one month after the first experiment (Metabolic Study #1), the participant completed the second experiment (Metabolic Study #2) which was exact in design and at the same time of day as the first study but the participant ingested the alternate food product. Prior to each of the metabolic studies, the participant ate a standardized research meal the night before the study and refrained from intense physical activity for 48 h. The morning of the metabolic studies, the participant arrived at the clinical research center after a ~10 h fast. A catheter was then placed in the participants' arm for blood sampling. Next, the participant underwent a baseline vastus lateralis skeletal muscle biopsy (0 min) using a modified version Bergström muscle biopsy technique [23]. Following the baseline muscle biopsy, the participant consumed either Cheddar cheese (65 g) or milk (370 mL; 2%; Fairlife) each amounting to 20 g of protein. The Cheddar cheese was processed at Glanbia Nutritionals, aged to one month, and frozen into batches distributed monthly by the sponsor as needed. The amino acid profile of low-fat Cheddar cheese and 2% Fairlife milk can be found in Supplemental Table S1. Subsequent muscle biopsies occurred 60 and 180 min on the same thigh after product ingestion which is an ideal timeframe to capture mTORC1 signaling and mRNA translational events following protein-enriched nutrient ingestion [24,25]. Blood sampling occurred before and periodically after ingestion of the products (up to 300 min). Blood samples were taken every 20 min during the first 3 h and then every 30 min for the last 2 h. Therefore, there were a total of 14 blood draws and 3 muscle biopsies for each Metabolic Study visit. The starting thigh for muscle biopsies for the first Metabolic Study was randomized for each subject and balanced with the second Metabolic Study (left leg then right or right leg then left). Muscle samples were frozen in liquid nitrogen (for immunoblotting and ribosomal profiling) or prepared in O.C.T. (Optimal Cutting Temperature) and frozen in liquid nitrogen-cooled isopentane for the immunohistochemical assessment.

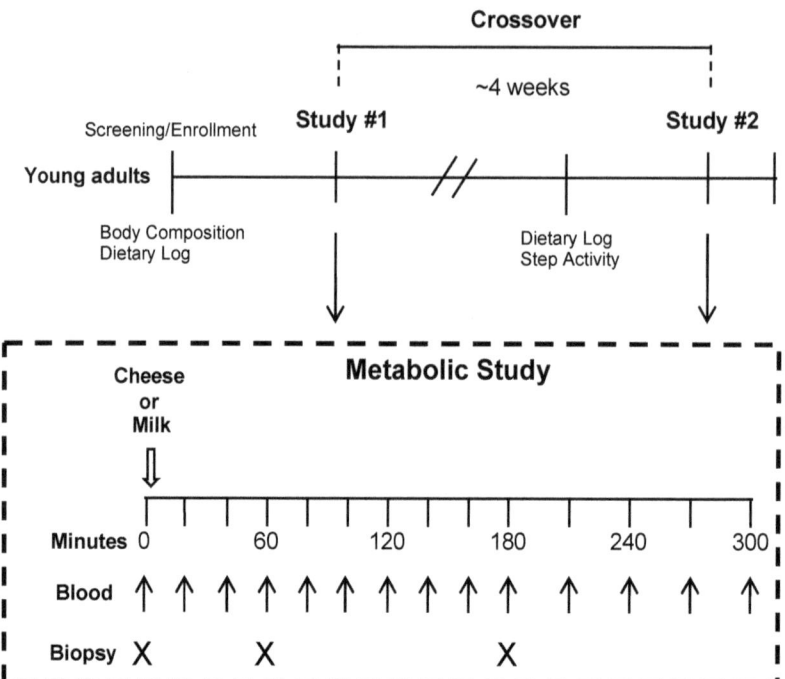

Figure 1. Overview of the crossover study experimental design.

Table 2. Nutrient content of experimental products.

	Cheddar Cheese	2% Fairlife Milk
Amount	65 g	370 mL
Protein (g)	20	20
Leucine (g)	1.97	1.98
Fat (g)	10	7.5
Carbohydrates (g)	0	9
Calories (kcal)	170	183

2.3. Blood Analyses

Blood samples were collected in EDTA (Ethylenediaminetetraacetic) vacutainer collection tubes and immediately placed on ice. Samples were centrifuged (2500 rpm, 10 min) and plasma was collected and frozen at −80 °C until later analysis. Plasma was processed for essential and non-essential amino acids using the EZ:Faast Amino Acid Kit (Phenomenex; Cat #KG0-7165) and analyzed using GCMS analysis in collaboration with the institution's Metabolomics Core. Essential amino acids included detection of leucine, isoleucine, valine, threonine, methionine, phenylalanine, lysine, histidine, and tryptophan. Non-essential amino acids included detection of alanine, glycine, serine, proline, asparagine, glutamate, glutamine, and tyrosine. Samples were also immediately assessed for glucose (YSI) at the time of the study and later assessed for insulin (Human Insulin ELISA, Millipore Sigma, Burlington, MA, USA; EZHI-14K) and non-esterified fatty acids (NEFA-HR; Wako Chemicals, Richmond, VA, USA) in replicate using commercially available kits. Insulin and free fatty acids were determined at select time points (baseline, 20, 40, 80, 140, 210, and 300 min).

2.4. Skeletal Muscle Immunoblotting

Approximately 30 mg of tissue at each biopsy time point for Cheddar cheese and milk interventions was homogenized 1:10 (wt/vol) using a glass tube and mechanically-driven pestle grinder in an ice-cold buffer containing 50 mM Tris (pH 7.5), 250 mM mannitol, 40 mM NaF, 5 mM pyrophosphate, 1 mM EDTA, 1 mM EGTA, and 1% Triton X-100 with a protease inhibitor cocktail. Homogenates were centrifuged for 10 min at 4 °C. After centrifugation, the supernatant was collected and the protein concentration was determined using a modified Bradford protein assay and measured by a spectrophotometer (EPOCH; BioTek, Winooski, VT, USA).

Thirty micrograms of protein from muscle homogenate was separated via polyacrylamide gel electrophoresis, transferred onto a polyvinylidene difluoride membrane (PVDF), and incubated with primary and secondary antibodies. PVDF Membranes were imaged on a ChemiDoc XRS (Bio-Rad, Hercules, CA, USA) and quantified with Image lab software (Bio-Rad). The primary antibodies were purchased from Cell Signaling Technology and were the following: phospho-S6K1, Thr389, 1:1000, #9205; phospho-ribosomal protein S6, RPS6, Ser240/244, 1:1000, #2215; phospho-AS160, Ser588, 1:1000, #8730; phospho-GSK-3β, Ser9, 1:1000, #9336; phospho-Akt, Ser473, 1:1000, #9271. Secondary antibody (HRP Anti-Rabbit, #SC2004, 1:2000) was purchased from Santa Cruz Biotechnology. Phosphorylation of these proteins were normalized to Ponceau-S staining and reported as fold change from baseline.

2.5. Skeletal Muscle Immunohistochemistry

Muscle was sectioned into 8 μm cross-sections, mounted on slides in −25 °C, then left to air-dry overnight, and stored at −20 °C. Immunofluorescent staining was used to detect mTORC1 (Cell Signaling Technology, #2983, 1:100), the lysosomes (LAMP2: Abcam, #ab25631, 1:100), and the membrane (WGA: Fisher Scientific, #W32466, 1:50) as demonstrated by others [26–29]. Briefly, tissue was fixed in acetone (10 min), and the following blocking steps were performed: (1) endogenous peroxidases: 3% H_2O_2 for 7 min, (2) Non-Specific Binding Sites: 5% goat serum, Vector Labs #S-1000 with 0.3% Triton-X for 1 h, and (3) Avidin/Biotin: Vector Labs #SP-2001 according to manufacturer's instructions. WGA was added (5 min), and mTOR and LAMP2 were incubated on the slide overnight. Secondary antibody for LAMP2 was performed using Alexa Fluor 488 Tyramide SuperBoost (Invitrogen, #B40932, according to manufacturer's instructions), while secondary for mTOR was on Cy3 (Jackson ImmunoResearch, #711-165-152; 1:500) for 1 h. Finally, slides were mounted, cover slipped (Vectashield with DAPI, Vector Labs, #H-1200), and stored in the fridge until imaged (within 1 month of staining).

Images were taken using a Leica SP8 White Light laser confocal microscope equipped with automated stage, and Nikon NIS-Elements multi-platform acquisition software. At least 9 images (16 bit) were taken at 40X/1.3 magnification with oil immersion, with each image capturing ~5 muscle fibers per image in high detail at each time point, analyzing a total of ~45 muscle fibers per subject per time point, for each product consumed. When looking at events detected above threshold (set with help of combinations of positive and negative controls) of mTOR and LAMP2, anything not within 80% of the average was not used. The number of objects/events per channel times the average area covered by each object gave us the total area per channel. As previously described [26], Mander's overlap coefficient of colocalization was employed (k1 for mTOR/LAMP2; k2 for mTOR/WGA) to quantify the cellular overlap of these proteins, and this was performed in NIS-Elements for mTOR/LAMP2 and mTOR/WGA.

2.6. Ribosomal Profiling

Muscle samples at each time point (0, 60, 180 min) from Cheddar cheese and milk studies were assessed from a subset of subjects (4 subjects, 2 M, 2 F). Traditional RNA-Seq captures total mRNA abundance within a tissue sample, while the emerging technique of Ribo-Seq allows the capture of ribosome protected fragments (RPF) measuring translational activity in a transcript-specific manner [30,31]. Polysome complexes were isolated, and unprotected mRNA digested with RNase I, and the ribosome protected mRNA footprints were analyzed by RNA-Sequencing methods as previously described by our group [22] with the exception that rRNA was removed from the RPF samples using the NEBNext rRNA Depletion kit and libraries were size selected by polyacrylamide gel electrophoresis on 6% native gels. Libraries were then sequenced on an Illumina Novaseq 6000 instrument. Raw sequence data can be obtained from the National Center for Biotechnology Information Gene Expression Omnibus repository entry GSE163279.

Uniquely mapping sequences were identified by alignments using bowtie to Reference Sequence database (RefSeq) mRNA entries obtained from the University of California, Santa Cruz browser (Hg38 human genome reference assembly) in which all mRNAs derived from the same gene were reduced to a single entry corresponding to the longest isoform. Normalization factors based on the trimmed mean of M-values were determined by using the calcNormFactors function of the Bioconductor package edgeR [32]. Dispersion estimates were obtained prior to likelihood ratio tests (glmFit and glmLRT functions of edgeR) to determine significance of the $\log_2$ fold change in RPFs or RNA for all transcripts with ≥ 1 count/million in all samples. Differences were considered significant if the false discovery rate was ≤ 0.05. Pearson's product-moment correlation coefficients were calculated.

Ingenuity Pathway Analysis was performed to determine significantly altered pathways informed by the translation changes at each time point for the two respective protein sources. mTOR pathway volcano plots used all of the molecules within the top 3 pathways ('EIF2 Signaling', 'Regulation of eIF4 and p70S6K Signaling', 'mTOR Signaling') in either cheese or milk for comparison, yielding presentation of the translation for 202 total transcripts, at 3 contrasts (60 vs. 0 min translation f.c.; 180 vs. 0 min translation f.c. and 180 vs. 60 min translation f.c.).

2.7. Statistical Analyses

Subject characteristics were compared between males and females using a t-test. Because there were no notable differences between males and females in major outcomes (i.e., blood amino acids), subjects were pooled and all comparisons (Plasma NEFA, Insulin, Amino Acids, Immunoblotting, and IHC colocalization) were analyzed using a 2-Way ANOVA with repeated measures for product and time. When appropriate after a significant interaction was detected, Sidak's multiple comparisons post-hoc test was used to identify differences from baseline or between protein products at a given time point. For all analyses, differences were considered statistically significant at $p < 0.05$. All statistical calculations and graphs were completed using GraphPad Prism (v8).

3. Results

3.1. Subject Characteristics

A total of 24 young adult participants completed both trials of this study. This was made up of 12 males and 12 females (Table 1). As expected, men had greater height, body weight, and had more lean mass than females ($p < 0.05$). The men also had less daily step activity than the females ($p < 0.05$). There were no differences between the sexes in age, BMI, fat mass, body fat % or daily protein intake.

3.2. Blood Insulin, Glucose and Non-Esterified Fatty Acids

Milk induced a rapid spike in insulin 20 min after ingestion (2-Way ANOVA: Time*Product Interaction, $p < 0.0001$; Sidak's multiple comparisons test, Milk different from baseline and from cheese at 20 and 40 min, $p < 0.0001$) while cheese consumption did not significantly change insulin at any time point (Figure 2A). Blood glucose decreased at 60 min following ingestion of either product, but this decrease occurred earlier for milk (40 min; Time*Product Interaction, $p < 0.0001$) and was significantly lower than cheese (Figure 2B). Similarly, NEFA levels decreased after ingestion of either Cheddar cheese or milk, (Time*Product Interaction, $p < 0.0001$), but this response was further decreased for milk compared to Cheddar cheese (Sidak's multiple comparisons test, Cheese vs. milk 40 min post, $p = 0.023$). Additionally, NEFA levels were significantly elevated in response to both protein sources by 300 min, in comparison to baseline NEFA values (Sidak's multiple comparisons test, Cheese: $p = 0.003$; Milk: $p = 0.011$) (Figure 2C).

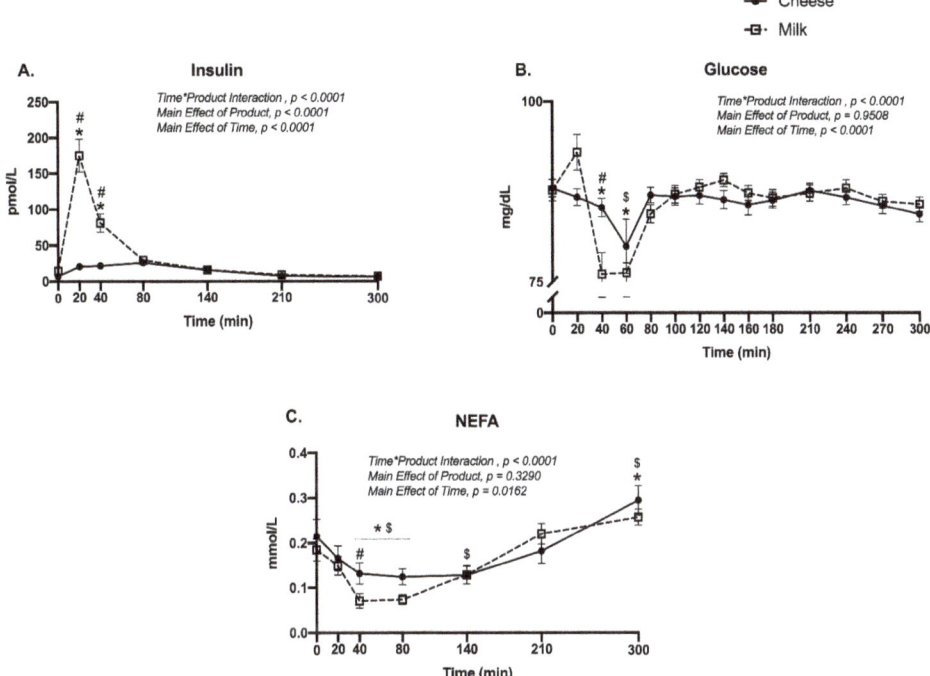

Figure 2. Plasma (A) insulin (pmol/L), (B) glucose (mg/dL), and (C) non-esterified fatty acids (mmol/L) in the fasted state (0 min) and over a 300 min time period following the ingestion of either cheese (solid line) or milk (dotted line) in men and women ($n = 24$). Different from baseline (0 min) for milk (*) and cheese ($), $p < 0.05$. #, Different between groups at the specific time point, $p < 0.05$.

3.3. Plasma Branched-Chain, Essential and Non-Essential Amino Acids

Total branched-chain amino acids (BCAAs) increased with different kinetics in response to ingestion of the respective products (Time*Product Interaction, $p < 0.0001$) (Figure 3A). After milk, BCAAs returned to baseline by 240 min post, and cheese maintained higher BCAA levels out to 270 min. Milk induced significantly higher BCAA levels than cheese from 20 to 60 min post ingestion, and decreased gradually towards baseline as cheese induced significantly higher BCAA in plasma than milk between 120 and 210 min (Sidak's, $p < 0.05$). Plasma leucine exhibited a similar response as total BCAAs (Time*Product Interaction, $p < 0.0001$) (Figure 3B), with both products increasing leucine levels out to 210 min and with cheese elevating leucine levels slightly longer to 240 min (Sidak's, $p < 0.05$). The leucine response occurred to a greater magnitude for milk from 20 to 60 min while cheese induced higher leucine levels (vs. milk) from 120 to 180 min. Plasma isoleucine (Time*Product Interaction, $p < 0.0001$) (Figure 3C) increased out to 160 min for milk while cheese increased isoleucine levels out to 240 min. Milk had a greater isoleucine response compared to cheese from 20 to 60 min while cheese had a greater plasma isoleucine response than milk from 120 to 210 min (Sidak's, $p < 0.05$). Plasma valine (Time*Product Interaction, $p < 0.0001$) (Figure 3D) increased over the 300 min time course for cheese and out to 270 min for milk. This response was greater for milk at 20–60 min while the cheese induced a greater valine level than milk from 120 to 210 min (Sidak's, $p < 0.05$). Total essential amino acids (EAA) increased above baseline for milk out to 180 min while cheese increased total EAA out to 300 min (Time*Product Interaction, $p < 0.0001$) (Figure 3E). Plasma EAA were higher for milk from 20 to 60 min (compared to cheese) while EAA were higher for cheese from 120 to 210 min (vs. milk). Non-Essential amino acids (NEAA) (Figure 3F) increased above baseline for milk from 20 to 100 min while NEAA were elevated above baseline from 40 to 180 min for cheese (Time*Product Interaction, $p < 0.0001$). Milk induced a greater NEAA response at 20–60 min while cheese induced a greater response than milk from 120 to 300 min (except at 270 min). Despite differences in amino acid kinetics between the products, the area under the curve over 5 h for total BCAA, leucine, isoleucine, valine, total EAA, and total NEAA were not different between cheese and milk products (Figure 3A–F).

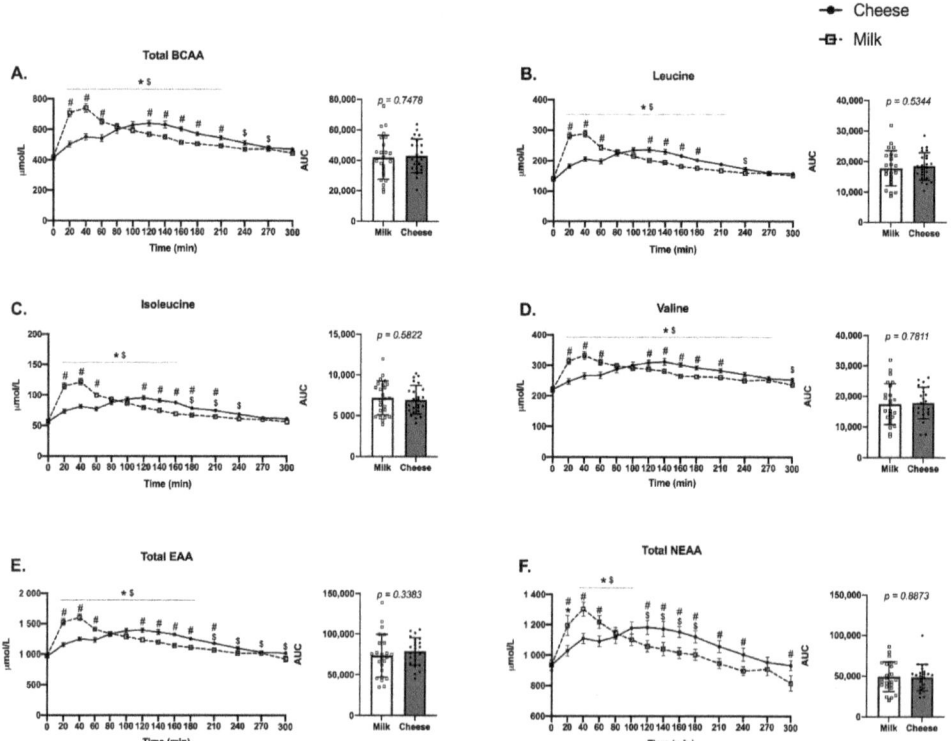

Figure 3. Plasma levels (μmol/L) of (**A**) branched-chain amino acids (Total BCAA), (**B**) leucine, (**C**) isoleucine, (**D**) valine, (**E**) essential amino acids (Total EAA), and (**F**) non-essential amino acids (Total NEAA) in the fasted state (0 min) and over a 300 min time period following the ingestion of either cheese (solid line) or milk (dotted line) in men and women ($n = 24$). Different from baseline (0 min) for milk (*) and cheese ($), $p < 0.05$. #, Different between groups at the specific time point, mboxemphp < 0.05. Units are in micromolar (μM). Note: Total EAA (E) does not include the BCAAs.

3.4. Muscle mTORC1 Signaling and Localization

Phosphorylated p70S6K(Thr389) (Time*Product Interaction, $p = 0.0005$) and phosphorylated rpS6(Ser240/244) (Time*Product Interaction, $p < 0.0001$) increased above baseline and were increased to a greater extent for milk at 60 min post ingestion compared to cheese (Sidak's multiple comparisons test, $p < 0.0001$ for p70S6K and rpS6K) (Figure 4A,B). Phosphorylated Akt(Ser473) was significantly elevated 60 min post ingestion as a result of cheese or milk with no difference between cheese and milk (2-Way ANOVA: Main Effect of Time, $p = 0.0097$) (Figure 4C). There were no significant differences detected for phosphorylated AS160(Ser588) or phosphorylated GSK-3β(Ser9) (Figure 4D,E). Figure 4F is representative immunoblotting images for the phosphorylated proteins.

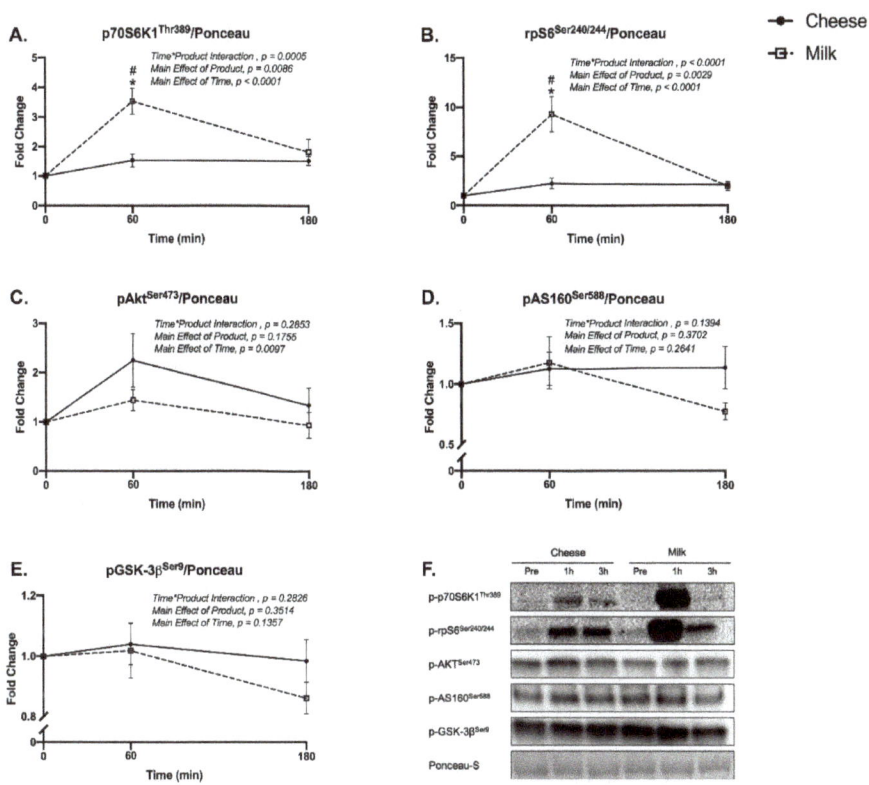

Figure 4. Skeletal muscle protein expression using immunoblotting for (A) p70S6K1(Thr389), (B) rpS6(Ser240/244), (C) Akt(Ser473), (D) AS160(Ser588), and (E) GSK-3β(Ser9) in the fasted state (0 min) and at 60 and 180 min following the ingestion of either cheese (solid line) or milk (dotted line) in men and women. Panels (A,B) are data for $n = 24$ while for (C–E) only $n = 8$ (4 M, 4 F) were analyzed. Panel (F) are representative images of immunoblotting. Phosphorylated protein levels were normalized to Ponceau-S. Different from baseline (0 min) for milk (*), $p < 0.05$. #, Different between groups at the specific time point, $p < 0.05$.

Using immunohistochemistry to fluorescently label and measure the spatial distribution of mTOR, we did not detect changes to the colocalization of mTOR with the lysosomal protein, LAMP2 (Figure 5A). However, mTOR colocalization with the sarcolemma (WGA) was different between groups at 60 and 180 min and increased at 180 min only after cheese ingestion (Time*Product Interaction, $p = 0.042$; Sidak's multiple comparisons test, $p = 0.003$) (Figure 5B). Representative images for DAPI, WGA, mTOR, LAMP2 and the overlay are found in Figure 5C.

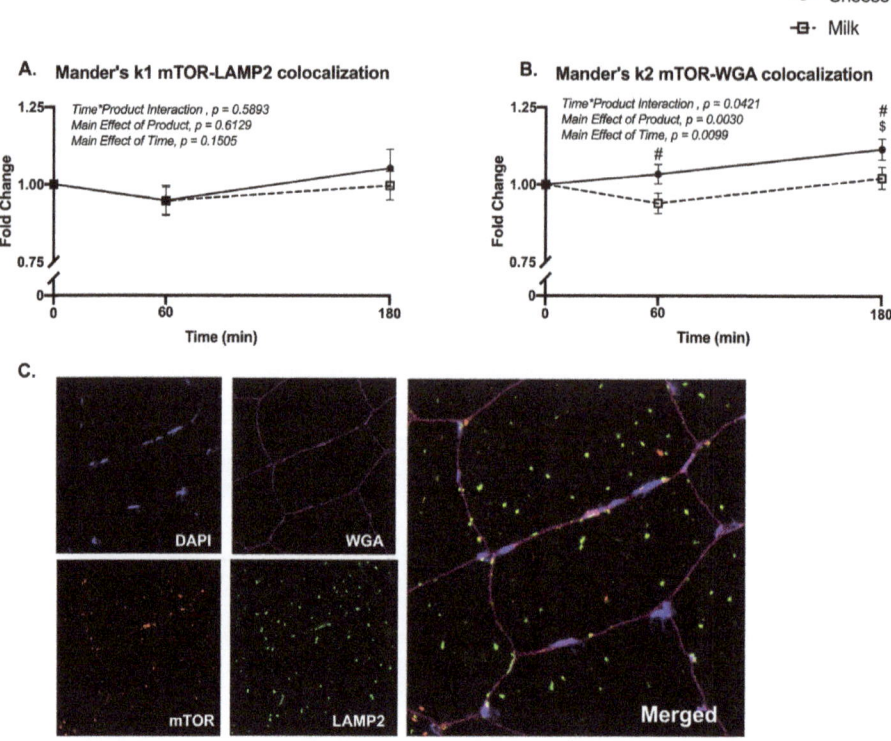

Figure 5. Skeletal muscle mTOR colocalization using immunohistochemistry. Panel (**A**) represents Mander's k1 mTOR-LAMP2 colocalization and (**B**) Mander's k2 mTOR-WGA colocalization at baseline (0 min) and at 60 and 180 min following the ingestion of either cheese (solid line) or milk (dotted line) in men and women ($n = 24$). Panel (**C**) are representative images using immunohistochemistry. Different from baseline (0 min) for cheese ($), $p < 0.05$. #, Different between groups at the specific time point, $p < 0.05$.

3.5. Ribosomal Profiling

A subset of subjects' muscle samples ($n = 4$) was used for ribosomal profiling. Ribosomal profiling captures ribosome protected mRNA fragments to measure active translation of specific transcripts using RNA sequencing libraries. Both cheese and milk altered the same top 3 Canonical Pathways related to mTORC1 signaling (IPA: EIF2 Signaling, Regulation of eIF4 and p70S6K Signaling, mTOR Signaling) (Figure 6A) at both 60 and 180 min, while only milk activated glucose metabolism-related pathways (Glycolysis I, Gluconeogenesis I) 60 min post ingestion. Next, we created a volcano plot for the significantly altered transcripts from within the top 3 Canonical Pathways for cheese and milk respectively, representative of all translation changes under control of mTORC1 signaling. As a result, we demonstrated a significant and dramatic milk-induced (in comparison to cheese) translational response from 0 to 60 min for these mTORC1 mediated molecules (Figure 6B). This response for milk was reduced at 0–180 min after ingestion while cheese-induced translation of mTORC1 molecules was maintained at similar levels as was observed at 60 min (Figure 6C). Moreover, translation changes across the 60–180 min time period (Figure 6D) highlight the observation that stimulation of mTORC1 pathway is reduced at 180 min after milk ingestion but persists with cheese.

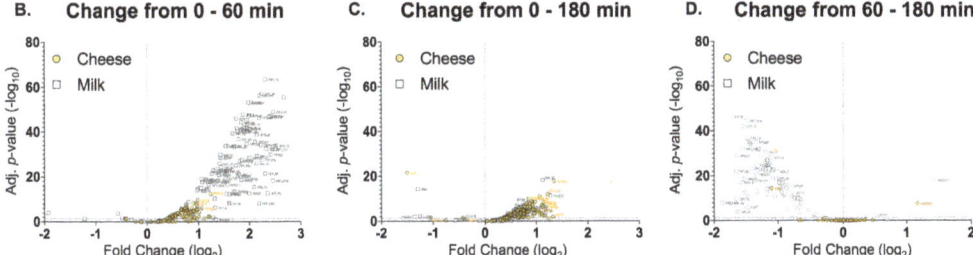

Figure 6. Skeletal muscle analysis of translated mRNAs assessed using ribosomal sequencing before and after cheese or milk ingestion in men and women (n = 4; 2 M, 2 F). (**A**) Canonical pathways identified by ingenuity pathway analysis for cheese and milk at 60 and 180 min and (**B**) Volcano plot of the change in translated mRNAs for cheese (orange squares) or milk (gray squares) from 0 to 60 min, (**C**) 0 to 180 min, and (**D**) 60 to 180 min post ingestion.

4. Discussion

Our current understanding of amino acid kinetics and subsequent skeletal muscle anabolism following protein intake has been informed by isolated protein sources (and often in liquid form) such as whey [33], casein [34], soy [5], and leucine-enriched EAAs [35]. Recently, the study of solid protein-enriched whole foods, of which the food matrix can greatly alter protein digestion and absorption kinetics and the subsequent muscle anabolic signature, is a valuable next step in studying the impact of dietary interventions on muscle health and disease [36]. The purpose of our study was to examine the response to 20 g of protein from Cheddar cheese on plasma amino acids, free fatty acids, insulin, and glucose and the subsequent skeletal muscle mTORC1 signaling and mRNA translational response. To better contextualize the results of Cheddar cheese ingestion with what is known in the field, we utilized a crossover design with comparison to milk, a highly studied protein source with a well-characterized absorption profile and muscle anabolic response [19,37]. The results from this study indicate that Cheddar cheese had a slow, yet persistent amino acid circulation appearance and subsequent skeletal muscle mTORC1 signaling and mRNA translation response when compared with the quick absorption and potent but short-lived mTORC1 stimulation induced by milk. At the studied dosage, Cheddar cheese did not induce a plasma insulinogenic or muscle glycemic response, a known effect of milk [19,38], suggesting Cheddar cheese may be an interesting food choice for dietary strategies geared to promote muscle protein anabolism yet requiring strict glycemic control.

The primary finding of this study was that consumption of Cheddar cheese (65 g) amounting to 20 g of protein promoted a delayed, yet sustained plasma amino acid concentration over 5 h, compared with the acute and potent appearance of circulating amino acids induced by milk proteins. Even though the amounts of protein and leucine were similar between the products, milk resulted in a more rapid and robust amino acid response likely driven by the whey protein component (20% whey, 80% casein in bovine milk vs. 100% casein in Cheddar cheese). Nonetheless, we found it interesting that there was no difference in the total circulating amino acids across the entire 5 h time course between cheese and milk. Thus, although the solid food matrix of cheese and protein composition may slow the digestion and absorption of protein and, subsequently, amino acid release into the circulation, when matched for protein, cheese and milk have similar total plasma amino acid availability. Casein hydrolysate, the form present in Cheddar cheese, has shown to result in a greater appearance in circulating levels of leucine compared with intact casein [4,17,39]. Though difficult to compare to a liquid casein beverage, the plasma leucine appearance data following 20 g Cheddar cheese protein ingestion demonstrated a slower plasma leucine appearance rate and magnitude compared to a similar amount of isolated casein hydrolysate [4] suggesting that the complex matrix of cheese may delay the release of amino acids into circulation [40]. It is currently unknown if a longer aged Cheddar cheese may speed the circulating appearance of amino acids. However, when compared with other solid, protein-dense foods, such as pork [6], cooked egg [14], and steak [13], Cheddar cheese aged to one month produced a similar plasma appearance, magnitude, and sustained amino acid availability response.

We also measured muscle mTORC1 activation using three different approaches with the cumulative result of these assessments demonstrating that anabolic signaling tracked closely with circulating amino acids for each product and with milk demonstrating a more robust mTORC1 signaling response early after intake (1 h). This is logical since essential amino acids, especially leucine, along with insulin, which also peaked prior to 1 h, are stimulators of mTORC1-mediated protein synthesis [41,42]. It is likely that the insulin response from milk, combined with the quickly absorbed leucine, synergized to enhance mTORC1 signaling as noted by the magnitude of p70S6K and rpS6 activation [42]. It is well known that anabolic cues such as insulin, mechanical stimulation, and amino acid ingestion stimulate mTORC1 and its downstream effectors (e.g., S6K1) to enhance translation initiation [43]. While our measurement of mTORC1 signaling was limited to a 3 h time course (based on other protein-dense whole food studies [6,13,14]), we found it noteworthy that mTORC1 activation following Cheddar cheese ingestion persisted at 3 h (and possibly beyond) in accordance with the plasma amino acid appearance and as supported by the mTORC1 localization data and the ribosomal profiling of translated mRNAs under control of mTORC1 signaling. It is unclear how a sustained circulation of amino acids following Cheddar cheese intake may impact muscle protein accretion. Whole foods that are slow digesting (in comparison to commonly studied dairy protein drinks) may have utility in sustaining the free amino acid pool so that they have a longer window to synergize with other anabolic cues such as exercise, or by offsetting protein breakdown to enhance net protein balance when combined with the acute stimulus of a faster digesting protein source [3,44]. For example, drinking a small glass of milk with cheese, may result in a greater net protein balance over several hours in comparison to a bolus of milk alone, because of the ceiling for acute anabolic activation and subsequent oxidation of excess amino acids (coined the 'muscle-full effect') thereby limiting the anabolic benefit of the beverage [38,44,45]. There is a similar underlying premise behind the recommendation of ingesting casein (a major component of cheese) prior to bedtime to enhance exercise adaptations [46–48].

Another interesting observation about the acute response to Cheddar cheese intake (in contrast to milk) in this study, was that cheese did not observably increase circulating insulin or translation of muscle mRNAs related to glycolytic pathways at any time point we measured after ingestion. Therefore, if a dietary intervention requires strict glycemic control, such as for individuals with diabetes [49] or requires adhering to a ketogenic diet [50], cheese may be a valuable protein food source to keep on the menu. The two most likely reasons for why cheese and milk ingestion had different circulating insulin responses, are that (a) milk contains higher levels of carbohydrates and (b) milk induced an early spike in circulating serum leucine (compared with cheese), which stimulates endogenous insulin release [51]. In addition to being less glycemic, the general public's health opinion of cheese should be re-examined since regular consumption of cheese does not appear to influence LDL or HDL levels despite the characteristically high fat content [52]. Though Cheddar cheese does incorporate a significant portion of its calories from fat, fat does not appear to influence muscle protein anabolism [53], and may even synergize with protein to promote a greater anabolic response [14,36]. However, because of the extra calories associated with fat as compared to other macronutrients, individuals who must restrict their calories may benefit from reduced fat cheese.

5. Conclusions

In summary, Cheddar cheese provided a slow and sustained appearance of circulating amino acids and subsequent activation of mTORC1 signaling when compared to milk matched for protein (and leucine) content. Also, Cheddar cheese at the amount consumed in this study did not noticeably increase circulating insulin or induce a muscle glycemic response in contrast with milk. Overall, low fat Cheddar cheese should be considered as a protein-dense food choice given its high leucine content, ability to sustain amino acid levels and promote protein anabolism and, especially, considering its low glycemic properties. Future studies are needed to examine muscle protein accretion in response to daily Cheddar cheese ingestion when combined with habitual exercise.

Supplementary Materials: The following are available online at https://www.mdpi.com/2072-6643/13/2/614/s1, Table S1: Amino acid composition of Cheddar cheese and milk.

Author Contributions: Conceptualization: E.D.B., L.S.W., M.J.D.; Methodology: M.J.D., P.T.R., M.J.B., E.D.B., L.S.W., M.T.H.; Clinical Execution and Data Collection: N.M.M.P.d.H., Z.S.M., P.T.R., J.J.K., A.I.M., J.J.P., L.M.B.; Writing: N.M.M.P.d.H., Z.S.M., M.J.D. All authors have read and agreed to the published version of the manuscript.

Funding: Funding for this study was provided by BUILD Dairy and Glanbia Nutritionals, partial clinical support by the National Center for Advancing Translational Sciences of the National Institutes of Health (UL1TR002538), and postdoctoral support from the Ruth L. Kirschstein National Research Service Award NIH 1T32HL139451.

Institutional Review Board Statement: Enrolled participants read and signed the informed consent document, which was approved by the University of Utah Institutional Review Board (IRB #110963) and in agreement with the Declaration of Helsinki.

Informed Consent Statement: Informed consent was obtained from all subjects involved in the study.

Data Availability Statement: Raw sequence data can be obtained from the National Center for Biotechnology Information Gene Expression Omnibus repository entry GSE163279.

Acknowledgments: We would like to thank J. Alan Maschek, University of Utah Metabolomics, Proteomics and Mass Spectrometry Cores, 15 N Medical Drive East, UT 84112, USA, for his contributions. We would also like to thank the clinical services core for their assistance with the blood sampling and muscle biopsies.

Conflicts of Interest: E.D.B. and L.S.W. are employed by Dairy West and Glanbia Nutritionals, respectively.

References

1. Atherton, P.J.; Etheridge, T.; Watt, P.W.; Wilkinson, D.; Selby, A.; Rankin, D.; Smith, K.; Rennie, M.J. Muscle full effect after oral protein: Time-dependent concordance and discordance between human muscle protein synthesis and mTORC1 signaling. *Am. J. Clin. Nutr.* **2010**, *92*, 1080–1088. [CrossRef]

2. Burd, N.A.; Yang, Y.; Moore, D.R.; Tang, J.E.; Tarnopolsky, M.A.; Phillips, S.M. Greater stimulation of myofibrillar protein synthesis with ingestion of whey protein isolate v. micellar casein at rest and after resistance exercise in elderly men. *Br. J. Nutr.* **2012**, *108*, 958–962. [CrossRef] [PubMed]
3. Churchward-Venne, T.A.; Breen, L.; Di Donato, D.M.; Hector, A.J.; Mitchell, C.J.; Moore, D.R.; Stellingwerff, T.; Breuille, D.; A Offord, E.; Baker, S.K.; et al. Leucine supplementation of a low-protein mixed macronutrient beverage enhances myofibrillar protein synthesis in young men: A double-blind, randomized trial. *Am. J. Clin. Nutr.* **2013**, *99*, 276–286. [CrossRef]
4. Pennings, B.; Boirie, Y.; Senden, J.M.G.; Gijsen, A.P.; Kuipers, H.; Van Loon, L.J.C. Whey protein stimulates postprandial muscle protein accretion more effectively than do casein and casein hydrolysate in older men. *Am. J. Clin. Nutr.* **2011**, *93*, 997–1005. [CrossRef]
5. Tang, J.E.; Moore, D.R.; Kujbida, G.W.; Tarnopolsky, M.A.; Phillips, S.M. Ingestion of whey hydrolysate, casein, or soy protein isolate: Effects on mixed muscle protein synthesis at rest and following resistance exercise in young men. *J. Appl. Physiol.* **2009**, *107*, 987–992. [CrossRef] [PubMed]
6. Beals, J.W.; Sukiennik, R.A.; Nallabelli, J.; Emmons, R.S.; Van Vliet, S.; Young, J.R.; Ulanov, A.V.; Li, Z.; Paluska, S.A.; De Lisio, M.; et al. Anabolic sensitivity of postprandial muscle protein synthesis to the ingestion of a protein-dense food is reduced in overweight and obese young adults. *Am. J. Clin. Nutr.* **2016**, *104*, 1014–1022. [CrossRef] [PubMed]
7. Burd, N.A.; Gorissen, S.H.; Van Vliet, S.; Snijders, T.; Van Loon, L.J. Differences in postprandial protein handling after beef compared with milk ingestion during postexercise recovery: A randomized controlled trial. *Am. J. Clin. Nutr.* **2015**, *102*, 828–836. [CrossRef] [PubMed]
8. Moore, D.R.; Robinson, M.J.; Fry, J.L.; Tang, J.E.; I Glover, E.; Wilkinson, S.B.; Prior, T.; A Tarnopolsky, M.; Phillips, S.M. Ingested protein dose response of muscle and albumin protein synthesis after resistance exercise in young men. *Am. J. Clin. Nutr.* **2008**, *89*, 161–168. [CrossRef]
9. Burke, L.M.; Winter, J.A.; Cameron-Smith, D.; Enslen, M.; Farnfield, M.; Decombaz, J. Effect of intake of different dietary protein sources on plasma amino Acid profiles at rest and after exercise. *Int. J. Sport Nutr. Exerc. Metab.* **2012**, *22*, 452–462. [CrossRef]
10. Beals, J.W.; Skinner, S.K.; McKenna, C.F.; Poozhikunnel, E.G.; Farooqi, S.A.; Van Vliet, S.; Martinez, I.G.; Ulanov, A.V.; Li, Z.; Paluska, S.A.; et al. Altered anabolic signalling and reduced stimulation of myofibrillar protein synthesis after feeding and resistance exercise in people with obesity. *J. Physiol.* **2018**, *596*, 5119–5133. [CrossRef] [PubMed]
11. Beals, J.W.; MacKenzie, R.W.; Van Vliet, S.; Skinner, S.K.; Pagni, B.A.; Niemiro, G.M.; Ulanov, A.V.; Li, Z.; Dilger, A.C.; Paluska, S.A.; et al. Protein-Rich Food Ingestion Stimulates Mitochondrial Protein Synthesis in Sedentary Young Adults of Different BMIs. *J. Clin. Endocrinol. Metab.* **2017**, *102*, 3415–3424. [CrossRef] [PubMed]
12. Robinson, M.J.; Burd, N.A.; Breen, L.; Rerecich, T.; Yang, Y.; Hector, A.J.; Baker, S.K.; Phillips, S.M. Dose-dependent responses of myofibrillar protein synthesis with beef ingestion are enhanced with resistance exercise in middle-aged men. *Appl. Physiol. Nutr. Metab.* **2013**, *38*, 120–125. [CrossRef]
13. Pennings, B.; Groen, B.B.L.; Van Dijk, J.-W.; De Lange, A.; Kiskini, A.; Kuklinski, M.; Senden, J.M.G.; Van Loon, L.J.C. Minced beef is more rapidly digested and absorbed than beef steak, resulting in greater postprandial protein retention in older men. *Am. J. Clin. Nutr.* **2013**, *98*, 121–128. [CrossRef] [PubMed]
14. Van Vliet, S.; Shy, E.L.; Sawan, S.A.; Beals, J.W.; West, D.W.; Skinner, S.K.; Ulanov, A.V.; Li, Z.; Paluska, S.A.; Parsons, C.M.; et al. Consumption of whole eggs promotes greater stimulation of postexercise muscle protein synthesis than consumption of isonitrogenous amounts of egg whites in young men. *Am. J. Clin. Nutr.* **2017**, *106*, 1401–1412. [CrossRef]
15. Aleman-Mateo, H.; Carreon, V.R.; Macias, L.; Astiazaran-Garcia, H.; Gallegos-Aguilar, A.C.; Enriquez, J.R. Nutrient-rich dairy proteins improve appendicular skeletal muscle mass and physical performance, and attenuate the loss of muscle strength in older men and women subjects: A single-blind randomized clinical trial. *Clin. Interv. Aging* **2014**, *9*, 1517–1525. [CrossRef]
16. vens, K.O.; Baumert, J.L.; Hutkins, R.L.; Taylor, S.L. Effect of proteolysis during Cheddar cheese aging on the detection of milk protein residues by ELISA. *J. Dairy Sci.* **2017**, *100*, 1629–1639. [CrossRef] [PubMed]
17. Koopman, R.; Crombach, N.; Gijsen, A.P.; Walrand, S.; Fauquant, J.; Kies, A.K.; Lemosquet, S.; Saris, W.H.M.; Boirie, Y.; Van Loon, L.J.C. Ingestion of a protein hydrolysate is accompanied by an accelerated in vivo digestion and absorption rate when compared with its intact protein. *Am. J. Clin. Nutr.* **2009**, *90*, 106–115. [CrossRef]
18. Haytowitz, D.; Ahuja, J.; Showell, B.; Somanchi, M.; Nickle, M.; Nyguyen, Q.; Pehrsson, P. *USDA National Nutrient Database for Standard Reference*; Release 28, released September 2015, slightly revised May 2016; US Department of Agriculture: Washington, DC, USA, 2015.
19. Van Vliet, S.; Beals, J.W.; Holwerda, A.M.; Emmons, R.S.; Goessens, J.P.; Paluska, S.A.; De Lisio, M.; Van Loon, L.J.C.; Burd, N.A. Time-dependent regulation of postprandial muscle protein synthesis rates after milk protein ingestion in young men. *J. Appl. Physiol.* **2019**, *127*, 1792–1801. [CrossRef]
20. Anthony, J.C.; Anthony, T.G.; Kimball, S.R.; Vary, T.C.; Jefferson, L.S. Orally administered leucine stimulates protein synthesis in skeletal muscle of postabsorptive rats in association with increased eIF4F formation. *J. Nutr.* **2000**, *130*, 139–145. [CrossRef] [PubMed]
21. Anthony, J.C.; Yoshizawa, F.; Anthony, T.G.; Vary, T.C.; Jefferson, L.S.; Kimball, S.R. Leucine stimulates trans-lation initiation in skeletal muscle of postabsorptive rats via a rapamycin-sensitive pathway. *J. Nutr.* **2000**, *130*, 2413–2419. [CrossRef]
22. Drummond, M.J.; Reidy, P.T.; Baird, L.M.; Dalley, B.K.; Howard, M.T. Leucine Differentially Regulates Gene-Specific Translation in Mouse Skeletal Muscle. *J. Nutr.* **2017**, *147*, 1616–1623. [CrossRef]

23. Bergstrom, J. Percutaneous needle biopsy of skeletal muscle in physiological and clinical research. *Scand. J. Clin. Lab. Investig.* **1975**, *35*, 609–616. [CrossRef]
24. Macnaughton, L.S.; Wardle, S.L.; Witard, O.C.; McGlory, C.; Hamilton, D.L.; Jeromson, S.; Lawrence, C.E.; Wallis, G.A.; Tipton, K.D. The response of muscle protein synthesis following whole-body resistance exercise is greater following 40 g than 20 g of ingested whey protein. *Physiol. Rep.* **2016**, *4*, e12893. [CrossRef]
25. Farnfield, M.M.; Breen, L.; Carey, K.A.; Garnham, A.; Cameron-Smith, D. Activation of mTOR signalling in young and old human skeletal muscle in response to combined resistance exercise and whey protein ingestion. *Appl. Physiol. Nutr. Metab.* **2012**, *37*, 21–30. [CrossRef]
26. Moro, T.; Brightwell, C.R.; Deer, R.R.; Graber, T.G.; Galvan, E.; Fry, C.S.; Volpi, E.; Rasmussen, B.B. Muscle Protein Anabolic Resistance to Essential Amino Acids Does not Occur in Healthy Older Adults Before or After Resistance Exercise Training. *J. Nutr.* **2018**, *148*, 900–909. [CrossRef] [PubMed]
27. D'Lugos, A.C.; Patel, S.H.; Ormsby, J.C.; Curtis, D.P.; Fry, C.S.; Carroll, C.C.; Dickinson, J.M. Prior acetaminophen consumption impacts the early adaptive cellular response of human skeletal muscle to resistance exercise. *J. Appl. Physiol.* **2018**, *124*, 1012–1024. [CrossRef]
28. Song, Z.; Moore, D.R.; Hodson, N.; Ward, C.; Dent, J.R.; O'Leary, M.F.; Shaw, A.M.; Hamilton, D.L.; Sarkar, S.; Gangloff, Y.-G.; et al. Resistance exercise initiates mechanistic target of rapamycin (mTOR) translocation and protein complex co-localisation in human skeletal muscle. *Sci. Rep.* **2017**, *7*, 1–14. [CrossRef] [PubMed]
29. Korolchuk, V.I.; Saiki, S.; Lichtenberg, M.; Siddiqi, F.H.; Roberts, E.A.; Imarisio, S.; Jahreiss, L.; Sarkar, S.; Futter, M.; Menzies, F.M.; et al. Lysosomal positioning coordinates cellular nutrient responses. *Nat. Cell Biol.* **2011**, *13*, 453–460. [CrossRef] [PubMed]
30. Dalley, B.K.; Baird, L.; Howard, M.T. Studying Selenoprotein mRNA Translation Using RNA-Seq and Ribosome Profiling. In *Selenoproteins*; Springer: New York, NY, USA, 2018; pp. 103–123.
31. Ingolia, N.T.; Ghaemmaghami, S.; Newman, J.R.S.; Weissman, J.S. Genome-Wide Analysis in Vivo of Translation with Nucleotide Resolution Using Ribosome Profiling. *Science* **2009**, *324*, 218–223. [CrossRef] [PubMed]
32. Robinson, M.D.; McCarthy, D.J.; Smyth, G.K. edgeR: A Bioconductor package for differential expression analysis of digital gene expression data. *Bioinformatics* **2009**, *26*, 139–140. [CrossRef]
33. Tipton, K.D.; Elliott, T.A.; Cree, M.G.; Aarsland, A.A.; Sanford, A.P.; Wolfe, R.R. Stimulation of net muscle protein synthesis by whey protein ingestion before and after exercise. *Am. J. Physiol.-Endocrinol. Metab.* **2007**, *292*, E71–E76. [CrossRef]
34. Tipton, K.D.; Elliott, T.A.; Cree, M.G.; Wolf, S.E.; Sanford, A.P.; Wolfe, R.R. Ingestion of Casein and Whey Proteins Result in Muscle Anabolism after Resistance Exercise. *Med. Sci. Sports Exerc.* **2004**, *36*, 2073–2081. [CrossRef]
35. Dreyer, H.C.; Drummond, M.J.; Pennings, B.; Fujita, S.; Glynn, E.L.; Chinkes, D.L.; Dhanani, S.; Volpi, E.; Rasmussen, B.B. Leucine-enriched essential amino acid and carbohydrate ingestion following resistance exercise enhances mTOR signaling and protein synthesis in human muscle. *Am. J. Physiol. Metab.* **2008**, *294*, E392–E400. [CrossRef]
36. Burd, N.A.; Beals, J.W.; Martinez, I.G.; Salvador, A.F.; Skinner, S.K. Food-First Approach to Enhance the Regulation of Post-exercise Skeletal Muscle Protein Synthesis and Remodeling. *Sports Med.* **2019**, *49*, 59–68. [CrossRef]
37. Mitchell, C.J.; McGregor, R.A.; D'Souza, R.F.; Thorstensen, E.B.; Markworth, J.F.; Fanning, A.C.; Poppitt, S.D.; Cameron-Smith, D. Consumption of Milk Protein or Whey Protein Results in a Similar Increase in Muscle Protein Synthesis in Middle Aged Men. *Nutrients* **2015**, *7*, 8685–8699. [CrossRef]
38. Witard, O.C.; Jackman, S.R.; Breen, L.; Smith, K.; Selby, A.; Tipton, K.D. Myofibrillar muscle protein synthesis rates subsequent to a meal in response to increasing doses of whey protein at rest and after resistance exercise. *Am. J. Clin. Nutr.* **2014**, *99*, 86–95. [CrossRef]
39. Dangin, M.; Boirie, Y.; Garcia-Rodenas, C.; Gachon, P.; Fauquant, J.; Callier, P.; Ballèvre, O.; Beaufrère, B. The digestion rate of protein is an independent regulating factor of postprandial protein retention. *Am. J. Physiol. Metab.* **2001**, *280*, E340–E348. [CrossRef]
40. Gorissen, S.H.M.; Burd, N.A.; Hamer, H.M.; Gijsen, A.P.; Groen, B.B.; Van Loon, L.J.C. Carbohydrate Coingestion Delays Dietary Protein Digestion and Absorption but Does Not Modulate Postprandial Muscle Protein Accretion. *J. Clin. Endocrinol. Metab.* **2014**, *99*, 2250–2258. [CrossRef]
41. Dickinson, J.M.; Fry, C.S.; Drummond, M.J.; Gundermann, D.M.; Walker, D.K.; Glynn, E.L.; Timmerman, K.L.; Dhanani, S.; Volpi, E.; Rasmussen, B.B. Mammalian Target of Rapamycin Complex 1 Activation Is Required for the Stimulation of Human Skeletal Muscle Protein Synthesis by Essential Amino Acids. *J. Nutr.* **2011**, *141*, 856–862. [CrossRef]
42. Vander Haar, E.; Lee, S.-I.; Bandhakavi, S.; Griffin, T.J.; Kim, D.-H. Insulin signalling to mTOR mediated by the Akt/PKB substrate PRAS40. *Nat. Cell Biol.* **2007**, *9*, 316–323. [CrossRef] [PubMed]
43. Bolster, D.R.; Jefferson, L.S.; Kimball, S.R. Regulation of protein synthesis associated with skeletal muscle hypertrophy by insulin-, amino acid-and exercise-induced signalling. *Proc. Nutr. Soc.* **2004**, *63*, 351–356. [CrossRef]
44. Mitchell, W.K.; Phillips, B.E.; Hill, I.; Greenhaff, P.L.; Lund, J.N.; Williams, J.P.; Rankin, D.; Wilkinson, D.J.; Smith, K.; Atherton, P.J. Human skeletal muscle is refractory to the anabolic effects of leucine during the postprandial muscle-full period in older men. *Clin. Sci.* **2017**, *131*, 2643–2653. [CrossRef]
45. Moore, D.R.; Churchward-Venne, T.A.; Witard, O.; Breen, L.; Burd, N.A.; Tipton, K.D.; Phillips, S.M. Protein Ingestion to Stimulate Myofibrillar Protein Synthesis Requires Greater Relative Protein Intakes in Healthy Older Versus Younger Men. *J. Gerontol. Ser. A* **2015**, *70*, 57–62. [CrossRef]

46. Trommelen, J.; Van Loon, L.J. Pre-sleep protein ingestion to improve the skeletal muscle adaptive response to exercise training. *Nutrients* **2016**, *8*, 763. [CrossRef]
47. Trommelen, J.; Kouw, I.W.K.; Holwerda, A.M.; Snijders, T.; Halson, S.L.; Rollo, I.; Verdijk, L.B.; Van Loon, L.J.C. Presleep dietary protein-derived amino acids are incorporated in myofibrillar protein during postexercise overnight recovery. *Am. J. Physiol. Metab.* **2018**, *314*, E457–E467. [CrossRef]
48. Snijders, T.; Res, P.T.; Smeets, J.S.J.; Van Vliet, S.; Van Kranenburg, J.; Maase, K.; Kies, A.K.; Verdijk, L.B.; Van Loon, L.J.C. Protein Ingestion before Sleep Increases Muscle Mass and Strength Gains during Prolonged Resistance-Type Exercise Training in Healthy Young Men. *J. Nutr.* **2015**, *145*, 1178–1184. [CrossRef]
49. Iribarren, C.; Karter, A.J.; Go, A.S.; Ferrara, A.; Liu, J.Y.; Sidney, S.; Selby, J.V. Glycemic Control and Heart Failure Among Adult Patients With Diabetes. *Circulation* **2001**, *103*, 2668–2673. [CrossRef]
50. Westman, E.C.; Yancy, W.S.; Mavropoulos, J.C.; Marquart, M.; McDuffie, J.R. The effect of a low-carbohydrate, ketogenic diet versus a low-glycemic index diet on glycemic control in type 2 diabetes mellitus. *Nutr. Metab.* **2008**, *5*, 36. [CrossRef]
51. Fajans, S.; Floyd, J., Jr.; Knopf, R.; Conn, J. (Eds.) Effect of amino acids and proteins on insulin secretion in man. In *Schering Symposium on Endocrinology, Berlin, May 26 to 27, 1967: Advances in the Biosciences*; Elsevier: Amsterdam, The Netherlands, 2016.
52. Raziani, F.; Tholstrup, T.; Kristensen, M.D.; Svanegaard, M.L.; Ritz, C.; Astrup, A.; Raben, A. High intake of regular-fat cheese compared with reduced-fat cheese does not affect LDL cholesterol or risk markers of the metabolic syndrome: A randomized controlled trial. *Am. J. Clin. Nutr.* **2016**, *104*, 973–981. [CrossRef]
53. Gorissen, S.H.; Burd, N.A.; Kramer, I.F.; Van Kranenburg, J.; Gijsen, A.P.; Rooyackers, O.; Van Loon, L.J. Co-ingesting milk fat with micellar casein does not affect postprandial protein handling in healthy older men. *Clin. Nutr.* **2017**, *36*, 429–437. [CrossRef]

MDPI
St. Alban-Anlage 66
4052 Basel
Switzerland
Tel. +41 61 683 77 34
Fax +41 61 302 89 18
www.mdpi.com

Nutrients Editorial Office
E-mail: nutrients@mdpi.com
www.mdpi.com/journal/nutrients